AF816074

# Environmental Earth Sciences

Series Editor:
James W. LaMoreaux

For further volumes:
http://www.springer.com/series/8394

Bartolomé Andreo · Francisco Carrasco
Juan José Durán · James W. LaMoreaux
(Editors)

# Advances in Research in Karst Media

Springer

Dr. Bartolomé Andreo
Centre of Hydrogeology
and Department of Geology
Faculty of Sciences
University of Málaga
Campus Universitario de Teatinos
29071 Málaga, Spain
Email: andreo@uma.es

Dr. Juan José Durán
Instituto Geológico y Minero de España
(Spanish Geological Survey)
C/ Ríos Rosas 23
28003 Madrid, Spain
Email: jjduran@igme.es

Dr. Francisco Carrasco
Centre of Hydrogeology
and Department of Geology
Faculty of Sciences
University of Málaga
Campus Universitario de Teatinos
29071 Málaga, Spain
Email: fcarrasco@uma.es

Dr. James W. LaMoreaux
P.E. LaMoreaux and Associates
2610 University Boulevard
Tuscaloosa, AL 35401
USA
Email: jlamoreaux@pela.com

*The first book of the Environmental Earth Sciences series*
Survival and Sustainability: Environment Concerns in the 21st Century
Gökcekus, H.; Türker, U.; LaMoreaux, J. W. (Eds.)
ISBN: 978-3-540-95990-8
Planned publication July 2010

ISBN 978-3-642-44091-5 ISBN 978-3-642-12486-0 (eBook)
DOI 10.1007/978-3-642-12486-0
Springer Heidelberg Dordrecht London New York



*Typesetting and production:* le-tex publishing services GmbH, Leipzig, Germany
*Cover design:* deblik, Berlin

Printed on acid-free paper

Springer is part of Springer Science+Business Media (www.springer.com)

# Presentation

Karst is a medium which has traditionally been the subject of hydrogeological research, given the abundant water resources that are stored in it. In many cases karst is the product of climatic and hydrological evolution in carbonate areas in recent periods of geological history. Karst contains key information on recent environmental changes. The action of water has generated a great range of karstic features that are part of our natural heritage and some of them form major tourist attractions (landscapes of natural parks, geosites and show caves, for example). Karst areas often serve as landscapes or as substrates for human activity. But karst is a highly fragile ecosystem and the exploitation of its resources or inappropriate land uses give rise to environmental problems (water pollution, subsidence, flooding, changes in the subterranean environment, etc.). Also civil engineering projects need special surveillance in karst areas.

Karstic outcrops cover approximately 12% of the terrestrial land surface and these areas have always attracted attention as a research field in which many disciplines are involved, not just geological sciences, but also physics, chemistry and biology. Recently, socio-economic science has also been closely involved with the karst medium, as many of these areas contain important natural resources (groundwater, caves, karst landscapes), for which proper management makes a significant contribution to local economies.

Karst medium has long been the object of scientific research. The results of these investigations have been presented and discussed at numerous congresses and meetings, especially in the second half of the 20th century. The threshold of the second decade of the 21st century seems a good time to reflect on the progress made in recent times and to set out some of the lines of research to pursue in the near future. This book is the result of the 4th International Symposium on Karst (ISKA-2010), held in Málaga (Spain).

ISKA-2010 is the last of the symposia on karst that have been held periodically in Málaga since 1992. These symposia are an international forum for scientific debate on the progress made in research into karst environments. The main objective of the 4th International Symposium on Karst (ISKA-2010) is to discuss and disseminate the latest trends in research into karst media on the basis of the results obtained with different methodologies in various worldwide areas.

This book contains 80 contributions grouped into four sections:

- Karst Hydrogeology
- Karst Geomorphology
- Engineering Geology in Karst
- Research on Caves

Most contributions are on Karst Hydrogeology, more than 60%, in connection with various topics: methods for groundwater recharge assessment, impact of climate change on karst aquifers, coastal aquifers, floods, karst groundwater flow, protection of karst aquifers, pollution and vulnerability in karst, thermal anomalies in carbonate aquifers, time series analysis, hydrochemistry, dye tracer and stable isotope applications, numerical modelling in karst, etc.

Concerning Karst Geomorphology (approximately 15% of total papers), the contributions deal with karst development in gypsum, wetlands, hypogene speleogenesis, sinkholes, travertines, fluviokarstic canyons and karst geosites.

Engineering Geology in Karst group contains about 10% of the published papers, concerning the following topics: motorways, dams, reservoirs, quarrying and mining, geophysical technical for mapping buried karst, karst risk assessment, and properties of the aeration zone in karst.

Finally, Research on Caves papers constitute the remaining15% of the works, related overall with mixing corrosion and speleogenetic processes, $CO_2$ sources and global carbon cycle in endokarst, speleothems and other deposits in caves, condensation and corrosion in caves, underground atmosphere and the colour in caves.

The number of authors involved in this issue (about 250) indicates the wide variety and representation of the selected papers from four continents (Europe, America, Asia and Africa) and over 20 countries.

The present book may be considered an update of karst hydrogeology landforms and cavities, described by numerous specialists who have investigated various aspects of these topics. It is a good example of investigations on karst systems in the last few years, and at the same time it provides an illustrative synthesis of the research tasks being performed out, among others, as part of the IGCP 513 project of the UNESCO, and IAH Karst Commission. This Project has been useful as a forum for the exchange of results and experience accumulated by researchers over recent years concerning karst-related issues. The result is a more integrated vision on karst, taking into account the diversity of geological and climatic contexts in the world. We hope this publication will be an interesting reference for all studying the karst medium.

# Acknowledgements

This book was made possible thanks to the dedicated, combined effort of the authors and members of the Scientific Committee as reviewers. The editors of this book offer to all of them our sincere gratitude. The members of the Scientific Committee composed of 40 international specialists from various karst domains are listed on following pages. They reviewed all the contributions, several times in some cases. Only manuscripts which met the criteria and quality required for the book were accepted.

The 4th International Symposium on Karst, Malaga 2010 has been organized by the Centre of Hydrogeology at the University of Malaga and the Spanish Geological Survey, in the framework of their "Advanced Hydrogeological Studies" partnership, with the cooperation of UNESCO and the IAH Karst Commission and the Spanish Group. We thank Nerja Cave Foundation for its continuous support of the series International Symposia on Karst; in fact, the two first symposia were held in Nerja. Furthermore, we must express our thanks to all the organisations that collaborated in organising, whose logotypes are included in following pages. Without these numerous collaborators, and without the support of all mentioned, it would not have been possible to publish this updated survey on karst.

This book is a contribution to projects IGCP 513 of UNESCO, CGL2008-06158 and CGL2008-04938 of Spanish Directorate of Research, to Integrated Actions HP2008-047 and DE2009-0060, and to P06-RNM-02161 project and Research Group RNM 308 of Andalusian Government. It is a publication related to the IAH Karst Commission.

# Scientific Committee

Contributions published in this book have been reviewed by the following Scientific Committee members:

Bartolomé Andreo (University of Málaga, Spain)
Tim Atkinson (University College, London, United Kingdom)
Michel Bakalowicz (University of Montpellier 2, France)
Bozidar Biondic (Institute of Geology, Zagreb, Croatia)
Ogden Bonacci (University of Split, Croatia)
Lhoussaine Bouchaou (University of Ibn Zohr, Agadir, Morocco)
Francisco Carrasco (University of Málaga, Spain)
Jose Antonio Crispim (University of Lisbon, Portugal)
Yuan Daoxian (Institute of Karst Geology, Guilin, China)
Nathalie Doerfliger (BRGM, France)
Wolfgang Dreybrodt (University of Bremen, Germany)
Juan Jose Durán (Spanish Geological Survey – IGME, Spain)
Cristophe Emblanch (University of Avignon, France)
Ian Fairchild (University of Birmingham, United Kingdom)
María Dolores Fidelibus (Polytechnic University, Bari, Italy)
Malcolm Field (Environmental Protection Agency, Washington, USA)
Stephen Foster (International Association of Hydrogeologists, United Kingdom)
Franci Gabrovsek (Karst Institute, Postojna, Slovenia)
Nico Goldscheider (University of Neuchatel, Switzerland/Technische Universität, München, Germany)
Chris Groves (Western Kentucky University, USA)
John Gunn (University of Birmingham, United Kingdom)
Heinz Hötzl (University of Karlsruhe, Germany)
Pierre Yves Jeannin (SISKA, Switzerland)
Alexander Klimchouck (National Academy of Science, Ukraine)
James LaMoreaux (P.E. LaMoreaux and Associates, Inc., USA)
Petar Milanovic (Serbia)

# Organizing Committee

The 4th International Symposium on Karst was organized by researchers from the Centre of Hydrogeology at the University of Malaga (CEHIUMA) and the Spanish Geological Survey (IGME), in the framework of their "Advanced Hydrogeological Studies" partnership. The Symposium was sponsored by UNESCO and IAH (Karst Commission and Spanish Group).

Bartolomé Andreo (CEHIUMA)
Francisco Carrasco Cantos (CEHIUMA)
Juan Jose Durán (IGME)
Pablo Jiménez Gavilán (CEHIUMA and INTECSA-INARSA)
Cristina Liñán Baena (CEHIUMA and Nerja Cave Foundation)
Juan Antonio López Geta (IGME)
Carlos Martínez Navarrete (IGME)
Pedro Agustín Robledo Ardila (IGME)
Iñaki Vadillo Pérez (CEHIUMA)

Organizing Secretariat: Ana Isabel Marin Guerrero (CEHIUMA)

**Organized and sponsorship by**

**In collaboration with**

# Contents

## Karst Hydrogeology

## Karst Geomorphology

---

**Engineering Geology in Karst**

---

## Research on Caves

# Karst Hydrogeology

# Sustainable Water Resources with Case Studies in Historic Areas in Egypt and Syria

J.W. LaMoreaux

**Abstract** The right to water use does not imply total sovereignty by some to the detriment of others over one of the most basic natural resources necessary to sustain life. Both developed and developing countries must find a workable balance for managing water resources. Water use decisions like those affecting Figeh Spring, Damascus, Syria, and Kharga Oases, Egypt, are controlled both by natural parameters, as well as human factors such as land use and population growth. Sustainable Water Development (SWD) is an internationally accepted program for balancing water use issues and goals among competing consumers.

## Introduction

Hydrogeology plays a major role in all aspects of environmental planning, execution and implementation. Without a safe sustainable water supply, life cannot exist and hopes for prosperity are limited. Much future world demand for water will be made up of the ground water component.

The United Nations has identified water resources as a top priority. Attention is particularly drawn to freshwater stress, which relates water withdrawal to percentage of water available. Based on the current rate of use, the United Nations is not optimistic about the global water outlook.

## Sustainable Water Development

Water quantity and water quality are inextricably linked; therefore, there needs to be rapid movement towards Sustainable Water Development (SWD) in both devel-

---

J.W. LaMoreaux
P.E. LaMoreaux and Associates, Inc., 1009 University Blvd., Tuscaloosa, AL 35401

B. Andreo et al. (Eds.), *Advances in Research in Karst Media* 
DOI 10.1007/978-3-642-12486-0, © Springer 2010

oped and developing countries. To address this issue, countries need to embrace sustainable water development. SWD is "the development of water in a manner in which [an] adequate supply of good quality water is sustained and the watercourse ecosystem is maintained for the uses of future generations" (Pichyakorn 2002).

SWD includes five specific elements:

(1) The right to use water
(2) The protection of water resources and prevention of water degradation
(3) The maintenance of water flow
(4) An ecosystem related approach
(5) The procedural elements to achieve sustainable development.

The right to use water involves comprehensive permitting with continuing review, strategic counseling, crisis management, creative dispute resolution, and enhanced relations with stakeholder and community groups. If there is the perception that one group of users is favored above another, the task of proper water management becomes more difficult.

Protection of water resources and prevention of water degradation includes addressing nonpoint sources of pollution on a national, regional and local basis; land use controls; integration of water and land management; and regulation of interbasin transfers.

Maintenance of water flow involves instream flows and environmental flows and may require appropriate controls. An ecosystem-related approach should not be limited only to watercourse mainstream or tributaries, but it should also incorporate terrestrial and marine environments interacting with it; promote health of the entire ecosystem; and utilize watershed management authorities.

To move toward SWD, freshwater needs to be managed in a holistic manner, or in other words, an ecosystem approach. Once a good scientific understanding of the nature, quantity and quality of available water resources has been gained, then proper planning for future water use is possible. Reliable scientific data must be the basis for political decisions in water policy. Developing and enacting regionally

**Fig. 1** Philip E. LaMoreaux in the Kharga Oases, 1962

appropriate regulations and water use policies are important aspects of the SWD approach.

"Management of water resources is holistic when it is done on a catchment or drainage basin basis. This includes both land and water resources, since land use can have significant impacts on freshwater and related ecosystems.... Thus water legislation should provide for a holistic, ecosystem approach to the management of water ..." (McCaffrey and Weber 2005).

Projects performed by P.E. LaMoreaux & Associates, Inc. (PELA) under the guidance of the late Philip E. LaMoreaux, past-president of the IAH (Fig. 1) in Kharga Oases in the Western Desert of Egypt (LaMoreax 1962a, 1962b, 1962c) and on Figeh Spring in Damascus, Syria (LaMoreaux 1989, 1992), will be used to illustrate this critical problem in the world today.

## Kharga Oases in Egypt

One of the longest continuous records of human use of groundwater in the world exists in Kharga Oases in Egypt. The present landscape has evolved over several million years of erosion, structure and deformation of beds. Geographic position and changes in climate over the past thousands of years have resulted in conditions favorable for inhabitation by man.

Groundwater reserves have been studied for long term socioeconomic development. The Nubian Sandstone forms a complex aquifer system and has been providing water to artesian wells and springs for several thousand years.

Kharga Oases are located in a topographic depression in the Western Desert of Egypt oriented north/south parallel to the Nile Valley (Fig. 2). The climate is arid, rainfall is sporadic and temperatures are extremely hot during the summer months. Mean annual rainfall is less than 1 mm (LaMoreaux et al. 1985).

The Nubian Sandstone (Fig. 3) there consists of a 600 m thick sequence of coarse clastic sediments of sandstone, sandy clay interbedded with shale, chalk and clay beds. Wells flow at about 800 liters per minute, with water temperatures of about 37 °C. The salinity of the water is remarkably variable, both laterally and vertically (LaMoreaux et al. 1985; El Ayouty et al. 1961).

Nearly all of the springs and wells originally in the Kharga Oases flowed (Fig. 4). Exploitation of groundwater from deeper and deeper wells for irrigation began about 1959. Artesian outflows in the springs and shallow wells declined as more and more closely spaced deep wells were drilled (Fig. 5). By 1975, many wells had ceased to flow (LaMoreaux et al. 1985).

Sustainability of water resources is related to the history of hydrogeology through the inhabitation of Kharga Oases. Caton-Thompson's work (1952) hypothesized that Paleolithic man migrated out of the desert plateau and open plains of Kharga as early as 25,000 BP. During dry seasons these groups relied on the availability of groundwater from mound springs, which are domed seeps of a rare wetland land type, more properly called a calcareous seepage fen. They occur where colder water that is rich in alkaline minerals flows out at the seepages in waves, and over time a roughly circular mound of successive deposits of minerals and mud builds up vertically in a rough circle around the water discharge point.

Prehistoric evidence exists of a post Pleistocene lake covering much of the Kharga Oases depression at a level of 70 m above sea level. Existence of good soils in these low areas allowed desert cultures to expand around them.

Evidence indicates that prehistoric man, and subsequently the Romans, used the springs in Kharga Oases for water supply. During Pharonic times, the oases were of little interest to Egyptian administrations until the very late Dynastic Period. Caravan routes crossed the area from Sudan to Libya. Stone temples, fortresses, and brick strongholds were built and remain as evidence of occupation by Persia, Greco/Roman, Roman, Coptic and subsequently Arab peoples.

Kharga inhabitants continue to use these ancient springs and wells, supplemented at present by groundwater from deep artesian wells penetrating the deeper Nubian

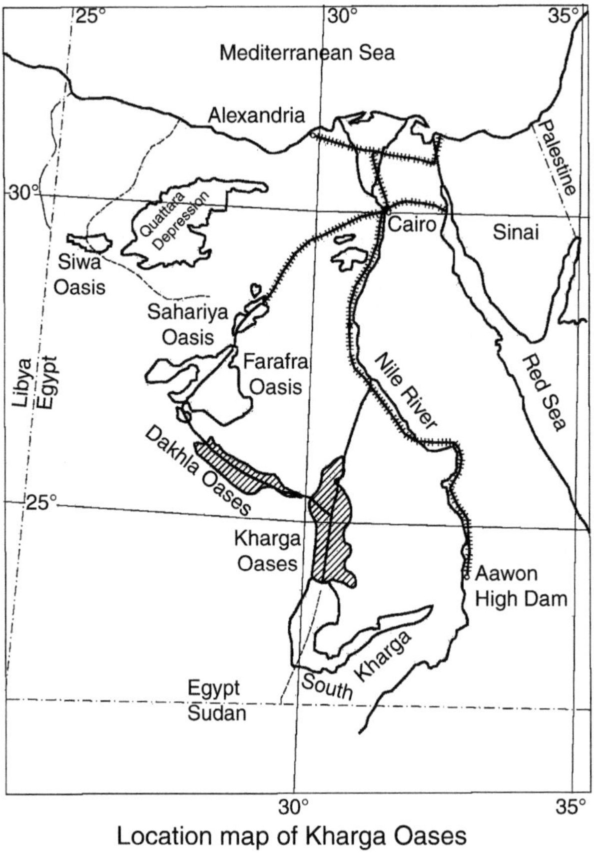


**Fig. 2** Location of Kharga Oases

sandstone layers. The challenge today is to manage the water resources in the area to allow continued usage in a prudent manner. Large reserves of groundwater remain in the Western Desert of Egypt. Reserves can be used effectively into the future but they must be developed cautiously and ideally combined with safe water reuse programs. Today water continues to flow at a reduced rate. Crop lands have a thin layer of salt, diminishing the productivity of the land.

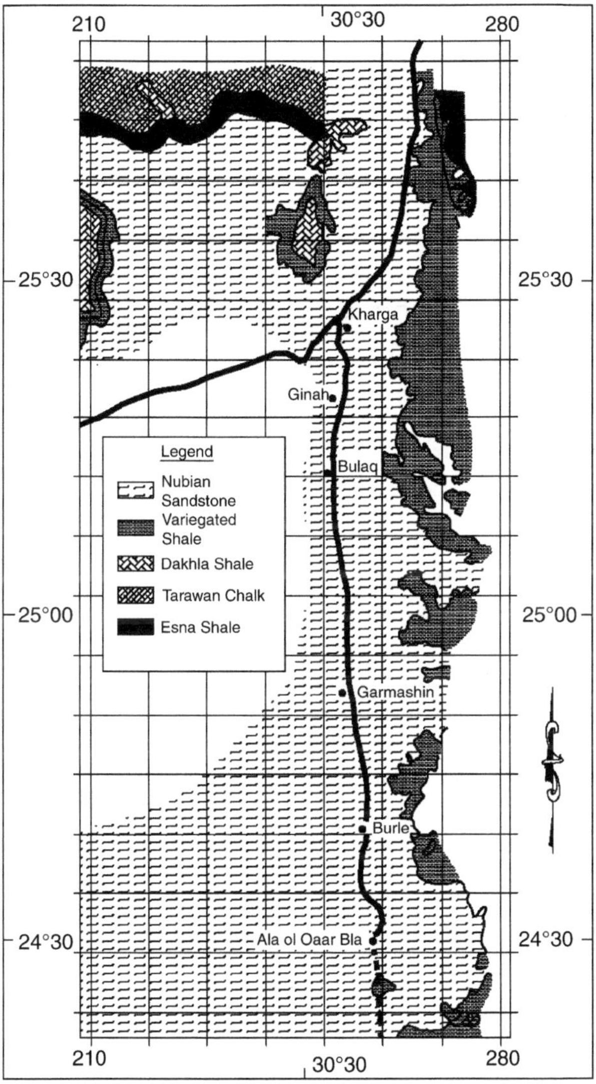


Generalized geologic map of the Kharga area

**Fig. 3** Generalized Geologic Map, Kharga

Fig. 4 Artesian Well in the Oases of Kharga, 1995

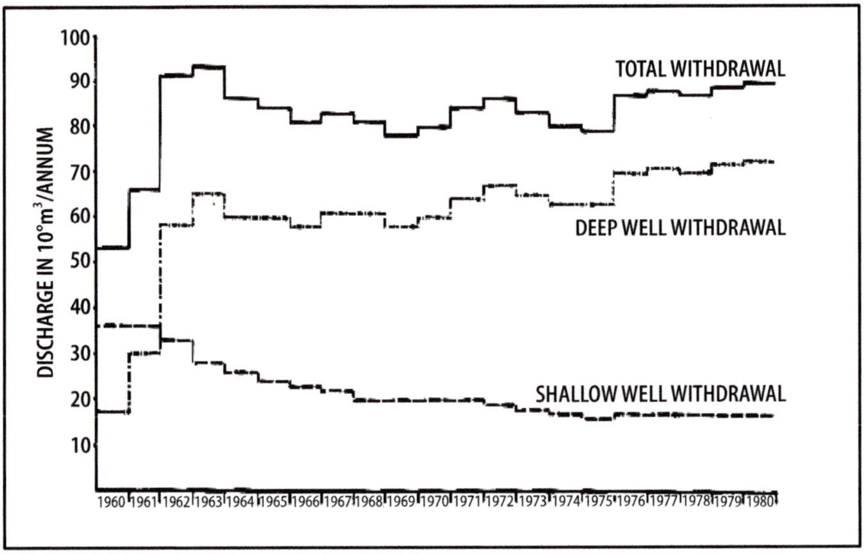


**Fig. 5** Annual withdrawal from shallow and and deep wells and total withdrawal, Kharga area, 1960–1980

The Nubian, being a complex geologic unit, comprises a very large storage system containing good quality water that moves through the sandstone beds very slowly (LaMoreaux 1964; LaMoreaux et al. 1985). This slow movement in a way comprises a safety factor. Even though these resources may have been misused in the past and artesian heads have been reduced drastically in some local areas, much water remains for future development. Applying the ideas of sustainable water resources and utilizing state of the art technologies available today will help to better manage this resource for future generations.

## Figeh Spring in Damascus, Syria

Figeh Spring in Syria (Fig. 6) is the third largest spring in the world, after Krasny Kutch (Red Spring) in Bashkortostan, Russia (Gareev 2004) and Silver Springs in Florida, USA (Wilson 1989). Figeh Spring has an average discharge of about $8.5\,m^3/s$, and the average temperature is about 14 °C. The waters are used for the supply of the city of Damascus, Syria. The spring was developed first by the Romans. Remains of Roman baths and of aqueducts leading south toward the city of Damascus are evidence of early development.

PELA performed hydrogeological studies in the vicinity of Figeh Spring to determine groundwater flow paths, recharge, storage, and discharge and the maximum reliable yield (LaMoreaux 1989; LaMoreaux et al. 1989). PELA's project provided information upon which to base pumpage to augment low-season flows from the spring which is the major water supply for the city of Damascus.

Studies performed include detailed surface geologic mapping, aerial photograph interpretation, well and spring inventory and a systematic water quality sampling program. Geologic structural work included mapping jointing, faulting, and folding and an analysis of their impact on groundwater movement.

Detailed quantitative studies were made on the large springs supplying the Damascus water supply with detailed monitoring during pumping. Pumping test results indicated that, with proper control and managed modifications at Figeh Spring, flow augmentation at an average annual rate of $4\,m^3/s$ is available to support the needs of Damascus during the low flow season. The reduction in storage is replaced first by rains and then by snowfall during the primary recharge season from November to January.

**Fig. 6** Figeh Spring, Syria: Discharge to the Barada River, 1988

Understanding the reservoir system from which water flows, the flow direction, and the rate of water withdrawn allows determination of the area around Figeh Spring in which environmental restraints should be imposed on development.

Three protection zones were identified with decreasing limitations:

(A) Total Protection
(B) Permit Restriction
(C) Permit Proper Development of light agriculture

Since PELA performed its studies in the 1980s, demands on water resources and arable soils have grown enormously due to high population growth rates. Water scarcity, over-extraction of groundwater and deteriorating water quality are common occurrences and are increasing over time. Water shortages in urban areas and limited access to clean drinking water are creating a water crisis.

In 2001, the population of the area had grown to about four million, irrigation use was increasing, and the average pumping rate was nearly 6 $m^3/s$. Water supplies were turned off on a rotating basis for between 16–20 hours a day during the summer months of 2001. Syria's Statistical Abstracts show that for the years between

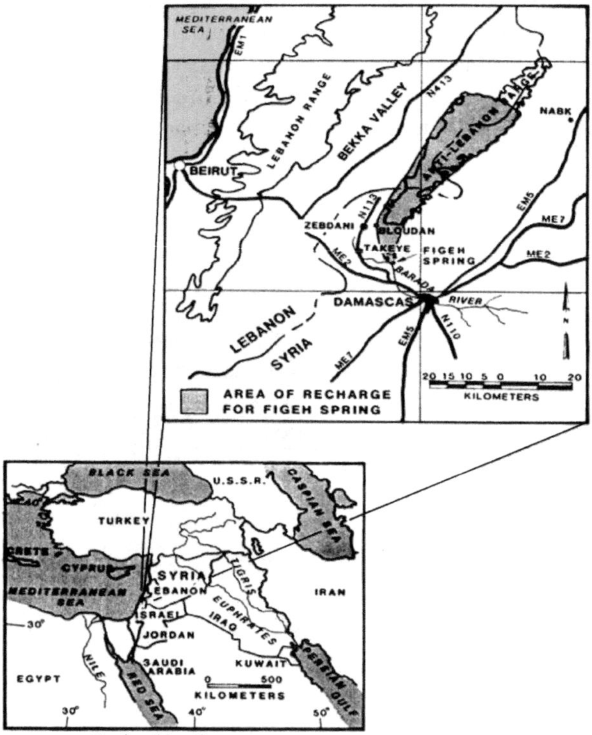


Regional location map

**Fig. 7** Location of Figeh Springs in Syria

1967 to 1984, average annual water flows for Figeh Spring fluctuated between a low of 4.227 $m^3/s$ and a high of 12.9 $m^3/s$. For the years between 1996 and 2000, average annual water flow recorded a low of 3.9 $m^3/s$ and a high of only 6.85 $m^3/s$ (Elhadj 2004).

Limited arable lands are endangered by intensification of land use and by increasing contamination, degradation and desertification. Socio-economic development of the region is limited by these factors and holds potential for even greater conflicts at the national and international level. The Federal Ministry for Economic Co-operation and Development, Germany (BMZ 2008), and the Arab Center for Studies of Arid Zones and Dry Lands (ACSAD 2008) are performing work in Syria to address these problems.

As it is difficult to estimate the results of management decisions in complex systems characterized by natural parameters such as those affecting Figeh Spring, as well as human factors such as land use and population growth, BMZ and ACSAD are developing a Decision Support System (DSS) involving visualizing and discussing the status quo and possible scenarios with relevant stakeholders.

The purpose of DSS is for stakeholders to be able to make joint decisions. In this way acceptable water allocations can be established and conflict between current water users can be reduced, which will help lead to more sustainable management of limited water resources for current and future years.

## Conclusions

Historically, sustainability of water resources has been a problem which has been difficult to manage without today's detailed knowledge of hydrogeological systems. History can be used as the basis for ongoing and future studies. Well planned investigations with quality control protocol and management provide better information upon which to base decisions.

New technologies allow more detailed investigations and the acquisition of accurate and verifiable data with which to make projections and model various scenarios. New management systems make it easier for stakeholders to be involved to minimize conflict and to collectively arrive at optimal solutions. This process embraces socio-economic development and sustainable water resources for future generations.

## References

Arab Center for Studies of Arid Zones and Dry Lands (2008) The Damascus Ghouta Pilot Study Area. http://www.acsad-bgr.org/

Bearden B (2009) Legal regime of water resources in Alabama. IN Proceedings of 23rd Annual Alabama Water Resources Conference and Symposium, Auburn University, Auburn, Alabama

Caton-Thompson G (1952) Kharga Oases in Prehistory. Athlone Press, University of London, 231 p

EL-Ayouty MK, Ezzat MA (1961) Hydrogeological Observations in the search for underground water in the Western Desert of Egypt, UAR. General Desert Development Authority, Cairo.

Elhadj E (2004) The Household Water Crisis in Syria's Greater Damascus Region. Occasional Paper 47, School of Oriental & African Studies Water Resources Group, King's College London, University of London

Federal Ministry for Economic Co-operation and Development, Germany (2008) Partner Country Information: Syrian Arab Republic. http://www.bmz.de/en/countries/partnercountries/syrien/zusammenarbeit.html#t4

Gareev EZ (2004) The Nature Geological Monuments of the Republic of Bashkortostan: Bashkir State University, Ufa, Taw, 296 p

LaMoreaux PE (1962a) A review of the New Valley Project, Western Desert of Egypt. Cairo, UAR Desert Institute, The General Desert Development Organization, Ministry of Reclamation, 18 p

LaMoreaux PE (1962b) Reconnaissance report and recommendations for ground-water investigations, Wadi el Natrun, Western Desert of Egypt. Cairo, UAR, Desert Institute, The General Desert Development Organization, Ministry of Reclamation, 56 p

LaMoreaux PE (1962c) A review of the New Valley Project, Western Desert of Egypt. Cairo, UAR, Desert Institute, The General Desert Development Organization, Ministry of Reclamation, 30 p

LaMoreaux PE (1964) A review of the New Valley Project, Western Desert of Egypt, with special reference to the development of a quantitative consultant's study. Cairo, UAR, Desert Institute, The General Desert Development Organization, Ministry of Reclamation, 8 p

LaMoreaux PE (1989) Environmental aspects of the development of Figeh Spring, Damascus, Syria. IN Beck, BF (ed) Proceedings of the Third Multidisciplinary Conference on Sinkholes and the Engineering and Environmental Impacts of Sinkholes and Karst, St. Petersburg Beach, Florida, Oct. 2–4, 1989; AA Balkema, Rotterdam, The Netherlands, 17–23

LaMoreaux PE (1992) Hydrogeotechnical hazards – Figeh Spring, Syria IN Proceedings of the 1992 National Conference of the American Institute of Hydrology

LaMoreaux PE, Hughes TH, Memon BA, Lineback NG (1989) Hydrogeologic assessment – Figeh Spring, Damascus, Syria. Environmental Geology and Water Sciences: Springer-Verlag, New York, New York, 13(2):73–127

LaMoreaux PE, Memon BA, Idris H (1985) Groundwater development, Kharga Oases, Western Desert of Egypt: A long-term environmental concern. Environmental Geology and Water Sciences, Springer-Verlag, New York, 7(3):129–146

Meinzer OE (1927) Large springs in the United States; U.S. Geological Survey Water-Supply Paper 557, 94 p

McCaffrey SC, Weber GS (2005) Guidebook for Policy and Legislative Development on Conservation and Sustainable Use of Freshwater Resources; United Nations Environment Programme, Nairobi, Kenya, 136 p

Pichyakorn, Bantita (2002) Sustainable Development and International Watercourse Agreements: The Mekong and the Rhine, International Union for the Conservation of Nature (Draft 30 June 2002)

Veni G, DuChene H, Crawford NC, Groves CG, Huppert GN, Kastning EH, Olson R, Wheeler BJ (2001) Living with Karst: a Fragile Foundation; American Geological Institute, Alexandria, VA 69 p

Wilson WL, Skiles, WC (1989) Partial Reclassification of first magnitude springs in Florida; IN Beck, BF (ed.) Proceedings of the Third Multidisciplinary Conference on Sinkholes and the Engineering and environmental Impacts of Karst, St. Petersburg Beach, FL

# Resources Assessment of a Small Karstic Mediterranean Aquifer (South-Eastern, Spain)

J.M. Andreu, P. Martínez-Santos, A. Pulido-Bosch, and E. García-Sánchez

**Abstract** Groundwater from small aquifer systems is frequently used for urban supply in southeastern Spain. Aquifers such as the Ventos system, located in Mediterranean semiarid environments, are sensitive to climatic and anthropogenic changes. Many of them have been severely depleted due to intensive pumping. Drawdowns in the Ventos aquifer amount to approximately 80 m over the last three decades. Adequate knowledge of groundwater resources is necessary for water planners and managers to guarantee suitable abstraction. This paper presents a methodology to estimate groundwater recharge in these kinds of quick-response semiarid karst aquifers. A distributed model has been used to evaluate the fraction of rainfall that ultimately results in aquifer recharge, as well as the correlation between the magnitude of rainfall events and infiltration rates. Modelling results are then compared with direct observations of the recharge processes and discussed to evaluate the implications of time scales.

## 1 Introduction

Groundwater resources contribute substantially to the supply of water in Alicante, southeastern Spain. The use of groundwater has increased significantly during the

J.M. Andreu
Dpto. de Ciencias de la Tierra y Medio Ambiente. Universidad de Alicante (Spain),
e-mail: Andreu.Rodes@ua.es

P. Martínez-Santos
Dpto. Geodinámica, Universidad Complutense de Madrid (Spain), e-mail: pemartin@geo.ucm.es

A. Pulido-Bosch
Dpto. de Hidrogeología y Química Analítica, Universidad de Almería (Spain),
e-mail: apulido@ual.es

E. García-Sánchez
Dpto. de Agroquímica y Medio Ambiente, Universidad Miguel Hernández (Spain),
e-mail: ernesto.garcia@umh.es

B. Andreo et al. (Eds.), *Advances in Research in Karst Media* 
DOI 10.1007/978-3-642-12486-0, 

last decades, largely due to an increasing population and to the development of the agricultural, tourism and industrial sectors. This has had an impact both on the quantity and quality of groundwater resources. Knowledge of groundwater resources is thus important to guarantee suitable abstraction levels and environmental flows.

The Alicante province consists of 144 councils, whose population amounts to two million inhabitants. Aquifers play a relevant role in local supply because groundwater is often the only water source. Thus, approximately 51.8% of urban supplies are met by groundwater. In turn, approximately 27% of these supplies come from aquifers located within the same (or neighbouring) municipalities (DCH 2007). Ventos is an example of this. It is a small karst unit which is solely exploited for urban supply purposes. Since the 1970s the Ventos aquifer provided drinking water to Agost, a 5,000-inhabitant town. Apart the importance of water for this small town, this system was specifically selected because the conceptual model seems quite simple and large datasets are available to develop and calibrate models.

Determining which of a wide variety of techniques is likely to provide reliable recharge in karst aquifer estimates is not a simple problem (Lenner et al. 1990; Scanlon et al. 2003). Recent times have witnessed an increasing interest on the part of the scientific community to comprehend the above the recharge processes. A big part of these efforts focuses on techniques and numerical models to calculate the magnitude of the recharge (Zagana et al. 2007; Andreo et al. 2008).

The aim of the present study is to quantify the resources in the Ventos aquifer and to establish the conceptual model of the system. A distributed model has been developed for this purpose.

## 2 Hydrogeological Characteristics

The region where the aquifer is located presents a typically Mediterranean climate, where long dry periods alternate with short wet sequences and hot dry summers follow short mild winters. Average temperature is 18.5 °C, and potential evapotranspiration estimates range between 870 and 1120 mm/yr. Mean precipitation is 275 mm/year, and presents marked seasonal variations. Most rainfall takes place in autumn.

Ventos aquifer is made up of Cretaceous limestones whose thickness ranges between 80–120 m (Fig. 1). The aquifer, about 7 $km^2$, is completely isolated from other aquifer formations. It presents a synclinal structure with its main axis tilted to the southwest.

Before the 1970s the aquifer behaved in an undisturbed manner. Rainfall infiltration is the only system input. Water follows a downward circulation path from the upper part of the mountain range to the area where Agost is located. Discharges originally took place through the Agost spring, which dried up some time after pumping from the aquifer began. Today, the sole output of the system is groundwater pumping (200,000 $m^3$/year). Pumping take place by one well, which works only a few hours a day.

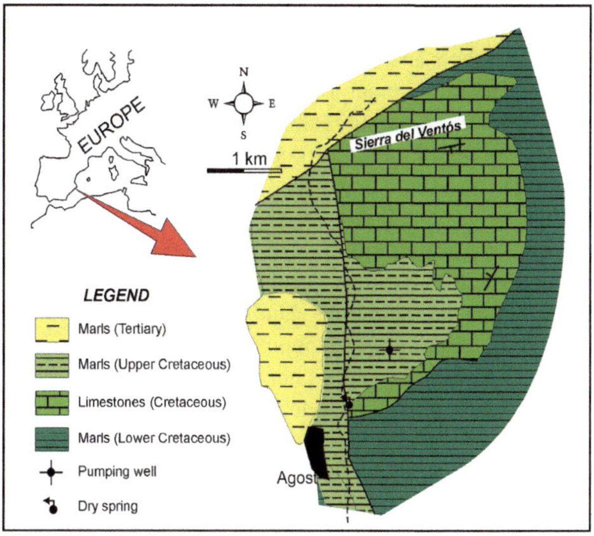


**Fig. 1** Geographical location and geological setting. Isolated pumping well and spring (nowadays dry) placement are shown

Though abstractions do not seem overly large, groundwater pumping has cause the level to drop by 80 m. Short-term changes in groundwater levels suggest that aquifer replenishment only responds to significant rainfall episodes. Thus, automated piezometric records reveal that rises in the water table are observed a few hours after each storm. These are followed by a slow recovery during the ensuing weeks, the replenishment rate gradually decreasing in magnitude over time (Touhami et al. 2008). On the other hand, it has been shown that similar rainfall values yielded significantly different recovery rates. This can be attributed to the joint influence of a variety of parameters, such as, frequency between rainfall events, precipitation intensity, soil moisture content and temperature.

## 3 Numerical Modelling

The numerical model is implemented in Visual Modflow Pro 4.1, one of the most widely available three-dimensional groundwater flow and solute transport simulation systems (WHI 2005). Using Modflow to estimate recharge presents two potential pitfalls, one mathematical and one physical. These refer to the numerical approach used to compute recharge, as well as to the peculiarities of karst systems.

Thus, parameter fitting is used to estimate recharge rates based on hydraulic heads, hydraulic conductivity and other parameters. Scanlon et al. (2002) however argue that recharge and hydraulic conductivity are highly correlated from the numerical viewpoint, and that inverse modeling estimates the ratio between those two parameters rather than their absolute values. In other words, several combinations of these two parameters may render a good adjustment between observed and calculated heads. Therefore, estimating recharge via inverse modeling may yield inaccurate results. This can be prevented by constraining the hydrodynamic parameters of the model, which in turn implies the need to count on sufficiently comprehensive field data.

On the other hand, the use of Modflow in karst environments remains controversial. This is because Modflow assumes porous flow, whilst karst systems may exhibit turbulent regimes. From the modelling standpoint this is tackled through defining an equivalent porous medium. Equivalent porous media models may be problematic in aquifers with well-developed conduit networks, but perform reasonably in those aquifers where karstification is poorly developed. This allows Modflow and other finite-difference distributed codes to provide sufficiently good results at the regional scale (Pulido-Bosch 1989; Scanlon et al. 2003; Martínez-Santos et al. 2005; Quinn et al. 2006).

The relatively simple behaviour of the system, together with its small size, allows for straightforward grid modelling. The aquifer is thus described as a single layer whose thickness varies with the relative elevation of the land surface and the impervious basement. Cells are uniform in size, covering an area of $50 \times 50$ m each. No-flow boundaries are prescribed in all directions. Active cells are defined for the aquifer area expected to remain saturated throughout the simulation process, and recharge is assumed uniform across the system. Hydraulic conductivity and specific yield are defined in agreement with pumping test data.

A steady-state simulation was carried out to replicate natural system conditions. This yields average spring discharges in the order of 0.5 $Mm^3$/year. Transient calibration ensued, taking into account the variability of pumping and recharge rates over time. Calibration considers monthly time steps, spanning a nine-year period (1999–2007).

Model calibration relies on observed groundwater levels, since the springs dried up long before the modelled interval. Figure 2 presents the calibration results. As shown, the model replicates the observed trend to a reasonable extent. Hydrodynamic parameters were modified for sensitivity analysis purposes. Storage coefficient was thus identified as the most sensitive parameter. Optimal correlation was obtained for storage values of 0.33 to 0.40%, the best adjustment standing at 0.36%. Doubling the existing estimate brings the $R^2$ coefficient to below 0.5, whereas halving it renders an $R^2$ in the order of 0.20. In both cases the trend remains recognizable, if considerably more detached from field observations.

The model suggests that correlation between rainfall and recharge is nonlinear. This confirms the hypothesis that temporal scales are crucial to reach more refined recharge estimations. From a conceptual standpoint the model assumes that recharge in the saturated zone is instantaneous. This ignores observed storage and transfer functions proper to the epikarst and vadose zones. As a result, the effect of individual recharge episodes is lost.

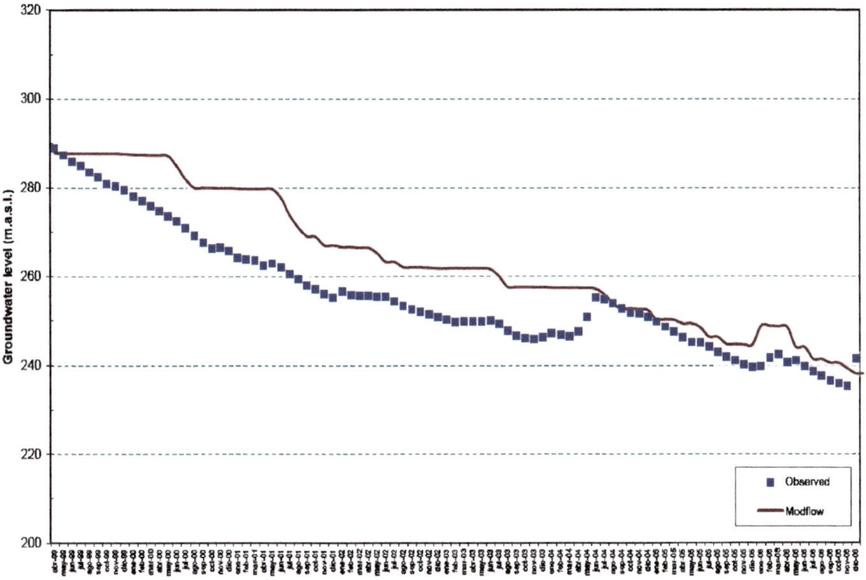


**Fig. 2** Observed and predicted piezometric levels during 1999–2006 period

# 4 Conclusions

Ventos aquifer is a small aquifer located in a Mediterranean climate, where present abstraction rates exceed several times their yearly resources. Knowledge of the resources has a crucial importance to water supply of small town of 5000 inhabitants. In order to estimate water balance, groundwater recharge to this aquifer has been presented and validated through the application of a distributed model (Modflow). The main results show that there is no linear relation between rainfall and recharge. Instead, the fraction of rainfall that results in recharge seems to increase exponentially with the magnitude of the precipitation event. Infiltration coefficients estimated vary widely, between 5 and 36%, and entail that the average yearly infiltration coefficient is less than 10%. Despite the limitations that stem from defining aquifers as equivalent porous media, this model provides an intuitive estimate of the system's behavior and replicates the observed trends reasonably well. This, together with direct observation of the system, suggests that this approach provides a sufficiently accurate representation of reality.

**Acknowledgements** This paper was developed under projects CGL2004-03627/HID (CYCIT), GV07/143 (Generalitat Valenciana) and IGCP-513 (UNESCO). The authors thank Diputación de Alicante and Ayuntamiento de Agost their interest and cooperation.

## References

Andreo B, Vías J, Durán JJ, Jiménez P et al (2008) Methodology for groundwater recharge assessment in carbonate aquifers: application to pilot sites in southern Spain. Hydrogeol J, 16: 911–925

DCH (2007) Mapa del Agua. Provincia de Alicante. Ed. Diputación de Alicante

Lenner DN, Issar AS, Simmers I (1990) Groundwater recharge: a guide to undersatanding and estimating natural recharge. International Contribution to Hydrogeology. IAH 8. Heise

Martínez-Santos P, Martínez Alfaro PE, Murillo JM (2005) A methodology to estimate the artificial recharge capacity of the Crestatx aquifer (Majorca, Spain). Environl Geol 47, 8: 1155–1161

Pulido-Bosch A (1989) Simulación del acuífero de Sierra Grossa (Valencia). Libro Hom. J. Porras, Hidrogeol y Rec Hidrául XIV: 301–313

Scanlon BR, Healy RW and Cook PG (2002) Choosing appropriate techniques to estimate groundwater recharge. Hydrogeol J 10, (1): 18–39

Scanlon BR, Mace RE, Barrett ME, et al (2003) Can we simulate regional groundwater flow in a karst system using equivalent porous media models? Case study, Barton Springs Edwards aquifer, USA. J Hydrol 276: 137–158

Tohuami I, Andreu JM, Pulido-Bosch et al (2008) Contribución de algunos eventos de recuperación al conocimiento de la recarga del acuífero del Ventós. Geotemas, 10: 1597–1600

WHI (2005) Visual MODFLOW Pro. User's Manual. Waterloo Hydrogeologic Inc. Waterloo

# Sensitivity Analysis of APLIS Method to Compute Spatial Variability of Karst Aquifers Recharge at the National Park of Viñales (Cuba)

H. Farfán, J.L. Corvea, and I. de Bustamante

**Abstract** In this contribution, sensitivity analysis of APLIS method is realized. APLIS method is based on a GIS analysis where the principal variables which influenced the recharge are taken into consideration, with a special emphasis on carbonate aquifers. The methodology is applied to carbonated aquifers of the tropical mountains at the National Park of Viñales. As the method has been developed from Mediterranean conditions, some modifications were required to improve the soil. The method shows acceptable results taking into account antecedent studies in the area. Results of sensitivity analysis of map removing show that the relative influence in the final recharge map is L>I>S>P>A. The single parameter sensitivity analysis shows some deviations between the empirical and the real weighting.

## 1 Introduction

Recharge assessment is one of the most important aspects in the evaluation and sustainable management of groundwater. In karst aquifers, recharge computing is more complex, and this aspect is conditioned by the heterogeneity and anisotropy of hydrological proprieties, as a direct consequence of the duality of the recharge. In such sense, two principal recharge types are defined: (1) Concentrate recharge and (2) Diffuse recharge. These principal recharge types have also been differentiated also into allogenic and autogenic ones.

H. Farfán, J.L. Corvea
Parque Nacional Viñales, Km 23, Carretera a Viñales, Viñales, Pinar del Río, Cuba,
e-mail: hfarfan@pnvinales.co.cu; direccion@pnvinales.co.cu

I. de Bustamante
Dpto. de Geología, Universidad de Alcalá, e-mail: irene.bustamante@uah.es

B. Andreo et al. (Eds.), *Advances in Research in Karst Media* 
DOI 10.1007/978-3-642-12486-0, 

In Cuba, over 65% of the Cuban land is affected by karst processes, approximately 76.9% of protected areas are found in these territories and 80% of the water resources come from karst areas. One of the most important karst areas of Cuba is the Viñales region, internationally known as one of the most spectacular exponents of tropical karst mountains. National Park of Viñales is a Protected Area with several scientific, historic, natural and aesthetic values. The area is localized in the center-oriental region of Sierra de los Órganos, with an exceptionally typical development in Jurassic-Cretaceous limestone rocks.

In this contribution, the autogenic groundwater recharge in the National Park of Viñales is evaluated by means of the APLIS method and is calibrated through the scarce studies related with the subject. The selection of the method is based on the possibility of mapping the spatial distribution of the recharge rate within aquifers, according to their particular characteristics, with a special emphasis on carbonate aquifers. On the other hand, sensitivity analysis with the APLIS method is realized by studying how imposed variations of the inputs affect the outputs of that analysis.

## 2 Methodology

### *2.1 APLIS Method*

APLIS is a parametric methodology which takes into account the principal variables that influenced the recharge: *Altitude*, *Slope*, *Lithology*, *Infiltration landform* and the *Soil*. As the method has developed in Mediterranean conditions, modifications related to the soil have been necessary. In Cuba, the prevalence of tropical stationary humid climatic condition and the *xerosols* are not present, because they are developed in Mediterranean dry climatic conditions. Other soils are not present in the area. For this reason, other soils with similar characteristics, have been substituted, taking into account the base criteria from their selection in the original method (Jaime et al., 2006). The final recharge map is computed by the expression (Andreo et al., 2008):

$$R = (A + P + 3 \cdot L + 2 \cdot I + S)/0.9 \tag{1}$$

Where: $R$: recharge (expressed in a percentage of precipitation); $A$: Altitude; $P$: Slope; $L$: Lithology; $I$: Infiltration landform and $S$: Soil. By dividing by 0.9, recharge rates are obtained that range from 8.88% to 88.8% of the rain that falls onto the surface of the aquifers.

### *2.2 Sensitivity Analysis*

A "*Map removal*" and "*Single parameter*" sensitivity analysis were performed. The concept of unique condition subarea is used in the analysis (Lodwik et al., 1990;

Gogu and Dassargues, 2000). Sensitivity of removing one or more maps can be expressed as (Lodwik et al., 1990):

$$S_{Xi} = \left| \frac{R_i}{N} - \frac{R_{Xi}}{n} \right| \tag{2}$$

Where: $S_{Xi}$: sensitivity (for the $i$th subarea) associated with the removal of one map; $R_i$: Recharge computed using Eq. 1; $R_{Xi}$: Recharge without considering the parameter $X$; $N$ and $n$: number of maps used to compute $R_i$ and $R_{Xi}$ respectively.

The variation index measures the effect removal of each parameter and was computed as (Gogu and Dassargues, 2000):

$$RX_i = \frac{R_i - R_{Xi}}{R_i} \cdot 100 \tag{3}$$

The effective weighting can be calculated for each subarea as (Napolitano and Fabbri, 1996):

$$W_{Xi} = \frac{X_{Ri} \cdot X_{Wi}}{R_i} \cdot 100 \tag{4}$$

Where $X_{Ri}$ and $X_{Wi}$ are, respectively, the rating and the weighting values for the parameter $X$.

# 3 Result and Discussions

## *3.1 Spatial Variability of Autogenic Recharge in National Park of Viñales. APLIS Method*

The map of recharge was made by overlaying the five factors determined in the original method, by a raster analysis in GIS (ArcView 3.3, ArcGIS 9.2). In the area, high recharge is extended by 4.69% of the National Park of Viñales. These areas of higher recharge are associated with development of preferential infiltration landforms in the karstified limestone. Recharges of autogenic concentrated remain prevalent in these areas.

Moderate recharges are the most represented in the area with 60.21%, and they are principally associated with karstified limestone, Low Mountain like mogotes and ruinform height. This area is characterized by a development of a typological variety of karren clustered in extensive karrenfield. The poljes and adjacent valleys have developed by the contact between karstified and non-karstified rocks, moderate recharge is also found. Recharge is principally diffuse. In the area, there is a prevalence of shales and silts, and low and very low recharges exist, in which case the distribution is 34.81 and 0.30% respectively.

Taking into account the antecedent work related with the recharge in the area, this application shows acceptable results. Farfán González (2006) valued by hydrologic

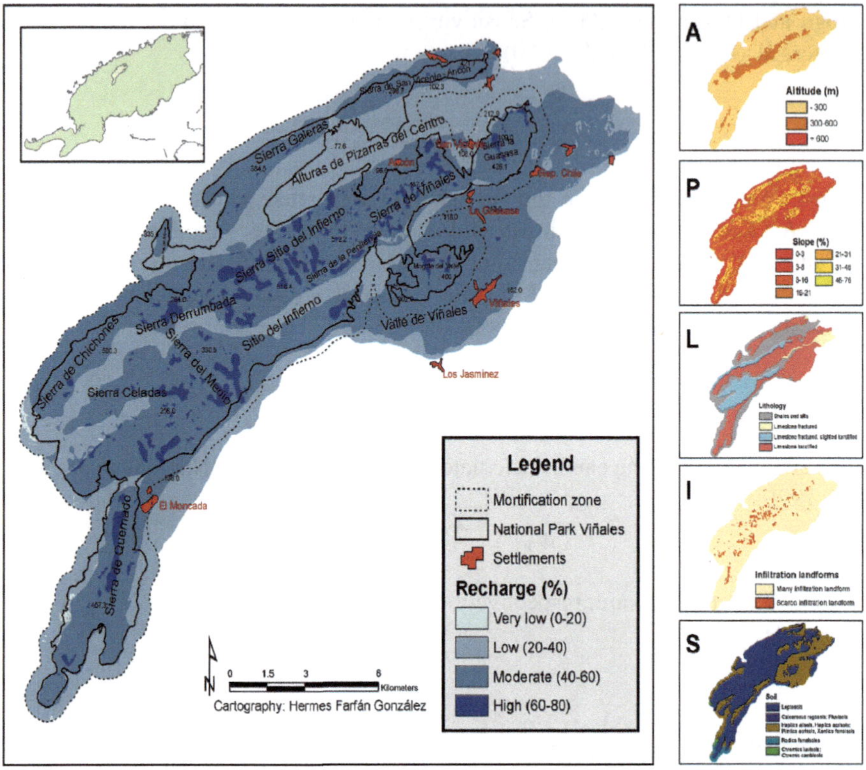


**Fig. 1** Recharge Map of the National Park of Viñales. APLIS method

balance method that the mean annual recharge in the territory is 47.4% of precipitations. In the Palmarito-Novillo-Pan de Azúcar karst systems, Farfán González (2005) computed an annual recharge mean of 56.7% by the mathematical model. In this study, Molerio León (2004) estimated the autogenic recharge in Sierra de Quemado by the chloride mass balance method, and he concluded that the capacity of natural recharge from the dolines over the systems can be as high as 90% of the mean annual rainfall, with a minimum of 70%, and a mean of 80% of precipitations.

## *3.2 Sensitivity Analysis of APLIS Model at the National Park of Viñales*

A result of map removal sensitivity analysis is shown in Table 1. The relative influence on the final recharge map is L>I>S>P>A. The low magnitude of the variation to remove the altitude parameter is expected, because in the area, he takes the minimum ratings and the range of variation is very low. The APLIS method is very sensible to remove the lithology and the infiltration landforms parameter, because

**Table 1** Statistics of map removal sensitivity analysis

| Parameter | Average | Minimum | Maximum | Range | Standard deviation |
|---|---|---|---|---|---|
| A | 0.83 | 0.00 | 1.67 | 1.67 | 1.179 |
| P | 1.17 | 0.00 | 2.33 | 2.33 | 1.650 |
| L | 3.56 | 0.00 | 7.11 | 7.11 | 5.028 |
| I | 2.39 | 0.00 | 4.78 | 4.78 | 3.379 |
| S | 1.33 | 0.00 | 2.67 | 2.67 | 1.886 |

these showed a high range of values. The spatial distribution of each value is another element which conditions the sensibility of each parameter.

The magnitude of the variation to map removal sensitivity of the Slope and Soil shows very similar sensibility values (35.761 and 36.619 respectively). The variation indexes of each parameter are always positive, which suggests that the exclusion of any parameter reduces the recharge which, therefore, increases the computation.

The single parameter sensitivity analysis shows some deviations between the empirical and real weighting. It is notable in the overvaluation of altitude parameter. The real importance of the slope, infiltration landform and soil are major in this particular analysis which was assigned in the original method. However, the lithology has approximately the same weighting. In the analysis, it is deducted that the result of recharge index is dominated by the infiltration landform parameter ($I$) with an average of 3.38 (42.206%) against the empirical weighting of 2 (25%). The spatial distribution and the relationship between five parameters are factors which determine the final real or effective weightings.

**Table 2** Statistics of single parameter sensitivity analysis

| Parameter | Empirical weighting | Empirical weighting (%) | Effective weighting (%) | | | | | Real weighting |
|---|---|---|---|---|---|---|---|---|
| | | | Average (%) | Minimum (%) | Maximum (%) | Range (%) | Standard deviation (%) | |
| A | 1 | 12.5 | 5.135 | 1.27 | 9 | 7.73 | 5.47 | 0.41 |
| P | 1 | 12.5 | 25.762 | 4.15 | 47.37 | 43.216 | 30.56 | 2.06 |
| L | 3 | 37.5 | 35.925 | 6 | 65.85 | 59.85 | 42.32 | 2.87 |
| I | 2 | 25 | 42.206 | 5 | 79.41 | 74.412 | 52.62 | 3.38 |
| S | 1 | 12.5 | 26.619 | 5.87 | 47.37 | 41.498 | 29.34 | 2.13 |

## 4 Conclusions

In this research, the APLIS method is applied to estimate the mean annual autogenic recharge expressed as a percentage of precipitation in the National Park of Viñales. From the application, it was necessary to modify the soil parameter, because the original method did not describe some of the soils' typologies in the area. Nevertheless, the application shows acceptable results, taking into account antecedent studies in the area because the method is developed considering the duality of the recharge in carbonate karst aquifers.

In National Park of Viñales, the map removal sensibility analyses shows the relative influence on the final recharge map which is L>I>S>P>A; it showed logical results. Map removal analysis showed that removing one of the parameters generates a notable variation in the resulting recharge map. In the single parameter sensitivity analysis, the altitude parameter is overvalued in the original method, and the effective weighting is higher from soil, slope and infiltration landforms parameters. The lithology showed approximately the same weighting.

The deviation between the empirical and real weighting is expected, because the effective weights calculated are the average of all real weight distributed spatially, and depend on the values of each parameter taken into account. To define an ideal weight is slightly effective and shows similar and not more efficient results. Nevertheless, sensitivity analysis is a useful tool to study how imposed perturbations (variations) of the inputs of geographical analysis affect the outputs of that analysis.

**Acknowledgements** The results were achieved due to the activities supported by the Management Plan of National Park Viñales, specially the investigation programs: "Hydrogeology and vulnerability of karst aquifers in the National Park Viñales" and "Integrated studies of physical environment for implementation of treatment residual water systems in communities of National Park Viñales", this last, in coordination with the University of Alcalá and IMDEA Agua Foundation.

## References

Andreo B, Vías J, Durán JJ, Jiménez P, López-Geta JA, Carrasco F (2008) Methodology for groundwater recharge assessment in carbonate aquifers: application to pilot sites in southern Spain. Hydrogeology Journal. 16 (5): 911–926

Farfán González H (2005) Modelación Matemática de Procesos Hidrológicos. Caso de Estudio: Sistema Cársico Palmarito-Novillo-Pan de Azúcar. Tesis Lic. Geografía. Fac. Geog. Univ. Habana. p 63

Farfán González H (2006) Potencial y recursos de agua explotable del Valle fluvio-cársico de Pan de Azúcar. Taller Ecovida. Delg. Prov. CITMA, PR. Libro de resúmenes

Gogu RC, Dassargues A (2000) Sensitivity analysis for the EPIK method of the vulnerability assessment in a small karstic aquifer, southern Belgium. Hydrogeology Journal. 8: 337–345

Jaime E., Luís M L., Olivera J (2006) Los suelos del Parque Nacional Viñales, Pinar del Río, Cuba. Condiciones genéticas y ambientales. Revista Cuadernos Geográficos (38). Instituto de Desarrollo Regional, Universidad de Granada: 195–205

Lodwik WA, Monson W and Svoboda L (1990) Attribute error and sensitivity analysis of map operations in Geographical Information Systems-suitability analysis. Int. J. Geog. Inf. Sist. 4: 413–428

Molerio León LF (2004) El enlace absorción-descarga de la Gran Caverna de Santo Tomás: evidencias derivadas de un ensayo con trazadores artificiales. Ing. Hidr. y Ambiental, XXV (3): 22–26 La Habana. Cuba

Napolitano P, Fabbri AG (1996) Single-parameter sensitivity analysis for aquifer vulnerability assessment using DRASTIC and SINTACS. HydroGIS 96: Application of Geographic Information Systems in Hydrology and Water Resources Management (Proceedings of the Vienna Conference, April 1996). IAHS Publ. no. 235. 559–556

# First Outcomes from New Approach in Assessing Recharge of Highly Karstified Terrains – Cases Examples from Montenegro

Milan Radulovic, Z. Stevanovic, and Micko Radulovic

**Abstract** The recharge of a karst aquifer, in terms of its quantity and spatial distribution depends on various natural factors, such as: climate, topography, vegetation, soil, geology etc. Selection of adequate method for assessing recharge in karst often represents matter of dispute. Multi-parameter methods using GIS tools have been recently successfully developed and applied in karstic terrains of Spain (Andreo et al. 2008) and Lebanon (Shaban et al. 2005). Specific local conditions as highly karstified terrains could additionally complicate such an assessment. The Montenegro karstified terrain, as southern part of External Dinarides, is characterized by very high precipitation rate, irregular seasonal distribution and absence of surface waters (high infiltration capacity of karst). Considering such conditions, an attempt to adapt existing knowledge and experiences and develop appropriate multi-parameter method for assessing spatial distribution of autogenous recharge has been made. Method KARSTLOP has been developed, applied and calibrated at catchments of several large karst springs of Montenegro. Obtained results in tested catchment areas indicate that with some further improvements the KARSTLOP method can support research of karstic aquifer in similar highly karstified terrains.

## 1 Introduction

Montenegro is situated in the southeastern part of Europe (Fig. 1). Its entire territory belongs to Alpine geo structure, to the Dinarides branch. The largest part of its

Milan Radulovic
Faculty of Civil Engineering, University of Montenegro, Cetinjski put bb, 81000 Podgorica, Montenegro, e-mail: radulovicmilan33@yahoo.com

Z. Stevanovic
Department of Hydrogeology, Faculty of Mining and Geology, University of Belgrade, Djusina 7, 11000 Belgrade, e-mail: zstev@eunet.rs

Micko Radulovic
Faculty of Civil Engineering, University of Montenegro, Cetinjski put bb, 81000 Podgorica, Montenegro, e-mail: mickor@ac.me

B. Andreo et al. (Eds.), *Advances in Research in Karst Media* 
DOI 10.1007/978-3-642-12486-0, 

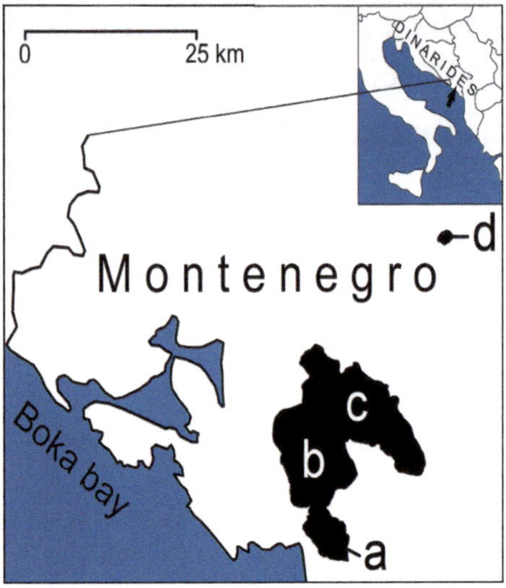


**Fig. 1** Geographical location of Montenegro and the four tested catchment areas: *a* Podgor springs; *b* Crnojevica springs; *c* Karuc springs; *d* Slatina springs

territory is highland. Carbonate rocks (limestone and dolomites) cover large parts of the area (over 60%) and are characterized by a very high degree of karstification.

External Dinarides karst is marked by specific characteristics, high infiltration capacity of karst being one of them. For example, the catchment area of the Bokakotor bay (covering the territory of 900 km$^2$), having the highest mean annual precipitation rate in the region (of approximately 2600 mm), has no permanent surface waters. High degree of karstification, lack of soil cover and poor vegetation, intensive precipitation, etc., have resulted in the most precipitation (over 70%) being infiltrated deep into the ground (Radulovic 2000).

Assessing spatial distribution of recharge in these karstified terrains represents a complex issue. The application of conventional methods (direct measuring on experimental sites, estimation of evapotranspiration by empirical equations, Darcy's law based methods, etc.) has not produced satisfactory results in these terrains. The most realistic recharge assessment is obtained by using the "black box" water budget method at the catchment areas, knowing input and output, but not all the other required system's parameters are known. However, in that case it is possible to calculate only the mean value for the entire catchment area, not including the spatial distribution. Very useful literary source for exploring groundwater recharge and standard measurement methods thereof is De Vries and Simmers (2002) and Scanlon et al. (2002).

In late 2007, research work started on developing a new method (KARSTLOP) which would provide the Map of spatial distribution of recharge in highly karstified terrains. Given the complexity of natural conditions, it is not possible to define its absolute terms. Therefore, the research has encompassed eight most important natural factors influencing recharge to obtain the most reliable assessment. Also, to facilitate the application process and increase the quality of the assessment, Remote Sensing and Geographic Information Systems (GIS) technologies have been used.

KARSTLOP is the regional method most convenient for generating maps in the scale of 1:25,000 to 1:100,000. Method is tested at catchments of four karst springs at the territory of Montenegro, the catchment areas of Podgor, Crnojevica, Karuc and Slatina springs (Fig. 1).

## 2 Methodology

KARSTLOP method is based on the GIS which facilitate spatial terrain analysis. Among earlier methods for groundwater recharge assessment in karst aquifers, it is important to mention APLIS method (Andreo et al. 2008) and the regional method developed by Shaban et al. (2005).

**Table 1** Parameters required and source of information for the preparation of a Recharge map following the KARSTLOP method

| Parameter | Description | Source |
|---|---|---|
| $K$ – Karstification ($K = 1$–5 points) | Karstification (surface and subsurface) is assessed on the basis of area of karst features and landforms per surface unit, distribution of a speleological objects and indirect parameters such as discharge amplitude of karst springs, velocity of artificial tracers and primary mineral saturation index of spring water. | Topographic maps, aerial and satellite images, speleological data, spring hydrographs, tracer tests results, hydrochemical data. |
| $A$ – Atmospheric conditions ($A = 1$–5 points) | Climatic conditions are mapped via altitude ($A_1$) and distribution of shadow areas ($A_2$). | Digital Elevation Model (DEM). |
| $R$ – Runoff ($R = 1$–5 points) | Distribution of permanent streams at the catchment area. It may indicate permeability of karstified terrains. | Topographic maps. |
| $S$ – Slope ($S = 1$–5 points) | Terrains with smaller slopes are described as being more suitable for groundwater recharge. | Digital Elevation Model (DEM). |
| $T$ – Tectonics ($T = 1$–5 points) | Two subfactors are being considered: density of faults $T_1$ and dip of strata $T_2$. | Geological maps, tectonics maps, aerial and satellite images. |
| $L$ – Lithology ($L = 0.7$–4.8 points) | $L$ map is obtained based on the following lithological data: the type of carbonate rock (subfactor a), bedding (subfactor b) and mineral-petrographic impurities (subfactor c). | Geological maps, lithological maps, borehole data, etc. |
| $O$ – Overlying layers ($O = 1$–5 points) | Type and thickness of geological cover and soil. | Pedological maps, geological maps, soil surveys, borehole data, geophisical data, etc. |
| $P$ – Plants ($P = 1$–5 points) | Type of vegetation cover. | Vegetation maps, landcover maps etc. |

KARSTLOP method has been developed for assessing autogenic recharge solely in karst terrains. The name is derived from the initial letters of selected factors: *Karstification* – $K$, *Atmospheric conditions* – $A$, *Runoff* – $R$, *Slope* – $S$, *Tectonics* – $T$, *Lithology* – $L$, *Overlying layers* – $O$ and *Plants* – $P$ (Table 1). The one of aim of developing such a method was to provide the most accurate categorization based on selected parameters to lessen the subjectivity in the application phase to the lowest possible level. Maps of factors show surfaces scored according to the established categorization (Radulovic 2009).

Final map showing the spatial distribution of mean annual recharge ($R_{\mathrm{ch}}$) expressed in percentages (%) is created from overlapping the maps of selected factors (Table 1) according to established methods and the presented algorithm.

## 3 First Application of the Method to Determine the Algorithm

The catchments of karst springs (Podgor, Crnojevica, Karuc, and Slatina springs) are selected due to availability of qualitative meteorological, hydrological, hydrogeological, hydrochemical and other data (Zivaljevic 1991; Zogovic 1992; Radulovic 1994; IJC 2001; etc.). Data on the amount of recharge obtained from previous investigations have been used for calibrating the method, i.e., for finding the most suited algorithm. One which best matchs with the results obtained in previous measurements at tested catchment areas has been chosen (Radulovic 2009):

$$R_{\mathrm{ch}} = 4 \times K + A + 4 \times R + 2 \times S + T + 4 \times L + 3 \times O + P \qquad (1)$$

By looking at algorithm it has been concluded that karstification, runoff and lithology have the most influence to recharge in these terrains. The recharge ($R_{\mathrm{ch}}$) is expressed as percentage (%) of precipitation. It is known that its rate is regularly between 60–80% in these terrains.

## 4 Results

The best matching between calculated and measured data has been obtained for the catchment area of Karuc springs. The recharge value of $R_{\mathrm{ch}} = 70.5\%$ obtained by application of other methods, mainly water budgeting and hydrologic analysis is almost equal to mean value for entire catchment of $R_{\mathrm{ch}} = 70.4\%$ obtained by using KARSTLOP method. Somewhat slightly weaker matching has been yielded for the catchment area of Slatina springs, with the measured value of $R_{\mathrm{ch}} = 69.7\%$ and calculated one $R_{\mathrm{ch}} = 66.3\%$. For the catchment area of Podgor springs, previously obtained result was $R_{\mathrm{ch}} = 76.2\%$, while KARSTLOP method yielded 71.5%. The biggest deviation has been noticed for the catchment area of Crnojevica springs, where earlier obtained result was $R_{\mathrm{ch}} = 76.1$ while newly calculated average value

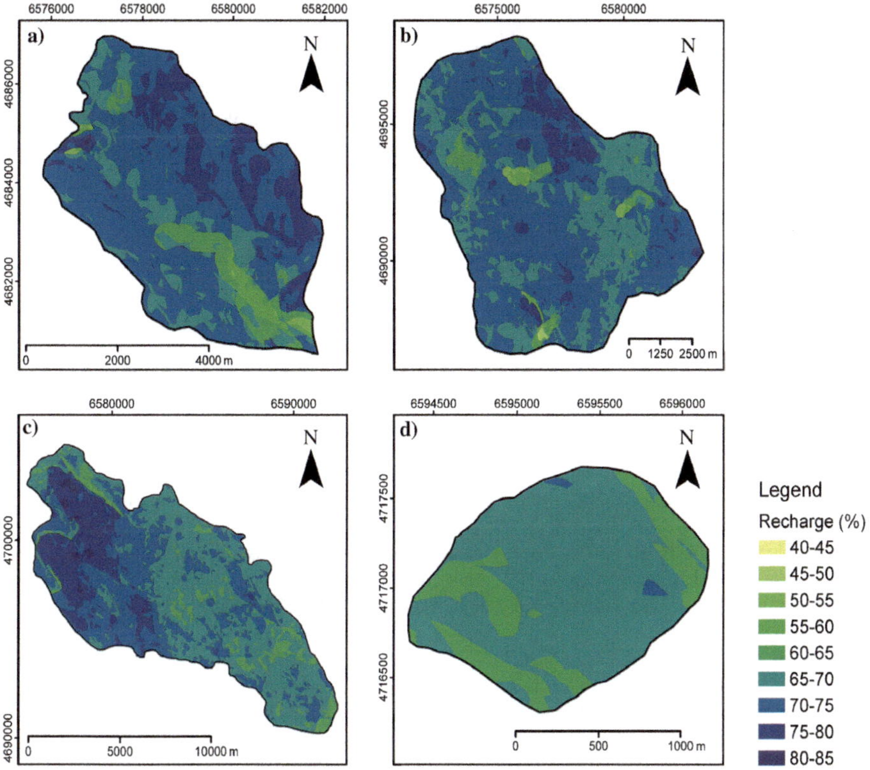


**Fig. 2** Maps of spatial distribution of recharge obtained by KARSTLOP method for catchments **a** Podgor springs; **b** Crnojevica springs; **c** Karuc springs; **d** Slatina springs

is 71.0% (Radulovic 2009). Therefore, deviation for selected catchment areas is between 0.1 and 5.10%. Figure 2 shows final maps of spatial distribution of recharge obtained by KARSTLOP method.

## 5 Conclusions

Method KARSTLOP has been applied and calibrated at catchments of several large karst springs of Montenegro. This method is aimed at mapping the spatial distribution of mean annual recharge of karstic aquifer in specific highly karstified terrains. The application of KARSTLOP method has provided a regional-scale map of spatial distribution of recharge (%). By comparing the method's outcomes with previously obtained results by conventional methods, it has been concluded that possible error in the assessment of mean annual recharge at the catchment area is approximately 5%, which is more than successful as an initial result. The final map as well as obtained layers can be used for addressing numerous practical karst water management issues such as: water supply (discharge assessment, spatial distribution of watersheds, etc.), transboundary water management, construction of underground objects, preliminary assessment of groundwater vulnerability, etc. In the forthcoming studies it will be necessary to provide further testing of the method at more catchment areas to highlight the method's possible shortcomings and take corrective actions if needed.

## References

Andreo B, Vías J, Durán JJ, Jiménez P, López-Geta JA, Carrasco F (2008) Methodology for groundwater recharge assessment in carbonate aquifers: application to pilot sites in southern Spain. Hydrogeol J, 16:911–925

De Vries JJ, Simmers I (2002) Groundwater recharge: an overview of processes and challenges. Hydrogeol J, 10:5–17

IJC (Institut Jaroslav Cerni) (2001) Water Master Plan of Montenegro. Government of Montenegro, Podgorica

Radulovic M (1994) Report on hydrogeological exploration of Sjenokos locality in Crmnica area – I phase (in Serbian). University of Montenegro, Podgorica

Radulovic M (2000) Karst hydrogeology of Montenegro (in Serbian). Geological Survey of Montenegro, Podgorica

Radulovic MM (2009) KARSTLOP method – Multiparameter analysis of karstic terrains potential for effective infiltration (in Serbian). Fac. Min. Geol., University of Belgrade, Belgrade

Scanlon BR, Healy RW, Cook PG (2002) Choosing appropriate techniques for quantifying groundwater recharge. Hydrogeol J, 10:18–39

Shaban A, Khawlie M, Abdallah C (2005) Use of remote sensing and GIS to determine recharge potential zones: the case of Occidental Lebanon. Hydrogeol J 14:433–443

Zivaljevic R (1991) Hydrological analysis of karst water flow – Crnojevica rijeka River case study (in Serbian). PhD Thesis, University of Montenegro, Podgorica

Zogovic D (1992) Final report of hydrogeological exploration of Karuc source (in Serbian). Energoprojekt, Beograd

# Climate Change Effects on Aquifer Recharge in a Glacierised Karst Aquifer System, Tsanfleuron-Sanetsch, Swiss Alps

V. Gremaud and N. Goldscheider

**Abstract** Alpine glaciers store considerable amounts of freshwater contributing to groundwater recharge in the warm season, but are rapidly retreating and many will disappear within 50 years. The Tsanfleuron-Sanetsch area in the Swiss Alps is an ideal site to study glacier-aquifer interactions; it consists of a vanishing glacier overlying a karst aquifer drained by a spring (Glarey) used for drinking water supply. Between the glacier and the moraine from 1855 (Little Ice Age), the karst surface is polished by ice flow; typical karrenfields are present below the moraine. Geologically, the area consists of folded Jurassic to Paleogene sedimentary rocks forming an anticlinorium limited by a narrow syncline. Relationships between stratigraphic and tectonic setting, recharge processes and aquifer drainage have been studied by means of tracer tests and hydrologic monitoring. The glacier's geometry was investigated by geophysical surveys, using radiomagnetotelluric (RMT): The estimated ice volume is $1.0 \times 108\,m^3$, corresponding to $0.92 \times 108\,m^3$ freshwater. Meltwater production displays diurnal and seasonal variability influencing the shape of tracer breakthrough curves and, thus, flow and transport in the aquifer. A preliminary prognosis of water availability when the glacier will have disappeared has been established, predicting more available water in winter, but shortages in long dry summer and autumn periods, due to the absence of meltwater and less estival rainfall. Increased irrigation water demand will aggravate water shortage.

## 1 Introduction

Rapidly retreating glaciers are clearly visible symptoms of climate changes in the Alps. Many glaciers have lost much of their volume since the maximum of the Little

V. Gremaud, N. Goldscheider
Centre for Hydrogeology, University of Neuchâtel, Rue Emile Argand 11, 2009 Neuchâtel, Switzerland, e-mail: vivian.gremaud@unine.ch; nico.goldscheider@unine.ch

B. Andreo et al. (Eds.), *Advances in Research in Karst Media*
DOI 10.1007/978-3-642-12486-0, © Springer 2010

Ice Age in the middle of the 19th century (Greene et al. 1999). Climate models predict warmer temperatures and more precipitation during winter, with a snowline at higher elevations, and less rain in summer. The loss of glaciers as intermediate water storage reservoirs will result in local or temporal water shortages, particularly during long dry summer periods (Seidel et al. 1998), with implications for drinking water supply, agriculture and hydropower (Viviroli and Weingartner 2004).

The Tsanfleuron-Sanetsch area in the Western Swiss Alps consists of a karst aquifer overlain and recharged in its upper part by a rapidly retreating glacier. At the lowest point, the Glarey spring drains the major part of the aquifer (Fig. 1). The spring is used for drinking water supply of a community (Conthey) and for irrigation. This site is ideal to study glacier-aquifer interrelations and impacts of climate-change on water resources.

The Tsanfleuron glacier was intensively studied by glaciologist (e.g. Chandler et al. 2008; Hubbard 2002; Hubbard and Hubbard 1998; Hubbard et al. 2000; Sharp et al. 1989). More recent studies investigate the relations between the geological structure and the drainage pattern in the karst aquifer (Gremaud et al. 2009) and glacier-aquifer relations (Gremaud and Goldscheider 2010).

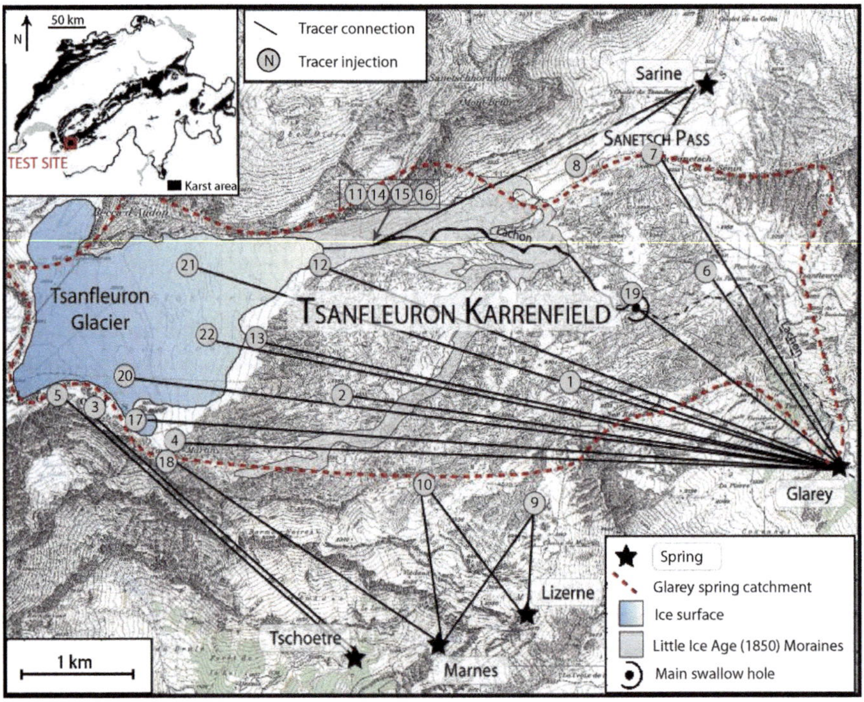


**Fig. 1** Location and general overview of the Tsanfleuron-Sanetsch glacier and karst aquifer system in the Western Swiss Alps. 22 tracer tests made it possible to delineate the recharge area of the Glarey spring (modified from Gremaud and Goldscheider 2010)

This article focuses on climate change impacts on the glacier and aquifer dynamics, and also presents estimations on glacier volume and meltwater availability in the near future.

## 2 Geologic, Hydrogeologic and Glaciological Setting

### *2.1 Geology*

The study area belongs to the Helvetic zone of the Alps and consists of Jurassic to Paleogene limestone, sandstone and marl. The Tsanfleuron glacier overlies 100–120 m thick Cretaceous limestone (Urgonian or Schrattenkalk) that outcrops on large parts of the land surface. This formation is a very karstifiable geologic unit in the Alps, due to its mechanical strength and high mineralogical purity (Goldscheider 2005). The underlying formation mainly consists of marl, about 100 m thick. The strata are folded, with SW-NE trending fold axes (Steck et al. 2001). The entire zone between the glacier and the Sanetsch pass is formed by a large and wide anticline dipping toward the NE with an angle of about 10°. The southward bordering narrow syncline forms the limit of the karst system.

### *2.2 Hydrogeology*

The hydrogeology of this karst system has been studied by means of multi-tracer tests with a total of 22 injections (Gremaud et al. 2009; Gremaud and Goldscheider 2010). The system is drained by five springs. The major part of the karst surface and glacier is drained by the Glarey spring in the SE (Fig. 1), with transit times of 5–57 h (peak times of tracer breakthrough curves) and recoveries of 5–80%. Water flow is parallel to the strata, from the crest of the wide anticline, downward its SE limb toward the narrow syncline, which collects all water and conducts it to the spring at 1550 m a.s.l. altitude. The other springs drain only marginal parts of the karrenfield. The water of the three springs in the S crosses the entire stratigraphic sequence, including several marl aquicludes, probably along deep fractures, with transit times of 15–32 h. The Sarine spring drains a small strip of karst limestone in the NE part of the area, with transit times around 12 h (Gremaud et al. 2009).

### *2.3 Glaciology*

Tsanfleuron glacier has a mean thickness of 35 m (Gremaud and Goldscheider 2010) and can be subdivided in two parts. A thick northern part filling a valley with a maximum thickness of 138 m, and a central and southern zone consisting of a thin layer of ice (locally only 15 m thick) on top of the aquifer (Fig. 2).

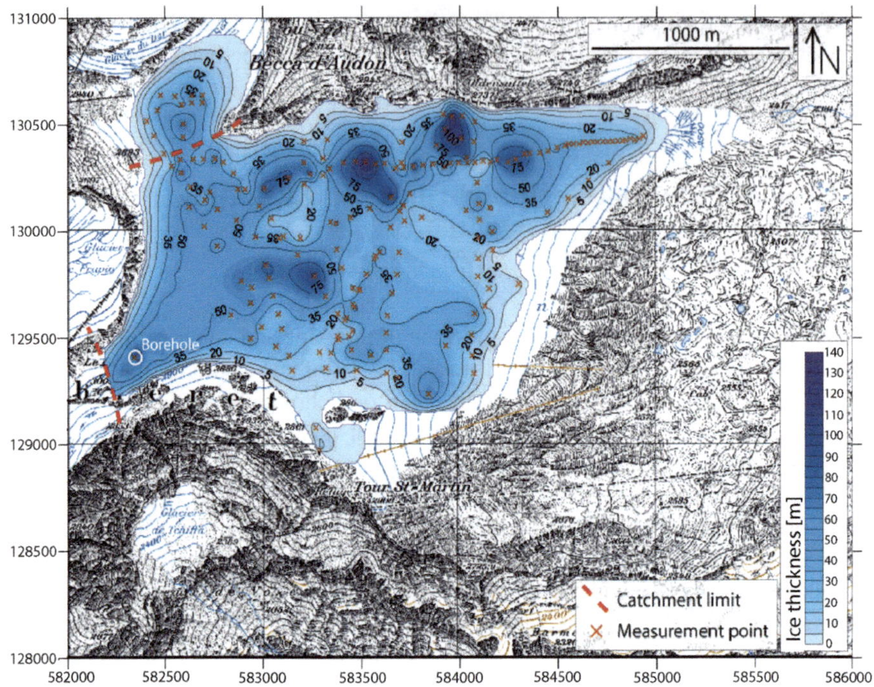


**Fig. 2** Ice-thickness map of the glacier (2007/2008), based on RMT measurements and a borehole. The thick northern part of the glacier forms a glacier tongue; the southern sector is a thin ice layer overlying limestone. The heterogeneous pattern of ice thickness reflects the uneven morphology below the glacier (Gremaud and Goldscheider 2010)

The land surface below the glacier is not plane but presents several wide depressions and mounts, with a subglacial morphology similar to the karst landscape in the glacier forefield. The ice volume was estimated at $1.0 \times 108\,m^3$, corresponding to $0.92 \times 108\,m^3$ freshwater available for aquifer recharge (Gremaud and Goldscheider 2010), assuming an ice/water density ratio of 0.92 (Benn 1998).

## 3 Outlook

Glaciers accumulate snow and ice in winter and release meltwater in summer. Under equilibrium conditions (no retreat or advancement), glaciers do not change much the annual water budget of a basin, but mainly influence the annual variability of flow by delivering meltwater during warm periods. However, data of the Swiss Glacier Monitoring Network (2009), satellite data (Paul et al. 2004) and models show that glaciers are shrinking and will continue to do so.

For Tsanfleuron glacier, field observations indicate that it loses about 15 m length and 1.5 m thickness per year. In 2008, the glacier covered a surface of $2.8\,km^2$, so

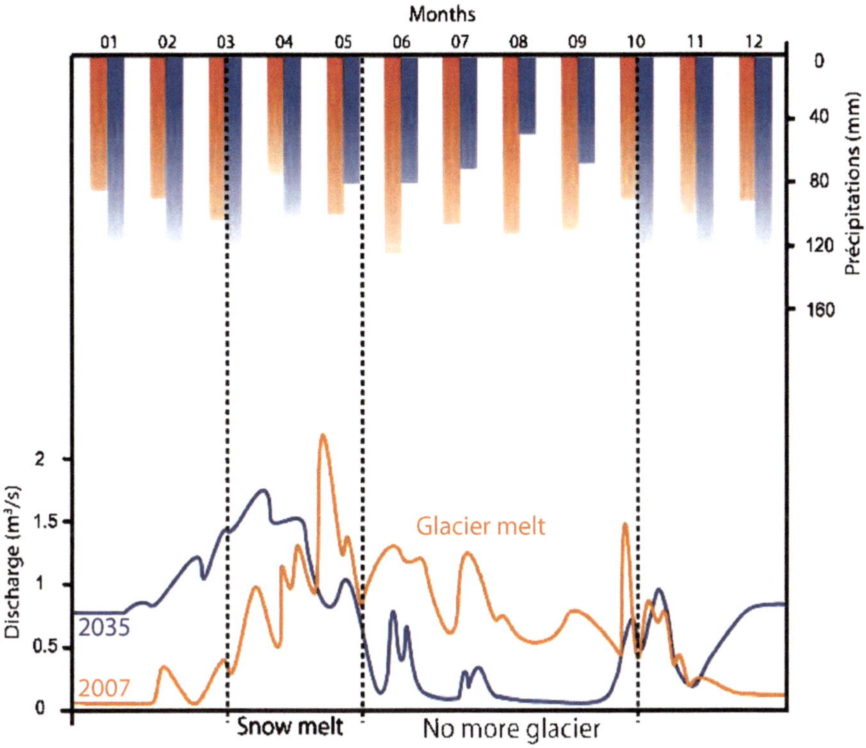


**Fig. 3** Comparison of hydrograph at Glarey spring in 2006 (available data) and 2035 (prognosis). Water availability will decrease during summer season, but increase during winter by more snowmelt and more rain at high elevation

the estimated ice loss is $4.2 \times 106\,m^3$/year, corresponding to a freshwater volume of $3.9 \times 106\,m^3$/year or a mean flow rate of 120 L/s (Gremaud and Goldscheider 2010). Evaporation losses from glaciers are small compared to meltwater production (Musy and Higy 2004). Thus, most of this flow is supposed to contribute to aquifer recharge. This quantity represents a transient surplus of water that is available now but will be missing when glacier will have disappeared (Fig. 3).

As a conclusion, water shortage has to be expected during long dry summer and autumn periods in the future, when at the same time, water demand for irrigation and others purposed will increase.

## References

Benn DI (1998) Glaciers and Glaciation. Arnold Publishers, 760 p

Chandler D, Hubbard B, Hubbard A, Murray T, Rippin D (2008) Optimising ice flow law parameters using borehole deformation measurements and numerical modelling. Geophysical Research Letters, 35(12)

Goldscheider N (2005) Fold structure and underground drainage pattern in the alpine karst system Hochifen-Gottesacker. Eclogae Geologicae Helvetiae, 98(1): 1–17

Greene AM, Broecker WS, Rind D (1999) Swiss glacier recession since the Little Ice Age: Reconciliation with climate records. Geophysical Research Letters, 26(13): 1909–1912

Gremaud V, Goldscheider N, Savoy L, Favre G, Masson H (2009) Geological structure, recharge processes and underground drainage of a glacierised karst aquifer system, Tsanfleuron-Sanetsch, Swiss Alps. Hydrogeology Journal, 17(8): 1833–1848

Gremaud V, Goldscheider N (2010) Geometry and drainage of a retreating glacier overlying and recharging a karst aquifer, Tsanfleuron-Sanetsch, Swiss Alps. Acta Carsologica, submitted

Hubbard B (2002) Direct measurement of basal motion at a hard-bedded, temperate glacier: Glacier de Tsanfleuron, Switzerland. Journal of Glaciology, 48(160): 1–8

Hubbard B, Hubbard A (1998) Bedrock surface roughness and the distribution of subglacially precipitated carbonate deposits: Implications for formation at Glacier de Tsanfleuron, Switzerland. Earth Surface Processes and Landforms, 23(3): 261–270

Hubbard B, Tison JL, Janssens L, Spiro B (2000) Ice-core evidence of the thickness and character of clear-facies basal ice: Glacier de Tsanfleuron, Switzerland. Journal of Glaciology, 46(152): 140–150

Musy A, Higy C (2004) Hydrologie 1, une science de la nature. Presses polytechniques et universitaires romandes, Lausanne, 309 p

Paul F, Kaab A, Maisch M, Kellenberger T, Haeberli W (2004) Rapid disintegration of Alpine glaciers observed with satellite data. Geophysical Research Letters, 31(21)

Seidel K, Ehrler C, Martinec J (1998) Effects of climate change on water resources and runoff in an Alpine basin. Hydrological Processes, 12(10–11): 1659–1669

Sharp M, Gemmell JC, Tison JL (1989) Structure and stability of the former subglacial drainage system of the Glacier de Tsanfleuron, Switzerland. Earth Surface Processes and Landforms, 14(2): 119–134

Steck A, Epard JL, Escher A, Gouffon Y, Masson H (2001) Carte tectonique des Alpes de Suisse occidentale et des régions avoisinantes 1:100000. Notice explicative, Service géologique national, Bern

Swiss Glacier Monitoring Network (2009) http://glaciology.ethz.ch/messnetz/index.html

Viviroli D, Weingartner R (2004) The hydrological significance of mountains: from regional to global scale. Hydrology and Earth System Sciences, 8(6): 1016–1029

# Recharge Enhancement and Protection of a Karst Aquifer in Central Texas

B.A. Smith, B.B. Hunt, and J. Beery

**Abstract** The Edwards Aquifer in central Texas, like many aquifers in the world, is threatened by over exploitation and contamination due to rapidly increasing populations in the aquifer recharge and contributing zones. Close to 2 million people depend on the Edwards Aquifer as their sole source of drinking water. Studies have shown that under extreme drought conditions, many wells would go dry and springflows would drop to such low rates that endangered species that live in the springs would be jeopardized. To address these concerns, a concrete vault was constructed over Antioch Cave that is located in the channel of Onion Creek. Large amounts of water enter this cave when water is flowing in the creek. However, during periods of stormwater flow, the water contains varying amounts of sediment, bacteria, and other contaminants. A system was constructed over the cave entrance to keep the contaminated water from entering the cave. This system also increases the amount of good quality water entering the aquifer. Water levels in the aquifer and springflow rates can be maintained at higher amounts because of this increased recharge. This will help protect wells from going dry and will provide more springflow for the endangered species at the springs.

## 1 Introduction

The Edwards Aquifer extends across a portion of central Texas that is 11,300 $km^2$ in area. The length of the aquifer is about 430 km (Fig. 1 inset). Lithologic units that make up the aquifer are mostly limestone and dolomite. Where the full thickness of the aquifer is present, thicknesses vary from about 90 m to 215 m. Water flows from source areas to springs and wells through an integrated system of conduits with additional flow through a matrix of primary and secondary voids between the conduits.

B.A. Smith, B.B. Hunt, J. Beery
Barton Springs/Edwards Aquifer Conservation District, 1124 Regal Row, Austin, Texas 78748, USA, e-mail: brians@bseacd.org; brianh@bseacd.org; joseph@bseacd.org

B. Andreo et al. (Eds.), *Advances in Research in Karst Media* 
DOI 10.1007/978-3-642-12486-0, 

Fig. 1 Location map of the Barton Springs segment of the Edwards Aquifer

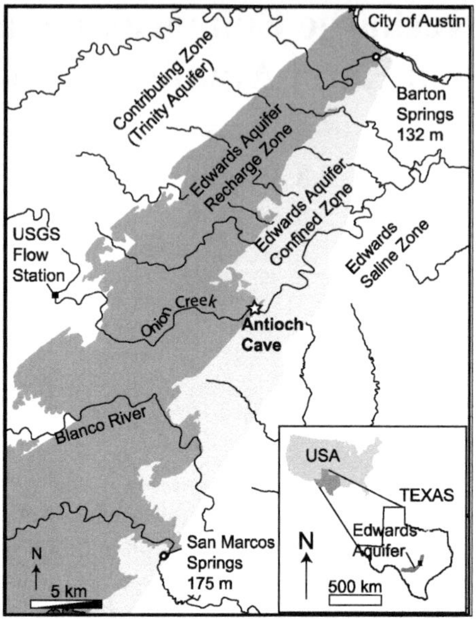


The majority of recharge occurs in stream channels that cross the recharge zone of the Edwards Aquifer. Recharge also occurs in upland areas with flow into caves and sinkholes, and by infiltration through soils. The development of the aquifer was influenced significantly by fracturing and normal faulting associated with the Balcones Fault Zone and dissolution of limestone and dolomite by infiltrating meteoric water (Sharp 1990).

## 2 Barton Springs Segment of the Edwards Aquifer

The Barton Springs segment of the Edwards Aquifer represents an area of about 400 $km^2$. This segment of the aquifer is bounded to the north by the Colorado River (Fig. 1). The main discharge point for groundwater in this segment is Barton Springs, located on Barton Creek about 1 km upstream of the confluence of Barton Creek and the Colorado River. Long-term average discharge from Barton Springs is 1.5 $m^3$/s. Highest recorded flow rates are 4.7 $m^3$/s. The lowest flow rate ever measured at Barton Springs is 0.3 $m^3$/s, which occurred in 1956 at the end of a 7-year drought. Groundwater withdrawals for domestic, industrial, commercial, and irrigation purposes are estimated to be about 9 $hm^3$/year. Approximately 60,000 people depend on the Barton Springs segment as their sole source of drinking water.

A pool has been constructed around the main outlet at Barton Springs. This pool and surrounding park serves as a major recreational attraction for the City of Austin. An endangered species of salamander lives in the four spring outlets that make up the Barton Springs complex. Degradation of water quality and loss of flow are the main threats to the continued survival of the salamanders.

## 3 Recharge Enhancement System at Antioch Cave

Earlier studies of recharge to the Barton Springs segment indicated that 85% of recharge occurs in the creek beds that cross the recharge zone (Slade et al. 1986). An additional 15% of recharge was estimated to occur in the upland areas. More recent studies based on detailed measurements of evapotranspiration, soil moisture, and runoff into recharge features such as caves and sinkholes, suggest that the amount of upland recharge is greater than that presented in the earlier studies (Hauwert 2009). Antioch Cave, situated within the channel of Onion Creek, is the most significant recharge feature in the Barton Springs segment. Because it is located in a stream channel, Antioch Cave has the largest capacity for recharge of any cave in the Barton Springs segment. Prior to the work performed for this study, the cave would periodically become plugged with rocks and debris carried into the cave by stormwater. When the cave was not plugged, stormwater would carry debris and contaminants into the cave and subsequently into the aquifer.

Water entering the cave has been traced to wells and Barton Springs using organic dyes. Under hydraulic conditions at the time of one tracer test, water travelled to Barton Springs in seven days, a straight-line distance of about 20 km (Fig. 2). Dye was also detected in wells located northeast, southwest, and northwest of the cave. This suggests radial flow from a groundwater mound developed when large amounts of water enter an aquifer in a limited area. A potentiometric surface map shows a significant groundwater mound that developed beneath Antioch Cave during wet conditions in 2002 (Fig. 2).

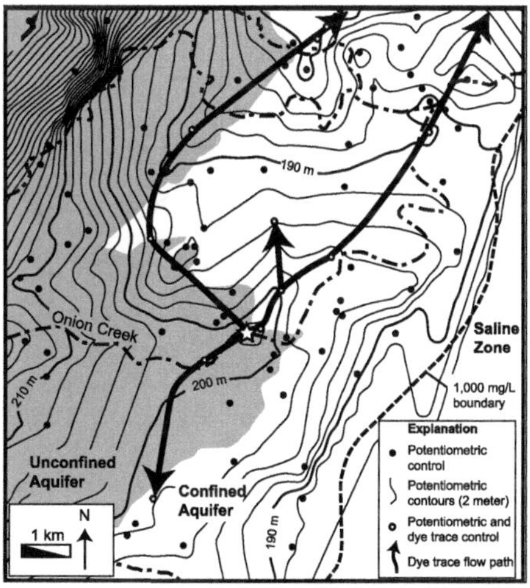


**Fig. 2** Potentiometric surface map in the vicinity of Antioch Cave (*star symbol*) and flow paths documented by tracer testing. *Hollow circles* represent wells in which dye was detected

In 1997, a concrete vault was constructed over the entrance to Antioch Cave (Fig. 3). A 0.9-m diameter butterfly valve was installed in the side of the vault. The valve would be manually closed to hold out contaminant-laden stormwater or opened to allow cleaner, baseflow water to enter the cave. In 2008, a second valve was installed in the vault. This valve can be automatically opened and closed depending on turbidity levels of water in the creek. In addition, an intake screen was installed to filter out sediment and debris from stormwater (Fig. 4). Automation of the process and filtration of stormwater is expected to significantly increase the efficiency of the system. This should increase the amount of good quality water entering the cave and recharging the aquifer.

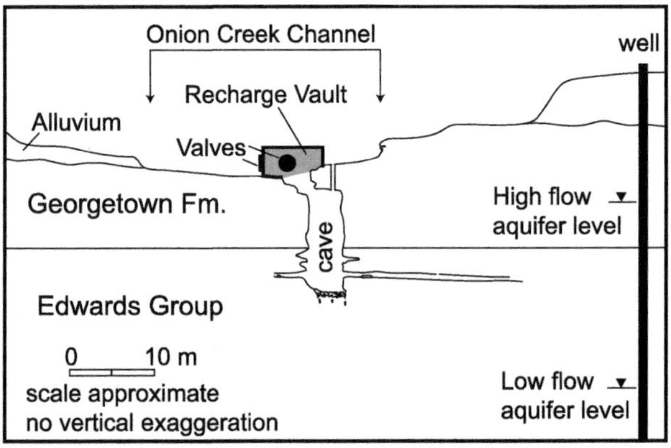


**Fig. 3** Schematic cross section through Antioch Cave. Direction of view is northwest (upstream)

**Fig. 4** Upstream view of the vault and screen constructed above the entrance to Antioch Cave. Length of intake screen is 10 m

## 4 Results of Flow Events at Antioch Cave

Following completion of the second phase of the project, an ongoing drought prevented testing of the system. In September and October 2009, adequate rains occurred that led to two flow events at Antioch Cave. Both events were small compared to other events in the past, but there was sufficient flow from each event that the automated valve was triggered and water-quality data were recorded by the monitoring instruments (Fig. 5). The September 29 flow event lasted only 16 hours. The highest flow rate through the automated valve was 2.4 cms. The October 27 flow event lasted 32 hours with similar rates of inflow.

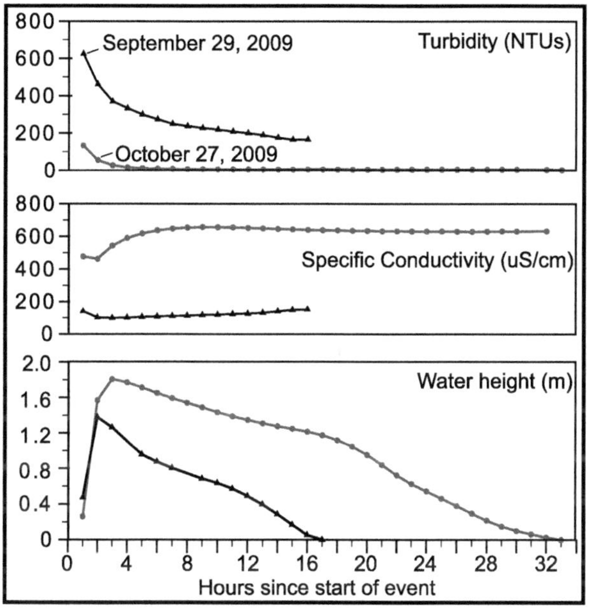


**Fig. 5** Charts of data from two storm events at Antioch Cave

## 5 Conclusions

During extended periods of above-average rainfall, flow at Antioch Cave can continue uninterrupted for as long as eight months. With extended periods of flow, the amount of recharge to the aquifer through Antioch Cave could be significant enough to increase the amount of water in storage which will help maintain water levels in wells and springflow at Barton Springs during periods of drought. If the enhanced recharge system at Antioch Cave functions at its maximum efficiency, it can potentially increase recharge to the aquifer by about 25% during periods of average rainfall. This additional water going into storage can reduce the impact of drought on wells and the endangered species at Barton Springs.

## References

Hauwert, NM (2009) Groundwater Flow and Recharge within the Barton Springs Segment of the Edwards Aquifer, Southern Travis County and Northern Hays Counties, Texas: PhD Dissertation, University of Texas at Austin, Texas. 328 p

Sharp J (1990) Stratigraphic, Geomorphic and Structural Controls of the Edwards Aquifer, Texas, USA: Simpson E, and Sharp J Jr, eds., Selected Papers on Hydrogeology: Heise, Hannover, Germany, International Association of Hydrogeologists. 1. 67–82

Slade RM Jr, Dorsey ME, Stewart S (1986) Hydrology and water quality of the Edwards Aquifer associated with Barton Springs in the Austin area, Texas. US Geological Survey Water-Resources Investigations Report 86-4036: 96

# Flow Potential Between Stacked Karst Aquifers in Central Texas, USA

B.A. Smith and B.B. Hunt

**Abstract** The Cretaceous-age Middle Trinity Aquifer in central Texas exhibits significant karst features where it is exposed at the surface. Where these units are overlain by the karstic Upper Trinity and Edwards Aquifers, the degree of karstification is not known, but groundwater in certain units of the Middle Trinity Aquifer is chemically similar to shallow groundwater in this area. Geochemical, potentiometric, and hydraulic conductivity data suggest that this deep groundwater is following pathways from the recharge areas that allow for rapid flow compared to other deep units that contain waters of significantly different chemistry. A 340-m deep multiport monitor well was installed in the Edwards (shallowest), and Upper and Middle Trinity Aquifers to better understand the vertical and horizontal relationships of these aquifers. The well was completed with 14 monitor zones allowing for groundwater sampling, hydraulic conductivity testing, and pressure measurements (potentiometric levels) in each zone. Data from this well suggest that karstification processes are active at depths of 300 m below land surface in a confined aquifer at least 25 km from the closest recharge areas.

## 1 Introduction

The majority of studies of the Edwards and Trinity Aquifers in central Texas are based on data collected from water-supply wells and very few scientifically-designed monitor wells. The purpose of this study was to collect data from discrete intervals of the Edwards and Trinity Aquifers (Fig. 1) that could provide insight to the potential for flow within and between the Edwards and Trinity Aquifers so that groundwater availability and quality issues for these aquifers may be better addressed.

B.A. Smith, B.B. Hunt
Barton Springs/Edwards Aquifer Conservation District 1124 Regal Row, Austin, Texas 78748, USA, e-mail: brians@bseacd.org; brianh@bseacd.org

B. Andreo et al. (Eds.), *Advances in Research in Karst Media*
DOI 10.1007/978-3-642-12486-0, © Springer 2010

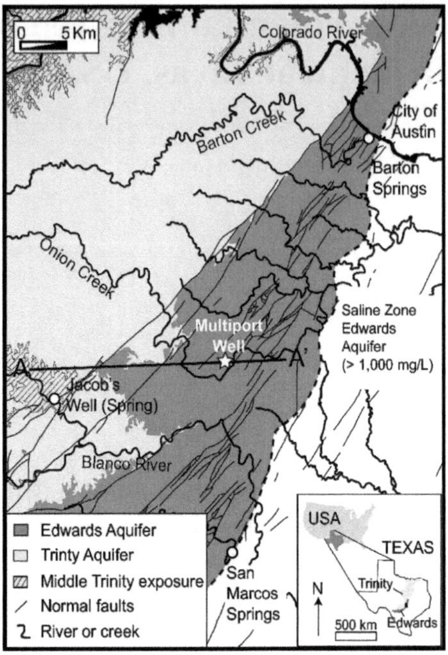


Fig. 1 Location map of study area

## 2 Geology

The geologic units that make up the Edwards Aquifer are mostly limestone and dolomite. The Trinity Aquifers are composed of (from stratigraphically highest to lowest) the Upper Glen Rose Limestone, Lower Glen Rose Limestone, Hensell Sand, Cow Creek Limestone, and the Hammett Shale (Fig. 2). The Upper and Lower

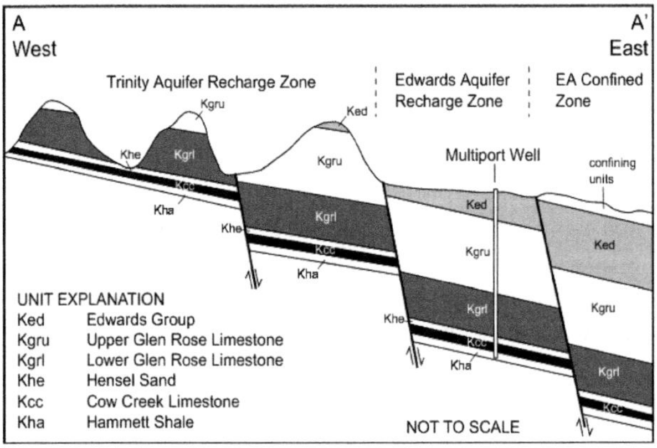


Fig. 2 Schematic cross section showing stratigraphic and structural relationships of geologic units

Glen Rose Limestones consist mostly of limestone, dolomite, shale, and marl. Some units of the Upper Glen Rose Limestone contain evaporates.

Studies of structures along the Balcones Fault Zone (BFZ) (Collins and Hovorka 1997) indicate that much of this area consists of southeast dipping, *en echelon*, normal faults with throws of as much as 260 m. Some of these faults are continuous over many kilometers, while others extend only a few kilometers or less. Fault blocks between the points where fault displacement decreases to zero are called relay ramps. The relay ramps transfer displacement from one fault to an adjacent fault (Collins 1995). Faulting along the BFZ plays a significant role in the development of the Edwards and Trinity Aquifers.

# 3 Methodology

Most aquifer studies rely on data that are obtained from wells that penetrate large sections of an aquifer, or multiple aquifers, or from wells with very limited completion intervals. Monitoring of many discrete intervals is needed to provide data

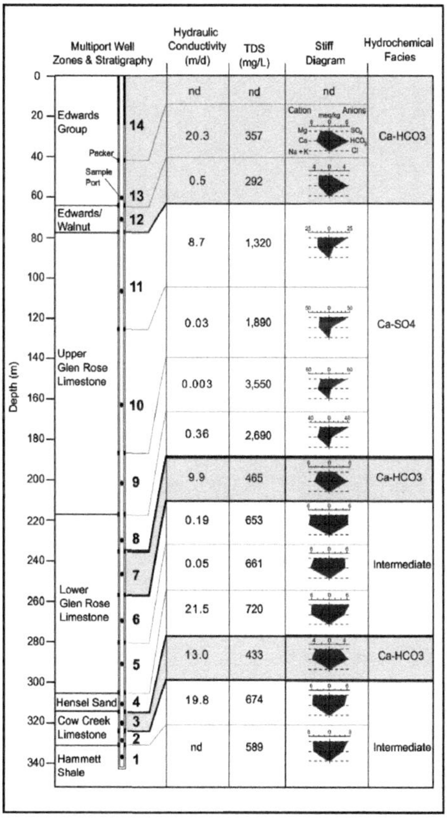


**Fig. 3** Construction diagram of multiport monitor well with stratigraphy, hydraulic conductivity data, geochemical data, and hydrochemical facies

that reflect the true complexity of most aquifers. To address these issues, a multiport well, using material manufactured by Westbay Instruments, Inc. (Schlumberger), was installed with 14 discrete monitoring zones in the Edwards, Upper Trinity, and Middle Trinity Aquifers (Figs. 2 and 3). The multiport system used in this study allows potentiometric and hydraulic conductivity measurements and groundwater sample collection from 14 discrete zones within a single borehole. A 12-cm diameter borehole was drilled to a depth of 340 m and equipped with permanent packers to isolate the 14 zones into separate hydrostratigraphic units of the Edwards and Trinity Aquifers. The average thickness of the multiport zones in the well is 22 m with the thickest zone at 60 m and the thinnest zone at 8 m (Fig. 3).

# 4 Results

Potentiometric data were collected over a 22-month period in all 14 zones in the multiport well (Fig. 4). The region was experiencing severe drought conditions during the first 20 months of data collection. Hydraulic heads in the uppermost five

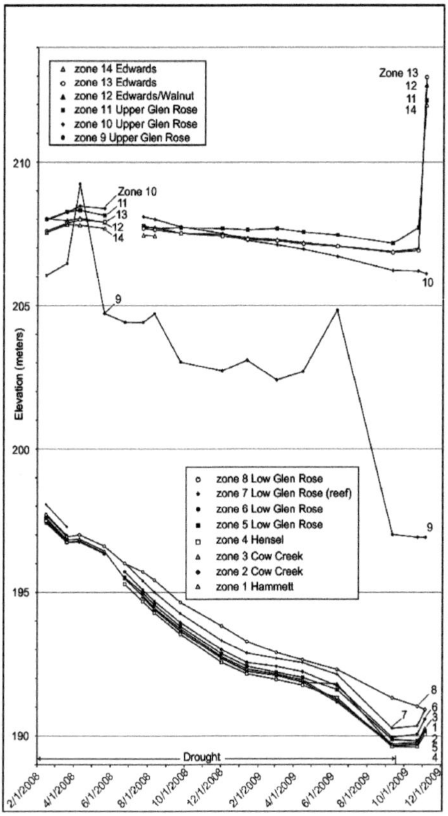


**Fig. 4** Potentiometric head data from multiport monitor well

zones were up to 22 m higher than hydraulic heads in the lowermost eight zones. Potentiometric data indicate a potential for flow from the upper zones consisting of Edwards and Upper Trinity units toward the lower zones composed of Middle Trinity units.

Groundwater samples collected from the multiport well in June 2009 were analyzed for total dissolved solids (TDS), major anions and cations, and certain metals. A summary of the results of these analyses are presented in Fig. 3. A review of select cation and anion concentrations shows that the samples can be divided into three distinct groupings (hydrochemical facies): Ca–$HCO_3$, Ca–$SO_4$, and an intermediate facies.

## 5 Interpretation

Differences in hydraulic head values show that there is a potential for flow between the various aquifer units, but pathways are needed for flow to actually take place. Considering that there are many low permeability beds in the upper and lower members of the Glen Rose Limestone (Ashworth 1983), vertical flow would likely occur only through faults and fractures. Declining heads in the Middle Trinity zones throughout most of the study period suggests a hydrologic regime that is well connected to a source area. The most likely recharge areas for the Middle Trinity Aquifer are about 25 km to the west of the multiport well where these units are exposed at the surface.

Significant geochemical differences between monitoring zones provide additional evidence that there is virtually no vertical flow through the formations. Hydraulic head values and geochemistry suggest that any flow along faults in this area is very limited.

Groundwater samples with the highest values of TDS and sulfate are associated with low permeability units of limestone, dolomite, marl, and shale with evaporites. Groundwater samples dominated by calcium and bicarbonate are associated with high permeability units of limestone and dolomite with few evaporites, if any. Samples with intermediate values of calcium and sulfate are associated with units with varied permeabilities that are predominantly limestone, dolomite, sand, silt, and shale.

Where fault displacement is greater than the thickness of high permeability units, flow through the units can be cut off if low permeability units are juxtaposed against high permeability units. However, many of these faults are relay-ramp structures that start at a point with no displacement then increase in displacement over the strike of the fault. Therefore, there is considerable lateral continuity of permeable units, even though some portions of the permeable units may be entirely offset by faults.

## 6 Conclusions

Significant differences in head and geochemical values suggest that there is little vertical flow between zones in the vicinity of the multiport well. The large differences in geochemical values between zones and the presence of distinct hydrochemical facies suggest that flow along faults in this area is small compared to horizontal flow in each zone. Extensive faulting associated with the BFZ might suggest that lateral flow would be fairly limited due to offsetting of permeable beds, however the presence of relay ramps between faults indicates that there is considerable lateral continuity of permeable beds.

Potentiometric, structural, and geochemical data support the interpretation that groundwater encountered in the permeable zones of the Middle Trinity is influenced more by lateral flow though distinct lithologies rather than flow along faults. Karst processes have affected these deep limestones by solutionally enlarging fractures and bedding planes.

## References

Ashworth, JB (1983) Ground-Water Availability of the Lower Cretaceous Formations in the Hill Country of South-Central Texas: Texas Department of Water Resources, Report 273, 172 p

Collins, EW (1995) Structural framework of the Edwards Aquifer, Balcones Fault Zone, central Texas: Gulf Coast Association of Geological Societies Transactions, vol. 45, pp. 135–142

Collins, EW, and Hovorka, SD (1997) Structure map of the San Antonio segment of the Edwards Aquifer and Balcones Fault Zone, south-central Texas: structural framework of a major limestone aquifer: Kinney, Uvalde, Medina, Bexar, Comal and Hays Counties: The University of Texas at Austin, Bureau of Economic Geology, Miscellaneous Map No. 38, scale 1:250,000, text 14 p

# Access to Coastal Karst Water Resources Through a Salinity Study During an Exceptional High-Water Event: the Case of Port-Miou (Cassis, SE France)

A. Tassy, B. Arfib, and E. Gilli

**Abstract** A conceptual model of the functioning of the complex coastal karst aquifer of Port-Miou (Cassis, SE France) is presented. The studied event consists in a high water discharge period recorded during December 2008 at Port-Miou spring. During this one week event, the discharge in the 20 m diameter saturated conduit had probably been in the range of 50 to 100 $m^3/s$. In spite of this huge spring water discharge, the salinity remained persistently above 2 g/L, while freshwater could have been expected by a simple dilution effect. A new conceptual model has been developed to better understand the remaining salinity by taking into account the dynamic behavior of the flow and the geological heterogeneities of the reservoir. It gives new insight for water management.

## 1 Introduction

For karstic aquifers, the sharing and the viability of groundwater resources depend on the dynamic flow of freshwater and saltwater, and on the spatial relation between the karstic conduits and the fractured rock matrix. The understanding of the hydraulic functioning of the coastal karst aquifer can improve knowledge of the conduits' geometry, and their influence on groundwater salinity. The objective of this study is to propose a new analytical method integrating the dynamic properties and the geological characteristics of the coastal aquifer. This method proposes to develop hydrogeological conceptual models that match the dynamic behavior of the

A. Tassy, B. Arfib
Université de Provence, Laboratoire de Géologie des Systèmes et des Réservoirs Carbonatés, EA 4229, Case 67, 3 place Victor Hugo, 13331 Marseille Cedex 3, France,
e-mail: auretassy@gmail.com; bruno.arfib@univ-provence.fr

E. Gilli
UMR 6012 "ESPACE" Université de Nice Sophia-Antipolis, 98 boulevard Édouard Herriot, BP 209, 06204 Nice Cedex, France, e-mail: e.gilli@wanadoo.fr


DOI 10.1007/978-3-642-12486-0, 

flow and take into account the geological uncertainties of this karstic system. The case study of the Port-Miou spring (Cassis, SE France) is used, a typical coastal brackish spring. An exceptional high-water event is studied.

## 2 Study Site and Methods

The Port-Miou spring is the outlet of a karst system that is developed in the fractured Urgonian limestone of the Lower Provence area and is an important drainage network of the Beausset Unit, from the Ste Baume mount to the sea (Gilli 2002) (Fig. 1). The cave is approximately horizontal over 2200 m long with an average diameter of 20 m. A vertical well, located at 2200 m from the coastal entrance, has been explored recently down to 179 m below the sea level. It is most likely that this cave is developed further both vertically and horizontally but remains unexplored for safety and technical reasons (Blavoux et al. 2004). The available database consists in speleonautic explorations and topographic survey since the 1950s, four years of in situ physical and chemical monitoring of groundwater, a geological and petrophysical survey of the outcrop in the area and well data. The Port-Miou site is of particular interest since a dam was constructed in the 1970s across the conduit, 500 m upstream from the coast (Potié and Ricour 1974; Ricour 1981). The dam aims at blocking the intruded sea water tongue inside the conduit directly from the mouth of the submarine spring. The continuous monitoring takes place upstream of the dam (Fig. 4), allowing studying the hydrodynamic behaviour of the aquifer. Data for a specific event is presented in Fig. 2, curves are smoothed with a six hours moving average.

**Fig. 1** Localisation of the studied area

## 3 Analysis of an Exceptional High-Water Event in the Coastal Karst of Port-Miou

Up to now the Port-Miou spring has always remained brackish, with a drop in salinity a few days after rainfall event due to discharge increase. In December 2008, an exceptional high-water event occurred at the Port-Miou spring giving access to data never monitored before. The studied event in this paper consists in a high water discharge period recorded from December 14th 2008 to December 24th 2008 at Port-Miou spring (Cassis, SE France). The water head increases up to 3 m, with an overflowing of the dam through the spillway, while no more than 1 m overpressure had been observed during the studied last four years. During this one week event, the discharge in the 20 m diameter saturated conduit had probably been in the range of 50 to 100 $m^3$/s. In spite of this huge spring water discharge, the salinity remained persistently above 2 g/L (Fig. 2), while freshwater could have been expected by a simple dilution effect by the freshwater contribution increase. During this period, the salinity and the temperature curves are tracking each other, showing the same trends. Salinity varies from 5.5 to 2 g/L, and temperature varies from 16 to 15 °C. Relative water level varies from 0 to 300 cm, corresponding to the overpressure in the conduit. It is possible to distinguish the following four intervals in the curve trends (Fig. 2):

A. This interval is showing the incoming of the flood at the dam characterized by a water level rising of almost 3 m. This overpressure allows the discharge to be estimated in the range of 50 to 100 $m^3$/s. This event is followed by a subsequent decrease of the water temperature and salinity. A shift of 14 hours occurs between the beginning of the flood, shown by the increase of the pressure, and the drop of salinity. The latter is preceded by the expulsion of 1,260,000 $m^3$ of brackish water with constant salinity, associated to a mean discharge of 25 $m^3$/s. This delay corresponds to the distance between the probe and the mixing zone inland.
B. This stage presents a wide sill of high water level corresponding to the maximum of the discharge.
C. The "C" interval is showing a wide sill of constant salinity and temperature, with a smaller duration than the previous pressure sill (stage B). The salinity reaches its minimum (2 g/L). However the water remains brackish, while freshwater could have been expected given the significant volume of the flooding.
D. The last stage presents the drop of the water level following the flood, and the progressive increase of temperature and salinity. It indicates the permanent sea water intrusion whenever the water head is decreasing.

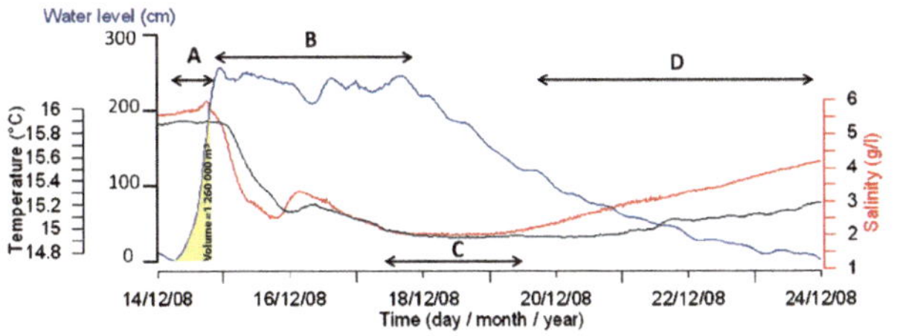


**Fig. 2** Continuous monitoring of salinity, temperature and relative water level of Port-Miou major flooding. Water-level and temperature curves are smoothed with a 6 hours moving average

## 4 Conceptual Modeling of the Coastal Aquifer Functioning

The Port-Miou karstic network is developed within a Cretaceous limestone formation (Urgonian) that appears well-bedded and fractured on the neighboring outcrops. Two conceptual models (1) and (2) are discussed taking into account the data given above (Sect. 3) and the geologic characteristics and uncertainties.

**Basic Conceptual Model (1).** The mechanism of salinization of a brackish karst spring is usually explained by two conceptual models (Fig. 3): 1) concentrate saline intrusion by connection of karst conduits, or 2) diffuse saline intrusion by matrix-conduit exchange (Arfib et al. 2007; Fleury et al. 2007). In these cases, salinity decreases with the discharge increase. For the two models, the salinity of the brackish spring is controlled by the difference of pressure head between the main conduit feeding the spring (conduit #1 in Fig. 3) and the connected medium containing the intruded seawater (karst conduits or matrix). According to this basic conceptual model, seawater entrance into a coastal karst conduit connected to a brackish spring is minimal in two cases: (1) a very strong discharge leading to a high pressure within the conduit prevents the seawater entrance, and results in freshwater at the coastal spring, or (2) a decrease in freshwater and seawater discharge and results in a drying up of the coastal spring. Data for the high-water event at the Port-Miou spring shows a persistent salinity. A rough calculation of the seawater contribution to the total discharge shows that during the minimal salinity period the discharge of seawater was maximal, superior to the one observed before the event (Table 1). This remaining salinity is not in accordance to the models described above.

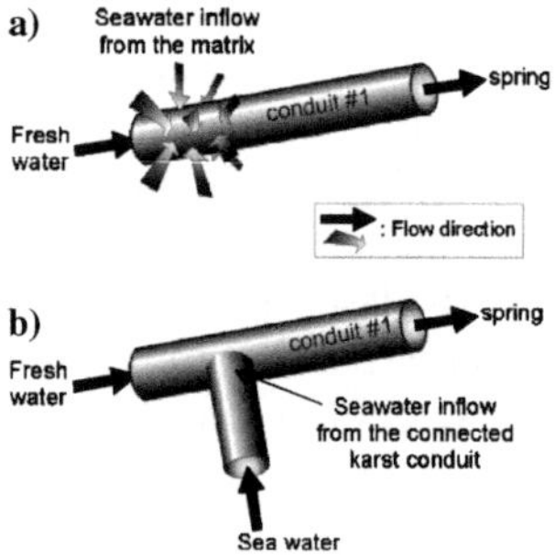


**Fig. 3** Conceptual model of seawater intrusion in a karst conduit feeding a brackish coastal spring. **a** diffuse saline intrusion by matrix-conduit exchange, **b** concentrate saline intrusion by connection of conduits

**Table 1** Rough calculation of the contribution of seawater to the total discharge at the brackish spring of Port-Miou in December 2008. (Salinity of the sea = 37 g/l)

| | Salinity (g/L) | Total discharge ($m^3/s$) | Discharge of seawater ($m^3/s$) |
|---|---|---|---|
| During the high-water event (Dec. 2008) | 2 | 50 | 2.7 |
| Before the high-water event | 5.5 | < 10 | < 1.5 |

**New Conceptual Model: Time Variable Head in the Matrix (2).** Outcrop surveys in the Cassis area indicate a very high density of open fractures in the Urgonian limestone enhanced locally by karstic dissolution. Joints between strata are also open at surface conditions (Jayet 2009). Fractures and joints form together a 3D network of well connected drains. According to this study, the fracture porosity is estimated between 10% and 20% at the scale of the studied outcrop (400 m). This estimation is based on analysis of fractures density and aperture on outcrop and photo. Available wells data in the area show that the freshwater aquifer around Port-Miou is limited to isolated aquifers. The limited freshwater aquifer combined to the high permeability of the Urgonian fractured aquifer allows the extensive penetration of seawater inland to be envisaged. A new conceptual model arises, taking into account the remaining salinity during the high-water event and the matrix contribution. The main conduit could be envisaged as a horizontal well draining a fractured aquifer saturated with salt water (Fig. 4a).This new conceptual model is based on a time variable head in the matrix with intruded seawater. The variable head generates a variation of the seawater entrance into the main conduit connected to the spring. During a high water discharge event (Fig. 4b), rainfall arrival has an impact on the head of the matrix.

The freshwater table increases over the saltwater wedge. Thus the head increases in the matrix, moving the saltwater around the deep karst conduit. For the Port-Miou case study, the increasing head within the conduit during the high-water event is measured at the dam around 3 m high. Thus the increasing head in the matrix must be higher than 3 m to allow an increasing inflow of seawater from the matrix to the conduit. This quite low value of 3 m can be reasonably quickly reached in the

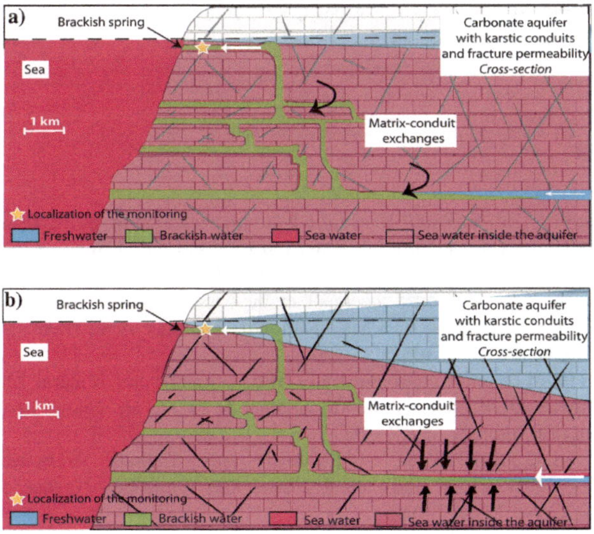


**Fig. 4** Hydrogeological conceptual model of Port-Miou aquifer functioning by time variable head in the matrix. **a** Low flow period; **b** High water discharge event

geological context of the area. The conceptual model presented in Fig. 4 does not exclude possible palaeoconduits inherited from Messinian karstic erosionnal period, but these are considered insignificant in term of storage capacity of sea water.

## 5 Conclusion

The objective of this study was to integrate the dynamic properties and the geological characteristics of the coastal aquifer. In this study, the method was applied to an exceptional flooding in a typical Mediterranean karst, with wide conduits and a deep development, linked to sea level variations. Persistent salinity of the water was observed, even during strong discharges. Two hydrogeological conceptual models, of the mixing were reviewed. The model (1), with a constant hydraulic head of the saltwater reservoir, does not predict the observed regular sea water discharge during flooding and lowest mean level. The model (2), with karsts conduits and a high fractured aquifer, matches the dynamic behavior of the flow by taking into account the geological heterogeneities of this reservoir. The strong fracturation of the surrounding aquifer supports the hypothesis of a great volume of seawater inland. The constant high residual salinity measured in the conduit is explained by the presence of sea water in the surrounding rocks, pushed by increasing head during high water events. It is therefore assumed that Port-Miou spring cannot be used as a freshwater supply, even using a hydraulic constrain.

**Acknowledgements** This work is part of KarstEau project, funded by the "Agence de l'Eau Rhône-Méditerranée-Corse", the "Conseil Général du Var", the "Conseil Général des Bouches-du-Rhône", and the "Région Provence-Alpes-Côte d'Azur". Thanks to Jean Borgomano for his interesting comments and discussions.

## References

Arfib B, Cavalera T, Gilli E (2006) Reach the hydraulic functioning of a coastal karstic aquifer with a conduit-matrix model. 3rd Salt Water Intrusion in Coastal Aquifers (SWICA) and 19th Salt Water Intrusion Meeting (SWIM), 24–29 Sept. 2006, Cagliari, Italy

Cavalera T (2007) Etude du fonctionnement et du bassin d'alimentation de la source sous-marine de Port-Miou (Cassis, Bouches-du-Rhône). Approche multicritère. Thèse de doctorat de l'Université de Provence. 403 p

Edmunds WM (1996) Bromine geochemistry of British groundwater, Mineral. Mag. 60 275–284

Fleury P, Bakalowicz M, Marsily de G (2007) Submarine springs and coastal karst aquifers: A review. Journal of Hydrology 339, 79–92

Gilli E (2002) Étude préalable sur le drainage des karsts littoraux. A/ Bouches-du-Rhône et Var; B/ Corse (Preliminary study about coastal karst drainage), Agence de l'Eau Rhône-Méditerranée-Corse, Lyon, 100 p

Jayet O (2009) Les faciès de fracturation dans les roches carbonatées: typologie, contrôles sédimentologiques et propriétés pétrophysiques (Crétacé inf. Bassin Sud Provençal). Mémoire de Master 2, Aix-Marseille Université

Potié L, Ricour J (1974) Etudes et captage de résurgences d'eau douce sous-marines. Ressources en eau pp 5–26

Ricour J (1981) Construction d'un barrage d'essais souterrain à Port-Miou (commune de Marseille) Bouches-du-Rhône. Compte-rendu de fin d'étude d'une recherche financée en partie par la Délégation générale à la recherche scientifique et technique, BRGM (Décision d'aide n° 76-7-1340), Mai 1981, 8 p

# Is the Term ‘Karst Aquifer’ Misleading?

J. Gunn

**Abstract** Carbonate rock sequences commonly provide significant quantities of groundwater but the term ‘aquifer’ is potentially misleading when applied to these sequences because, in common usage, it implies a single body of groundwater in which water moves down a hydraulic gradient towards a specific surface outlet. Such aquifers are often modelled as an equivalent porous medium (EPM) in which the outlet is assumed to be a seepage face. These conditions rarely apply in carbonate rock sequences due to the development by dissolution of channel networks. In a $\sim$8 km$^2$ block of carbonate rocks around the town of Buxton (United Kingdom) the degree of interconnectedness of the channel networks discharging at springs and intersected at wells varies to such an extent that there is clearly no single aquifer and it would be impossible to model the carbonate block as a simple EPM aquifer.

## 1 Introduction

The importance of carbonate rocks as water sources is widely acknowledged as is the fact that “The distinctive surface and subterranean features that are a hallmark of karst result from rock dissolution by natural waters along pathways provided by the geological structure” (Ford and Williams 2007, p. 1). The processes of rock dissolution are now well understood and numerical models have been developed that simulate the development of permeability in carbonate rocks (see reviews in Worthington and Gunn, 2009; Worthington and Ford 2009). The general conclusion of these models is that large-aperture pathways with increases in hydraulic conductivity of several orders of magnitude can evolve in 103 to 106 years in most natural-gradient situations. These pathways form self-organized channel networks

J. Gunn
Limestone Research Group, GEES (School of Geography, Earth and Environmental Sciences), University of Birmingham, Edgbaston, Birmingham B15 2TT, UK, e-mail: j.gunn.1@bham.ac.uk

B. Andreo et al. (Eds.), *Advances in Research in Karst Media*
DOI 10.1007/978-3-642-12486-0, © Springer 2010

with a continuum of apertures that are typically in the range of mm to cm but can reach dimensions sufficiently large for human exploration, at which point they can be referred to as caves. To date most models have considered the development of dissolution channels as an essentially random process but field observations demonstrate that this is not the case. One reason for this is that certain parts of a carbonate rock sequence (inception horizons) have a greater susceptibility to dissolution, and to channel forming processes, by virtue of physical, lithological and/or chemical deviation from the predominant carbonate facies within the sequence. The validity and potential usefulness of the inception horizon concept, which was developed in the 1990s (Lowe and Gunn 1997), have recently been confirmed by quantitative analysis (Filipponi et al. 2009). Notwithstanding the many advances, hydrogeologists continue to have problems understanding the manner in which carbonates function as aquifers and the effects of karstification. Part of the problem may lie with the term 'aquifer', which commonly implies a single body of groundwater in which water moves down a hydraulic gradient towards a specific surface outlet. Such aquifers are commonly modelled as an equivalent porous medium (EPM) in which the outlet is assumed to be a seepage face. These conditions rarely apply in carbonate rock sequences due to the development by dissolution of channel networks. The term 'carbonate aquifer' has been used to cover a very wide range of hydrogeological conditions, ranging from what have been described as fractured rock sequences (with the implication that there has been little development of channel networks) to rock sequences containing caves. The term 'karst aquifer' has sometimes been used in a restricted sense for a rock sequence containing "solution conduits in which a turbulent flow regime occurs" (Atkinson and Smart 1981, 182–183) and sometimes in a broader sense to describe a rock sequence in which "the permeability structure evolved as a consequence of dissolution by the fluid" (Huntoon 1995, 343). Worthington and Ford (2009, 334) supported the broader definition and noted that if this is used, "then most aquifers in carbonate rocks are karstic". They also argued that it was not important to achieve concensus on a definition for the term 'karstic aquifer' provided it is recognized that flow through carbonate rocks leads to the formation of self-organized channel networks and rapid groundwater flow. The formation of self-organized channels with is rapid groundwater flow is not disputed. However, it is debatable whether a carbonate rock sequence containing discrete channel networks can be considered as an aquifer, at least in the common usage of the term. The argument is developed in the context of carbonate rocks around the town of Buxton in the Derbyshire Peak District, UK.

## 2 The Buxton Karst

Buxton lies in a topographical basin and straddles the boundary between a carbonate-dominated Visean age sequence, the Peak Limestone Group, and overlying sandstones and mudstones of the Namurian age Millstone Grit Group (Fig. 1). The Millstone Grit Group strata crop out to the north and west of the town and the lime-

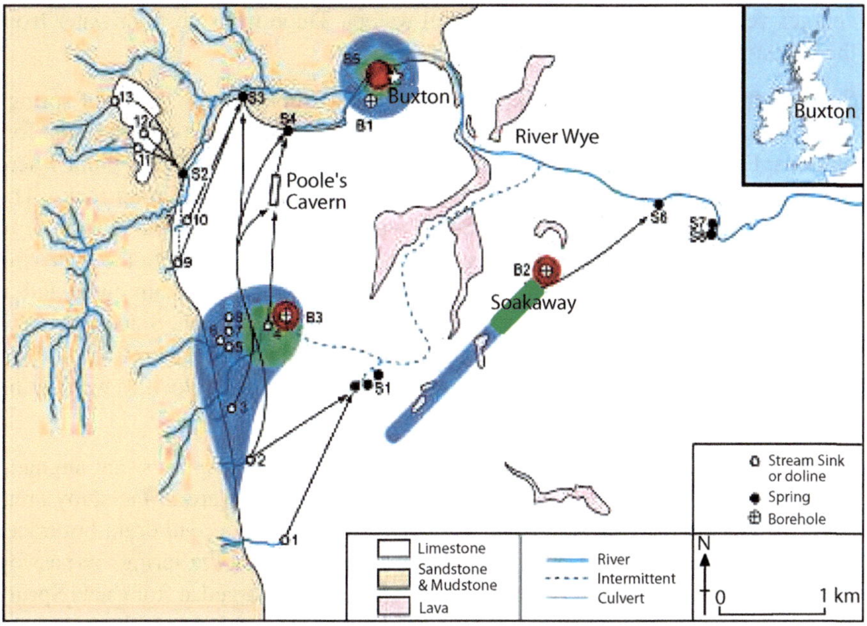


**Fig. 1** The Buxton karst. *Lines* show links between tracer injection and tracer recovery points. 1 Leap Edge Swallet; 2 Turncliffe Swallet and Anthony Hill Shakeholes; 3 Jakes Hole; 4 Borehole Swallet; 5 Axe Hole; 6 Plunge Hole; 7 Perseverance Pot; 8 Virgin Pot; 9 Can Holes 1; 10 Can Holes 2; 11 Shay Lodge 2; 12 Shay Lodge 3; 13 Shay Lodge 1; S1 Brook Bottom; S2 Dog Holes; S3 Otter Hole; S4 Wye Head; S5 St Anne's Well; S6 Devonshire Arms; S7 Rockhead; S8 Kidtor Spring; B1 Buxton Hydro Borehole; B2 Staden Borehole; B3 Stanley Moor Borehole. Environment Agency Source Protection Zones (SPZ): *Red* = inner, *Green* = outer; *Blue* = total catchment

stones dip steeply beneath them. To the east the carbonate outcrop continues for some 20 km as an up to 2000 m-thick sequence of bedded shelf limestones of which only the uppermost 600 m are exposed. Further details of the geological setting are given in Banks et al. (2009). The area is drained by the River Wye, which has its headwaters on Millstone Grit Group strata to the southwest of Buxton. The river has no perennial surface tributaries to the east of Buxton but the discharge increases steadily downstream due to inputs from springs and through its bed.

This paper examines the hydrogeology of an 8 km$^2$ area bounded to the west and south by overlying non-carbonate rocks and to the north by the River Wye. The eastern boundary is arbitrary as the carbonate sequence extends for a further 20 km. The block receives allogenic recharge from streams that sink soon after crossing onto the limestone. Autogenic recharge is mainly dispersed through a 1–2 m cover of superficial deposits but a small number of dolines provide concentrated recharge. There are three types of groundwater discharge: spring groups that discharge water partly derived from concentrated allogenic recharge; springs that discharge waters derived entirely from dispersed autogenic recharge and St Anne's Well which discharges very long residence time thermal waters. Three wells abstract water from the limestones.

**Springs that Receive Allogenic Recharge.** The Brook Bottom group of springs emerge close to an inception horizon at the boundary between two limestone formations. Dog Holes Spring has a single outlet that is very close to the point where the limestone dips westwards beneath overlying Millstone Grit Group rocks. The Otter Hole group has two perennial outlets which are about 80 m apart and close to the point at which the limestone dips towards the northwest beneath the overlying Millstone Grit Group rocks. The Wye Head springs comprise a main outlet and at least five other much smaller discrete outlets along a 100 m reach. Since 1984 over 30 natural gradient water tracing experiments have been undertaken from sinking streams to these springs (summarised on Fig. 1). Straight line velocities were in the range 10–120 m/h.

**Springs Fed Entirely by Autogenic Recharge.** The many springs that augment the River Wye are mostly fed by autogenic recharge. Three are in the study area. The Devonshire Arms Spring was originally thought to be a road drain but tracer injected into a soakaway 1000 m to the southwest emerged at the spring, by-passing the Staden Borehole (Fig. 1). The chemistry of water discharged at Rockhead Spring has been remarkably consistent since 1977 which suggests a long residence time as does the failure to recover any tracer injected in the area. Kidtor Spring is only a few metres from Rockhead Spring and was initially thought to be a distributary. However, water chemistry analyses have shown them to be distinct and Kidtor Spring also shows occasional evidence of pollution, possibly from a septic tank system located 400 m to the southwest.

**The St Anne's Well Thermal Spring Complex.** St Anne's Well is thought to comprise a single deep source with multiple outlets, including at least two major fissures in the underlying limestone. The spring has a constant temperature of 27 °C and the water is several thousand years old. Barker et al. (2000) suggest that recharge is from the limestone outcrop and that the water circulates to a depth > 1 km. Gunn et al. (2006) obtained a similar depth but suggest that the thermal water has migrated from deep sandstone aquifers below high ground to the west before discharging through fissured limestone at a topographic low. In the 1990s local SPZ were calculated using a saturated thickness of 75 m, a porosity of 1%, and 50 and 400 day time of travel for the Inner and Outer zones. The results were circles with radii of 68 and 193 m implying hydraulic conductivities of 1.36 and 0.48 m/d.

**Wells.** The Buxton Hydro Borehole (118 m deep) was driven through the bottom of a 20.4 m deep hand-dug well. Borehole water has high Ca–$HCO_3$ and no faecal bacteria (= longer residence time) whereas well water contains faecal bacteria and has lower Ca–$HCO_3$ but higher N and P (= shorter residence time and contamination from urban recharge). Both the shallow and the deep waters are chemically totally different from the water discharged by the nearby St Anne's Well. Pumping tests (up to 176 $m^3$/d) gave drawdowns up to 20.7 m. AquiferWin32 was used to fit

Theis confined solutions to the 5-minute drawdown data. The transmissivity was 194 $m^2$/day. Neuman curve analysis yielded a transmissivity of 125 $m^2$/d.

The Staden Borehole (150 m deep) is in an area of dispersed autogenic recharge. Water levels are 60–82 m below ground level (bgl). Two soakaways lie 150 m southwest of the borehole and within a SPZ derived in the 1990s using an EPM model (Gunn, 2007). Injected tracer was not recovered from the borehole but took less than 7 days to reach the Devonshire Arms Spring ~1000 m away, giving a velocity of at least 110 m/d. The borehole was dry when drilled and explosive was detonated at the base to improve the fracturing. Following the detonation, water entered the borehole and rose up it for ~75 m, suggesting that a channel had been intercepted. At the time of the tracing experiment the water level in the borehole was > 70 m bgl and the tracer probably followed a lateral route past the borehole rather than descending to the 'water table'. Hence, the SPZ has no basis.

In the 1990s a SPZ was derived for the Stanley Moor Borehole (177 m deep) using an EPM model. The inner zone (50 day travel time) was a circle with a radius 107 m. The water sinking at Borehole Swallet which is within this circle has been traced to the Wye Head Resurgences via Poole's Cavern and when pumping was taking place from the Stanley Moor borehole the stream flowing through the cave reduced markedly in volume. This indicates a degree of connectivity between the sinking streams and the borehole that renders the SPZ almost meaningless.

## 3 Discussion

From a surface geomorphological perspective the Buxton karst is relatively simple, with a 'classic' combination of sinking streams, dolines, dry valleys, caves and springs. However, the hydrogeology is extremely complex. Consideration of water tracing, water chemistry and pumping test data, together with the geological context suggests that the carbonate rock sequence is best conceptualised as a block containing multiple channel networks at different levels that have evolved as a function of hydraulic gradients interacting with lithology and structure. Some channels have permanent bifurcations (the Stanley Moor sinks always flow to both Otter Hole and Wye Head springs). Others have switches that are dependent on water levels (water from Borehole Swallet only flows through Poole's Cavern during high water levels). Springs that are in close proximity (Dog Holes and Otter Hole) have separate catchments whereas Otter Hole and Brook Bottom springs (> 2 km apart) have part of their catchment in common. Channels at different elevations cross without mixing (the Staden soakaways flow to the Devonshire Arms spring passing through the 'surface' catchments of Staden Borehole and Rockhead Spring, receiving recharge from the former but probably not the latter). The Buxton Hydro and Staden boreholes intercept a deeper channel network that drains broadly southeastwards, whereas St Anne's Well discharges water from the west that has been to even greater depth and rises without mixing with local waters.

The level of complexity is such that it is hard to conceive of this area as a single aquifer (ie a homogenous water body) and it would be impossible to model the carbonate block as a simple EPM aquifer. SPZ derived using EPM models bear no relation to reality.

**Acknowledgements** This paper incorporates data from research partly undertaken by former students (notably Joanne Gilman and Anita Edmans) and was assisted by other students and colleagues at Huddersfield University. Some of the work was commissioned by Rockhead Mineral Water Ltd, Buxton Hydro Spring Ltd, High Peak Borough Council and the Environment Agency. The text has benefitted greatly from discussions with Dr David Lowe and Dr Steve Worthington. Thanks are due to Kevin Burkhill (University of Birmingham) for drafting Fig. 1.

## References

Atkinson TC, Smart PL (1981) Artificial tracers in hydrogeology. In: A Survey of British Hydrogeology 1980, 173–190. London: Royal Society

Banks VJ, Gunn J, Lowe DJ (2009) Stratigraphical influences on the limestone hydrogeology of the Wye catchment, Derbyshire. Q Jnl of Eng Geol and Hydrogeol 42, 211–225

Barker JA, Downing RA, Gray DA, Findlay J, Kellaway GA, Parker RH, Rollin KE (2000) Hydrogeothermal studies in the United Kingdom. Q Jnl of Eng Geol and Hydrogeol 33:41–58

Filipponi M, Jeannin P-Y, Tacher L. (2009) Evidence of inception horizons in karst conduit networks. Geomorphology 106 86–99

Ford D C and Williams P W. (2007). Karst hydrogeology and geomorphology. Wiley

Gunn J (2007) Contributory area definition for groundwater source protection and hazard mitigation in carbonate aquifers. In: Parise, M and Gunn, J (eds.) (2007). Natural and anthropogenic hazards in karst areas: recognition, analysis and mitigation. Geol Soc Sp Pub 279, 97–109

Gunn J, Bottrell SH, Lowe DJ, Worthington SRH (2006) Deep groundwater flow and geochemical processes in limestone aquifers. Hydrogeology Journal, 14, 868–881

Huntoon PW (1995) Is it appropriate to apply porous media groundwater circulation models to karstic aquifers? In: A.I. El-Kadi (ed.), Groundwater Models for Resources Analysis and Management, 339–358. Boca Raton, Florida. Lewis Publishers

Lowe DJ, Gunn J (1997) Carbonate speleogenesis: An inception horizon hypothesis. Acta Carsologica, 26(2), 457–488

Worthington SRH and Ford DC (2009) Self-Organized Permeability in Carbonate Aquifers. Ground Water 47, 326–336

Worthington SRH and Gunn J (2009) Hydrogeology of carbonate aquifers: a short history. Ground Water 47, 462–467

# Evidence of an Early Phreatic and Confined Karst Phase in Minervois, South of France

A. Nou, S. Pistre, V. Borrell, C. Batiot-Guilhe, and M. Bakalowicz

**Abstract** The Minervois karst, South of France, developed in the Eocene carbonate formation, constitutes a regional karst system divided in several subsystems. The Lower Karst Subsystem developing at the base of the series is formed by small to medium size conduits with a circular cross section, organized in mazes and cut off by the canyons. It constitutes the earliest cave development, earlier than the canyons cutting the plateaus. Confined conditions account for the shape and organization of the conduits, which development precedes the down-bending separating Minervois from the Carcassonne basin.

## 1 Introduction

The Minervois karst located in the Languedoc-Roussillon, South of France has been considered by the Rhone, Mediterranean and Corsica water authority (AE-RMC 2009) as a high potential water body of regional interest to be developed for the increasing water demand of the region. Compared to other Mediterranean karst systems mainly developed in Jurassic and Cretaceous limestones, it extends through the Eocene carbonate formation. It is a multirank karst system only known and described as a geomorphic object (Le Coz 1967) which hydrogeological settings are presently badly known. The present work is a first attempt for a better understanding of the hydrogeological settings of this regional karst system.

## 2 Hydrology Settings

The Cesse River, 54 km long, is the main river crossing the Minervois karst system (Fig. 1). It has its source in the south side of the Montagne Noire, near Ferrals-les-

A. Nou, S. Pistre, V. Borrell, C. Batiot-Guilhe, M. Bakalowicz
HydroSciences Montpellier, 300 av. Emile Jeanbrau, 34090 Montpellier, France,
e-mail: nou@msem.univ-montp2.fr

B. Andreo et al. (Eds.), *Advances in Research in Karst Media*
DOI 10.1007/978-3-642-12486-0, © Springer 2010

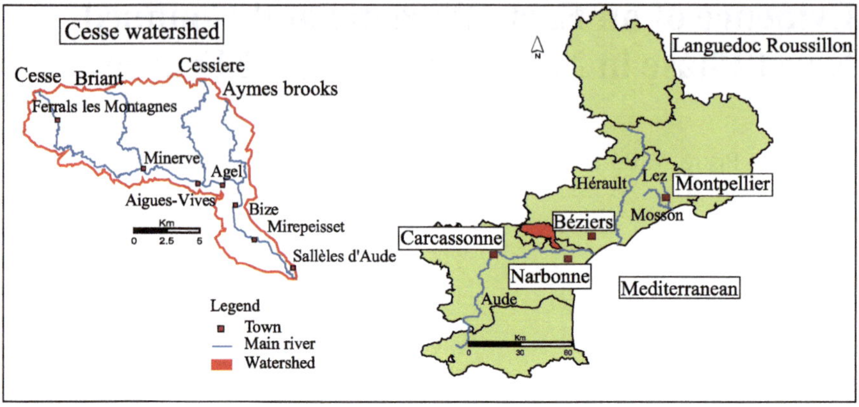


**Fig. 1** Location of Cesse watershed

Montagnes (450 m a.s.l.) and runs off on impermeable formations in its upstream course. It is the last left bank tributary of the Aude River, with a watershed total area of 275 $km^2$. During low stage, the Cesse River is feeded by two permanent springs, Saint Pierre and Autheze springs. It has three tributaries, the Briant, the Cessiere and the Aymes brooks and flows through the villages of Minerve, Aigues-Vives, Agel, Bize and Mirepeisset before joining the Aude River.

The watershed is subject to a Mediterranean climate, with an autumn and winter rainy season alternating with a long dry season. The annual precipitations are unequally distributed, with more than 1500 mm on the northern part, and only 350 mm on the area, and a mean annual rainfall of 620 mm (Yvroux 2001). The soil cover depends on the nature of rock and on the rainfall. The northern area of the watershed is mainly covered by oak and chestnut forest, while on the southern part, the limestone plateaus are covered by dense and thorny scrubland. The intermediate valley, between Agel and Bize, is planted with vines surrounded by rocky slopes with pine forest. The lower valley develops on the alluvial plain occupied by vines and orchards.

## 3 Geology Settings

The Cesse watershed (Fig. 2) develops on two major geological formations:

- In the North the Monts de Pardailhan, the south side of the Montagne Noire, are made of primary terrains (Bertolini 1980) beginning with the Marcory sandstones of lower Cambrian overlaid with massive dolomite and dolomitic limestone, overlaid by a carbonate sandstone formation (middle Cambrian), the Coulouma schists (upper Cambrian), quartz sandstones (lower Ordovician), Cassagnole flysch (middle Ordovician), Abéouradou flysch (upper Ordovician), and finally the so-called “griotte” limestone (Devonian).

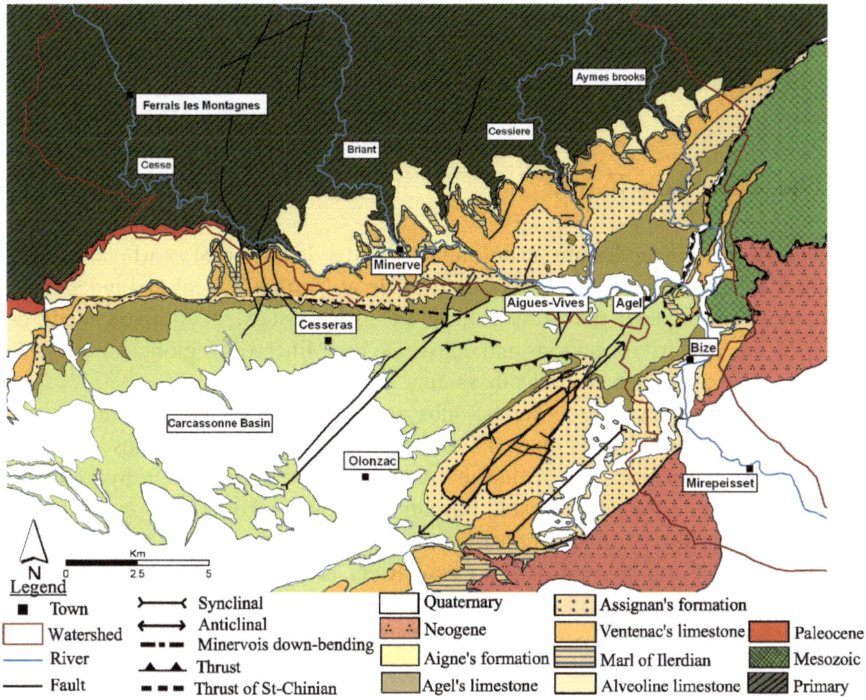


**Fig. 2** Main geological structures of the Minervois area

- In their southern part the Minervois limestone plateaus develop at North of the Aigne-Tourouzelle sedimentary basin, part of the Carcassonne basin. This tertiary formation is transgressive upon the Cambrian dolomite with an unconformable contact oriented 10° S. It begins with silt sands and freshwater silts of Montian age. It is overlaid with the Thanetian Montolieu lacustrine limestone, and the Sparnacian freshwater silt complex (Yvroux 2001).

The Eocene series begins with the lower and middle Ilerdian composed of Alveoline marine limestones, overlaid with the middle and upper Ilerdian pyritic blue marls. The upper Ilerdian is represented by oyster sandstones which indicate a transition between the previous marine formations and the following continental formations. It continues with the Cuisian Ventenac lacustrine limestones, the upper Cuisian Assignan detritic formation composed of sandstones and marls, and the Lutetian Agel lacustrine limestones with lignite. The tertiary formation ends with the Bartonian Aigne marls and conglomerate (Yvroux 2001).

From the structural point of view, the Paleozoic formations in the North are intensely folded and faulted (Bertolini 1980). The tertiary formations are structured in a series of anticlines and synclines dipping towards the North-West in the southern area of the Minervois (Chaigne 1964; Genna 1990). The Pyrenean tectonic phase created the main down-bending bordering the Montagne Noire range at North

(Genna 1990), combined with fracturing and faulting oriented N20-40, N90-110 and N140-160. At South this down-bending buries all ante-Oligocene formations underneath the recent continental sediments of the Carcassonne basin.

## 4 The Minervois Karst

Karst developed in the Cambrian dolomite as well as in the tertiary carbonate formations. The karst systems in Cambrian dolomite at North are local and independent from those developed in the tertiary limestones. The most extended cave systems develop in the Tertiary formations and constitute the Minervois regional karst system which can be divided in several subsystems.

From West to East, three karstic subsystems may be considered: the first located in the Alveoline limestones; the second in the Ventenac limestones and Assignan formation; and the third within the Agel limestones. They seem hydraulically connected despite the impermeable formations between them (Yvroux 2001). The Aigne impermeable formation constitutes the roof of this regional karst.

Minerve and the canyon of the Cesse River are very famous for the natural bridges, remains of karst conduits. The Minerve cave ("grotte d'Aldène") is another famous karst feature, known for its phosphate sediments and prehistorical remains (Ambert et al. 2005). The walls of the canyons of Cesse and Briant Rivers show several levels of karst conduits. The Minerve cave, the most important one, develops on several stories corresponding to successive karst stages. It contains thick sediments partly originating from Paleozoic rocks carried by surface rivers.

The site of this study, located in Minerve area, showed the presence of particular conduits at the bottom of the Tertiary limestone series in the Alveoline limestone. These small (dm) to medium (m) size conduits are organized in maze. Their cross section is sub-circular, sometimes elongated along joints or bedding planes. These conduits look randomly distributed all along the canyon walls, on both sides, at the bottom of the Alveoline limestone. They are free of fluvial sediments, while some of them are uniformly coated with calcite crust on their walls. The most interesting forms are observed at the base of limestone, on contact with Cambrian dolomites and at the base of the cliffs in the Briant and Cesse canyons close to their confluence. They do not seem to develop extensively in the Cambrian dolomite. It is named here as the "Lower Karst Subsystem" (LKS).

## 5 Interpretation and Discussion

Three main characteristics differentiate the LKS from the upper ones: i) the sub-circular shape and the small size of the conduits; ii) their organization in maze; and iii) their distribution on both sides of the canyons. The Upper Karst Subsystems (UKS) are obviously classical, with both vadose and phreatic features, organized in drainage systems from the plateaus towards the canyons. Their multi-storey organization is the consequence of their evolution related to the deepening of the canyons during the regional uplift (Gèze 1979).

The LKS is only formed with phreatic conduits which obviously developed before the entrenchment of the plateaus by rivers. It is then older than the canyons and constitutes the earliest cave development in the tertiary carbonate formations. Specific conditions are required (Ford and Williams 2007) for their phreatic development: a slow mass transfer in a confined aquifer. That means that in the Minervois area the limestone was at that time capped with the impermeable Aigne formation and its cover. It then formed a confined aquifer, certainly connected to aquifer formations buried beneath the Carcassonne basin.

This type of karst is well known and described in the United States in particular in the Black Hills karst, North Dakota (Bakalowicz et al. 1987; Palmer and Palmer 2000) and elsewhere in Europe, known in gypsum (Klimchouk 2000). It is related to sedimentary basins latter uplifted. The conduit systems developed either in epigenic conditions from groundwater infiltrated on the carbonate outcrops, or in hypogenic conditions from deep, hydrothermal circulations, enriched or not in deep $CO_2$. In the Minervois, there is no field evidence for a hypogenic origin of the LKS. The preferential development of the caves in the limestone above the Cambrian dolomite could actually be accounted for an epigenic more than hypogenic origin. Otherwise solution processes would have concerned the Cambrian dolomite as intensely as the tertiary limestone. This type of caves, never described in France, is poorly documented in Europe, only in gypsum formations (Klimchouk 2000). The most likely reason is that sedimentary basins containing carbonate formations and recently uplifted are not common.

This earliest karst phase could be related to the installation of the Minervois down-bending which created a semi-impervious barrier to groundwater flows (Genna 1990). This structural effect could have locally conditioned the conduit system allowing a slow mass transfer probably due to an upwards leakage through the confining layers in the Carcassonne basin.

Because there is no field evidence for connections between LKS and UKS, and because UKS is a multirank karst, it is likely that the LKS developed in phreatic conditions. The regional uplift then allowed the surface rivers to progressively entrench the limestone plateaus and to exhume the LKS. The resulting relief contributed to the development of a classical karst, the UKS, which recorded the different phases of uplifting by creating several stories. More detailed field data on the local hydraulic influence of the Minervois down-bending and the structure of the different cave systems are necessary for a better understanding of karst development and its connections with the aquifer formations of the Carcassonne basin.

## 6 Conclusions

The Minervois karst plateaus with the canyon entrenchments offer an interesting opportunity for observing different stages of karst development. The LKS, the earlier stage, presents characteristics definitely different from the UKS, indicating that it formed under confined conditions, before the opening of the canyons. Its development seems to be related to the functioning of the down-bending bordering the Carcassonne basin. It is likely that such a karst conduit system of the same type may exist at depth in the tertiary limestones of the basin, constituting a confined aquifer with a high karstic porosity and permeability. Because these limestone formations are all connected thanks to the high and dense fracturing related to the down-bending, the UKS certainly act as the recharge zone of the deep carbonate aquifer of the basin.

Improving the knowledge of the LKS and its development phase would allow moving towards a better understanding of kartogenesis and overall functioning of the regional Minervois karst, an essential condition for a sustainable exploitation and an efficient protection of the regional groundwater resource.

## References

AE-RMC, Agence de l'Eau Rhône-Méditerranée-Corse (2009) SDAGE vol 3_09. http://sierm.eaurmc.fr/sdage/sdage.php

Ambert P et al (2005) Attribution des gravures paléolithiques de la grotte d'Aldène (Cesseras, Hérault) à l'Aurignacien par la datation des remplissages géologiques. Comptes rendus Palevol Volume 4 Issue 3: 275–284

Bakalowicz M, Ford DC, Palmer A et al (1987) Thermal genesis of dissolution caves in the Black Hills, South Dakota. Geology Society of America Bulletin 99: 729–738

Bertolini P (1980) Etude des aquifères karstiques de la région "Saint-Pons, Minerve", structure des magasins, organisation des écoulements souterrains. Thèse 3ème cycle, Université Montpellier II

Chaigne M (1964) Contribution à l'étude stratigraphique et sédimentologique de secteur d'Aigne-Torouzelle (bassin tertiaire de Carcassonne). Thèse 3ème Cycle, Bordeaux

Ford DC, Williams PW (2007) Karst hydrogeology and geomorphology. Wiley, London

Genna A (1990) Déformation hydroplastique et circulation de fluide dans un contexte compressif d'avant-chaîne (bassin nord-pyréneen de Carcassonne, Aude-Hérault, France). Géologie de la France 2: 21–56

Gèze B (1979) Languedoc méditerranéen. Montagne Noire. Guides géologiques régionaux. Masson, Paris

Klimchouk A (2000) Speleogenesis of Great Gypsum Mazes in the Western Ukraine; in: Klimchouk A, Ford D.C, Palmer A, Dreybrodt W. Speleogenesis: Evolution of karst aquifers: Huntsville, Alabama. National Speleological Society 261–273

Le Coz J (1967) Aspects du Quaternaire languedocien, les niveaux de la Cesse et de l'Orbieu (Aude). Bulletin de la Société Languedocienne de Géographie 1 (2): 127–146

Palmer A, Palmer M (2000). Speleogenesis of The Black Hills maze caves, South Dakota, USA; in: Klimchouk A, Ford D.C, Palmer A, Dreybrodt W. Speleogenesis: Evolution of karst aquifers: Huntsville, Alabama. National Speleological Society 274–281

Yvroux M (2001) L'aquifère karstique de Pouzols-Minervois (Aude, France). Synthèse hydrogéologique et données nouvelles. Conseil Général de l'Aude

# Characteristics of Heterogeneous Shimbar Karstic Systems in Southwest Iran

N. Kalantari, M.R. Keshavarzi, F. Hamidzadeh, and M. Sahebdel

**Abstract** The study area of intensely tectonised rocks consists mainly of Asmari limestone formation (lower Miocene) and the impermeable Pabdeh-Gurpi formation (Oligocene) which lies beneath the karstic rocks and acts as a hydraulic barrier. The bare sedimentary formations generally crop out and in situ soil and regolith cover the limestone rock. The main structural features are anticlines, faults and fractures. The structural (fault) and geomorphological (polje) features have dissected the study area into three distinct karstic systems. Understanding a karstic horizon requires understanding many aspects of the system. This includes structural analysis, geomorphologic, hydrodynamic and geochemical approaches. Therefore, to achieve the goal a comprehensive investigation including structural geology, geomorphology, hydrogeology and hydrochemistry were carried out in the Shimbar karstic area. Hydrodynamic characteristics perceived at the outlet of major springs consisting of Dareh-e-Anari (DA), Abshekalo (AS) and Sarhoni (SH) revealed poorly and well developed karstified systems. The collected data indicated that ponor and polje are operating as underground transporting systems and convey drained water from two smaller reservoirs (AS and SH) into the main reservoir (DA). The principal aim of this research was to determine: 1) flow nature to make out prevail aquifer condition, 2) reservoirs dynamic potential and 3) chemical characteristics of the discontinuous aquifers.

## 1 Introduction

The Shimbar karstic aquifers system, which is focus of this investigation, lies within 49°30′ to 49°40′ E and 32°14′ to 32°25′ N and occupies an area of 325 km$^2$ (Fig. 1). It is a part of the Zagros structural belt in the north east of Khuzestan province in the

N. Kalantari
Geology Department Shahid Chamran University Ahvaz Iran, e-mail: nkalantari@hotmail.com

B. Andreo et al. (Eds.), *Advances in Research in Karst Media*
DOI 10.1007/978-3-642-12486-0, © Springer 2010

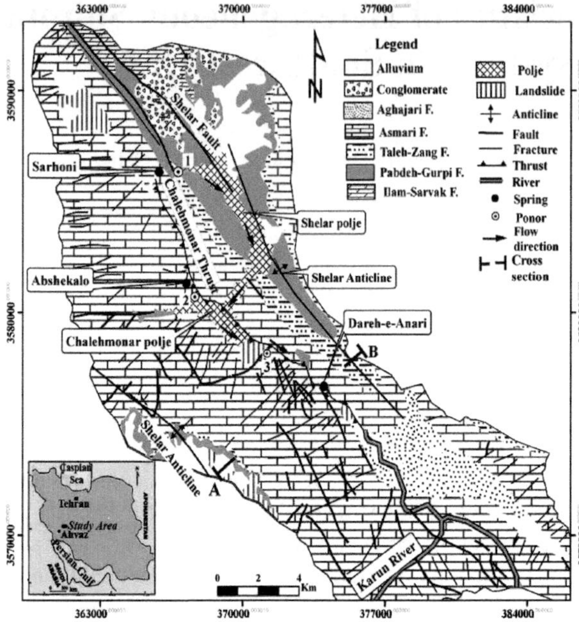


**Fig. 1** Location and hydrogeological map of the study area

southwest of Iran. The area experiences temperate climate and temperatures range from –4 to 46 °C in winter and summer respectively. The average annual rainfall exceeds 1000 mm/year, resulting from intense precipitation of short duration during winter and spring (October to March). This type of climate is prevails in the mountainous parts of Khuzestan province.

Carbonate formations crop out over 23% of the Zagros range of Iran (Raeisi and Kowsar 1977) and tectonic activity has a great impact on the karst processes. Local and regional faults (Fig. 1) influence karstification development, and as a whole govern hydrogeological situation in the study area. Apart from minor faults, two great thrust faults with dozens meters displacement (Fig. 2) are responsible for present landscape and general condition of the area. Lithological and geological structure studies and the position of the springs (DA, SH and AS) indicate rising up

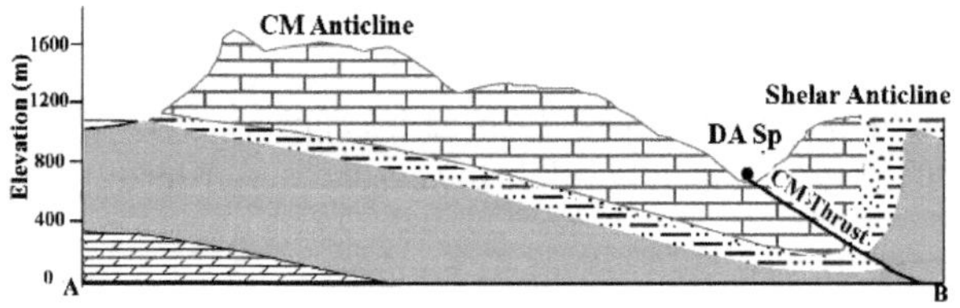


**Fig. 2** Geological cross section along line AB on Fig. 1

of the springs from three different catchments. The catchment area of DA and the ephemeral SH springs are marked by a density of longitudinal and lateral fractures, and these govern water input and movement. Small fissures and voids feed the AS spring. Probably regolith development is retarding fracture expansion in the spring watershed.

Karst catchments are usually characterized by the occurrence of large number of dolines and other karstic features which readily conduct surface water underground (Bogli 1980; Bonacci 1987; Kalantari 2002). Despite scattered development of ponors and grouts, the most distinguished karstic geomorphological feature is polje (Fig. 3) and these karst depressions are mostly situated in the central part of the area. The moderately permeable nature of poljes deposits control water movement underground and makeup significant pathways for channeling water in the direction of DA spring. In addition to this, the main tectokarstic features are intense fracturing and small and large scale faults, which also influence the discharge of surface water into groundwaters. In other words, the poljes and Chalehmonar (CM) fault are operating as groundwater collector and collector channel respectively and ultimately feeding DA reservoir.

**Fig. 3** Shows Chalehmonar Polje

# 2 Results and Discussion

## 2.1 *Hydraulic Behavior of Poljes*

The morphology of the poljes (CM and Shelar) indicated that these karstic features (Figs. 1 and 3) were formed as a result of tectonic activities. These two interconnected surface depressions are almost a system with a great basin, underground drainage, steep slopes and end with a small output channel in the east of the CM polje. In rainy periods, water from the larger polje (Shelar) flows into the smaller one. The permeable nature of polje sediments permits water to sink, but in the time of heavy rainfall, they flood for long periods because the infiltration and outlet capacities are too low as compared to inflow water. In dry periods, the spring waters are sinking, either through highly permeable soil or ponors into these poljes. Groundwater finds such zones favorable to pass through. General knowledge of the hydrogeological status of the area has revealed that groundwater movement takes place at the base of these interconnected poljes. Once diffuse and probable conduit flow reaches the impermeable bedrock, the sediment interface concentrates along CM fault trend at the CM polje and leads toward the DA karstic basin.

## 2.2 Flow Characteristics

The karst springs in the study area represent natural exits for groundwater. Springs appear at the contact between fractured Asmari limestone rock and low permeability Pabdeh formation (DA and SH) and in the Asmari formation itself (AS). The karstic springs can be classified in terms of discharge into high and low discharge. The high discharge springs with considerable variation (DA and SH) and low discharge spring with limited variation (AS) reflect the geometrical nature of the karstic systems.

In karstic areas the flow can be ascertained using spring hydrograph measurements (Bonacci 1993) to compile karst spring discharge curves. The same discharge curves are also used to determine flow characteristics and hydrodynamic nature of the karstic system in the area. Based on aforementioned explanation and as shown in Figs. 4 and 5, hydrographs for DA, SH and AS springs were prepared. It is evident that the flow considerably varies with time for DA and SH springs exhibiting turbulent flow (Fig. 5) while the AS spring (Fig. 4) does not depict variation with rainfall. Factors controlling conduit flow of springs include presence of faults (CM and Shelar), highly fractured rocks, and considerable thickness of karstic rocks. In addition, water movement at the floor of poljes plays a significant role, such as the DA spring catchment with its, remarkable discharge.

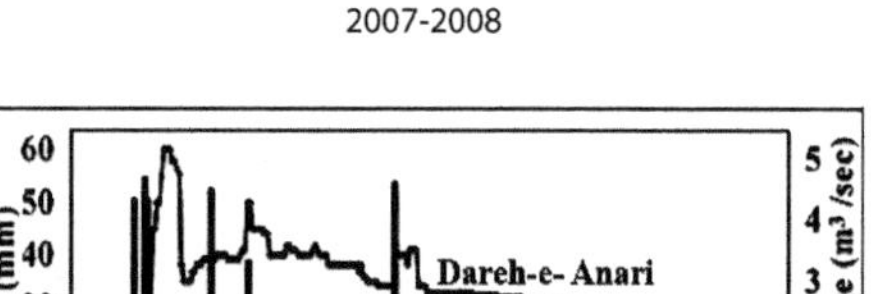


**Fig. 4** Abshekalo spring hydrograph

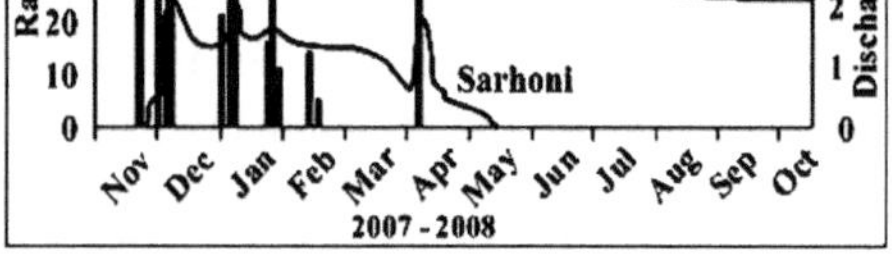


**Fig. 5** Dareh-e-Anari and Sarhoni spring hydrographs

## 2.3 Reservoir Dynamic Potential

The catchment area and altitude exit of the springs are given in Table 1. It is evident from Table 1 that catchment area of the DA spring is considerably larger with respect to other springs, and this is influencing outflow discharge of the spring.

The turbulent and diffuse flow show development of large fissures to conduits and small fractures to fissures in the catchment areas of DA, SH and AS springs respectively. The discharge curves of springs, which are plotted from daily and weekly gauge measurement installed at each spring's location, indicate maximum and minimum discharge of the springs. As shown in Figs. 4 and 5, the discharge of DA and AS springs are sustainable throughout the year, while the SH spring flows only for few months and dries up after rainfall. The same curves have been used for estimation of dynamic potential of springs and as depicted in Table 1, the DA storage potential is appreciable.

**Table 1** The physical characteristics of the springs

| Spring | Type | Altitude (m) | Catchment area (km$^2$) | Discharge (m$^3$) | Dynamic storage (MCM/Annum) |
|---|---|---|---|---|---|
| DA | Conduit | 550 | 154 | 0.7–5 | 81 |
| SH | Conduit | 874 | 19 | 0–1.7 | 10 |
| AS | Diffuse | 950 | 5 | 0.04–0.05 | 1.07 |

## 2.4 Hydrochemical Characteristics

During the study periods, samples were collected monthly from September 2008 to December 2009 (Table 2) to find out relationship among various parts of the karstic area. The electrical conductivity (EC) of springs water ranges from 416 to 507 and 435 to 472 µs/cm prior and post rainfall respectively, and the dominant water type is $HCO_3$-Ca-Mg. The values of EC, bicarbonate and calcium in DA as

**Table 2** Physico-chemical characteristics of representative samples (mg/l), EC (µs/cm)

| Spring | Date | *T* (°C) | EC | pH | HCO3$^-$ | SO4$^{-2}$ | Cl$^-$ | Ca$^{+2}$ | Mg$^{+2}$ | Na$^+$ | SIc |
|---|---|---|---|---|---|---|---|---|---|---|---|
| DA | Nov 2007 | 18.1 | 507 | 7.1 | 286 | 10.0 | 15.9 | 61.7 | 24.4 | 3.4 | 0.17 |
| DA | March 2008 | 20.4 | 387 | 7.4 | 250 | 7.2 | 12.7 | 51.7 | 25 | 2.7 | 0.26 |
| SH | Nov 2007 | 19 | 288 | 7.4 | 194 | 9.12 | 8.86 | 68.13 | 2.8 | 1.84 | 0.20 |
| SH | March 2008 | 17 | 258 | 7.5 | 185 | 7.2 | 9.5 | 51.3 | 11 | 1.8 | 0.18 |
| AS | Nov 2007 | 19.4 | 416 | 8.2 | 238 | 6.7 | 18.7 | 54.1 | 20.2 | 1.8 | 0.18 |
| AS | March 2008 | 28 | 351 | 8.1 | 210 | 17.7 | 12.4 | 37.2 | 29.4 | 1.8 | 0.5 |
| Rain | Nov 2007 | 20 | 133 | 8.9 | 37 | 15.6 | 4.0 | 19.8 | 2.4 | 0.8 | 0.7 |

compared to even AS diffuse spring, imply probable hybrid origin of this spring. Plot of the EC versus $HCO_3$ and Ca indicated a linear pattern, while in other springs such a pattern has not been observed. In cluster analysis, the DA spring falls into a diffuse spring group, denoting the heterogeneous character of this spring. The increase in concentration of ions and composite diagram results reveal water rock interaction during groundwater movement from various catchments towards DA spring through ponors (1 and 2), poljes and faults. On the other hand, the increase in concentration indicates flow gaining by the DA spring from other springs and mingling nature of water in DA.

## 3 Conclusion

One of the main factors for karst development in the area is tectonic activity and occurrence of lineaments in various scales. Structural processes are responsible for the three distinct karstic zones and polje development. The hydraulic connection between springs is governed by fault, poljes and ponors. The frequency of longitudinal and lateral fractures trending N30W and N45E respectively in SH and DA catchments are significant for aquifers recharge. These fractures are responsible for quick response of aquifers to events. Due to various sources of recharge and the large catchment area, DA flows throughout the year while SH discharge ends within a short period after rainfall. The fracture development in the AS tract is negligible and its recharge and discharge are strongly related to open spaces and small fissures. This spring flows throughout the year with minor discharge fluctuation.

The high TDS and enhancement in chloride concentration of the DA samples with respect to AS diffuse spring could confirm the blended character of DA spring and its hydraulic connection with CM polje. This polje collects water from the surrounding water bodies and transports into the DA catchment. The coefficient variations, composite diagram and cluster analysis also supporting mixing processes of DA water composition and capturing of its catchment from neighboring reservoirs.

## References

Ashjari J and Raeisi E (2007) Lithological control on water chemistry in karstic aquifers of the Zagros Range, Iran. Cave and Karst Science, Vol. 33, 111–118

Bogli A (1980) Karst hydrology and physical speleology. Springer, Berlin Heidelberg New York

Bonnacci O (1987) Karst hydrology. Springer, Berlin Heidelberg New York

Bonnacci O (1993) Karst spring hydrographs as indicators of karst aquifers. Hydrology Science Journal, Vol. 38, 51–62

Kalantari N (2002) Groundwater tracing in the Posht-e –Naz karstic area in northern Iran. Conference on evolution of karst from prekarst to cessation, Postojna Slovenia, 375–380

Raeisi E and Kowsar N (1997) Development of Shahpour Cave, southern Iran. Cave and Karst Science, Vol. 24, 27–34

# Application of Wavelet Correlation Analysis to the Karst Spring of Fuenmayor. San Julián de Banzo, Huesca, Spain

D. Chinarro, J.A. Cuchí, and J.L. Villarroel

**Abstract** The Fuenmayor spring has continuously been monitored for identification purposes, to study the behavior, and optimize the management of a karst system. Previous studies have examined the model through classical spectral analysis and system identification techniques under a linearity hypothesis. Since a karst area is generally accepted as a non-linear and dynamic system, different approaches, from wavelet paradigm to high order statistically stable wavelet techniques, are discussed. The results obtained from spectral and polyspectral wavelet analysis shed light on karst system non-linearity, to better understand its behavior.

## 1 Introduction

Fourier analysis has been extensively used to study the karst systems, since it is usually accepted that its hydrological response is non-linear (Bakalowicz 2005). Under this context, wavelets were introduced by the French school and expanded to other karst systems (Andreo et al. 2006; Labat et al. 1999; Larocque et al. 1998; Massei 2006; Mathevet et al. 2004). In karst hydrology, signal fluctuations are statically non-stationary and can vary in a short time, showing periodicity and abrupt changes, and frequently involve transient processes. A single input event affects all frequencies in Fourier transform. However, wavelet transform expands time series into time frequency space, and can reveal temporary behavior at local time.

D. Chinarro, J.A. Cuchí, J.L. Villarroel
Group of Technologies in hostile Environments (GTE), University of Zaragoza, I3A, C/ Mar a de Luna 3, 50018-Zaragoza, Spain, e-mail: chinarro@unizar.es; cuchi@unizar.es; jlvilla@unizar.es

B. Andreo et al. (Eds.), *Advances in Research in Karst Media* 
DOI 10.1007/978-3-642-12486-0, 

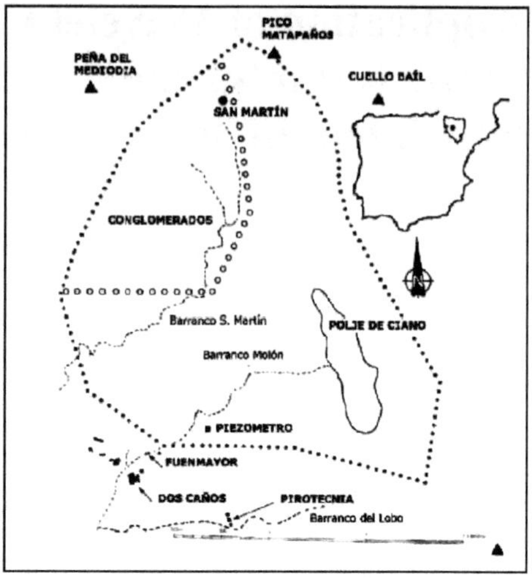


**Fig. 1** Location map of Fuenmayor spring

The present study applies wavelets techniques to Fuenmayor Spring. This karst system is located at San Julian de Banzo, Huesca, Spain, in the south Pyrenean Sierras Exteriores range (Fig. 1). The recharge area, with clear Mediterranean exokarst features, is around $10\,\mathrm{km}^2$ and opens on the limestone of the Guara Formation of the Middle Eocene (Cuchí 2002).

For several years, an automatic monitoring station measured hourly discharge, rainfall, electrical conductivity of the water, air and water temperature. A first rainfall-discharge transfer equation was given by Cuchí and Villarroel (2002) and Villarroel and Cuchí (2003). The spring's response to a severe drought was analyzed (Cuchí et al. 2005) through an extended study using classical spectral analysis and system identification techniques, and under the hypothesis of a linear and time-invariant response of the system.

## 2 Methods

The *continuous wavelet transform* $W\Psi_\tau x$ of a function $x(t)$, is defined as (Daubechies 1990):

$$W_{\Psi,x} = \int_{-\infty}^{+\infty} x(t)\Psi^*(t)\mathrm{d}t\ ; \qquad \Psi_{s,\tau} = \frac{1}{\sqrt{s}}\Psi\left(\frac{t-\tau}{s}\right), \tag{1}$$

where $\Psi^*$ is the complex conjugate. The mother wavelet $\Psi(t)$ is normalized with

$$||\Psi||^2 = \int_{-\infty}^{\infty} |\Psi(t)|^2\mathrm{d}t = 1$$

and can be scaled and translated by modifying $s$ and $\tau$.

*Wavelet power spectrum* (WPS) is defined as autocorrelation function of the wavelet transformation of the signal. The expression

$$P_{2x}(s,\tau) = W_x(s,\tau) * W_x^*(s,\tau) = |W_x(s,\tau)|^2$$

describes the WPS (a representation of the time-varying frequency content) of the signal $x(t)$ at a certain time $\tau$ on a scale $s$.

The *wavelet transform* applied to a finite length time series inevitably suffers from border distortions due to the fact that it involves points away from the ends. The affected region is the *cone of influence* (COI) of the wavelet spectrum. The edge effect increases with scale and reduce the effective length of the data series.

*Cross wavelet spectrum* (XWT) between $x(t)$ and $y(t)$ is defined as (Meyer 1992):

$$P_{xy}(s,\tau) = W_{\Psi,x}^*(s,\tau)W_{\Psi,y}(s,\tau)\,, \quad \text{where } P_{xy}(s,\tau)$$

can be complex.

The significance test is carried out at every point in the time/scale plane, to check whether the wavelet power exceeded a certain critical value corresponding to the chosen *significance level.*

A kind of normalized wavelet cross-spectrum is the *wavelet coherence* (WTC):

$$\Theta_i(s,\tau) = \left|P_{xy}(s,\tau)\right| / \left[P_2x(s,\tau)P_2y(s,\tau)\right]^{1/2}\,. \tag{2}$$

The WTC determines both the absolute magnitude, defined between 0 (no coherence) and 1 (100% coherence), and the phase shift between the series as a function of the frequency/timescale and time.

*Higher Order Spectrum* (HOS) uses the wavelet power bispectrum to better estimate parameters in noisy situations or to clarify nonlinearity features. The *wavelet bispectrum* is defined:

$$P_{3x}(s_1,s_2) = \int_T W_x(s_1,\tau)W_x(s_2,\tau)W_x^*(s,\tau)\mathrm{d}\tau \tag{3}$$

and wavelet cross bispectrum is:

$$P_{2xy}(s_1,s_2) = \int_T W_x(s_1,\tau)W_x(s_2,\tau)W_y^*(s,\tau)\mathrm{d}\tau\,.$$

*Wavelet bicoherence* is a normalized version of the bispectrum and is a useful tool for classication and quantization of non-linearity in time series

$$\Theta_{3x}(s_1,s_2) = |P_{3x}(s_1,s_2)| / \left[P_{2x}(s_1+s_2)P_{2x}(s_1)P_{2x}(s_2)\right]^{1/2}\,. \tag{4}$$

## 3 Application to Fuenmayor Spring

The rainfall and discharge signals are depicted on Fig. 2 and cover about three years with a sampling period of one hour, including dry and wet seasons. The sequence of

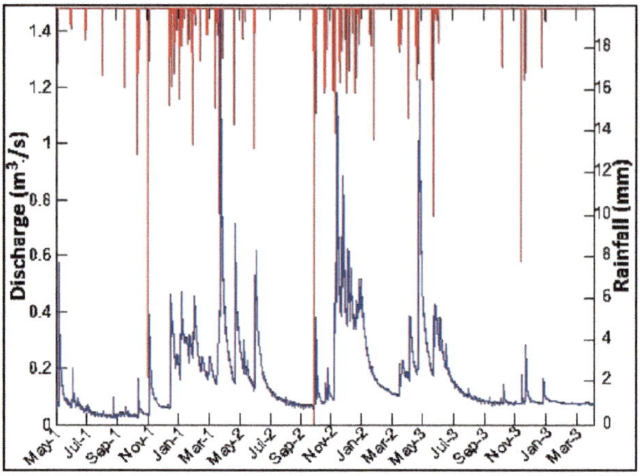


**Fig. 2** Rainfall (*upper plot*) and discharge (*lower plot*) in Fuenmayor karst system from 2002 to 2005

intensive precipitation in certain periods and dry intervals constitutes the structure of the rainfall time series. The *effective rainfall* is estimated by subtracting evapotranspiration (Eagleman method) from the total rain, and the discharge of the spring is measured by flumes.

The selection of the wavelet basis depends on circumstance and the type of time series. In this analysis, as commonly used in hydrology, Morlet wavelet is the basis function for the transforms.

The wavelet transforms (1) on Fig. 3 has been plotted with software provided by C. Torrence and Compo (1998) and Aslak Grinsted et al. (2004). By visual inspection, some features are perceptible:

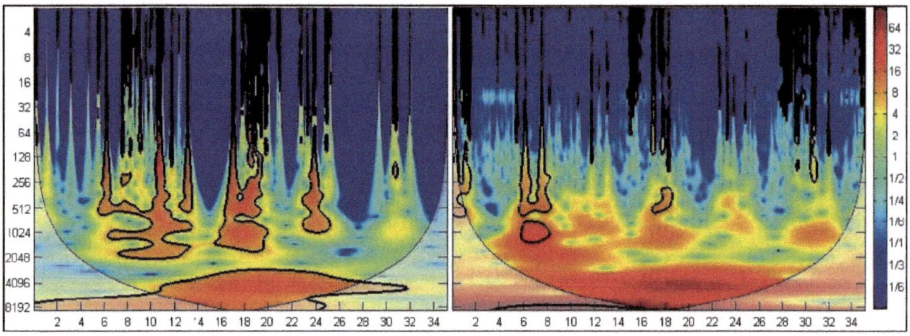

**Fig. 3** *Left*: Rainfall power spectrum. *Right*: Discharge power spectrum. The *y axis* is the wavelet scale (hours). The *x axis* is time (months). *Curved lines* on either side indicate the cone of influence where edge effects become important. The *thick black line* indicates significance level (established at 5%) for power spectrum. Spectral strength is shown by *colors* ranging from *deep blue* (weak) to *deep red* (strong)

- The spectral plot shows a half-annual cycle, typical in the Mediterranean weather, given by the continuous colored band at nearly 4096 hours (about 6 months).
- In rainy seasons (6–13, 16–20, 23–25 month intervals), the rain reveals a richer spectrum with frequency components in the 200–2000 hour band (about 8–80 days) . In dry intervals these components are absent.
- The plot shows rainfall events as small isolated zones in the high frequency band.

Figure 4 (top graph) presents the XWT of both signals, with regions in time frequency space with large common spectral power, showing a consistent phase relationship. Therefore, it suggests causality between both time series. Since the Morlet wavelet is complex, the XMT amplitude $|P_{xy}(s,\tau)|^2$ is large when $P_{2x}$ and $P_{2y}$ are large at closeness of scales (frequencies) and around the same time, regardless of the local phase difference. The 5% significance level is plotted with a black thick line (Fig. 4).

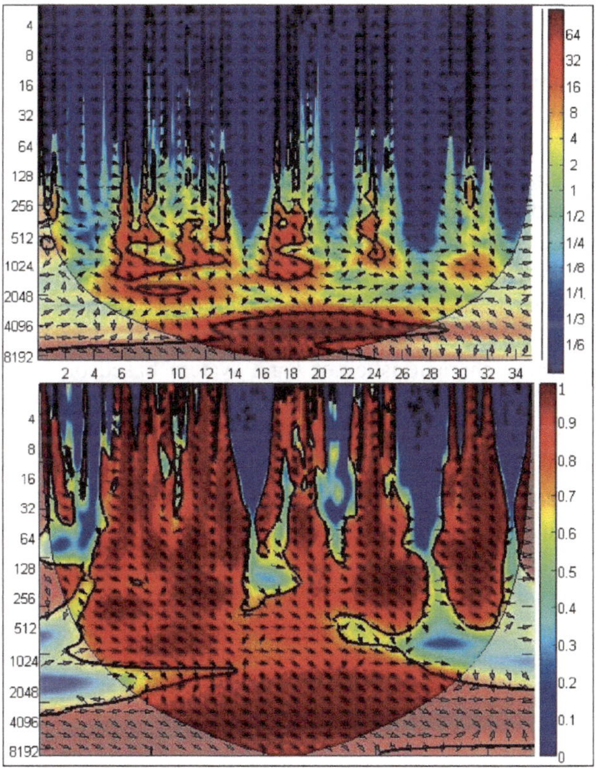


**Fig. 4** XWT (*left*) and WTC (*right*) of rainfall-discharge. The *left axis* is the wavelet scale (hours). The *bottom axis* is time (months). Shared spectral strength is shown by *colors* ranging from *deep blue* (weak) to *deep red* (strong). The *thick contour* encloses regions of greater than 95% confidence for a red-noise process with a lag-1 coefficient of 0.7. Sections with *arrows point to right and below*, indicate that the rainfall leads the discharge by a relative phase

As expected, *wavelet coherence* (2) on Fig. 4 (bottom) shows wide areas of strong correlation between the input and output signals, with values close to 1. However, there are zones of very low coherence inside the COI. This suggests there are two behaviors of the system.

- The coherent one happens when the system reacts in consequence of the rain.
- A non-coherent behavior occurs when it rains after a dry season (0–6,15–18, 28–30 months), because the soil and epikarst zones need to be drenched before the effective rainfall has some influence on the discharge. This fact also shows the nonlinearities of the system.

To assert the linearity of Fuenmayor time series, the Hinich test (1982), based on calculating the normalized bispectrum ((3) and (4)), has been performed. The linearity hypothesis is rejected, because the estimated interquartile ranges (482.80), based on seven samples, is much larger than the theoretical value (44.97).

## 4 Conclusions

The resolution advantage of the wavelet transform permits detection of changes in nonstationary signals, with exceptional localization both in time and frequency domains. Cross wavelet power spectrum gives a quantified indication of the similarity or strong relation of rainfall-discharge, independently of the energy amplitude of the signals, and explains some hydrological cycle phenomena. By a thorough wavelet coherence analysis, certain incoherent behavior reveals the nonlinearities of the karst system, which is confirmed by assessing the phase coupling.

This study clearly reinforces the assumed stationary case in Fourier-based spectral analysis and correlation. In addition, it shows a new approach in the analysis of the nonlinear characteristics of karst system based on a combination of wavelet transform with third-order statistics. Therefore, wavelet spectral analysis and wavelet bispectral analysis unveils a non-linear rainfall-discharge interaction and karst features that otherwise were concealed in the Fourier-based discourse. These results permit extending the analysis to large-scale experiments, considering different parameters as conductivity or water temperature, and for other karst systems.

**Acknowledgements** This research is supported financially by Confederación Hidrográfica del Ebro and Instituto de Estudios Altoaragoneses. The authors thank Parque Natural de Guara and Ayuntamiento de Huesca for facilitating and supporting the activities in the study area.

## References

Andreo B, Jiménez P, Durán J, Carrasco F, Vadillo I, Mangin A (2006) Climatic and hydrological variations during the last 117–166 years in the south of the Iberian peninsula, from spectral and correlation analyses and continuous wavelets analyses. Journal of Hydrology 324, 24–39

Bakalowicz M (2005) Karst groundwater: a challenge for new resources. Hydrogeology journal 160, 13–14

Cuchí J, Villarroel J (2002) An alisis de respuesta de Fuenmayor (San Julián de Banzo, Huesca), primeros resultados. Geogaceta pp. 71–74

Cuchí J, Villarroel J, Carceller T, Azcón A (2005) Comportamiento del manantial de Fuenmayor (San Julián de Banzo) durante la sequía de 2005. Publicaciones del Instituto Geologico y Minero de Espana. Serie: Hidrogeología y Aguas Subterraneas, pp. 135–143

Cuchí JA, Villarroel JL, Manso J (2002) Caracteristicas del manantial de Fuenmayor, San Julián de Banzo, Huesca. Geogaceta 31, 75–78

Daubechies I (1990) The wavelet transform time-frequency localization and signal analysis. IEEE Trans. Inform. Theory pp. 961–1004

Grinsted A, Moore JC, Jevrejeva S (2004) Application of the cross wavelet transform and wavelet coherence to geophysical time series. Nonlinear Processes in Geophysics 11, 561–566, URL http://www.glaciology.net/Home/PDFs

Hinich M (1982) Testing for gaussianity and linearity of a stationary time series. Journal of Time Series Analysis 3, 169–176

Labat D, Ababou R, Mangin A (1999) Analyse en ondelettes en hydrologie karstique. Geoscience 329, 881–887

Larocque M, Mangin A, Razack M, Banton D (1998) Contribution of correlation and spectral analyses to the regional study of a large karst aquifer (Charente, France). Journal of Hydrology 205, 217–231

Massei N, Dupont J, Mahler B, Laignel B, Fournier M, Valdes D, Ogier S (2006) Investigation transport properties and turbidity dynamics of a karst aquifer using correlation, spectral, and wavelet analysis. Journal of Hydrology 329, 244–257

Mathevet T, Lepiller M, Mangin A (2004) Application of time-series analyses to the hydrological functioning of an alpine karstic system: the case of Bange-l'eau-Morte. Hydrology and Earth System Sciences 8(6) 1051–1064

Meyer Y (1992) Wavelets and operators. Cambridge University Press, Cambridge

Torrence C, Compo GP (1998) A practical guide to wavelet analysis. Bull. America Meteorol. Society 79, 61–78, URL http://paos.colorado.edu/research/wavelets/

Villarroel JL, Cuchí JA (2004) Estudio cualitativo de la respuesta, de Mayo de 2002 a Abril 2003, del manantial karstico de Fuenmayor (San Julian de Banzo, Huesca) a la lluvia y a la temperatura atmosferica. Bolet n Geologico y Minero pp. 237–246

# Recession Curve Analysis to Constrain Rainfall-Discharge Model Parameterisation

F. Moussu, V. Plagnes, L. Oudin, and H. Bendjoudi

**Abstract** This paper explores the conceptual reservoir model applied to karst aquifers. A methodology is proponed to facilitate model parameter optimisation by improving overall parameter identifiability. This approach uses recession curve analysis to determine a parameter often included in the optimisation process: the baseflow coefficient. The proposed methodology was applied to six French karst systems representing a relatively large spectrum of karst system types. The efficiency of the model's two versions (the recession coefficient as a free parameter and the recession coefficient obtained from recession curve analysis) was similar. In addition, the analysis of the variability of the model's parameters shows that fixing the baseflow coefficient provides better identifiability of the model's other parameters during the optimisation procedure.

## 1 Introduction

Identifying relevant patterns in the data or observations is the basis for testing hypotheses on underlying processes in hydrological science (Sivapalan 2005). One such pattern or signature is the recession curve, which represents the gradual depletion of discharge during periods with little or no precipitation. The recession curve contains valuable information on storage properties and aquifer characteristics (Mangin 1975; Bonacci 1993; Padilla et al. 1994). Besides providing key information on a system's behavioral patterns, recession analysis can assist in determining the routing parameters of rainfall-discharge models through stepwise calibration procedures. In these procedures, some model parameters are calibrated one after another to mimic different hydrograph patterns, including the recession curve. These calibration procedures can be seen as the counterpart of fully automatic single-step, single-criterion optimization techniques widely used in hydrological modelling. The

F. Moussu
UPMC Univ Paris 06, UMR 7619 Sisyphe, Case 105, 4 place Jussieu, 75005 Paris, France,
e-mail: Francois.moussu@upmc.fr

B. Andreo et al. (Eds.), *Advances in Research in Karst Media* 
DOI 10.1007/978-3-642-12486-0, 

stepwise calibration procedure aims at constraining these parameter values on the basis of a process-oriented interpretation of the role of each parameter. A number of previous studies have shown the benefits that could be gleaned from this stepwise procedure in model calibration (Harlin 1991; Zhang and Lindstrom 1997; Hogue et al. 2000; Shamir et al. 2005). As for modelling karst systems, the recession curve coefficients are considered so crucial that they are often fixed to the value of the baseflow coefficient obtained by recession analysis (Fleury et al. 2007; Jukic and Denic-Jukic 2009). This study was designed to test the differences between a fully automatic calibration procedure and a stepwise calibration procedure that takes the baseflow coefficients obtained from recession curve analysis explicitly into account. The differences are analyzed on the basis of the efficiency of a model applied on six karstic systems.

## 2 Materials and Methods

Table 1 presents the drainage area, mean and maximum discharge, annual precipitation and mean elevation of the six karst systems. Note that the systems are rather different in terms of these characteristics, e.g., the drainage area ranges from 13 km$^2$ (the Baget system) to 1115 km$^2$ (Fontaine de Vaucluse).

The rainfall-discharge model used was developed by Fleury et al. (2007) to study the Fontaine du Vaucluse karst system. Such conceptual rainfall-discharge models are widely used to model karst systems (Le Moine et al. 2008 for a review). However, the model's structure is flexible and generic enough to ensure good transferability to other karst systems. This model will be noted hereafter KDM for "karstic devoted model". Ttwo versions of the KDM (Fig. 1) were used in the analysis: 1) the KDM-4p having four free parameters, including the baseflow coefficient of the slow response reservoir and 2) KDM-3p being similar to the KDM-4p but for which the baseflow coefficient is predetermined from recession curve analysis. The free parameters of both models were obtained by an automatic calibration procedure: the SCE-UA algorithm (Duan et al. 1992) using a modified Nash and Sutcliffe (1970) coefficient (see Sect. 3).

**Table 1** Main characteristics of the six karst systems studied

| | Drainage area (km$^2$) | Mean annual discharge $Q$ (m$^3$/s) | Max of measured discharge $Q_{max}$ (m$^3$/s) | Mean annual precipitation (mm) | Mean elevation (m) NGF | Available data |
|---|---|---|---|---|---|---|
| Gervanne | 200 | 1.7 | 21 | 1340 | 760 | 1995–2005 |
| Baget | 13 | 0.48 | 9.8 | 1640 | 950 | 1973–1992 |
| Vaucluse | 1115 | 18 | 80 | 960 | 870 | 1996–2006 |
| Homède | 27.3 | 0.4 | 5.2 | 1040 | 800 | 1996–2005 |
| Durzon | 115 | 1.7 | 35 | 1022 | 800 | 1996–2005 |
| Boundoulaou | 17.1 | 0.21 | 12.6 | 924 | 800 | 1999–2005 |

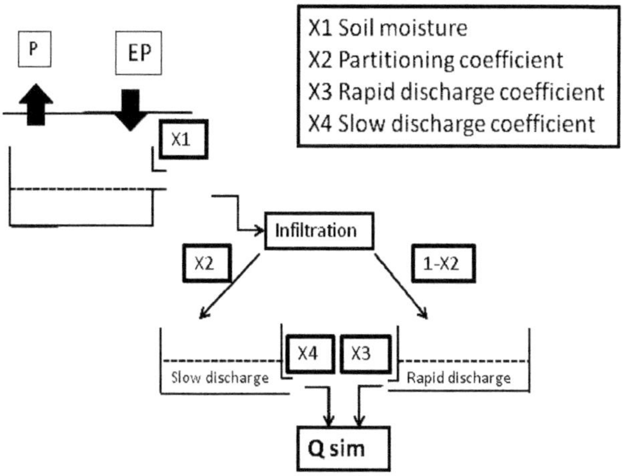


**Fig. 1** Schematic representation of the KDM model structure (see Fleury et al. 2007 for greater detail)

The baseflow coefficient $\alpha$ is determined from the hydrograph analysis of several recession events following the methodology proposed by Mangin (1975). The value of the baseflow coefficient $\alpha$ is given in days$^{-1}$ and is highly variable among the karst systems studied. Table 2 presents the values of the mean baseflow coefficients of the six systems studied. These coefficients are used as the value of the parameter X4 for the KDM-3p (and thus only X1, X2 and X3 are calibrated by the automatic optimisation procedure).

**Table 2** Main results of the recession analysis performed on the six karst systems

| | Mean baseflow coefficient $\alpha$ (day$^{-1}$) | Number of recessions studied | Mean duration of the infiltration delay (day) | Mean duration of the recession (day) | Mean baseflow coefficient $\alpha$ (day$^{-1}$) obtained by automatic calibration |
|---|---|---|---|---|---|
| Gervanne | 0.0200 | 3 | 35.6 | 91 | 0.136 |
| Baget | 0.0076 | 3 | 42.6 | 163 | 0.281 |
| Vaucluse | 0.0060 | 3 | 80 | 150 | 0.052 |
| Homède | 0.0064 | 3 | 42.3 | 165 | 0.118 |
| Durzon | 0.0028 | 3 | 28.3 | 172 | 0.131 |
| Boundoulaou | 0.0088 | 4 | 35.5 | 100 | 0.085 |

## 3 Results

First efficiencies of the KDM-4p and the KDM-3p were compared. The model was tested on the six karst systems studied, using the standard split sample test scheme

(Klemeš 1986). This means that the record period was divided into two sub-periods. Then the model was calibrated on the first and tested in validation mode on the other period, followed by the calibration on the second sub-period, and the model efficiency was finally assessed on the first sub-period. The Nash and Sutcliffe efficiency (NSE) criterion computed on $\sqrt{Q}$ was used to assess the model's efficiency since this criterion was shown by to be a good compromise between several alternative objective functions (Oudin et al. 2006). Efficiency was compared in terms of discharge simulations on the six systems studied using the two models. Figure 2 presents the values of the NSE($\sqrt{Q}$) obtained on the six karst systems in validation mode by the KDM-4p and the KDM-3p models; it must be remembered that for the former the baseflow coefficient (X4) was calibrated, whereas for the later it was determined from preliminary recession curve analysis. Whatever system was considered, the efficiency of the two models was quite similar. Two systems (Gervanne and Durzon) were particularly poorly modelled, with NSE$\sqrt{(Q)}$ values under 0.60. For these two systems, the different values obtained on the two sub-periods suggest that the calibration of the models is affected by the hydroclimatic specificities of one of the two sub-periods. The best modelling results were obtained for the Fontaine de Vaucluse system with NSE($\sqrt{Q}$) values around 0.90. The other three systems (Boundoulaou, Baget and Homède) presented relatively similar NSE($\sqrt{Q}$) values around 0.70.

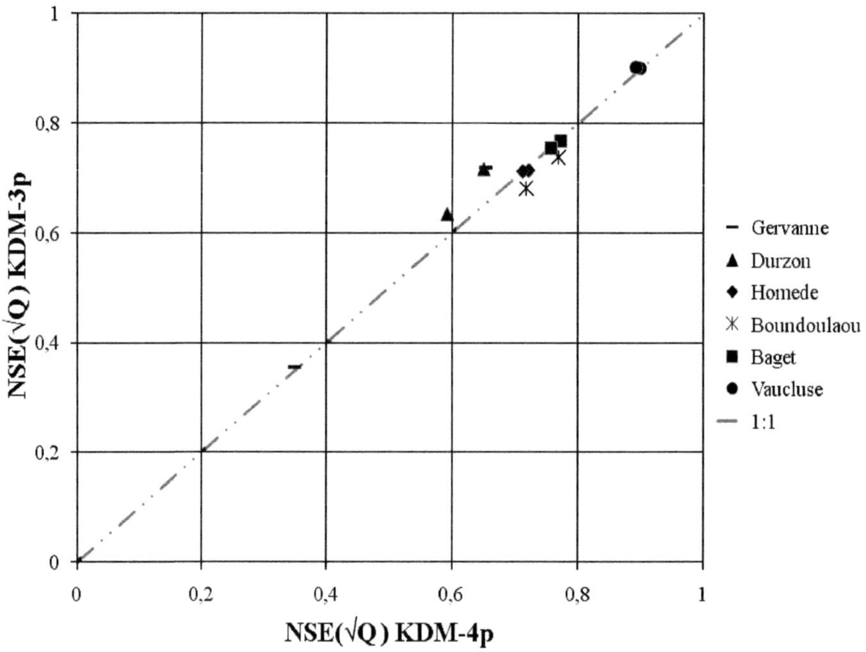


**Fig. 2** NSE($\sqrt{Q}$) values obtained by the two versions of the model when tested on the six karst systems. Results are shown for the two sub-periods in validation mode

The baseflow coefficient obtained from automatic calibration gave quite different values from those obtained from recession analysis (Table 2, last column): the baseflow coefficient obtained by calibration is greater than the one estimated from recession analysis. This difference could be explained by the high connection of the three parameters X2, X3 and X4 in the KDM-4p. This suggests that the KDM-4p is likely over-parameterised, i.e., the degree of freedom is too high in the parameter's optimum values. Therefore, the KDM-3p is particularly advantageous since it potentially constrains the parameterisation of the model.

## 4 Conclusions

Recession curve analysis was used to improve parameter consistency and identifiability in a karst system model calibration. The methodology compares the model's performance and parameter variability with and without using the baseflow coefficient obtained from recession curve analysis.

Efficiency of the KDM-3p is shown to be similar to the KDM-4p for the six karst systems examined in this study, meaning that giving the model more physically meaningful parameters does not affect its overall efficiency over the systems studied. Interestingly, the value of the X4 parameter obtained by automatic calibration is much higher than the value obtained by recession analysis. Indeed, the four-parameter version of the KDM seems to suffer from the interactions existing between the three routing parameters (X2, X3 and X4).

**Acknowledgements** We would like to thank Alain Mangin for the Baget streamflow data and Laurent Danneville from Parc des Grand Causes for making data available. We also thank Meteo France for rainfall data and the Hydro database for the steamflow data.

## References

Bonacci O (1993) Karst springs hydrographs as indicators of karst aquifers. Hydrological Sciences Journal, 38(1): 51–62

Duan Q, Sorooshian S, Gupta V (1992) Effective and efficient global optimization for conceptual rainfall-runoff models. Water Resour. Res., 28(4): 1015–1031

Fleury P, Plagnes V, Bakalowicz M (2007) Modelling of the functioning of karst aquifers with a reservoir model: Application to Fontaine de Vaucluse (South of France). Journal of Hydrology, 345(1–2): 38

Harlin J (1991) Development of a process oriented calibration scheme for the HBV hydrological model. Nordic Hydrol., 22(1): 15–36

Hogue TS, Sorooshian S, Gupta H, Holz A, Braatz D (2000) A multistep automatic calibration scheme for river forecasting models. Journal of Hydrometeorology, 1(6): 524–542

Jukic D, Denic-Jukic V (2009) Groundwater balance estimation in karst by using a conceptual rainfall-runoff model. Journal of Hydrology, 373(3–4): 302–315

Klemeš V (1986) Operational testing of hydrological simulation models. Hydrological Sciences Journal, 31(1): 13–24

Le Moine N, Andréassian V, Mathevet T (2008) Confronting surface- and groundwater balances on the La Rochefoucauld-Touvre karstic system (Charente, France). Water Resources Research, 44: W03403, doi:10.1029/2007WR005984

Mangin A (1975) Contribution à l'étude hydrodynamique des aquifères karstiques, Thesis, Université de Dijon, 422 pp

Nash JE, Sutcliffe JV (1970) River flow forecasting through conceptual models part I – A discussion of principles. Journal of Hydrology, 10(3): 282

Oudin L, Andréassian V, Mathevet T, Perrin C, Michel C (2006) Dynamic averaging of rainfall-runoff model simulations from complementary model parameterizations. Water Resour. Res., 42: W07410, doi:10.1029/2005WR004636

Padilla A, Pulido-Bosch, A, Mangin A (1994). Relative importance of baseflow and quickflow from hydrographs of karst spring. Ground Water, 32(2): 267–277

Shamir E, Imam B, Gupta HV, Sorooshian S (2005) Application of temporal streamflow descriptors in hydrologic model parameter estimation. Water Resour. Res., 41: W06021, doi:10.1029/2004WR003409

Sivapalan M (2005) Pattern, process and function: elements of a unified theory of hydrology at the catchment scale. Encyclop. Hydrol. Sci, 1186: 1190–1192

Zhang XN, Lindstrom G (1997) Development of an automatic calibration scheme for the HBV hydrological model. Hydrological Processes, 11(12): 1671–1682

# Time Series Analysis of Saraw Springs – SE of Sulaimaniya, Iraqi Kurdistan Region

S.S. Ali and Z. Stevanovic

**Abstract** Saraw Springs are located 45 km southeast of Sulaimaniya city of the Iraqi Kurdistan Region. This group of three vauclusian springs is recharged by infiltration through Cretaceous and Jurassic limestone outcrops. During three years of drought (1999–2001), one of the springs located at a few meters higher elevation than the others dried up. It became again active after the heavy rains at the end of December 2001. Based on the daily discharges measurements from October 2004 until the end of June 2006, a high correlation between discharge rates of the springs has been observed which indicates the prevailing role of long term reserves. The gentle declining of the auto-correlogram with no inter-subsistence of any peaks during that period indicates a type of homogeneity in the fissured systems, with no significant double porosity. The cross-correlogram shows that the stochastic connection between precipitation and discharge rate is weakly synchronized. This confirms dominant discharge through small fissures of the aquifer. The late reactions of the bigger channels can be forecasted according to the highly significant correlation of the first two days. In accordance with the obtained recession coefficients the average total dynamic reserves for the three springs equals $22 \times 10^6$ $m^3$ per 277 days.

## 1 Introduction

Karst springs represent imperative potential resources for water supply and irrigation in Iraqi Kurdistan Region. Sharazoor-Piramagroon basin as one of the most important hydrogeological basins in the northeastern part of Iraq is characterized by the existence of a number of karstic and karstic-fissured springs with appre-

S.S. Ali
Department of Geology, College of Science, University of Sulaimani, Sulaimaniya, Iraqi Kurdistan Region, Iraq, e-mail: salahsulus@yahoo.com

Z. Stevanovic
Department of Hydrogeology, FMG, Belgrade University, Belgrade, Serbia, e-mail: zstev@EUnet.rs

B. Andreo et al. (Eds.), *Advances in Research in Karst Media*
DOI 10.1007/978-3-642-12486-0, 

ciable discharge (Ali 2007; Ali et al. 2007, 2009). Saraw group springs as one of the largest karstic springs in the mentioned basin consists of three discharge points namely Sarchawi Saraw, Qalabo and Qurena (Fig. 1). These springs are draining Cretaceous and Jurassic limestones, and all have a vauclusian (artesian) character. The three springs are most probably controlled by the outlets of the same karstic system named by Stevanovic and Iurkiewicz (2004) the Saraw-Galal system due to the hydraulic connection of these springs with karstified rocks in nearby Galal valley.

During the three years of drought (1999, 2000 and 2001) Sarchaway Saraw, which is located at a few meters higher elevation than the others dried up. It suddenly became active after the heavy rains at the end of December 2001 (Stevanovic and Iurkiewicz 2004).

The abrupt spring increasing in one day from 50 to 600–700 l/s validates the hypothesis of an overflow type spring draining the same karst system regularly discharged through lower positioned springs of Qalabo and Qurena. The large discharge and fluctuations are the result of the fast transferred recharge. The spring returned to its normal yield during March, 2002 with maximum discharges at the end of April, 2002.

The waters of Saraw are diverted into two irrigation channels in addition to a pipeline for the gravity water supply of Saraw district in the vicinity of Said Sadiq town. The relatively high elevation makes the spring water a potential source for further utilization and development, mainly for drip and spray irrigation.

**Fig. 1** Location map of the studied area

## 2 Geology and Hydrogeology

Saraw group springs are located in the High Folded Zone, which is related to Western Zagros Fold–Thrusted Belt, created by Laramide and post Laramide Orogenies (Buday 1980, Buday and Jassim 1987). The northern and northeastern boundary of the basin coincides with the boundary between High Folded and Imbricated Zones (Buday 1980). In this basin, the anticlines and synclines are high and tight, in most cases they are overturned toward southwest due to stress of overriding Iranian plate. The springs are emerging from the southeastern plunge of Warsaka anticline forty five km SE of Sulaimaniya.

Kometan Formation (L. Cretaceous) comprises the main and top aquifer of these springs. It is composed of well-bedded, white or grey pelagic limestone, with a total thickness of 200 m. They overlay impervious Shiranish and Tanjero Formations (U. Cretaceous) which act as natural obstacle (barrier) for the accumulation of groundwater and caused creation of contact type springs. The deeper aquifer parts and furthermost catchment consists of Jurassic age rocks. The Saraw springs are predominantly recharged by the infiltration of precipitation within the Galal valley and the areas that surround Kora Kazhaw Mt. The annual precipitation in normal years approaches 900 mm. Part of the groundwater from the karstic aquifers underlying the Galal valley flows into the intergranular aquifer, towards the Said Sadiq spring or possibly south of the study area (Fig. 2).

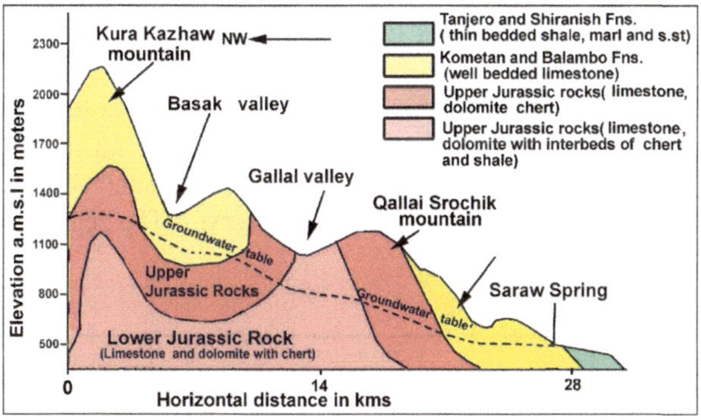


**Fig. 2** Geological cross section passing through Kura Kazhaw Mountain

## 3 Spring Discharge Regime

Figure 3 shows the hydrograph of the three springs based on the daily discharge measurements from October 2004 until the end of June 2006.

The three springs react similarly to rainfall events. The feeding of the reservoir starts with a sudden increase in the reserve due to the fast transfer of water

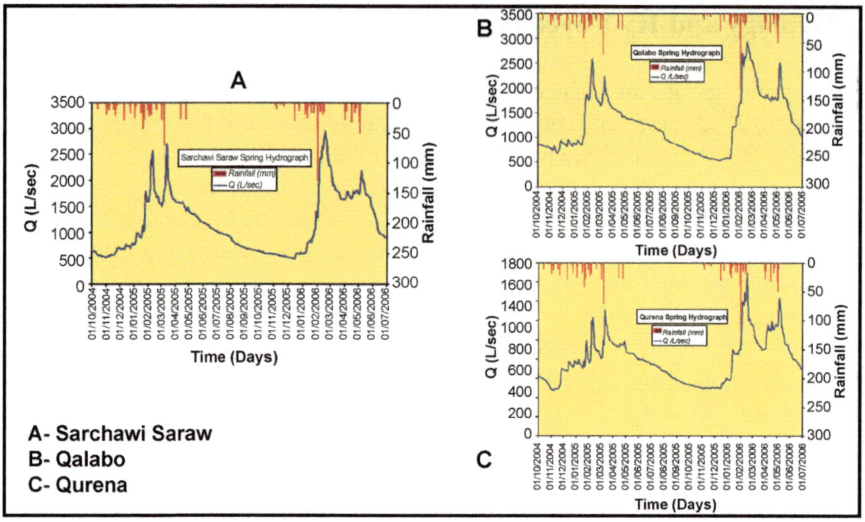


**Fig. 3** Hydrograph of Saraw group springs, for the period of Oct. 2004–July 2006

through the Jurassic and Cretaceous karstic aquifer. Auto-correlational and cross-correlational analyses give insight into the structure of the karst system, and the characteristics of the discharge regime and discharge rate (Kresic 1997). Auto-correlogram and cross-correlogram were made daily for a period of ninety nine days.

From the auto-correlogram of the three springs (Fig. 4) a high correlation between discharge rates of the springs could be observed with a statistical significance of 80, 75 and 88 days for Sarchaway Saraw, Qalabo and Qurena respectively, which indicates the prevailing role of long term reserves. The gentle declining of the

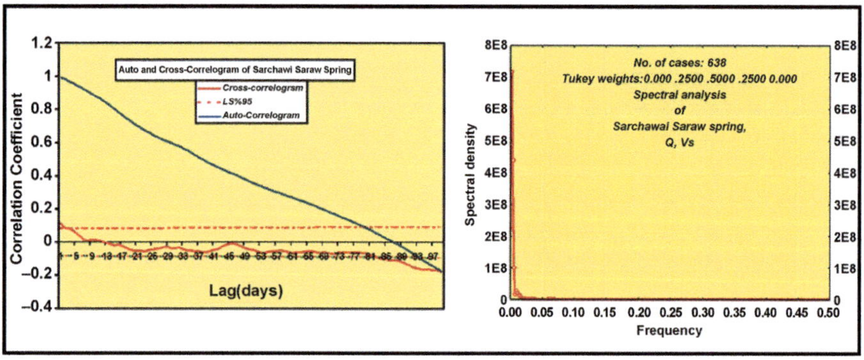


**Fig. 4** Auto and cross-correlational analysis of Sarchawai Saraw spring (*left*) and spectral density analysis, Oct.2004–July 2006

auto-correlogram with no inter-subsistence of any peaks during that period indicates a type of homogeneity in the fissured systems with no significant double porosity.

The cross-correlogram of the three springs shows an almost similar shape, which again confirms the hydraulic connection and the dominancy of one main source of karstic aquifer for the three springs.

The cross-correlogram shows that the stochastic connection between precipitation and discharge rate is weakly synchronized. This confirms dominant discharge through small fissures of the aquifer. Relatively quick response by the discharge to the precipitation was observed in the first few steps.

The late reactions of the bigger channels can be forecasted according to the highly significant correlation of the first two days. This correlation is repeated with a descending correlation coefficient throughout the seasonal period, but with less statistical significance. Behavior of the frequency–density curve is noticeable, which substantiates the annual cyclicity nature of the system (Fig. 4).

**Table 1** Characteristic data recorded for Saraw springs

| Spring name | $Q_{max}$ (L/s) | $Q_{min}$ (L/s) | $Q_{av}$ (L/s) | $Q_{max}/Q_{min}$ |
|---|---|---|---|---|
| Sarchaway Saraw | 2956 | 495 | 1160 | 5.97 |
| Qalabo | 2930 | 530 | 1280 | 5.52 |
| Qurena | 1700 | 477 | 793 | 3.56 |
| Total | 7586 | 1502 | 3233 | 5.01 |

# 4 Application of Maillet Recession Coefficient

The three springs react to the rainfall events similarly, and follow a similar way of discharging. Relatively rapid emptying during the end of the wet season with a steep recession curve within an average of twenty days is followed by a gradual gentle discharge of the potential reserve inside the fissures of the aquifer under descending pressure. The recession coefficients have been determined based on Maillet recession curve. Moreover the estimated dynamic reserves are shown in the table below. The average total dynamic reserves for the three springs equals $22.1 \times 10^6$ $m^3$ per 277 days. This is equivalent to average 0.92 $m^3/s$. Theoretically approximately four years are required for the springs to be exhausted totally without any recharge.

**Table 2** Recession coefficients for Saraw springs

| Spring Name | $\alpha_1$ | $\alpha_2$ | $V_{total} \times 10^6$ $m^3$ |
|---|---|---|---|
| Sarchaway Saraw | 0.037 Per 22 days | 0.0047 Per 255 days | 24.1 |
| Qalabo | 0.02 Per 16 days | 0.0043 Per 262 days | 24.65 |
| Qurena | 0.023 Per 12 days | 0.0026 Per 264 days | 17.59 |
| Total | 0.026 Per 16.6 days | 0.0038 Per 260.3 days | 22.1 Per 276.9 days |

## 5 Conclusions

1. The main aquifers of the Saraw group of springs are Kometan and Jurassic limestone formations. The non-karstic rocks consist of part of Lower Jurassic and Balambo formations. They act more or less as aquitard, which supply the spring during the end of water year.
2. The localization of the spring is due to topography, structure and stratigraphy. The spring is located at the lowest elevation and contact of karstified Kometan rocks with the impervious Shiranish formation.
3. The main characteristics of the water regime are cyclic variations and significant aquifer storage capacity.
4. The cross-correlogram shows that the stochastic connection between precipitation and discharge rate is weakly synchronized. This confirms dominant discharge through small fissures of the aquifer. Relatively quick response by the discharge to the precipitation was observed in the first few steps.
5. The Jurassic carbonate sediments extend beneath the clastic sediments to the south. This means that the karstic aquifer may be also discharged to a certain degree by subterranean flow outside the study area.
6. The springs are of strategic importance for water supply and irrigation in the study area of Sharazoor Plain. Currently, the water is tapped out and by gravity channels diverted to many downstream consumers. The possibility to regulate springs and tap more waters over the critical summer – autumn months exists. The most favorable regulation option appears to be over pumping of deeper channels of this vauclusian type spring. Conversely, it is risky to approach the groundwater development through the wells due to an exceptional heterogeneity of the karstic aquifer and problem of controlling extraction from possibly a great number of the wells.

## References

Ali SS (2007) Geology and Hydrogeology of Sharazoor-Piramagroon Basin in Sulaimani area, NE Iraq. PhD Thesis, FMG, Belgrade University, 330p

Ali SS, Stevanovic Z, Al-Jabbari MH (2009) Auto and Cross Correlation Analysis of Bestansur spring – Sharazoor Plain-Iraqi Kurdistan Region – Proceedings of 3rd sci. conf. of the College of Science, 24 to 26 March 2009, University of Baghdad, Baghdad

Ali SS, Stevanovic Z, Jemcov I (2007) The mechanism and influence on karstic spring flow: Sarchinar spring case example NE Iraq, 4th Conference hydrogeology, ecology, monitoring, and management of groundwater in karst terrains, Feb. 27–28, 2007, Florida

Buday T (1980) Regional Geology of Iraq: Vol. 1, Stratigraphy, I.I. Kassab and S.Z. Jassim (Eds) D. G. Geo Survey. Min. Invest. Publication. 445p

Buday T, Jassim SZ (1987) Regional geology of Iraq: Tectonis, Magmatism, and metamorphism. II. Kassab, MJ Abbas (Eds), Baghdad, 445p

Energoprojekt Beograd (1981) Kaolos dam project. Definite planning report. Vol. 2, Book 2, Part 2. Min. of Irrigation, State organization of dams, Iraq. 70p

Kresic N (1997) Quantitative solutions in hydrogeology and groundwater modeling. Lewis publishers. CRC. 461p

Stevanovic Z, Iurkiewicz A (2004) Hydrogeology of Northern Iraq, Vol. 2 General hydrogeology and aquifer systems, Spec. Edition TCES, FAO, 175p

# Analysis of Spring Discharges During Droughts: Example from Karst Systems of Southern Italy

F. Fiorillo and F.M. Guadagno

**Abstract** Analyses of spring discharges of a large karst system (Campania, Southern Italy) have been carried out to quantify the relation of rainfall and droughts. Due to a Mediterranean climate and hydrogeological conditions of the aquifers, spring hydrographs are characterised generally by one annual smoothed peak during spring season and a negligible quick flow component. Insufficient recharge due to poor annual rainfall can cause continuous decreasing discharge at springs, producing flat hydrographs (with no peak). Flat spring hydrographs induces a reduction in discharge during the following year as well. Thus, aquifers present a memory effect, which amplifies periods of poor rainfall. Droughts also appear to be induced by consecutive years with lower than average rainfall. Due to a long historical series and the specific karst spring regime, a flat hydrograph can be forecast as early as winter, providing a useful tool for water management.

## 1 Introduction

Analyses of time series of some spring discharges have been carried out to quantify the relation between rainfall and droughts. Data come from the karst system of Picentini mountains, Campania, where the length of the time series available allows investigation into some hydrogeological, hydrological and climate features.

Local karst springs supply the main aqueducts of Southern Italy, and have a strategic importance since Roman Age. Differently from other hydrological risks (flood, landslides) specific procedures to forecast groundwater droughts are generally missing in Southern Italy. Thanks to long hydrological time series (rainfall, temperature, spring discharges), some probabilistic approaches have been carried

F. Fiorillo
Dipartimento di Studi Geologici e Ambientali, Università degli Studi del Sannio, Via dei Mulini 59/A, 82100 Benevento, Italy, e-mail: francesco.fiorillo@unisannio.it

B. Andreo et al. (Eds.), *Advances in Research in Karst Media* 
DOI 10.1007/978-3-642-12486-0, 

out to forecast droughts in karst environment (Fiorillo 2009), providing useful tools for water resource management.

## 2 Hydrogeological Features and Meteorological Data

The karst springs are located along the northern boundary of the Picentini Mountains, a large karst system in the Campania region of Italy (Fig. 1a–b). Outcropping rocks primarily belong to the calcareous and calcareous-dolomite series (Late Triassic–Miocene), are 2500 m thick, heavily fractured and faulted and frequently reduced to breccias. Pyroclastic deposits of Somma-Vesuvius activity mantle the slopes, mainly in the western sector (Mt. Terminio) which have an important role in the infiltration of water into the karst substratum below. The calcareous-dolomite series is tectonically bounded by terrigenous deposits, constituting complex argillaceous (Paleocene) and flysch sequences (Miocene).

The powerful springs of the Picentini Mountain belong to three main groups (Fig. 1b), known as the Serino, Cassano and Caposele springs, and their main hydrological characteristics are shown in Table 1 and Fig. 1c.

A prevalently Mediterranean climate is present, characterised by dry and warm summers and a wet period that occurs during autumn, winter and spring. Figure 1c shows the distribution of the monthly rainfall reduced by the amount of evapotranspiration computed by the method of Thornthwaite and Mather (1957).

Runoff can be considered almost null in the karst area analysed, and the difference between monthly rainfall and monthly potential evapotranspiration appears very similar to the amount of effective infiltration.

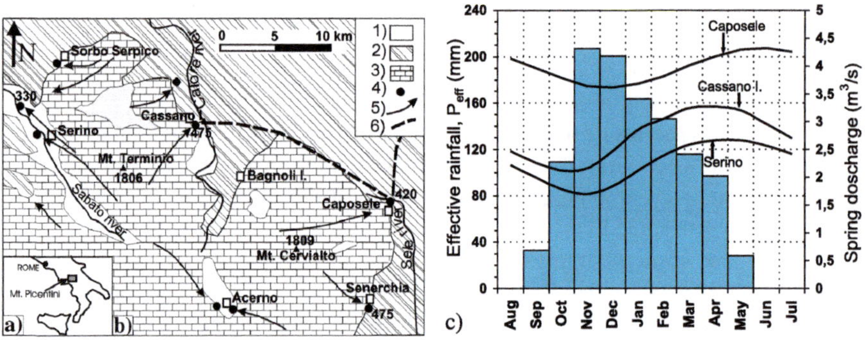


**Fig. 1** **a** Southern Italian peninsula. **b** Hydrogeological map of the Northern Picentini Mountains (modified from Celico, 1978). 1) Covering deposits (Quaternary); 2) argillaceous complex and Flysch sequences (Paleogene–Miocene); 3) calcareous-dolomite series (Jurassic–Miocene); 4) main spring; 5) groundwater flow direction; 6) Pugliese Aqueduct. **c** Mean monthly rainfall reduced by the evapotranspiration (mean of stations at Serino, 351 m a.s.l. and Montevergine, 1270 m a.s.l.) and mean monthly spring discharges of the Caposele, Cassano and Serino groups (period 1930–2006, except Cassano spring discharges, 1965–2006)

Figure 1c shows the historical monthly mean discharges from the spring groups. In each group of springs, the maximum discharge occurred several months after the maximum rainfall, whereas the minimum occurred during the wettest period. Spring hydrographs present a smoothed shape, with absence or a very low component of quick flow, and can be associated to *diffuse* karst system flow (Bonacci 1993). During droughts, hydrographs can have a continuous decreasing trend during the hydrological year.

**Table 1** Main characteristics of the spring groups (period 1930–2006): $\mu$, mean; $\sigma$, standard deviation; $Q_{10}$, 10th percentile; $Q_{90}$, 90th percentile

| Spring group | Elevation $M$ (a.s.l.) | Discharge $Q$ ($m^3/s$) | | | | Recharge area ($km^2$) |
|---|---|---|---|---|---|---|
| | | $\mu$ | $\sigma$ | $Q_{90}$ | $Q_{10}$ | |
| Serino | 330–380 | 2.25 | 0,33 | 2.64 | 1.81 | 67 |
| Cassano I. | 473–476 | 2.65 | 0.68 | 3.55 | 1.79 | 78 |
| Caposele | 420 | 3.96 | 0.59 | 4.69 | 3.15 | 103 |

## 3 Hydrological Time Series and Droughts

Annual mean series of the Serino, Caposele and Cassano spring discharge are shown in Fig. 2a. Spring discharges show a very similar trend, with drier and wetter years being well correlated between each series. The minimum of 1946, 1949, and 2002 are present in the all series; the spring hydrographs obtained after 1987 all clearly show a decrease in discharge up to 1993. During this period, many water resource management problems occurred in Southern Italy due to the reduced water supply.

All of the spring discharges reached their maximum value following the Irpinia earthquakes in November 1980 (Ms = 6.9), which caused an anomalous increase in spring discharge.

The annual rainfall recorded at the Serino rain gauge (Fig. 2b) appears to be well correlated with the annual mean spring discharge series (Fiorillo 2009). The two droughts of 1949 and 2002 are explained by the two lowest historical values of annual rainfall, and the continuous rainfall period 1987–1993 constantly below the mean explains the relative lowering of the spring discharge in the same period.

During any hydrological year, spring hydrographs characterised by a constant decrease or without a clear flood denote insufficient recharge of the karst system. These flat spring hydrographs can be defined as indicators of drought, as described by Fiorillo (2009) in these karst environments. Based on this hydrogeological criterion, any drought has a duration of about one year, beginning in the autumn and ending the following autumn. In addition, flat spring hydrograph can characterise two or more consecutive years, indicating a typical multiyear drought. Fiorillo and Guadagno (2010) showed the relation between the meteorological drought defined by the Standard Precipitation Index (SPI of McKee et al. 1993) and the groundwater drought defined by the flat spring hydrographs.

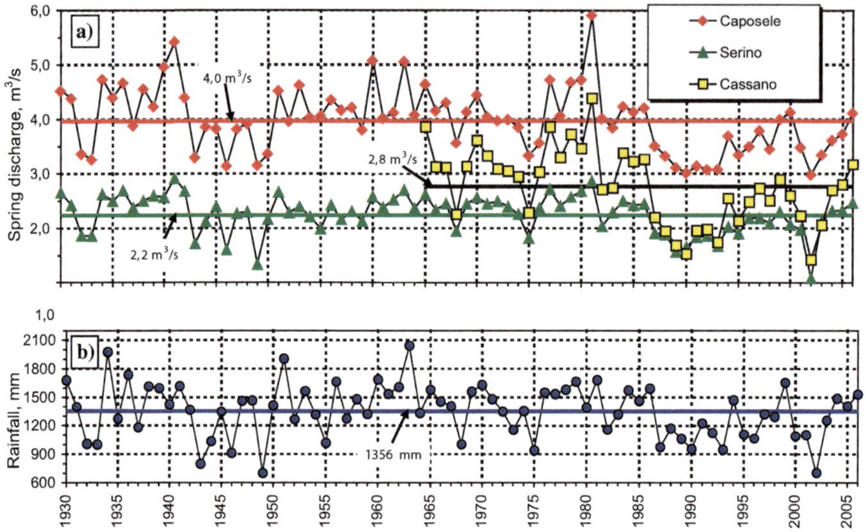


**Fig. 2** **a** Annual mean series (November–October) of the Caposele, Serino and Cassano spring discharge; **b** Annual rainfall (September–August) of Serino rain gauge. Mean of the 1930–2006 period is also shown

**Annual Scale Analysis.** Autocorrelation of the annual mean spring discharge and annual rainfall series were conducted to determine if there is any link between consecutive values (Fig. 3a). For a value of $k = 1$, the coefficient of autocorrelation, $r(k)$, is 0.68, 0.62 and 0.51 for the Cassano, Caposele and Serino springs, respectively, and is very low (0.23) for the rainfall series. Because the value of $r(k)$ is high (specially for Cassano and Caposele annual mean spring discharge), each year appears connected to the previous one. For these springs, the dependence occurs mainly after droughts, where the effect of poor annual rainfall have been found to last for several years in the Caposele and Cassano springs (Fig. 3b).

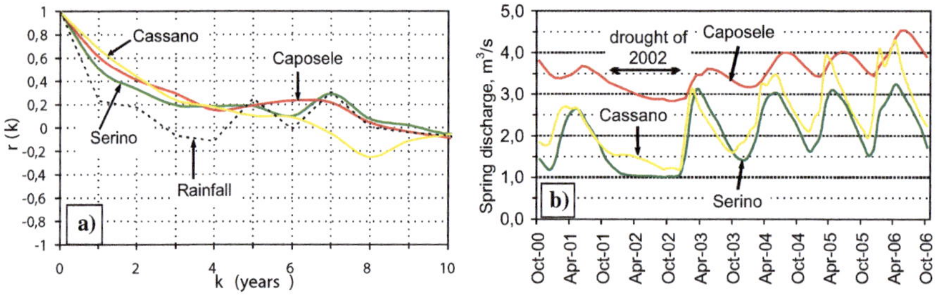


**Fig. 3** **a** Correlogram of the annual mean spring discharge series and annual rainfall (period 1965–2006). **b** Spring discharges for the period of October 2000 to October 2006, showing the progressive increasing of the Cassano and Caposele spring discharges after the 2002 drought (Fiorillo 2009)

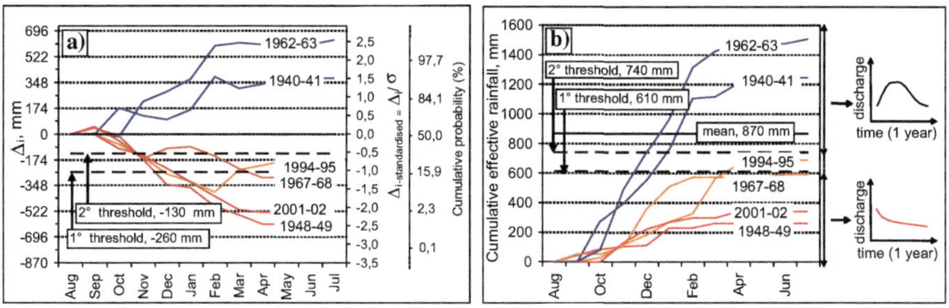


**Fig. 4** **a** Trend of the index $\Delta_i$ for some hydrological years (Fiorillo 2009). The *right-hand axis* is normalised by the standard deviation, $\sigma$, and gives the cumulative probability for normal distribution. **b** trend of the cumulative effective rainfall for the same years shown in **a**

**Monthly Scale Analysis.** The relationship rainfall–spring discharge has been investigated for each hydrological year. The combined effect of increasing evapotranspiration and decreasing rainfall towards the summer season results in the effective rainfall after April being negligible. Therefore, the rainfall that occurs up to March–April of each hydrological year recharges karst aquifers, and subsequent rainfall, up to September–October, generally does not recharge the aquifers. To highlight the characteristics of rainfall in a Mediterranean climate, Fiorillo et al. (2007) introduced the cumulative effective rainfall index:

$$\Delta_i = (P_i - E_\mathrm{p}) - (P_\mathrm{m} - E_\mathrm{p}) + \Delta_{(i-1)}\,,$$

where $P_i$ is the monthly rainfall, $P_\mathrm{m}$ is the mean monthly rainfall and $E_\mathrm{p}$ is the monthly potential evapotranspiration; when $P < E_\mathrm{p}$, $P - E_\mathrm{p} = 0$ is fixed. This index splits the cumulative effective rainfall into two primary groups: (i) negative values of $\Delta_i$, define the deficit of effective rainfall with respect to the mean, and (ii) positive values of $\Delta_i$, define the surplus of effective rainfall with respect to the mean.

Figure 4a shows the trend of the index $\Delta_i$ for several wet and dry hydrological years, and in Fig. 4b the cumulative effective rainfall for the same hydrological years have been plotted. The lower paths are associated with poor rainfall, which causes a weak rising or a continuous decreasing of spring discharge during the hydrological years. During these years, recharge is unable to exceed the discharge from the aquifer. This results in a poor aquifer response, which is reflected as a flat spring hydrograph. For the Caposele spring, these years have been delimited empirically by threshold lines, with values below $\Delta_i = -260$ mm producing flat hydrographs and values above the threshold of $\Delta_i = -130$ mm producing hydrographs that indicate flooding. Values $-260\,\mathrm{mm} < \Delta_i < -130\,\mathrm{mm}$ produce an uncertain type of hydrograph, depending mainly on the annual rainfall of the antecedent year.

To avoid a drought, the cumulative rainfall has to reach or exceeds the threshold given in Fig. 4. In particular, for a fixed probability, this amount of cumulative rainfall diminishes over time (from September to June) and can be evaluated by statistical analysis (Fiorillo 2009).

## 4 Conclusions

Karst spring regimes appear to be strictly connected to climate regime and controlled by the geological and hydrogeological features of the aquifer. Hydrographs of karst springs primarily reflect the amount of cumulative rainfall that has occurred from the beginning of the rainy season. During years in which a flat spring hydrograph is produced, the expected increase in spring discharge in autumn does not occur and flow is often diminished until the following autumn. This results in the prolonged aquifer emptying, and the resulting deficit of storage reduces discharge during the following year. This aquifer behaviour indicates the presence of a memory effect of antecedent drought that appears to be higher for springs characterised by highest discharge, such as the Caposele spring. Due to continuous adaptation of the water volume storage into the karst aquifer to the long term climate trend, two or more consecutive years with rainfall below the mean can cause a progressive emptying of the aquifers. In such cases, flat spring hydrographs are also produced, as shown by hydrographs corresponding to the 1987–1993 period.

The role of the negative trend in annual rainfall and positive trend of temperature recorded from the late '80 (Fiorillo et al. 2007; Fiorillo & Guadagno 2010), suggest that future storage in the aquifer could diminish, and that droughts characterised by a flat spring hydrograph could be more frequent than in the past.

**Acknowledgements** Authors are grateful to Acquedotto Pugliese and ARIN water companies for the spring discharge data.

## References

Bonacci O (1993) Karst spring hydrographs as indicators of karst aquifers. Hydrological Sciences Journal 38: 51–62

Celico P (1978) Schema idrogeologico dell'Appennino carbonatico centro-meridionale. Memorie e Note Istituto di Geologia Applicata, Napoli 14: 1–43

Fiorillo F (2009) Spring hydrographs as indicators of droughts in karst environment. J. Hydrology 373: 290–301

Fiorillo F, Esposito L, Guadagno FM (2007) Analyses and forecast of water resources in an ultra-centenarian spring discharge series from Serino (Southern Italy). J. Hydrology 336: 125–138

Fiorillo F, Guadagno FM (2010) Karst spring discharges analysis in relation to drought periods, using the SPI. Water Resource Management (in press, 10.1007/s11269-009-9528-9)

McKee TB, Doesken NJ, Kleist J (1993). *The relationship of drought frequency and duration to time scales*. 8th Conference on Applied Climatology, Anaheim, CA. American Meteorological Society, Boston; 179–184

Thornthwaite CW, Mather JR (1957) Instructions and tables for computing potential evapotranspiration and the water balance. Publication In Climatology 10, Drexel Institute Of Technology, Centerton, New Jersey

# Use of Combined Recession Curves Analyses of Neighbouring Karstic Springs to Reveal Karstification Degree of Groundwater Springing Routes

P. Malík and S. Vojtková

**Abstract** In the southern part of the Slovenský Kras Mts., positioned on the territory of Hungarian–Slovak transboundary karstic hydrogeological structure of Dolný vrch/Alsóhegy, several karstic springs occur. Hydrograph recession curves from nine gauged springs on the Slovak part of the transboundary aquifer were analyzed. Individual springs were classified by typical groundwater depletion hydrograph into categories by typical depletion equations. These were also assigned to the different lithostratigraphic limestone types occurring in the springing areas and assumed recharge areas. Previous hydrogeological mapping had first indicated a suspicion on the possible linking of groundwater from the two individual springs, to be only the two surface appearances of the same springing groundwater in the neighbouring springs. By coupling the outflowing volumes of the two springs together, and analyzing the resulting new time series, new classification of subregimes could be done. Surprising new facts were revealed about the analysed pairs of springs, which primarily were not recognized as connected branches of one spring. Summation of discharged volumes into one entity lead in two cases to discovery of apparent presence of turbulent flow subregimes on the recession curves. Two branches of one outflow from the recharge area exist there; one of them usually represents the lower branch with "baseflow" and another, the upper one, is the overflow spring. This also influences the aspects of groundwater-sensitivity-to-pollution, and also gives unexpected hopes for further speleological investigations nearby.

---

P. Malík
ŠGÚDŠ – Geological Survey of Slovak Republic, Mlynská dolina 1, 817 04 Bratislava, Slovakia, e-mail: peter.malik@geology.sk

S. Vojtková
Veternicová 22, 841 05 Bratislava, Slovakia, e-mail: s.vojtkova@gmail.com

B. Andreo et al. (Eds.), *Advances in Research in Karst Media* 
DOI 10.1007/978-3-642-12486-0, 

# 1 Geographical and Hydrogeological Background

Karstic area of the Slovenský Kras Mts. (southern Slovakia) contains also the Slovak/Hungarian transboundary karstic hydrogeological structure of Dolný vrch/Alsóhegy (Fig. 1). The total area of this structure is 46.70 $km^2$, the Slovak part outcrops on the 21.63 $km^2$. The structure itself is formed mainly by Wetterstein type of limestones. Relatively impervious rocks, delineating the karstic structure itself – the Lower Triassic shales, sandstones and claystones form the underbed/barrier to the groundwater flow, on which the majority of springs appear. The springs are on the Slovak part of this transboundary karstic aquifer mentioned together with ID (number), under which they are registered by the Slovak Hydrometeorological Institute (SHMÚ), the institution that regularly monitors the discharge of the springs on weekly and in some cases on daily bases since the 1960s (Malík & Vojtková 2006).

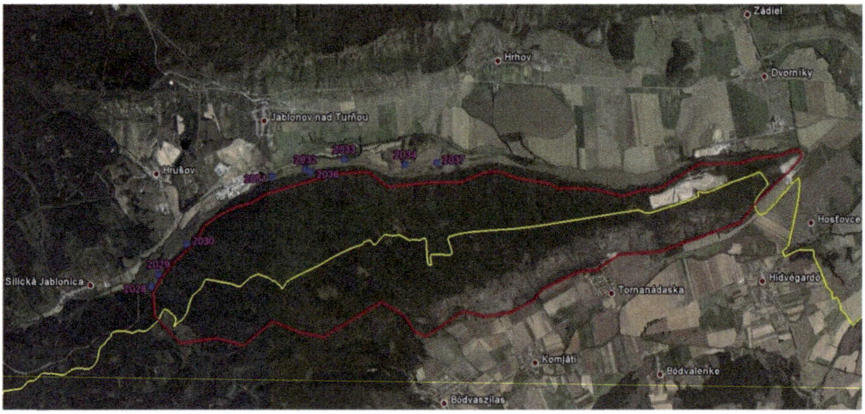


**Fig. 1** Location of the Dolný vrch/Alsóhegy transboundary Slovak/Hungarian hydrogeological structure in Europe. Springs monitored by the Slovak Hydrometeorological Institute in the NW part of the structure are *blue points*, structure delineation in *dark violet line*, state boundaries in *yellow line*. The coordinates of the centre of the area are N 48.5710° and E 20.7267°

# 2 Recession Curves Analyses

In this study, simple Maillet's exponential description ($Q_t = Q_o \cdot e^{-\alpha \cdot t}$) of laminar subregimes is used (Maillet 1905). Using only the Reynold's number (usually technically difficult) it is possible to distinguish the presence of laminar or turbulent flow of water (groundwater/flow in karst channels). Kullman (1990) offers a simplified method of turbulent flow identification. The linear turbulent model ($Q_t = Q_o \cdot (1 - \beta \cdot t)$) for turbulent flow supposed for karstic channels as described by Kullman (1990, 2000) is used within this study. Several both laminar and turbulent subregimes may exist in one aquifer, and its discharge can be described by superposition of several appropriate equations. Hydrograph recession curves can be used for analyses of type and properties of a karstic aquifer (Kullman 2000), as well as for estimation of regional karstification degree and groundwater sensitivity to pollution (Malík 2007). The basis of this method, which is used for the assessment of the degree of the rock disruption and karstification, were described by Kullman

**Table 1** Equations of recession curves of springs monitored in the Dolný vrch/Alsóhegy area

| SHMÚ No. (Fig. 1) | Site and spring's name | Final discharge equation |
|---|---|---|
| 2028 | Hrušov – Studňa pri hradnom buku | $Q_t = 1.5 \cdot e^{-0.008 \cdot t} + 25.0 \cdot e^{-0.055 \cdot t}$ |
| 2029 | Hrušov – Čierna studňa | $Q_t = 0.35 \cdot e^{-0.0031 \cdot t} + 8.0 \cdot e^{-0.047 \cdot t}$ |
| 2030 | Hrušov – Jarček | $Q_t = 0.80 \cdot e^{-0.003 \cdot t} + 13.0 \cdot (1 - 0.02 \cdot t) + 22 \cdot (1 - 0.12 \cdot t)$ |
| 2032 | Jablonov nad Turňou – Köszörű | $Q_t = 17.0 \cdot e^{-0.005 \cdot t} + 20.0 \cdot e^{-0.011 \cdot t}$ |
| 2033 | Jablonov nad Turňou – Tapolca | $Q_t = 24 \cdot e^{-0.01 \cdot t} + 70 \cdot (1 - 0.0147 \cdot t) + 160 \cdot (1 - 0.045 \cdot t) + 790 \cdot (1 - 0.14 \cdot t)$ |
| 2034 | Hrhov – Körkút | $Q_t = 0.15 \cdot e^{-0.008 \cdot t} + 1.1 \cdot e^{-0.045 \cdot t}$ |
| 2036 | Jablonov nad Turňou – Köszörű občasný | $Q_t = 30.0 \cdot (1 - 0.0109 \cdot t) + 120.0 \cdot (1 - 0.016 \cdot t) + 146.0 \cdot (1 - 0.058 \cdot t)$ |
| 2037 | Hrhov – Žmaň | $Q_t = 2.7 \cdot e^{-0.024 \cdot t}$ |
| 2051 | Jablonov nad Turňou – Za mostom | $Q_t = 3.0 \cdot e^{-0.036 \cdot t} + 12.0 \cdot e^{-0.18 \cdot t}$ |
| 2032 + 2036 | Jablonov nad Turňou – Köszörű + Köszörű občasný | $Q_t = 12.0 \cdot e^{-0.003 \cdot t} + 130.0 \cdot e^{-0.05 \cdot t} + 500.0 \cdot (1 - 0.04 \cdot t)$ |
| 2028 + 2029 | Hrušov – Čierna studňa + Studňa pri hradnom buku | $Q_t = 2.0 \cdot e^{-0.0055 \cdot t} + 5.0 \cdot e^{-0.033 \cdot t} + 14.0 \cdot (1 - 0.07 \cdot t)$ |

**Table 2** Karstification degree of the monitored springs in the Dolný vrch/Alsóhegy hydrogeological structure

| SHMÚ No. (Fig. 1) | Site and spring's name | Karstification degree |
|---|---|---|
| 2034 | Hrhov – Körkút | 3.7 |
| 2037 | Hrhov – Žmaň | 2.3 |
| 2028 | Hrušov – Studňa pri hradnom buku | 3.7 |
| 2029 | Hrušov – Čierna studňa | 3.0 |
| 2030 | Hrušov – Jarček | 8.0 |
| 2032 | Jablonov nad Turňou – Köszörű | 3.7 |
| 2033 | Jablonov nad Turňou – Tapolca | 6.0 |
| 2036 | Jablonov nad Turňou – Köszörű občasný | 10.0 |
| 2051 | Jablonov nad Turňou – Za mostom | 4.7 |
| 2032 + 2036 | Jablonov nad Turňou – Köszörű + Köszörű občasný | 5.5 |
| 2028 + 2029 | Hrušov – Čierna studňa + Studňa pri hradnom buku | 5.5 |

(1990, 2000). Karstification degree is understood as a level of endogenous karst development, which – for the phreatic and epiphreatic zone – can influence the mode of groundwater upwelling from the karst system (Malík 2007).

In the case of Dolný vrch/Alsóhegy hydrogeological structure, the previous analyses of springs' recession curves had revealed presence of several discharge subregimes, including turbulent ones. In the Table 1, recession curves analyses undertaken on individual springs are shown as resulting equations.

Differences in character of individual depletion hydrographs enabled also the assessment of the anticipated extent of karstification of the rock environment between groundwater infiltration in the catchment area and its springing (outflow) in the individual springs (Table 1). According to Malík (2007), the karstification degree there is within the range of 2.3 to 10.0 (Table 2).

In the same way, for the purposes of regional comparison, the mean ranking can be linked to certain aquifer types, based on the results of individual spring's evaluation. In this way, Wetterstein limestones aquifer can be characterized by "4.7" degree as rock environment with anticipated existence of crushed water-bearing zones (e.g. fault zones) or by dense network of open small fissures in combination with simple, partly (perennially) phreatic conduit system of considerable extent (e.g. with open karstified fault in the vadose zone; Kullman 1990).

## 3 Discharge Coupling

Previous hydrogeological mapping had first indicated the possible linking of groundwater from the two individual springs as only two surface appearances of the same springing groundwater in the neighbouring outlets. In the case of the northern part of the Dolný vrch/Alsóhegy hydrogeological structure, springs 2032 Köszörű + 2036 Köszörű občasný near Jablonov nad Turňou and 2028 Studňa pri hradnom buku + 2029 Čierna studňa near Hrušov seemed to be close enough to start further investigations. Counting the outflowing volumes of these two pairs of springs together, and analyzing resulting new time series, new classification of subregimes could be done (Fig. 2, Tables 1 and 2 – lower part). Outflows from the aquifer in the area of spring's outlets (2032 + 2036) and (2028 + 2029) are both characterised by the "5.5" degree. Regime of groundwater discharge is mathematically expressed by superposition of two exponential and one linear equations (Table 1, last rows). Groundwater regime type is a combination of one sub-regime with turbulent flow (with short-term influence) and two sub-regimes with laminar groundwater flow. Discharge coefficients in individual sub-regimes are high (Table 1), which indicates very high permeability and low storativity of rock environment. Discontinuities in the rock environment can be described as affected by extensive deformation, with a majority of open, medium size, both not karstified and karstified fissures, in the phreatic zone of the fissure karst groundwater reservoir and with smaller influence of connected conduits.

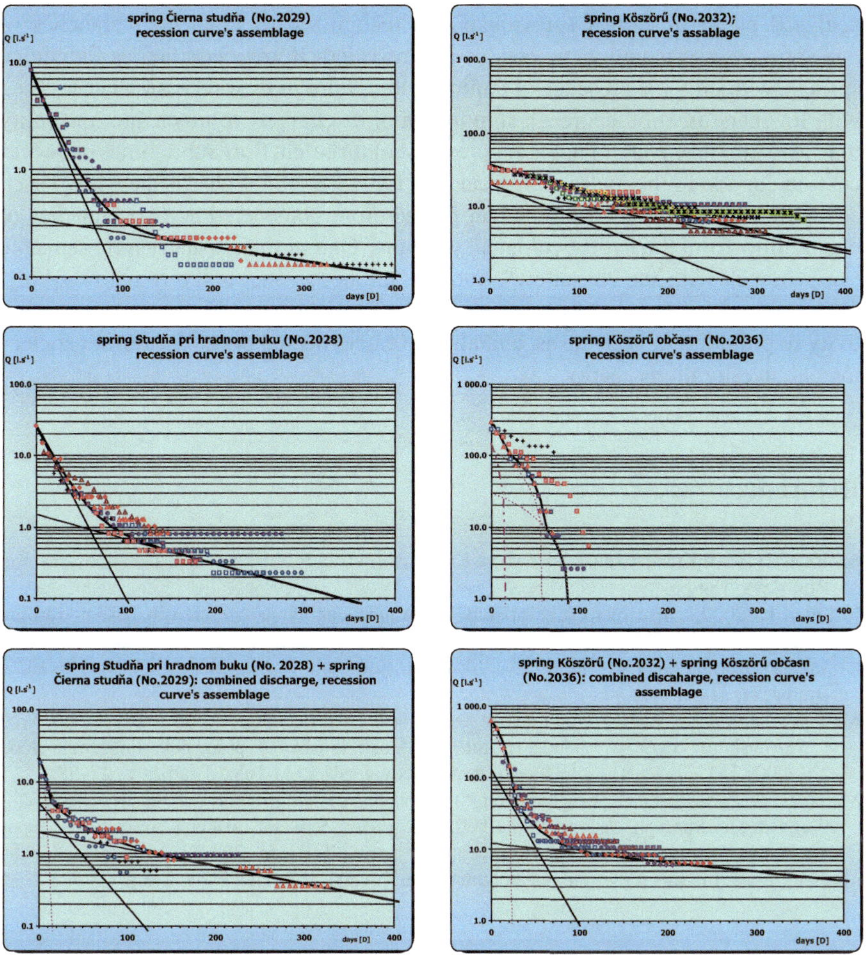


**Fig. 2** Recession curves of the individual outflows and of the coupled discharges (total sum of discharges) analysed for the cases of springs Studňa pri hradnom buku (No. 2028), spring Čierna studňa (No. 2029), Köszörű (No. 2032) and Köszörű občasný (No. 2036), situated in the NW part of the Dolný vrch/Alsóhegy transboundary Slovak/Hungarian hydrogeological structure. Individual decreasing discharge series are drawn by different marking, the final evaluation was done on the base assemblage of individual recession curves from the different dry periods. *Thick lines* represent the final assessment of recession curve, based on equations shown in Table 1. *Thin lines* represent estimated subregimes. Both equations and their representation by lines were created manually by help of MS Excel macro developed in Geological Survey of Slovak Republic, using the "maximum allowed accession"

# 4 Conclusions

Recession curves analyses of discharge time series observed on individual springs can be used for estimation of regional karstification degree. In some cases of analysed pairs of springs (previously not recognized as connected branches of one spring), surprisingly new facts can be revealed. Being classified as individual springs, with only the presence of laminar subregimes in their recession curves (i.e., with low karstification degrees), summation of discharged volumes into one entity may lead into discovery of apparent presence of turbulent flow subregimes on the recession curves. In the hydrogeological structure of Dolný vrch/Alsóhegy, two such cases were found. These springs can be therefore considered as two branches of one outflow from the hydrogeological structure. One of them usually represents the lower branch with the "baseflow", another, the upper one, is the overflow spring. These complex relationships bring more light in the aspects of groundwater sensitivity to pollution in such cases, and also give unexpected hopes for further speleological investigations nearby.

## References

Kullman E (1990) Krasovo–puklinové vody. Karst-fissure waters. Geologický ústav Dionýza Štúra, Bratislava, 184 pp.

Kullman E (2000) Nové metodické prístupy k riešeniu ochrany a ochranných pásiem zdrojov podzemných vôd v horninových prostrediach s krasovo – puklinovou priepustnosťou [in Slovak]. Podzemná voda VI. / 2000 č. 2, Slovenská asociácia hydrogeológov, ISSN 1335-1052, Bratislava, 31–41

Malík P, Vojtková S (2006) Hodnotenie zraniteľnosti hydrogeologickej štruktúry Dolný vrch v Slovenskom krase na základe režimu výdatnosti krasových prameňov. Podzemná voda XII./2006 č. 1, Slovenská asociácia hydrogeológov, ISSN 1335-1052, Bratislava, 27–38

Malík P (2007) Assessment of regional karstification degree and groundwater sensitivity to pollution using hydrograph analysis in the Velka Fatra Mts., Slovakia. Water Resources and Environmental Problems in Karst. Environmental Geology (2007), 51, 707–711

Maillet E (1905) Essais d'hydraulique souterraine et fluviale. Hermann Paris, 218 p.

# Problems of Flood and Drought in a Typical Peak Cluster Depression Karst Area (SW China)

F. Guo and G. Jiang

**Abstract** Flood and drought are typical hazards in karst areas, especially in the peak cluster depression karst area of Southwest China. In this study, some karst areas prone to flooding and drought area are mentioned; and the flooding and drought problem in Mumei subterranean river system in Yunnan Province is discussed in detail. There are multiple reasons for the problems, including uneven precipitation in a year, a complicated underground river system, intensive human engineering activities, deforestation and unreasonable land use. Some solutions to the karst flood and drought cycles could be returning farmland to forest and some engineering measures. Because flood and drought threaten the limited land resources and also human life, they should have more attention.

## 1 Introduction

Flood and drought are the most outstanding natural hazards in the world. People have paid much attention to these (Jones 1990; Liu et al. 2005; Singh 2001). But most of studies focused on non-karst areas. In China, flooding in karst terranes is frequent, as is drought. Together, these natural hazards cause damage to property, businesses and life. In the karst mountain areas, besides line distribution of surface rivers, another landform which is easily impacted is a blind valley, or a trough valley in long strip distribution and depression which can be either erose (jagged) or nearly rotundity. This negative landform acts as a separate hydro-unit which collects

F. Guo
Karst Dynamics Laboratory, Institute of Karst Geology, MLR, Guilin 541004, People's Republic of China, e-mail: gfkarst@126.com

G. Jiang
The International Research Center on Karst under the Auspices of UNESCO, Guilin 541004, People's Republic of China, e-mail: bmnxz@126.com

B. Andreo et al. (Eds.), *Advances in Research in Karst Media* 
DOI 10.1007/978-3-642-12486-0, 

surface water. Pooling of water enters to the underground through sinkholes developed in depressions, then discharges to an adjacent river via the conduit. Flooding occurs when a sinkhole's drainage capacity is not adequate to transfer the storm water runoff to the underground. Because depressions contain the most concentrated areas of local people and farmland distribution, but are rather flood-prone, the importance of this issue can not be underestimated. Floods can lead to the formation of cover-collapse sinkholes and groundwater pollution. Human beings' activities may increase the runoff rate to a sinkhole so that the rate of recharge exceeds the acceptance capacity of the sinkhole drain (Zhou 2007).

Drought is another hazard in karst areas. As a result of extensive distribution of karst surface features, such as karst windows and sinkholes, rain water usually discharges the underground system quickly. And because the infiltration coefficient is large, rain water easily leaks underground. So in dry seasons, it is difficult to find drinking and domestic water for the local people. They obtain water from far away by carrying containers on their shoulders or by horse. When the land resources are too dry, the people can not feed themselves, which contributes to reoccuring poverty.

## 2 Flood and Drought in Karst Area of Southwest China

Flood and drought are serious phenomena in the karst area of Southwest China, especially in the peak cluster area in Guizhou Province, Yunan Province and Guangxi Zhuang Autonomous Region (Li and Wang 2001). For example, about 94% of the areas are occupied by carbonate rocks in the east Mashan County, Guangxi Autonomous Region (Li et al. 2008). The karst depression and valley are well developed and flooding occurs quite often. According to the survey, there are more than 170 depressions with flooding, and the area frequently submerged by water is 1400 ha, accounting for 74% of the 189 thousand ha. Banwen subterranean river system. One of the largest underground rivers, Hongshuihe River in Guangxi, is often flooded. One flooding event in 1993 resulted in a flooded area of 12.35 km$^2$ and affected 6671 farmers, with a direct economic loss of more than 700 million yuan (Xie and Pei 2002). Drought is also an outstanding problem. According to statistics in 1997, Yunnan, Guizhou, Guangxi Province with 800 million people has severe drinking water problems.

## 3 Flood and Drought in Mumei Subterranean River System

The Mumei subterranean stream is situated in the boundary of Guangnan and Funing county, Yunnan province, China. The catchment covers an area of 308 km$^2$. The discharge point of Mumei subterranean stream is located in Mumei village, Babao town. The Mumei catchment is characterized by a humid subtropical climate, and is heavily influenced by the monsoon regime of Southwest China. Two distinct seasons can be observed in the area: the dry winter lasting from November to April, and the rainy summer from May to October. The yearly mean temperature is 17 °C in the study area, the mean annual precipitation is 1410 mm.

The hazards of flood and drought are frequent, occuring seasonally almost every year. As everybody know, groundwater in karst is plentiful, whilst there is a shortage of surface water. The subterranean river system has characteristics of surface river and underground river alternation. When the conduit can not discharge the runoff timely, groundwater will backflow from karst windows or resurge from portals connected with the conduit. Although there are usually sinkhole in the depression which can discharge the water, the depression will submerge due to the rate of backflow being larger than that of discharge. The submerge time is related to rain features such as intensity or duration. Table 1 shows the submerge time and depth of some typical depressions in the Mumei subterranean river system.

**Table 1** Submerged areas and depth of some typical flood depressions

| No. | Flooding area ($m^2$) | Flooding duration | Water depth (m) |
|---|---|---|---|
| 1 | $32.66 \times 10^3$ | May to November | 20 |
| 2 | $367.32 \times 10^3$ | submerging all the year | 10 |
| 3 | $390.45 \times 10^3$ | April to August each year | 5 |
| 4 | $102.94 \times 10^3$ | one month after heavy rain | 10 |
| 5 | $188.72 \times 10^3$ | May to August each year | 15 |
| 6 | $100.91 \times 10^3$ | several months in a year | 2 |
| 7 | $75.80 \times 10^3$ | several times each year, 7–15 days once | 12 |
| 8 | $8.83 \times 10^3$ | April to November | 10 |
| 9 | $31.30 \times 10^3$ | May to August each year | 40 |
| 10 | $75.44 \times 10^3$ | May to August each year | 10 |

Drought always occurs in dry seasons. All of water tanks and epikarst springs are dry. People have to get water from karst windows which still have water. Because the transportation is poor, it may take a day to get water from the farthest water sources. In some place a karst window will supply 40 villages from the surrounding area. Drought brings to reduction of output of crops. Many people can only live by relying on almsgiving.

## 4 Cause of Flood and Drought

**Meteorological Factors.** Although there is plenty of rain in the Mumei subterranean river drainage area, it is uneven. About 83% of precipitation is distributed between May and October, and only 17% falls in the dry fall and winter season. The amount of precipitation has a tendency to decrease from south to north spatially.

**Hydrogeological Conditions.** The Mumei subterranean stream lies in the slope area of Yunnan-Guizhou altiplano and Guangxi plain, with highland parts in the west and south, and with lowland in the east and north. The highest peak has an altitude of 1,851 m (a.s.l.). The altitude of the Mumei discharge point is 1150 m. Surface water is normally absent in the study area. Only several intermittent streams emerge in the mid-section or in the downstream part of the subterranean water-courses. The system has a complex conduit network. Four tributary conduits (I, II, III and IV) of the Mumei subterranean stream have been identified, which show a parallel channeling in the catchment. Water in conduit I, II, and III is mainly discharged by the Mumei outlet. In the rainy season the water is discharged at Jiaba intermittent stream synchronously (Guo et al. 2009). The average discharge of Mumei subterranean stream is 5.37 $m^3$/s.

Jiaba depression is one of the easily waterlogged depressions in the system. Because it is a converged area for several conduits with plenty of water outlets, it is very important to understand the groundwater status of it when the distribution of groundwater in the basin is discussed. Jiaba depression is divided into two parts, namely Shangba and Xiaba. Jiaba village lies in Shangba. From south to north along Shangba depression, there are eight water points including No. 1 to No. 8. All of water gathers into the open flow, and it flows to the north continually. It accepts three water points when passing Xiaba depression, which is No. 9, No. 10 and No. 11.

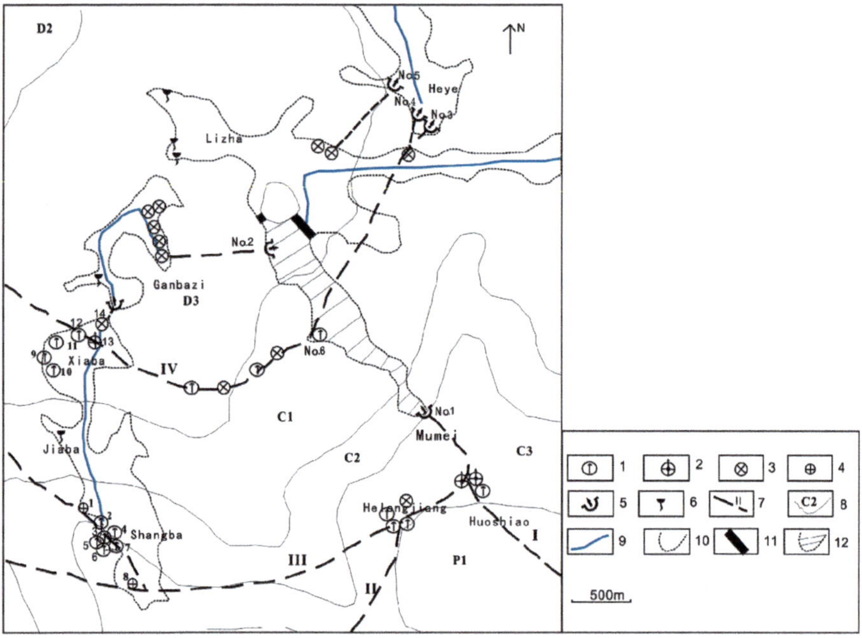


**Fig. 1** Groundwater distribution in Jiaba-Mumei. 1. spill hole; 2. karst window; 3. sinkhole; 4. swallow-spill hole; 5. underground river outlet; 6. seasonal spring; 7. conduit; 8. stratum boundary; 9. surface water; 10. karst depression; 11 reservoir dam; 12. Babao reservoir

Then it turns into groundwater between Jiaba and Ganbazi depression. After that it outcrops in Ganbazi depression, then it enters into ground in the other side of the depression. Finally it discharges at No. 2 of Mumei outlet (Fig. 1). The discharge, water chemistry and physical properties were monitored to better understand the relationship among them. They showed that there were some similar features in water chemistry of water point in Shangba, and those in Xiaba too. But there are different character between Shangba and Xiaba, indicating they have different sources.

Another question is how both muddy water and clear water discharge happens. Some of water points in Jiaba have clear water, while some emerge as muddy water. Two water points outcrop closely with different water colors, what is the reason? The possible cause is that they belonging to different sources, or maybe some independence layer in one underground river system receives more direct runoff. Upward layers are closer to ground surface, and with the characteristics of surface and underground river interphased, groundwater may take a large bedload of sediment. Lower layers further away from surface have less probability of pollution. For example, water in Shangba is clear, and has fish, indicating water source is far and it is a real groundwater in saturated zone. Water in Dapingzi (No. 6, No. 7) has a short duration time, with large seasonal variation, and is often muddy, indicating it is groundwater in seasonal variation zone. Water systems in Shangba and Dapingzi have no contact relationship although their discharge points are closely positioned, further showing the characteristic of heterogeneity of karst development in the vertical direction.

**Human Factors.** Intensive human engineering activities, deforestation and unreasonable land use, are the direct cause of serious damage to the ecological environment, which leads to vegetation reduction, enhanced surface runoff, and serious soil erosion. The formation of the mountain torrents of rainfall brings mud sands as waste deposition, with silting in the mouth of sinkhole, resulting in reduction of flow cross sections, lowered discharge capacity of conduits, and then flooding. In recent years, the affection of ecological destruction has gradually increased, and durations of floods are more prolonged. The rock desertification in Mumei underground river basin is serious. According to an investigation from 2003 to 2004, out of 308 $km^2$ in Mumei, there are 231.9 $km^2$ of moderate and severe rock desertification, 47.8 $km^2$ of mild degree rock desertification, accounting for 75.2% and 15.5% of the total area respectively (Jiang et al. 2005).

## 5 Solutions to Karst Flood and Drought

Solutions to karst flooding requires knowledge of the unique characteristics of karst systems and regulatory control of human being's activities in sinkhole areas. There is an intimate relation between surface water and groundwater in karst aquifers. A flood-prone depression is not an isolated geologic feature, rather a source or sink that connects to the whole karst system.

**Returning Farmland to Forest.** Vegetation can conserve water. It reduces surface water runoff and water from blind valley and depressions. Returning farmland to forests or grasslands work should be speeded up in the steep slopes above 25°. Through returning farmland to forests or grasslands, karst source material which leads to pipeline blockage can be reduced. Resultingly, the epikarst springs will retain more usable water, which will be a good measure in solving the problems of drought.

**Engineering Measures.** Before the implementation of flood disaster management, hydrogeology investigations and evaluations should be done. Delineation of the karst basin may require a comprehensive investigation using specialized techniques such as tracer tests, geo- or hydrogeo-physics, and long-term groundwater monitoring. After systematic analysis of the causes of karst floods, effective and economical measures may be proposed. Expanding the groundwater import, such as creating additional sinkholes or swallow holes by artificial blasting, may accelerate the discharge rate in depressions. The most favorable topography and geological conditions of the terrain will need to be selected for excavating drainage tunnels. Adjusting the water flow between different drainage areas can also control or alleviate the flooding. For example, Tingzhai karst blind valley in Xianfeng county, Hubei Province, China, was made to adjust the water to Tangyan river drainage area by excavating a drainage tunnel. Its creation solved the long-term problem of devastating flood (Su et al. 2008). For the Mumei system, flood water can be collected in a selcted depression, which will alleviate water content pressure of other depressions. If a mini-type reservoir can be built in the west of the drainage area to collect allogenic water, it will have two good effects, one is reducing flood flow, the second is intercepting mud and sand to better protect water quality for irrigation.

## 6 Conclusions

Flood and drought are key environmental problems in karst areas of southwest China, which results from water imbalance between recharge and discharge. The unique geology and hydrology background of karst determinate this. Human being's activities further increase the probability and severity of flood and drought in some areas. Some measures such as engineering projects often can be used to change this situation, but they should be based on understanding the characteristics of karst aquifers. Locations for the incident depressions to control flooding should not be considered as single independent areas; they should consider the whole karst system as part of the scope of work.

**Acknowledgements** This project is funded by the National Natural Science Foundation (40802063) and Guangxi Natural Science Foundation (0991096). We would like to thank the reviewers for their constructive comments to further improve the manuscript.

## References

Guo F, Jiang GH, Lin YS (2009) The heterogeneity of water resources in a karst peak cluster depression area. Global Groundwater Resources and Management, General Symposium: Hydrogeology, Oslo (Norway). Scientific Publishers (India), Jodhpur, pp. 123–131

Jiang GH, Guo F, Hou MF (2005) The Characteristics and Rebuild of Rock Desertification in Mumei Underground River Basin. Journal of Guangxi Academy of Sciences, 21(2):95–98 (in Chinese)

Jones AR (1990) Zoobenthic variability associated with a flood and drought in the Hawkesbury estuary, New South Wales: some consequences for environmental monitoring. Environmental Monitoring and Assessment, 14:185–195

Li QS, Li ZhL, Pei JG et al. (2008) Characteristics of waterlogging and management planning in karst depress ions and valley in the east Mashan. Carsologica Sinica, 27(4):359–365 (in Chinese)

Li TY, Wang SJ (2001) Analysis of Flood Aggravation in Guizhou Karst Depression. Bulletin of Soil and Water Conservation, 21(3):1–4 (in Chinese)

Liu YB, Batelaan O, De Smedt F. et al. (2005) Test of a distributed modelling approach to predict flood flows in the karst Suoimuoi catchment in Vietnam. Environ Geol, 48:931–940

Singh CV (2001) Probabilities and distribution of monsoon rainfall in normal flood and drought years over India. Meteorol Atmos Phys, 78:205–214

Su Ch, Chen HB, Dong BW (2008) Genetic Mechanism Anaslysis and Controlling Suggestions on Flood Disaster in Karst Area. Resources Environment & Engineering, 22(2):193–196 (in Chinese)

Xie YQ, Pei JG (2002) Waterlogging in underground river system of Banwen. Journal of Geological Hazards and Environment Preservation, 13(4):9–13 (in Chinese)

Zhou WF (2007) Drainage and flooding in karst terranes. Environ Geol, 51:963–973

# Influence of Great Flood on the Functioning of Karst Aquifer: Example of the Fontaine de Vaucluse Karst System (SE France)

C. Danquigny, C. Emblanch, T. Blondel, B. Garry, A. Roch, and C. Sudre

**Abstract** In the experimental catchment area of the Fontaine de Vaucluse, a long term monitoring has been completed. It enables the study of the system functioning with variable hydrological conditions. Coupling flow rates and $Mg^{2+}$ concentrations permits highlighting the particular contribution of water with great $Mg^{2+}$ concentration during low water periods. It appears that these high $Mg^{2+}$ concentrations are linked not only to the low flow rate itself, but also to flow rate of previous flood periods. Thus, the average residence time of discharged water during low flow period, which is usually linked to the severity of the low water level, seems to be influenced by previous flood period too. Great floods and related high pressures could lead to circulation of hardly mobilized water with very high residence time.

## 1 Introduction

Natural tracing is a useful tool for the study of karst aquifers (Blavoux 1992; Plagne 1997). However, variations of natural tracer concentration can have several origins that make the interpretation more complicated. Depending on the studied system, the relevance of the selected marker should be checked for example with lithology, studied structure, and climate system inertia (Bakalowicz 1995). In developed karstic system, where the magnesium distribution is homogeneous, $Mg^{2+}$ is a good indicator of the water residence time because of its low dissolution kinetic (Mudry 1987; Plagnes 1997; Batiot 2003; Celle-Jeanton 2003). Regarding the Fontaine de Vaucluse (FdV) catchment area, magnesium is homogeneously distributed within

C. Danquigny, C. Emblanch, T. Blondel, B. Garry, A. Roch
UMR 1114 INRA-UAPV "EMMAH", UAPV, Faculté des Sciences Exactes et de la Nature, 33 rue Louis Pasteur, 84000 Avignon, France, e-mail: christophe.emblanch@univ-avignon.fr

C. Sudre
Laboratoire Souterrain Bas-Bruit de Rustrel Pays d'Apt (LSBB), UNS/CNRS/OCA, La Grande Combe, 84400 Rustrel, France

B. Andreo et al. (Eds.), *Advances in Research in Karst Media* 115
DOI 10.1007/978-3-642-12486-0, © Springer 2010

the carbonate matrix (Masse 1968). At the outlet, $Mg^{2+}$ concentrations can vary from 1.7 to 9 mg/L with maximum levels during the low flow period. This amplitude is large enough to be very significant. In addition, magnesium from rainfall and snow is negligible (Celle 2000). Magnesium is a good marker of residence time on the karst system of the FdV (Mudry 1987). Commonly, the more severe the low flow conditions, the more mobilized becomes the water, with increased residence time. Therefore, it is reasonable to assume that $Mg^{2+}$ concentration increases with the importance of low flow water. However, some specific behaviours have been identified (Emblanch 1998). Significant increases of $Mg^{2+}$ concentrations during flood rises were observed. This information, coupled with other tracers ($^{13}$CTDIC and T°), indicates that changes in mineralization observed during the rising flood of April 1995 are due to arrivals of water with long residence time in the saturated zone. Consequently, variations of $Mg^{2+}$ concentration could be related to particular phenomena. This may be characteristic of the aquifer, but also specific hydrogeological conditions.

This study documents hydrodynamic and hydrochemical data measured during 11 discontinuous hydrological cycles at the FdV. The aim is to highlight phenomena and different hydrogeological conditions which could explain the $Mg^{2+}$ concentrations measured at the outlet of karst system. The focus is to better understanding the high concentrations of magnesium during low flow period.

## 2 Material and Methods

The FdV is located in the South-Eastern of France. It is the only discharge point of a karst aquifer system (Fig. 1) which is peculiar for the surface of its recharge area (1130 km$^2$) and its mean flow rate of 20 m$^3$/s. Another significant characteristic of this karst system is its approximately 800 m thick unsaturated zone. Such a great

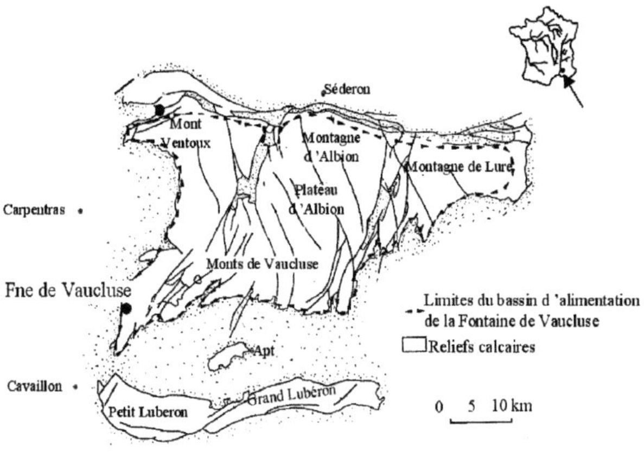


**Fig. 1** Catchment area of the FdV spring

layer is due to the presence of a nearly 1500 meters thick massive and continuous reef limestone (Masse 1968).

The FdV flow rate is monitored every day. However, $Mg^{2+}$ concentration has been measured regularly only over 4 periods: 1981/1982, 1994/1995, 1999/2001 and since 2003. Among these data, only 10 hydrological cycles end with low flow rate

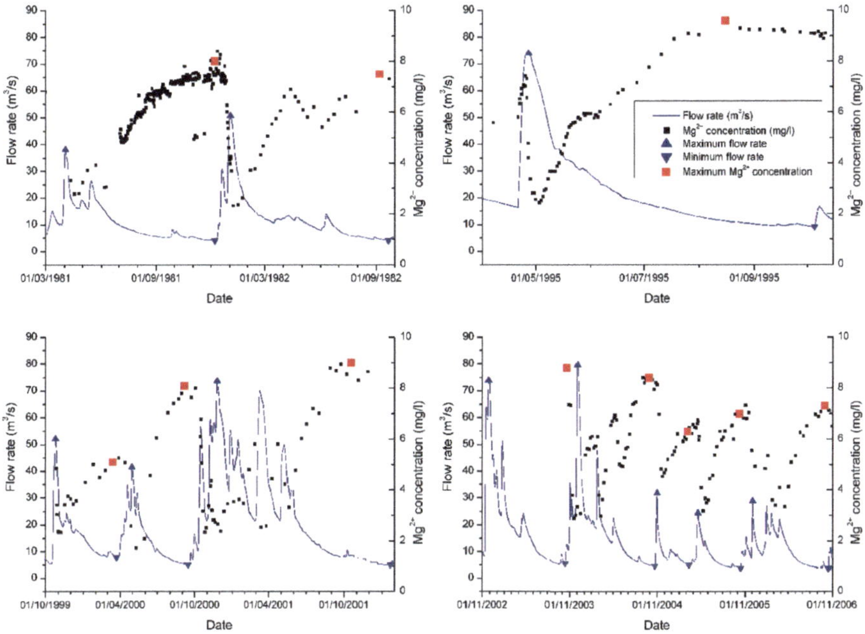


**Fig. 2** Flow rate and $Mg^{2+}$ concentration measured at the FdV outlet

**Table 1** Maxima $Mg^{2+}$ concentrations and extreme flow rate at the FdV outlet

| Time period | Maximum flow rate $Q_{max}$ (m³/s) | | Maximum $Mg^{2+}$ concentration $Mg_{max}$ (mg/L) | | Minimum flow rate $Q_{min}$ (m³/s) | |
|---|---|---|---|---|---|---|
| | Date | Flow rate | Date | Conc. | Date | Flow rate |
| 1981 | 03/04/1981 | 38.1 | 08/12/1981 | 8.0 | 08/12/1981 | 4.17 |
| 1982 | 03/01/1982 | 50.7 | 07/09/1982 | 7.5 | 21/09/1982 | 4.19 |
| 1995 | 27/04/1995 | 74.0 | 16/08/1995 | 9.6 | 05/10/1995 | 9.07 |
| 1999/2000 | 26/10/1999 | 52.0 | 15/03/2000 | 5.1 | 23/03/2000 | 7.93 |
| 2000 | 30/04/2000 | 41.6 | 06/09/2000 | 8.1 | 15/09/2000 | 5.23 |
| 2001/2002 | 26/11/2000 | 73.7 | 19/10/2001 | 9.0 | 22/01/2002 | 5.22 |
| *2002/2003* | *28/11/2002* | 74.0 | *22/10/2003* | $\geq$ *8.8* | *15/10/2003* | 5.77 |
| 2003/2004 | 06/12/2003 | 79.7 | 29/09/2004 | 8.4 | 24/10/2004 | 4.87 |
| 2004/2005 | 03/11/2004 | 31.8 | 10/03/2005 | 6.3 | 15/03/2005 | 5.23 |
| 2005 | 21/04/2005 | 24.5 | 12/10/2005 | 7.0 | 17/10/2005 | 3.91 |
| 2005/2006 | 05/12/2005 | 28.8 | 04/10/2006 | 7.3 | 17/10/2006 | 3.73 |

below $10\,m^3/s$ (Fig. 2). In addition, an eleventh cycle, 2002/2003, has been selected for this study. Despite a unique value of $Mg^{2+}$ concentration (in italics in Table 1), measured around the low flow period, this cycle is considered important because of its particularly hard flood period. This concentration value has been considered carefully. Each period of interest is bounded by the daily maximum and minimum flow rates, $Q_{max}$ and $Q_{min}$, i.e., respectively the daily flow rate of flood peak and the minimum daily flow rate of the following low water period (Fig. 2). For each cycle, the maximum $Mg^{2+}$ concentration, $Mg_{max}$, has been identified (Fig. 2). Values and corresponding dates are shown in Table 1.

## 3 Results

For the 11 hydrological cycles with a low flow rate ($Q_{min}$) less than $10\,m^3/s$, two groups can be identified disregarding the season when they occurred:

- Four "extreme" cycles (1995, 2001/2002, 2002/2003 and 2003/2004 – in bold in Table 2) show maximum flow rate ($Q_{max}$) very high, characteristic of a decadal flood (71 to $88\,m^3/s$) and close to the known maximum flow ($85\,m^3/s$); these particular hydrodynamic conditions are followed by a low flow period with the highest $Mg^{2+}$ concentrations measured at the FdV (near or above 9 mg/L);
- Seven "ordinary" cycles have a maximum flow rate ($Q_{max}$) less than $52\,m^3/s$ and maximum $Mg^{2+}$ concentrations between 5.1 and 8.1 mg/L.

However, for the "extreme" cycles, the highest maximum concentrations of $Mg^{2+}$ ($Mg_{max} > 8.4$ mg/L) appear during low flow periods that are not particularly severe ($9.07\,m^3/s > Q_{min} > 4.87\,m^3/s$). In order to understand the relationships between the maximum $Mg^{2+}$ concentration during the low water period ($Mg_{max}$), the maximum flow rate of the previous flood period ($Q_{max}$) and the low water flow rate itself ($Q_{min}$), correlation coefficients between $Mg_{max}$ and, respectively, $Q_{max}$, $Q_{min}$ and ($Q_{max} - Q_{min}$) have been calculated. Calculations have been done firstly for the whole cycles and secondly by considering only the "ordinary" cycles ($Q_{max} < 52\,m^3/s$). The results are presented in Table 2.

First, if the whole cycles are considered, the usual correlation between low water flow rates and $Mg^{2+}$ concentrations during this period, commonly representative of the increased contribution of reserves during very low water period, is not verified ($CC_{all} = 0.08$). However, if the "extreme" cycles are not considered, this correlation is actually verified ($CC_{classic} = -0.76$): $Mg^{2+}$ concentration increases with the importance of low water flow rate ($Q_{min}$). The calculation of the correlation coefficient between $Mg_{max}$ and $Q_{max}$ (or $Q_{max} - Q_{min}$) for all cycles considered ($Q_{min} < 10m^3/s$) reveals a positive correlation ($CC_{all} = 0.61$) between the maximum $Mg^{2+}$ concentration during low water period and the previous flooding. This correlation does not exist if only "ordinary" cycles are considered ($CC_{classic} = -0.22$). Thus, it seems that harder the flood is, the greater the $Mg^{2+}$ concentration during the following low flow period. Regardless of the severe character of the low flow, these very high concentrations signify the arrival at the outlet

**Table 2** Correlation coefficients between $Mg^{2+}$ concentration and extreme flow rate

| Time period | Maximum $Mg^{2+}$ concentration $Mg_{max}$ (mg/L) | Maximum flow rate $Q_{max}$ ($m^3$/s) | Minimum flow rate $Q_{min}$ ($m^3$/s) | $Q_{max} - Q_{min}$ |
|---|---|---|---|---|
| 1981 | 8.0 | 38.1 | 4.17 | 33.93 |
| 1982 | 7.5 | 50.7 | 4.19 | 46.51 |
| **1995** | **9.6** | **74.0** | **9.07** | **64.93** |
| 1999/2000 | 5.1 | 52.0 | 7.93 | 44.07 |
| 2000 | 8.1 | 41.6 | 5.23 | 36.37 |
| **2001/2002** | **9.0** | **73.7** | **5.22** | **68.48** |
| **2002/2003** | **≥ 8.8** | **74.0** | **5.77** | **68.23** |
| **2003/2004** | **8.4** | **79.7** | **4.87** | **74.83** |
| 2004/2005 | 6.3 | 31.8 | 5.23 | 26.57 |
| 2005 | 7.0 | 24.5 | 3.91 | 20.59 |
| 2005/2006 | 7.3 | 28.8 | 3.73 | 25.07 |
| Correlation coefficient for all cycles $CC_{all}$ | | 0.61 | 0.08 | 0.63 |
| Correlation coefficient for cycles without high max. flow rate $CC_{classic}$ | | –0.22 | –0.76 | –0.12 |

of particularly old water. Apparently, this water is not mobilized during "ordinary" cycles, i.e., without previous hard flooding.

## 4 Discussion

For all considered cycles, the relatively high $Mg^{2+}$ concentrations measured during the low flow periods confirm the contribution of water with long residence time to the low water flow. In recent years, the recurrence of extreme hydrological events enables a better characterization of these involved reserves. By distinguishing two groups of hydrological cycles, i.e., "extreme" cycles and "ordinary" cycles, it is now possible to identify the solicitation of two types of reserves during the low flow period. These solicitations seem to be independent of the seasons because both "extreme" and "ordinary" cycles include some different periods of the year. If the low flow period is ordinary, i.e., is not preceded by a major flood, $Mg^{2+}$ concentrations during the low flow period reflect the contribution of water that has a relatively long residence time but is regularly mobilized. The occurrence of an extreme flood, which implies strong pressure in the aquifer, leads to unusually high $Mg^{2+}$ concentrations during consecutive low flow period. High $Mg^{2+}$ concentrations reflect the contribution of water with very long residence time. This water is not involved during "ordinary" cycles and is therefore hardly mobilized. Detection at the outlet of this particular water can indirectly enables to characterize highly capacitive parts of the aquifer. In these parts, flow of water should only be caused by great water pressure produced by major floods. The volumes involved may be large enough to impact the usually mobilized reserve.

In a changing climate scenario involving a multiplication of extreme hydrological events, the impact on reserves of karst systems may be dramatic. Indeed, if hardly mobilized reserves are regularly involved, then the karstification can develop quickly. Consequently, these highly capacitive parts of the aquifer could be lost. A significant decrease of the reserves of karst aquifers is ultimately possible.

## 5 Conclusions

This study shows once more the relevance of natural tracing, but also the complexity of karst aquifers. Hardly mobilized water can impact significantly the chemical quality of reserves more easily and is usually mobilized during the low flow period. This implies significant flows from the capacitive part of the aquifer involved. In this part, the flow is probably due to strong water pressure implied by major floods. According to the hydrological conditions of the studied cycles, two types of reserves were identified. Two questions arise: are there many types of reserves with discrete degrees or difficulties to mobilize water, and for example, a pressure threshold effect that helps to mobilize one or other of these reserves? Or does a continuum of reserves exist? A relative desynchronization of the $Mg^{2+}$ signal during the low water period was also revealed: it is more the previous flooding conditions than the low flow itself that influence the $Mg^{2+}$ concentration. All these observations have an impact on the karstification in a changing climate scenario. The risk of reactivation of karstification processes is possible on systems that were thought to be at equilibrium or near equilibrium. This implies a risk of "drainage improvements" and reduction of reserves and/or low water flow rates. These results demonstrate once again the need to establish observatories in the long term (several decades) of karst systems.

## References


Bakalowicz M (1995) La zone d'infiltration des aquifères karstiques. Méthodes d'étude. Structure et fonctionnement. Hydrogéol 4:3–21

Blavoux B, Mudry J, Puig J M (1992) Bilan, fonctionnement et protection du système karstique de la Fontaine de Vaucluse (S-E de la France). Geodyn Acta (Paris) 5:3:153–172

Batiot C, Emblanch C, Blavoux B (2003) Carbone organique total et magnésium: deux traceurs complémentaires du temps de séjour dans l'aquifère karstique. Académie des sciences, éditions scientifiques Elsevier:205–214

Celle-Jeanton H (2000) Caractérisation des précipitations sur le pourtour de la méditerranée occidentale, approche isotopique et chimique. PhD Thesis. Univ. Avignon

Celle-Jeanton H, Charmoille A, Emblanch C, Mudry J (2003) Contribrution of time tracers to understand the role of unsatured zone. Geophys res let 30:6:5p

Emblanch C, Puig J M, Zuppi G M, Mudry J, Blavoux B (1998) Comportement particulier lors des montées de crues dans les aquifères karstiques, mise en évidence d'une double fracturation et/ou de circulation profonde : Exemple de la Fontaine de Vaucluse. Eclogae Geol Helv 92:251–257

Masse JP (1968) L'Urgonien de Sault (Vaucluse). Bull Soc Géol de France 4:495–496

Mudry J (1987) Apport du traçage physico-chimique naturel à la connaissance hydrocinématique des aquifères carbonatés. Mémoire n°4 Univ. Besançon, 382p
Plagnes V (1997) Le transport de matière dans les aquifères karstiques. 6th conf on limeston hydrology and fissured media, La Chaux de Fond, Suisse, 14-18/08/97:179–183
Trcek B (2008) Flow and solute transport monitoring in the karst aquifer in SW Slovenia. Environmental geology 55:2:269–276

# Localisation of a Reactive Transport Zone in a Saturated Karstic Conduit Deduced from Natural and Artificial Tracer Tests

S. Binet, A. Joodi, E. Joigneaux, P. Albéric, and A. Gutierrez

**Abstract** For groundwater resources managers, flow modeling is a useful tool to investigate sustainable scenarios of water use. However, in karstic aquifers, the quality of scenarios is limited by the difficulties of locating and describing the position, geometry and possible time evolution of conduits. The location of conduits in the karstic aquifer of the "Val d'Orléans" (France) were defined using 200 boreholes, surface collapses and 24 artificial tracer tests, which facilitated the development of a simplified conceptual model of flow in the saturated conduits and the surrounding rocks. 68 logs present voids $> 50$ cm and locate a highly porous zone around 80 m a.s.l. with voids that average 3.5 meter in diameter. In this saturated conduit, 1D quantitative interpretation of artificial tracer tests validate the proposed conceptual model of a saturated conduit under pressure, with an efficient section about $10\,m^2$ an input flow about $3.1\,m^3/s$ with $2.9\,m^3/s$ flowing from the conduit toward the surrounding rock before arriving at the Loiret Spring. The conceptual model of flow and the previous water chemical analysis show that the transported elements in the groundwater react and dissolve carbonate rocks, mainly inside the conduit, and that this may increase the diameter of the conduit zone by an estimated 40 cm in 100 years.

## 1 Introduction

For groundwater resource managers, groundwater flow modeling is an essential tool for investigating proposed sustainable scenarios of water uses. Applied to karstic aquifers, the modeler is confronted by the duality of flows, with the major role of

S. Binet, A. Joodi, E. Joigneaux, P. Albéric
Institut des Sciences de la Terre d'Orléans – UMR 6113 Université d'Orléans, CNRS/INSU, Université François Rabelais – Tours Campus Géosciences, 1A rue de la Férollerie, 45071 Orléans Cedex 2, France, e-mail: stephane.binet@univ-orleans.fr

E. Joigneaux, A. Gutierrez
Bureau des Ressources Géologique et Minières, 3 avenue Claude-Guillemin, 45060 Orléans Cedex 2, France

B. Andreo et al. (Eds.), *Advances in Research in Karst Media*
DOI 10.1007/978-3-642-12486-0, © Springer 2010

the permeable conduits as the water transfer compartment and the important role of the surrounding rock as the water storage compartment (Kiraly et al. 1995). This duality is enhanced by the existence of shear stress in water (Joodi et al. 2009) and by turbulent flows (Cheng and Chen 2004) that may take place in the conduits. Thus in most karstic groundwater flow models, accuracy is limited by the difficulties of describing the location and the geometry of the conduits. Furthermore, long term accuracy of models is limited by rock dissolution, that creates porosity change and may modify the conduit/rock relationship.

The paper uses a well known conduit flow system (The Val d'Orleans karst system), through 200 boreholes and 28 artificial tracer tests, to deduce a conceptual model of conduits in this karstic system. An average diameter approximation method is proposed for describing the conduit, with the aim of inferring their exact geometry. The conceptual model is validated using a 1D artificial tracer test model. In this chemically active karst system, the proposed conceptual model uses the chemical reactions identified by Albéric and Lepiller (1998) to estimate a change rate of the conduit diameter.

## 2 Hydrogeological Setting

The Val d'Orléans is considered as a major depression in the bed of the Loire river, 37 km long and from 4 to 7 km wide (Fig. 1).

The karst aquifer is hosted within an Oligocene carbonate lacustrine deposit called the limestone of Beauce. The Loire River feeds more than 80% of the wa-

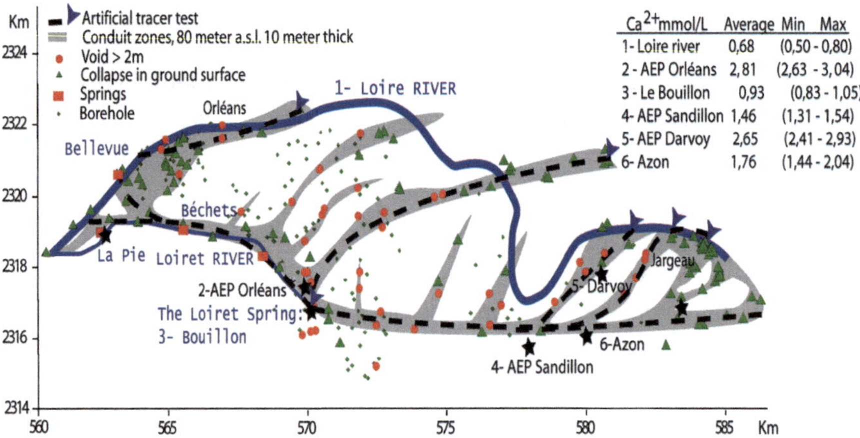


**Fig. 1** Map of the Val d'Orléans Aquifer with *in grey* the zone with a 3 to 40% probability of finding a conduit. This conduit zone was interpreted from observed voids in the boreholes (*circles*), collapses on the surface (*triangles*) and tracer tests (*dashed lines*). The calcium concentrations in water are presented in the table for 6 boreholes in and around the conduit (average data from 1970 to 2009)

ter hosted in the carbonated karstic aquifer (from 15 to 100 $m^3/s$). The water runs from Jargeau through the under pressure karst networks toward several springs of the Loiret River: the Bouillon, Béchets, Bellevue, and the Pie springs (Fig. 1) (Zunino 1979; Chéry 1983; Lepiller 2006). The Loiret springs are considered as the main emergence of the water lost close to Jargeau in the Loire River (from 0.1 to 5 $m^3/s$). In this area the conduits have been explored by a speleologist diver and the total flow rate in this section of the conduit is about 2 to 10 $m^3/s$ (personal communication from Mr Boismoreau). In Albéric and Lepiller (1998), the existence of calcite dissolution was demonstrated in this system by the oxidation of riverine organic matter. The magnitude of the reaction is controlled by dissolved $O_2$, nitrate for organic matter oxidation and by the release of $Ca^{2+}$ for calcite dissolution (Fig. 1). Numerous surface collapses were observed around Orléans.

# 3 Methods

The location of the conduits is deduced from two databases (borehole logs and surface collapses realized by BRGM, the French geological survey (http://infoterre.brgm.fr/ and http://www.bdcavite.net/). In the Val d'Orléans aquifer, 200 borehole logs were analyzed to extract information about underground karstic voids. The location $(X, Y)$ of voids encountered are presented in Fig. 1. Results are presented using the frequency curves in percent (Fig. 2). To refine to location of conduits, the 147 collapses observed in the surface (cavity database) were considered to be activated by the presence of an underground active conduit (triangle in Fig. 1). 24 artificial tracer tests (Lepiller 2006; Joodi et al. 2009) demonstrate that observed voids are a part of the karstic conduit (dashed lines in Fig. 1). To validate the average geometrical data deduced from the boreholes, an artifical tracer test was carried out on October 24th 2009 during a low water period. 1 kg of uranine was injected in the Loire River under the Jargeau Bridge. Recovery was recorded at the Bouillon Spring (Fig. 3).

A 1D model is applied to the tracer test results. If the conduits are considered as an equivalent conduit with a section $A$ ($m^2$) and a porosity $n_f$ (%), a 1D solution of the advection/dispersion/reaction equation can be applied for a Dirac injection type where the concentration $c(x,t)$ is described by the injected mass $\Delta M$ (kg), an average flow velocity $u$ (m/s) and the longitudinal diffusion coefficient $D_L$ (m). Usually, $\lambda$ ($s^{-1}$) is a decay constant. Calculation is resolved with a new software program, TRAC (download possible from the BRGM web site www.brgm.fr/trac).

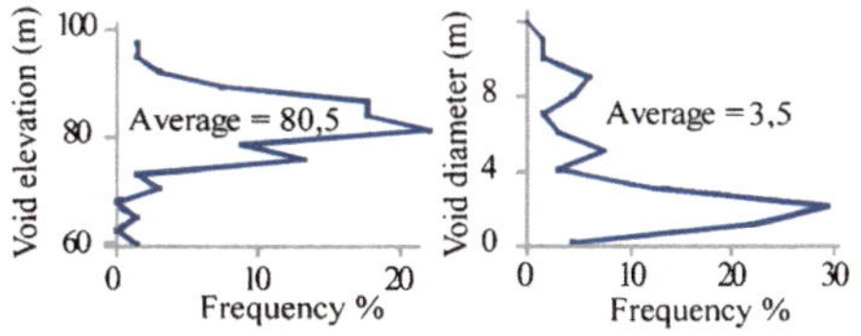


**Fig. 2** Frequency distribution of voids observed in the boreholes related to **a** the elevation, **b** their diameters

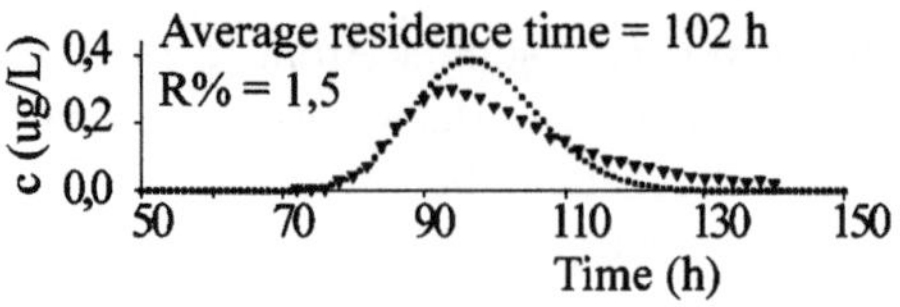


**Fig. 3** Artificial tracer test recovery curve at the Loiret spring realized in October 2009 (*triangle*) and best fit from 1D solution with $x = 13$ km; $D = 55$ m; $An_{\mathrm{f}} = 10\,\mathrm{m}^2$; $\Delta M = 1$ kg; $u = 3.6$ cm/s; $\lambda = 1.25 \times 10^{-5}\,\mathrm{s}^{-1}$ (*square*)

If the flow velocity ($u$) in the conduit and the leak flow rate ($Q_{\mathrm{m}}$) from the conduit to the hosting rock are constant with $x$ and $t$, an analogy can be used where the decay phenomenon $\lambda$ ($\mathrm{s}^{-1}$) becomes a leakage constant describing the amount of tracer flowing toward the hosted rock. $Q_{\mathrm{m}}$ can be calculated with $\lambda$ ($\mathrm{s}^{-1}$). $Q_{\mathrm{m}} = \lambda Q_{\mathrm{f}}\mathrm{d}t$ and $Q_{\mathrm{o}} = Q_{\mathrm{m}} + Q_{\mathrm{f}}$. Water chemistry has been compiled from previous studies (Chéry 1983; Alberic and Lepiller 1998) (Fig. 1). Following Albéric and Lepiller (1998), calcium is used to estimate the amount of dissolved calcite between two points.

# 4 Results

## 4.1 Conduit Location and Geometry

On the 220 boreholes logs, 68 logs present voids $> 50$ cm. These are interpreted as karstic features and are presented in Fig. 1. The whole data set enables a 100 meter wide zone to be drawn where there is a high probability of finding a conduit (Fig. 1). This zone includes voids, collapses and the probable path between the tracer injections areas and the recovery points. These zones have two preferential flow directions: NE/SW and E/W in the system. Figure 2 shows that average elevation of the voids are around 80 meters above sea level, or about 10 to 20 meter below the ground surface. In this zone, voids have an average diameter of 3.5 meter diameter. An effective cross-section of about $10\,\mathrm{m}^2$ is estimated, assuming a circular shape. The conduit is located in a zone around 70–80 meter a.s.l. (10 meter thick). This zone is considered as the zone with a high probability of containing a conduit.

## 4.2 Validation Using a Quantitative Tracer Test Interpretation

Using (1) and the average previous values ($A * n_{\mathrm{f}} = 10\,\mathrm{m}^2$) it is possible to predict the recovery curve at the Loiret Springs ($x = 13$ km).

$$c(x,t) = \frac{\Delta M}{2An_{\mathrm{f}}\sqrt{\pi D_{\mathrm{L}} t}} \exp\left[-\frac{(x-ut)^2}{4D_{\mathrm{L}}t}\right] \exp(-\lambda t) \quad \text{(Sauty et al. 1992)} \qquad (1)$$

During low water period, the flow rate in this zone (Loiret springs + underground conduit) is estimated around $0.4\,m^3/s$ (with a $10\,m^2$ of efficient flow section an average flow velocity about 3.6 cm/s). The only unknown parameters are the diffusion coefficient and the leakage constant. Best fit gives $D = 55\,m$ and $\lambda = 1.25 \times 10^{-5}\,s^{-1}$ and is presented Fig. 3. The model results reproduce the average residence time and the recovery rate and validate the possibility of using the geometrical observations realized in boreholes to describe the conduits at the aquifer scale. The shape of the curve cannot be fitted with this 1D model. A significant physical mechanism taking place in the conduit may be neglected, such as shear stress in water created by tortuosity of the conduit zone as described by Joodi et al. 2009, or by turbulent flows (Cheng and Chen 2004).

### *4.3 Water Chemistry Around the Conduit Zone*

Water chemistry analyses since 1970 for six points located within or outside the conduit zone (Fig. 1) show that the calcium concentration increases between the Loire river and the Bouillon Spring (on average for more than 100 analysis) by about 0.5 mmol/L. The water from the Bouillon is characteristic of the conduit flows. All the boreholes around the conduit zone show higher calcium concentration (2 mmol/L) (Chéry 1983; Albéric and Lepiller 1998).

## 5 Discussion

The geological data, the voids characterization and the tracer tests model from the boreholes enables a conceptual model of flows in these conduit zones to be proposed (Fig. 4) Overall, the conduits are under pressure and water flow is toward the surrounding rock. The conduit zone is 20 meters under the ground surface in the limestone of Beauce. The input flow rate ($Q_o$) for the 2009 low water period

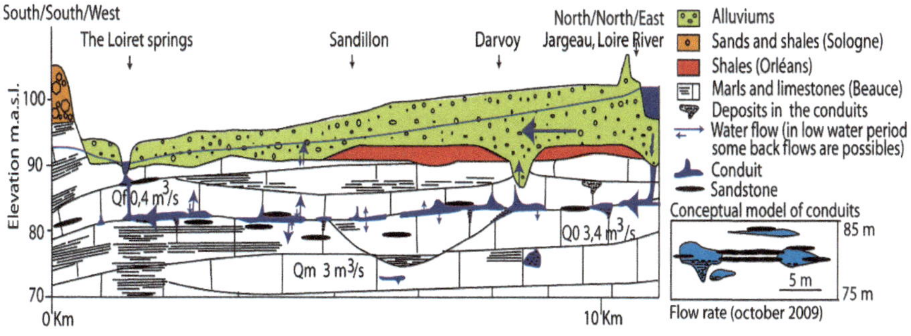


**Fig. 4** Cross-section of the Val d'Orléans

can be estimated around 3.4 $m^3/s$ for this conduit with 3 $m^3/s$ flowing from the conduit toward the surrounding rock, before arriving at the Loiret spring. 0.4 $m^3/s$ flow through the conduit zone located around the Loiret springs.

According to this conceptual model from Lepiller (2006) and Joodi et al. (2009), most of the time the drain is under pressure and feeds the surrounding rock. Thus dissolution takes place mainly inside the conduit. Calcium concentration increases with the distance from the conduit zone. This may explain the 147 ground surface collapses observed in this area. Using the 0.25 mmol/L of calcium observed between Loire River and Loiret Spring (Fig. 1) and taking calcite density about 2.7, a conduit about 13 km long, a 10 $m^2$ effective flow section, the diameter of the conduit may increase about 40 cm in 100 years. Dissolution seems to be a significant process to understand flow rate change from the last 100 years. The magnitude order estimated here needs to be thoroughly studied to improve the long term groundwater management.

## 6 Conclusions

The conduits in the karstic aquifer of the "Val d'Orléans" (France) were located in a 100 meter wide highly porous zone around 80 m a.s.l. with an average diameter of 3.5 meters. 1D quantitative interpretation of artificial tracer tests validate the proposed geometrical model of the conduit, with 3 $m^3/s$ flowing from the conduit toward the surrounding rock. The conduit flows are under pressure and water chemistry analysis shows that the transported elements in the groundwater react and dissolve carbonate rocks, mainly inside the conduit, where conduit diameter can increase about 40 cm/100 years. This opens questions about the porosity changes and long-term stability of the water flow rate in the Loiret River.

**Acknowledgements** This work was founded by the "Région Centre" through the TRAC project.

## References

Albéric P, Lepiller M (1998) Oxydation de la matière organique dans un système hydrologique karstique alimenté par les pertes fluviales (Loiret, France). Water Resources 32:2051–2064

Joodi A, Sizaret S, Binet S, Bruand A, Albéric P (2009) Development of a Darcy–Brinkman model to simulate water flow and tracer transport in a heterogeneous karstic aquifer (Val d'Orléans, France) Hydrogeology Journal on line

Kiraly L, Perrochet P, Rossier Y (1995). Effect of the epikarst on the hydrograph of karst springs: a numerical approach. Bulletin du Centre d'Hydrogéologie 14:199–220

Cheng JM, Chen CX (2004) An integrated linear/non-linear flow model for the conduit-fissure-pore media in the karst triple void aquifer system. Environmental Geology 47:163–174

Chery JL (1983) Etude hydrochimique d'un aquifère karstique alimenté par perte de cours d'eau (la Loire). Ph.D. Thesis, Orleans University

Lepiller (2006) Hydrogéologie du Val d'Orléans. In Les aquifères de France. ed. BRGM – AIH 2006, 956 p

Sauty JP, Kinzelbach W, Voss A (1992) CATTI – Computer Aided Tracer Test Interpretation. BRGM report

Zunino (1979) Contribution à l'étude hydrogéologique du Val d'Orléans. Ph.D. Thesis, Orleans University

# The Characteristics of Groundwater Flow in Karst Aquifers During Long Lasting Low Flow Conditions, Example from SW Slovenia

N. Ravbar, M. Petrič, and J. Kogovšek

**Abstract** As it is expected that global warming will induce changes in hydrological regimes, the presented study has been conducted to understand how karst groundwater resources may be affected during long-term dry periods. A multi-tracer test with three injection points aimed at identifying the dynamics of underground drainage through a well-developed system of karst channels and through vadose zone. Results show low apparent dominant flow velocities (through well developed karst conduits between 5.9 and 22.8 m/h, through vadose zone 3.6 m/h), multiple peaked and extended breakthrough curves. These findings are relevant for groundwater quality monitoring, protection and management, as they infer that contaminant transport in karst, at least in the studied area, may be retarded and long-lasting.

## 1 Introduction

The incidence of increased precipitation intensity and variability provoked by global warming is expected as a consequence in the large-scale hydrological cycle. This will have implications for water quantity and quality in many areas and affect freshwater dependant ecosystems and several socio-economic activities (Kundzewicz et al. 2008).

Due to their specific nature, i.e., rapid infiltration rates and underground water flow, highly controlled by heterogeneous permeability, karst aquifer systems are greatly dependant on respective hydrological conditions. Therefore these landscapes are very susceptible to climatic stresses and have the potential to be strongly impacted by freshwater shortfalls or floods (White 2002).

A specific study has been conducted to better understand the aquifer behaviour, recharge and transport processes under extreme hydrological conditions. This paper presents preliminary results of a multi-tracer test with three injections carried out during long lasting low water conditions, which is a part of a holistic research of variability of groundwater flow under different hydrologic conditions.

N. Ravbar, M. Petrič, J. Kogovšek
Karst Research Institute SRC SASA, Titov trg 2, 6230 Postojna, Slovenia,
e-mail: natasa.ravbar@zrc-sazu.si; petric@zrc-sazu.si; kogovsek@zrc-sazu.si

B. Andreo et al. (Eds.), *Advances in Research in Karst Media*
DOI 10.1007/978-3-642-12486-0, © Springer 2010

The study focuses on the springs at the Polje of Planina and their catchment in SW Slovenia (Fig. 1). At the southern margin of the Polje two efficient karst springs occur: Unica and Malenščica, the latter being an important regional drinking water source.

The Unica springs out of a large underground conduit Planinska Jama, where two subsurface rivers confluence (Rak and Pivka branches). After a few hundred meters of surface flow Unica and Malenščica join, cross the Polje and sink on its W and NW margins. The total discharges of the Unica River range between 1.08 and 100 $m^3/s$; with an average of 20.95 $m^3/s$ (Gospodarič and Habič 1976).

The catchment of the observed springs covers about 746 $km^2$ and comprises autogenic and allogenic recharge areas. The majority of the catchment belongs to the Javorniki karst plateau composed of Jurassic and Cretaceous limestones. Westwards the Pivka valley is covered by very poorly permeable Eocene flysch draining the surface network of the Pivka River that sinks at the NE edge of the valley and flows towards the Polje of Planina. The underground current of the Pivka River runs through very well developed karst conduits of Postojnska Jama cave system.

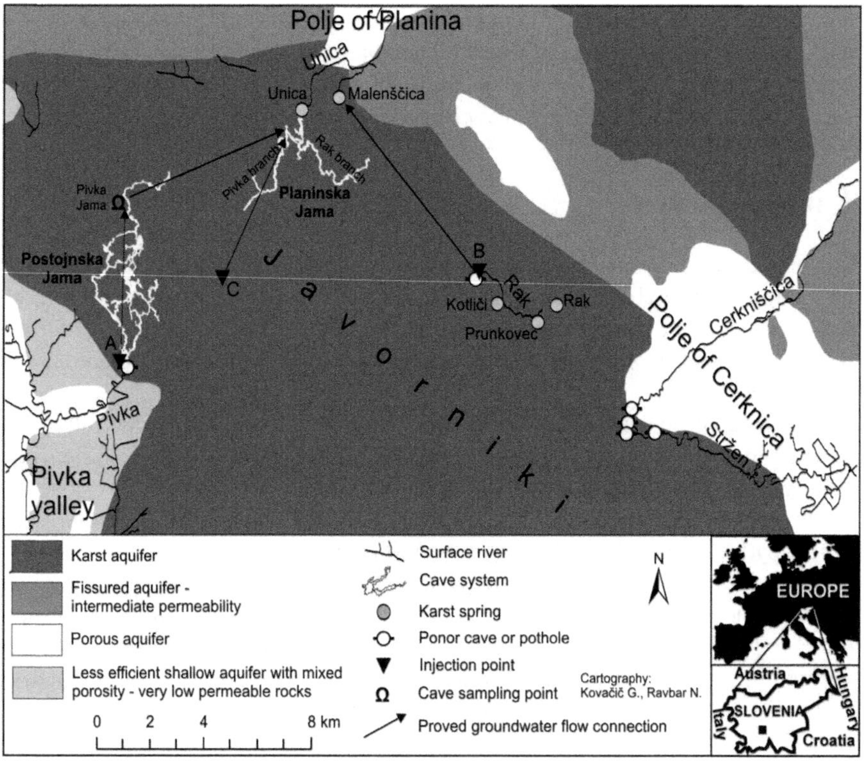


**Fig. 1** Hydrogeological settings of the studied area. The underground flow connections proved by multi-tracer test carried out in June 2009 are shown

Pivka Jama, among several cave system entrances, presents the ultimate access to the underground Pivka River course.

Eastern edge of the Javorniki plateau is bordered by the neotectonic strike-slip fault zone, along which a chain of karst poljes has developed. Catchment area of the studied springs comprises sinking rivers flowing through the poljes, among which the Polje of Cerknica is the largest, gathering surface waters also from the Triassic dolomite.

The hydrogeology and underground connections of the studied karst system has been researched in several studies in the past. Lately tracer tests have been conducted for various other purposes (e.g. Kogovšek et al. 1999; Kogovšek and Šebela 2004; Petrič and Šebela 2005; Gabrovšek et al. 2010). However, with the exception of Gams (1970) several decades ago, none of these tracings have been performed during extremely low water conditions.

## 2 Methods

To study the characteristics of groundwater flow during low flow conditions, three artificial tracer injections were done on 2 June 2009. The injection sites were selected representing different recharge pathways: 2 kg of Naphtionate was injected at the Pivka River ponor (injection point A), 500 g of Amidorhodamine G was injected at the Rak River ponor (injection point B) and 500 g of Uranine was injected in karren on the karst surface (injection point C).

At the Malenščica and Unica springs samples were taken by automatic sampler ISCO 6700, and at the Pivka and Rak branches manually. All samples were analysed in the Karst Research Institute laboratory (Postojna) with a luminescence spectrometer (Perkin Elmer LS 30).

For monitoring prior and during the tracer test, precipitation was measured near Postojna by rain-gauge Onset RG-M in 30-minute intervals. At the Malenščica and Unica springs water level was measured in the same interval with ISCO 6700-750 Area-velocity flow module and Gealog-S data-logger. The discharges were calculated based on the stage-discharge curves, which were prepared by the Environmental Agency of the Republic of Slovenia. In Pivka Jama samples were taken manually.

## 3 Results and Conclusions

The weather conditions from spring to autumn 2009 allowed observation of the behaviour of karst aquifers during an extraordinary long-term low water period that lasted for several months. From May to October the precipitation amount reached about 347 mm in total, which was only 52% of the 1961–90 period average amount. Consequently the groundwater level was very low. At the time of injection Malenščica spring discharged 3.3 $m^3/s$ and Unica spring 2.8 $m^3/s$. Discharges at both springs were gradually lowering, with some short periods of increased values in response to rain events, and reached minimum during the course of the tracer test (Malenščica 1.08 $m^3/s$, Unica 0.04 $m^3/s$; Fig. 2).

A milder rainy event followed the injection, but a more intense precipitation did not occur until three weeks after the injection. At the Malenščica spring only Amidorhodamine G was detected (Fig. 2; Table 1). It was first detected very sporadically, forming small individual peaks of breakthrough curve (BTC), but the more continuous BTC was observed from 23 June on. Also then the BTC formed very unbalanced shape with the maximum concentration at 0.6 μg/L. Altogether 79.7% of Amidorhodamine G has been recovered at the Malenščica spring. The dye was also detected at a few samples from the Rak branch and Unica spring. However, the tracer detection was very oscillatory, which makes the connections very uncertain.

Continuous and uniformly shaped Naphtionate BTCs were observed along the Pivka underground course. In Pivka Jama was the tracer first detected on 8 June with maximal concentration of 14.8 μg/L on the next day. In Pivka branch of the Planinska Jama and at Unica spring the tracer was not detected until 28 and 30 June, respectively. On 11 July the maximum concentration of 2.8 μg/L was detected in Pivka branch and of 1.5 μg/L at Unica spring. At all three sampling points, secondary peaks were observed in lower concentrations after the subsequent rainy events. The tracer was almost entirely recovered.

Uranine was only detected in Pivka branch and at Unica spring from 9 July on. The maximum concentrations of 0.2 μg/L (Pivka branch) and 0.1 μg/L (Unica spring) were reached. Tracer recovery was 54.3%.

Despite being site-specific, the performed multi-tracer test demonstrated some general characteristics of underground drainage in karst aquifer through different recharge pathways during long-lasting dry period:

1. Transport velocities were very low. Previous study conducted at similar hydrological conditions (Gams 1970) showed even lower values (might also be due to weaker technical feasibilities). Studies at mean low flow conditions showed that maximum flow velocities were for about four times higher, and in a well-developed conduit network (Postojnska Jama) even ten times higher. Several other studies showed considerable variations of tracer transport as a function of respective hydrologic conditions (e.g. Pronk et al. 2007; Göppert and Goldscheider 2008).
2. Transfer of tracers was accelerated by subsequent raining, causing either first tracer detection or multi-modal and broader BTCs.
3. Longitudinal dispersion along the flow path was observed at sampling points along the underground river course.
4. Multi-modal BTCs resulted also from diffuse infiltration, sediment or soil retention, multiple flow paths and/or water stagnation; (due to temporal storage in the unsaturated zone, hydraulic restrictions, such as syphons and pools in conduit flow, obstructions by channel narrowing and rockfalls, etc.
5. After long-term monitoring satisfactory tracer recoveries were observed, tracing under low flow conditions may also result in substantially lower recoveries (Ravbar and Goldscheider 2007).

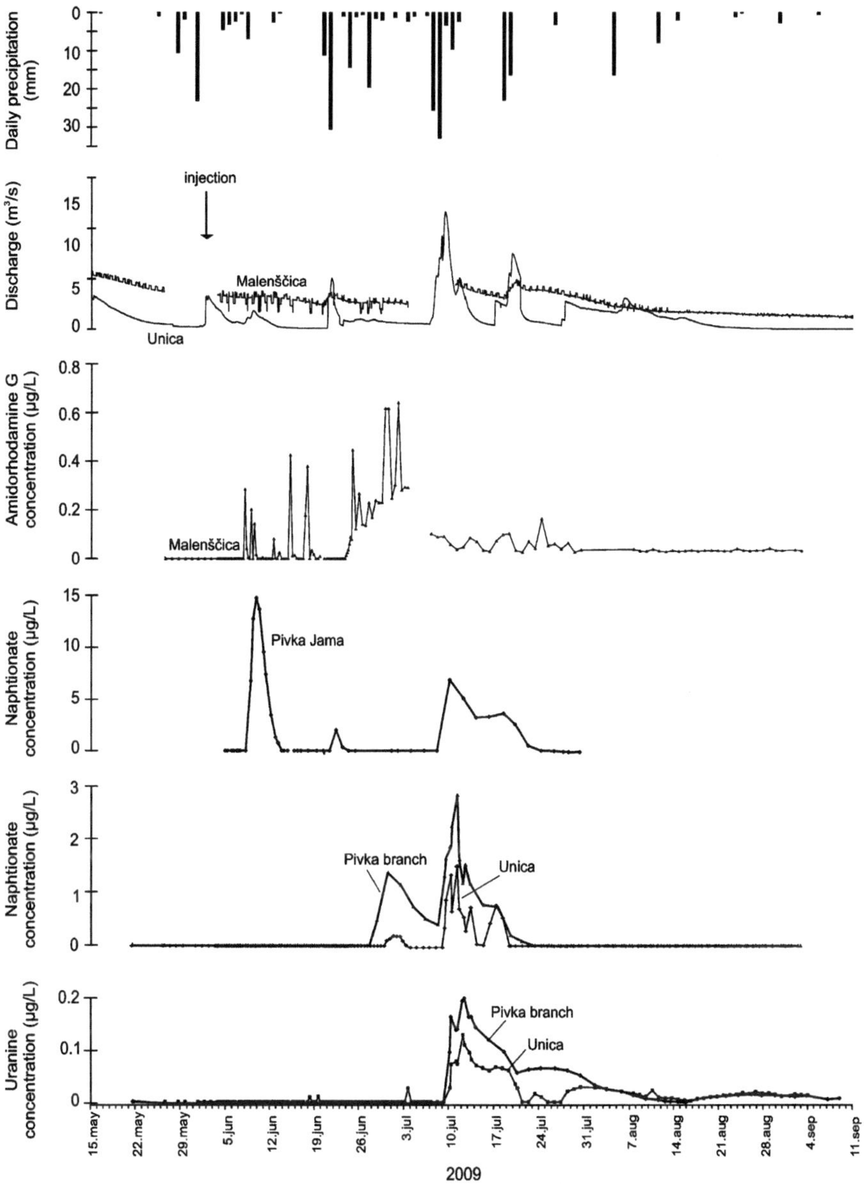


**Fig. 2** Hydrological conditions and breakthrough curves recorded at the observed springs. Abrupt changes in discharge values at Malenščica spring are due to supply capture

The expected effects of climate change in Europe are increased occurrence of hot and dry summers in the future. The obtained results are relevant for groundwater quality monitoring and for water suppliers. The findings indicate that the dynamics

**Table 1** Summarized flow velocities during low flow conditions

| Injection point | Sampling point | Distance (m) | $v_{max}$ (m/h) | $v_{dom}$ (m/h) |
|---|---|---|---|---|
| A | Pivka Jama | 3850 | 26 | 22.8 |
| | Pivka branch | 5390 | 8.5 | 5.8 |
| | Unica spring | 6100 | 6.7 | 6.5 |
| B | Malenščica spring | 4060 | 32.2 | 5.9 |
| C | Pivka branch | 3370 | 3.8 | 3.6 |
| | Unica spring | 4100 | 4.6 | 4.4 |

of contaminant transport in karst, at least in the studied area, may be retarded, but persistent. It is important that these findings are considered when planning protection and management.

**Acknowledgements** The study has been conducted in the frame of the Slovenian Research Agency Project No. Z6-0156, and supported by the World Federation of Scientists and the Slovenian Science Foundation. The authors also acknowledge the help of Juan Antonio Barberá Fornell (Malaga) and Gregor Kovačič (Koper) for their valuable assistance.

## References

Gabrovšek F, Kogovšek J, Kovačič G, Petrič M, Ravbar N, Turk J (2010) Recent results of tracer tests in the catchment of the Unica River (SW Slovenia). Acta carsologica 39, published online

Gams I (1970) Maksimiranost kraških podzemeljskih pretokov na primeru ozemlja med Cerkniškim in Planinskim poljem. Acta carsologica 5:171–187, in Slovene

Gospodarič R, Habič P (1976) Underground water tracing: Investigations in Slovenia 1972–1975. Third International Symposium of Underground Water Tracing (3. SUWT), Ljubljana, Bled

Göppert N, Goldscheider N (2008) Solute and colloid transport in karst conduits under low- and high-flow conditions. Ground Water 46, 1:61–68

Kogovšek J, Šebela S (2004) Water tracing through the vadose zone above Postojnska Jama, Slovenia. Environmental Geology 45, 7:992–1001

Kogovšek J, Knez M, Mihevc A, Petrič M, Slabe T, Šebela S (1999) Military training area in Kras (Slovenia). Environmental Geology 38, 1:69–76

Kundzewicz ZW, Mata LJ, Arnell NW, Doll P, Jimenez B, Miller K, Oki T, Sen Z, Shiklomanov I (2008) The implications of projected climate change for freshwater resources and their management. Hydrol. Sci. J. 53:3–10

Petrič M, Šebela S (2005) Hydrogeological research as a basis for the preparation of the plan of monitoring groundwater contamination – a case study of the Stara vas landfill near Postojna (SW Slovenia). Acta carsologica 34, 2:489–505

Pronk M, Goldscheider N, Zopfi J (2007) Particle-size distribution as indicator for fecal bacteria contamination of drinking water from karst springs. Environmental Science and Technology 41:8400–8405

Ravbar N, Goldscheider N (2007) Proposed methodology of vulnerability and contamination risk mapping for the protection of karst aquifers in Slovenia. Acta carsologica 36, 3:397–411

White WB (2002) Karst hydrology: recent developments and open questions. Engineering Geology 65, 2:85–105

# Tracer Investigations of the Nature and Structure of Subsurface Voids in Mildly Karstic Aquifers: an Example from the English Chalk

L. Maurice, T.C. Atkinson, J.A. Barker, A.T. Williams, A. Farrant, and A. Gallagher

**Abstract** The Chalk in England is a mildly karstified fractured limestone with high matrix porosity. Chalk landscapes are predominantly a fluviokarst with an extensive dry valley network and with stream sinks and dolines developed in zones close to overlying strata. The nature and extent of subsurface dissolutional voids is poorly understood and field tests using artificial tracers were carried out to investigate their structure. Tracer testing from stream sinks demonstrated rapid groundwater flow (up to 5 km/day) confirming previous Chalk tracer studies. However tracer attenuation was variable with extremely low tracer recoveries at some sites indicating that not all flowpaths fed by stream sinks comprise fully connected conduit systems. A conceptual model of flow along multiple flowpaths comprising a distributary network around stream sinks and a tributary network around springs with variable connectivity between the two is presented. The proposed conceptual model of flowpath structure in the Chalk resembles previous models of the early stages of speleogenesis. It is suggested that groundwater flow in the Chalk and other mildly karstic aquifers occurs within complex small scale dissolutional networks, in which there is a high degree of connectivity between larger conduits, fissures, and a primary fracture network. Advection from karstic channels into smaller voids results in high attenuation of solutes and particulates providing a degree of protection to groundwater outlets that is not seen in more highly karstic aquifers.

---

L. Maurice, A.T. Williams, A. Farrant, A. Gallagher
British Geological Survey, e-mail: loma@bgs.ac.uk

T.C. Atkinson, J.A. Barker
University College London

J.A. Barker
Southampton University

B. Andreo et al. (Eds.), *Advances in Research in Karst Media*
DOI 10.1007/978-3-642-12486-0, © Springer 2010

# 1 Introduction

The Cretaceous Chalk is a pure microporous limestone with a high matrix porosity of ~30 to 45% (MacDonald and Allen 2001). Storage in the matrix is high, but permeability is extremely low. Groundwater flow is predominantly through fractures, which are often solutionally enlarged to form fissures, or small karst conduits. Small-scale surface karst features (dolines and stream sinks) are commonly associated with the Chalk, especially at the boundary with the overlying Palaeogene strata (Banks et al. 1995). More than 100 sinking streams have been observed in the roughly 400 $km^2$ Pang-Lambourn catchment where this study was carried out (Fig. 1). In highly karstic aquifers sinking streams generally feed conduit flowpaths along which groundwater flow is rapid and attenuation is low (Atkinson 1977; Ford and Williams 2007). Previous tracer studies in chalk have found rapid flow of several km/d combined with low attenuation (Atkinson and Smith 1974; Barnes et al. Barnes 1999; Massei et al. 2006). The objective of this study is to explore whether the large numbers of sinking streams in the English Chalk all feed similar flowpaths.

# 2 Methods

Natural gradient tracer tests with injections into stream sinks in the Chalk were carried out over distances of 1.3 to 5.1 km (Fig. 1). The stream sinks with the highest flows were selected. The tests were carried out following rainfall when the sinking streams have flows of roughly 1 to 5 L/s. They are dry at other times. A qualitative test was carried out using the optical brightener Photine C with monitoring at 13 spring, borehole and river sites using cotton detectors changed every one to three days later observed under ultra violet light for the presence of the tracer. Six injections of 25 g to 1 kg were carried out.

Semi-quantitative tests were carried out from four sinking streams in two catchments using bacteriophage tracers (*Serratia marcescens* and *Enterobacter cloacae*). One litre of $1 \times 10^{12}$ pfu m/l was injected at each site. Samples were collected by automatic water sampler at springs and downstream monitoring points in nearby rivers (initially every hour with decreasing sample frequency) and analysed by a specialist microbiological laboratory. There was no background presence of the bacteriophages in the springs, but there was an occasional low background presence in rivers. Flow was measured at the monitoring sites to enable tracer recoveries to be estimated. The flow at the Blue Pool was ~200 L/s during all tracer tests, and that at Bagnor springs was ~25 L/s.

# 3 Results

Photine C optical brightener injected at Mirams Copse stream sink was not detected at any of the monitoring sites, including the Blue Pool spring (Fig. 1). *Ser-*

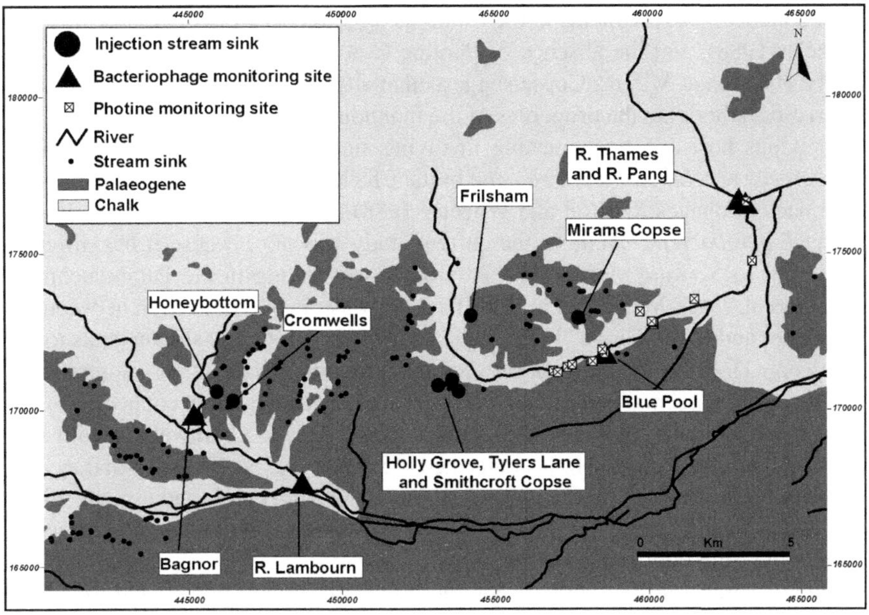


**Fig. 1** Geology, stream sinks and tracer testing sites in the Pang and Lambourn catchments in Southern England

*ratia marcescens* bacteriophage was injected into Mirams Copse stream sink and Cromwells stream sink (Fig. 1). Tracer from Mirams Copse was detected 1.6 km away at the Blue Pool spring and the groundwater velocity was 4.1 km/d. Tracer recovery was $< 0.00005\%$. Because bacteriophages can be injected at much higher concentrations than dyes they are much more detectable, and the low bacteriophage recovery suggests that around 100 kg of Photine would have been needed to produce a positive result. Tracer from Cromwells was detected at Bagnor springs 1.3 km away and the groundwater velocity was $> 0.6$ km/d. Tracer recovery was estimated as $< 0.000007\%$. In both these tests the *Serratia marcescens* bacteriophage was near the detection limit. The *Enterobacter cloacae* injected into Frilsham and Honeybottom sinks was not detected at any of the monitoring points.

## 4 Discussion

Tracer testing between Mirams Copse sink and the Blue Pool and Cromwells sink and Bagnor springs demonstrated rapid groundwater flow combined with very high tracer attenuation. Bacteriophages are known to undergo attenuation due to sorption and die off and have lower recoveries than simultaneously injected dye tracers (Rossi et al., 1998). Therefore the high attenuation observed during these tests may be due to the tracer properties. However, larger recoveries of 0.1 to 0.5% have been obtained in experiments in the Chalk using these bacteriophages (Skilton and Wheeler 1988), and the absence of Photine C at the Blue Pool spring following a 1 kg injection at Mirams Copse suggests that attenuation is due to flowpath characteristics rather than the properties of the individual tracers.

Previous borehole tracer testing involving simultaneous injection of *Serratia marcescens* and *Enterobacter cloacae* in the Chalk indicated similar recoveries for both bacteriophages (Skilton and Wheeler 1988). This suggests that the failure to detect *Enterobacter cloacae* in the current study was not because it has different properties to *Serratia marcescens*. Previous tracer testing in the Pang catchment (Banks et al. 1995; Maurice et al. 2006) has demonstrated a connection between the Blue Pool and Holly Grove, Tylers Lane and Smithcroft Copse stream sinks to the west (Fig. 1). Frilsham sink lies between this area and Mirams Copse stream sink, and therefore it seems likely that Frilsham sink is hydrologically connected to the Blue Pool. Similarly, Honeybottom stream sink is near to Cromwells stream sink and might also be expected to be connected to Bagnor springs (Fig. 1). No tracer injected at these sites was detected at any of the monitoring points and it seems likely that this is because the *Enterobacter cloacae* was attenuated to below the detection limit at the Blue Pool and Bagnor springs.

Previous tracer testing between Smithcroft Copse stream sink and the Blue Pool in similar hydrological conditions demonstrated rapid groundwater velocities of 5 km/d and tracer recoveries of 20–25% indicating relatively low attenuation (Maurice et al. 2006). This study suggested that the travel time was too rapid for significant diffusion of tracer into the porous matrix and that the loss of 75–80% of the tracer was probably due to dispersion of tracer from the main conduit flowpath into small voids where double porosity diffusion occurs. The results of the new tracer tests which demonstrate rapid groundwater flow combined with high attenuation (and perhaps total attenuation of tracer along some flowpaths fed by stream sinks)

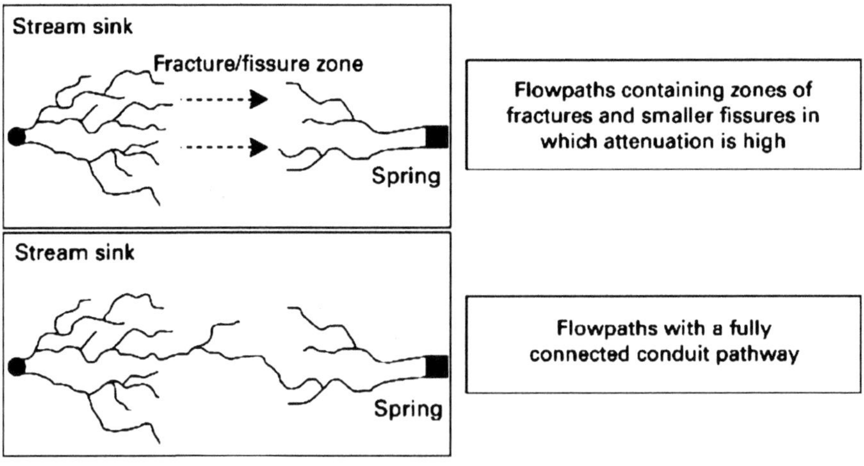


**Fig. 2** Conceptual model of flowpaths between stream sinks and springs in the Chalk

supports the theory that groundwater flow in the Chalk (and perhaps other mildly karstic aquifers) occurs along multiple flowpaths. The proposed conceptual model of flowpath structure in the Chalk resembles previous models of the early stages of speleogenesis (Ford and Williams 2007), with a network of solutionally enlarged fissures and small conduits developed around stream sinks and springs, but only limited connectivity between them (Fig. 2). In tests with lower attenuation a higher proportion of tracer travels along a fully connected conduit network. In tracer tests with high attenuation more tracer is dispersed into smaller voids, and it is conjectured that at some sites there is no connected conduit network and the tracer passes through an area of fractures and smaller fissures which causes total tracer attenuation between the stream sink and the groundwater outlet.

## 5 Conclusions

Tracer testing has demonstrated groundwater flow between stream sinks and springs in the Chalk that is often as rapid as that observed in highly karstic aquifers. However, in contrast to many karstic flowpaths in which rapid flow is combined with very low attenuation, tracer attenuation in these Chalk tests is more variable and can be extremely high. The tracer testing carried out in this study was from the most developed stream sinks and therefore it is possible that groundwater flowpaths fed by many stream sinks in Chalk are highly attenuating. Results suggest that groundwater flow in the Chalk and other mildly karstic aquifers occurs within complex small scale dissolutional networks, in which there is a high degree of connectivity between larger conduits, fissures, and a primary fracture network. Advection from karstic channels into smaller voids results in high attenuation of solutes and particulates providing a degree of protection to groundwater outlets that is not seen in more highly karstic aquifers.

**Acknowledgements** This work is funded by Natural Environment Research Council grant NER/T/S/2001/00956. This paper is published with the permission of the Executive Director, British Geological Survey (BGS).

## References

Atkinson TC (1977) Diffuse flow and conduit flow in limestone terrain in the Mendip Hills, Somerset (Great Britain). Journal of Hydrology 35, 93–110

Atkinson TC, Smith DI (1974) Rapid groundwater flow in fissures in the Chalk: An example from South Hampshire. Quarterly Journal of Engineering Geology 7, 197–205

Banks D, Davies C, Davies W (1995) The Chalk as a karstic aquifer: evidence from a tracer test at Stanford Dingley, Berkshire, UK. Quarterly Journal of Engineering Geology 28, S31–S38

Barnes S (1999) Karstic groundwater flow characteristics in the Cretaceous Chalk aquifer, Northern Ireland Quarterly Journal of Engineering Geology 32, 55–68

Ford DC, Williams P (2007) Karst hydrogeology and geomorphology. John Wiley and Sons Ltd. 562 pp

Massei N, Wang HQ, Field MS, Dupont JP, Rodet J (2006) Interpreting tracer breakthrough tailing in a conduit-dominated karstic aquifer. Hydrogeology Journal 14 (6), 1431–2174

MacDonald AM, Allen DJ (2001) Aquifer properties of the Chalk of England. Quarterly Journal of Engineering Geology 34, 371–384

Maurice L, Atkinson TA, Barker JA, Bloomfield JP, Farrant AR, Williams AT (2006) Karstic behaviour of groundwater in the English Chalk. Journal of Hydrology 330, 53–62

Rossi P, Döerfliger N, Kennedy K, Müller I, Aragno M (1998) Bacteriophages as surface and groundwater tracers. Hydrology and Earth System Sciences 2, 101–110

Skilton H, Wheeler D (1988) Bacteriophage tracer experiments in groundwater. Journal of Applied Bacteriology 65, 387–395

# Transit Time Environmental Tracing from Dissolved Organic Matter Fluorescence Properties in Karstic Aquifers. Application to Different Flows of Fontaine de Vaucluse Experimental Basin (SE France)

T. Blondel, C. Emblanch, Y. Dudal, C. Batiot-Guilhe, Y. Travi, and S. Gaffet

**Abstract** For about ten years, environmental tracing development using Dissolved Organic Matter (DOM) has been the subject of several studies. Particularly, the use of characterization techniques, like fluorescence Emission-Excitation Matrices (EEM), has enabled identification and monitoring of DOM sources within mainland or marine hydrosystems. Hydrogeologists have already shown the significance of Total Organic Carbon (TOC) used as fast seepage tracer in karstic aquifers. The aim of this study consists in using DOM fluorescence signals to develop a transit time quantitative tracer in heterogeneous hydrosystems. The Low-Noise Underground Laboratory of 'Rustrel – Pays d'Apt' (France) offers a special access to different kinds of unstructured karstic flows, cutting randomly through the fault network of 'Fontaine de Vaucluse' vadose zone. The hydrochemical monitoring of these flows since 2002 has allowed their behaviour to be well known, and therefore to calibrate a relationship between a fluorescence index (Humification IndeX) and their mean transit time. Finally, this relationship has been tested on two springs of 'Vaucluse' plateaus, giving good transit time estimations for hydrosystems which do not present mixture between recent and pluriannual waters.

---

T. Blondel, C. Emblanch, Y. Travi
UMR 1114 INRA-UAPV "EMMAH", UAPV, Faculté des Sciences Exactes et de la Nature, 33 rue Louis Pasteur, 84000 Avignon, France,
e-mail: christophe.emblanch@univ-avignon.fr

Y. Dudal
UMR "Biogéochimie du Sol et de la Rhizosphère", INRA-SupAgro, 34060 Montpellier, France

C. Batiot-Guilhe
Laboratoire HydroSciences, Université de Montpellier II, Maison des Sciences de l'Eau, 34095 Montpellier, France

S. Gaffet
UMR "Géoazur", CNRS-Université de Nice-Sophia-Antipolis Laboratoire Souterrain à Bas Bruit (LSBB), 84400 Rustrel – Pays d'Apt, France

B. Andreo et al. (Eds.), *Advances in Research in Karst Media* 
DOI 10.1007/978-3-642-12486-0, 

# 1 Introduction

Artificial and environmental tracings have been used to obtain transit time estimation, including dye and salt injections, dissolved gases, dissolved element contents (Silica [$SiO_2$], Magnesium [$Mg^{2+}$], ...), isotopes ($^{18}O$, $^{14}C$) and bacteriophage (Lapworth et al. 2008). Total Organic Carbon (TOC) is a particularly relevant tracer for studying water transit within karstic environment (Batiot 2002). However, in this hydrosystem type, all of these tracers often give only either specific information between injection point and monitored spring for specific hydric conditions, or either a comparative approach of transit time of different waters, or they are put in the wrong because of system complexity.

Most of the soil-borne Dissolved Organic Matter (DOM) can provide a fluorescence signal (Stedmon et al. 2003). For some years, fluorescence has been used extensively in the water sciences, showing some great potential to decipher DOM fractions, and to monitor and understand DOM transformation in aquatic systems (Hudson et al. 2007). Recent studies have highlighted the use of fluorescence proprieties of DOM to discriminate between different sources of DOM, and as a natural tracer to understand water movement (Baker and Genty 1999; Lapworth et al. 2008). This paper outlines another use of fluorescence properties of DOM as a possible transit time tracer for highly complex karstic flows and tests this hypothesis for two hydrosystems of 'Vaucluse' plateau.

# 2 Presentation of Site

The 'Fontaine de Vaucluse' spring (FV) is the largest karstic outlet in France. It is the only outlet of a carbonated formation, without organic matter (Masse 1968), situated in the north-western part of Provence. Some perennial small outlets in the limestone plateau can be found. Among them (Fig. 1a), two were investigated: FV, and a perennial small spring, 'St Trinit' (TRI), located in the middle of limestone plateau. Different studies have been carried out on these discharge points (Batiot 2002), and their general flow characteristics are well known.

Located to north of Rustrel-Pays d'Apt (Vaucluse, France), the Low-Noise Underground Laboratory (LSBB) has become an experimental laboratory to investigate the UZ of Mediterranean karst aquifers. The tunnel is located in the UZ of the 'Fontaine de Vaucluse' catchment area, and cuts randomly through the karstic network. Five significant flows can be found in the gallery (Fig. 1b): four are perennial (A, B, C and D) and one is intermittent (GAS). All correspond to drip water points observed at several faults zone levels (Garry et al. 2008): points A and B are located in a fractured zone, point C in a jointed zone, point D is found at the base of epikarst and point GAS in a faulted area showing an important karstification (Fig. 1b). This diversity of fracturations and karstification degrees allows the study of different kinds of flows in the UZ, which have different transit times. After a prospecting at the surface (Fig. 1c), two groups of soil are found: a lithosol covered

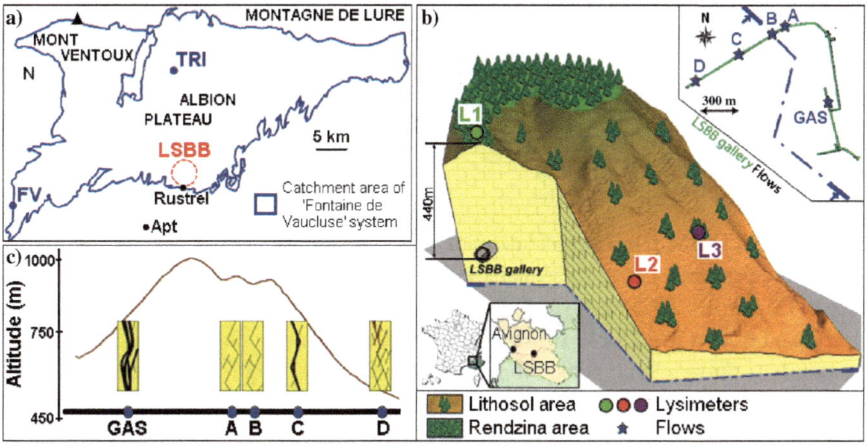


**Fig. 1** **a** Localisation of studied springs and LSBB in 'Fontaine de Vaucluse' catchment area, **b** karstification pattern above LSBB flows, **c** LSBB studied area pattern

by a more or less scattered scrubland, and a rendzina covered by helm oak forest. Cultivated soil and other anthropogenic organic sources are absent in the studied site. Here, soil is therefore the only organic source.

## 3 Materials and Methods

All materials were cleaned and rinsed with osmosis water. Glassware was heated to 500 °C to pyrolize manufactured organic matter. In Fig. 1c, sampling was performed on (i) the different drip water points of LSBB, when they had sufficient water flow, and (ii) three permanently installed lysimeters, representative of prospected pedological conditions: one in rendzina (L1), one in lithosol under herbaceous vegetation (L2) and one in lithosol under shrubby vegetation (L3). Soil samples were leached for few minutes on field lysimeters to obtain a total leachate volume of 500 mL, by using osmosis water. Finally, all samples were filtered through 0.45 μm fiberglass filters (Pall Corporation), priorly heated to 500 °C.

A Perkin-Elmer LS 55 luminescence spectrometer was used to acquire (i) Excitation-Emission Matrices (EEM), and (ii) necessary emission spectra to calculate the Humidification IndeX (HIX) of Zsolnay et al. (1999). The slit widths were 10 nm for the excitation and emission monochromators. EEM were obtained by subsequently scanning from 215 to 530 nm, every 0.5 nm, and by increasing the excitation wavelength by 15 nm from 215 to 530 nm. Emission spectra were scanned with the same emission wavelength range and with an excitation wavelength of 254 nm. All water samples showed DOC content lower than 2 mg/L and fluorescence analysis are in the linear range for all components of emission spectra, precluding the need for absorbance correction (Ohno 2002). This linearity was verified prior to fluorescence acquisitions for all soil leachate samples.

# 4 Results and Discussion

**Fluorescent DOM Characterization of LSBB Samples.** Examples of general fluorescence properties of soil leachate and LSBB groundwater samples collected between December 2005 and April 2008 are illustrated in Fig. 2.

It can be seen from these EEMs that there are four main fluorophores. Their corresponding fluorescence peaks, together with the wavelength ranges of their maxima and their assignments, are given in Table 1, and are in accordance with what is reported in the literature by several authors.

Transit time tracing problematics must consider those molecules which are present in both soil leachate samples and groundwaters, and have a sufficiently slow evolution to be observed all along their transfer in the hydrosystem. After laboratory simulations (Blondel 2008), fluorophores, corresponding to $\alpha'$-peak, are seen to be (i) much less quickly adsorbed, and (ii) more resistant to biocenosis, than ones of $\alpha$-peak. The $\alpha'$-peak evolution is examined in more detail.

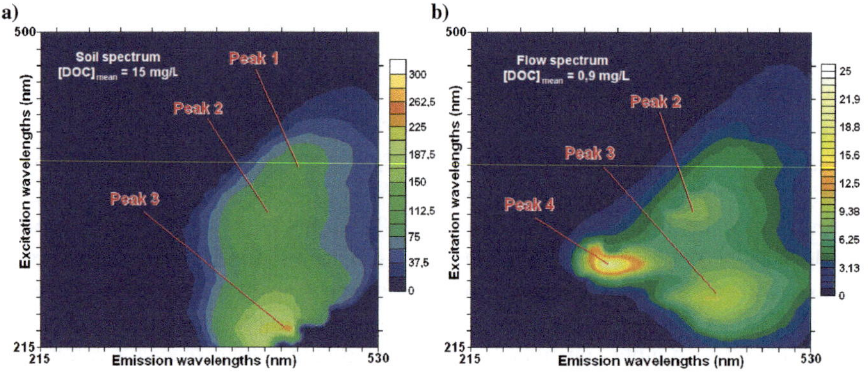


**Fig. 2** Examples of **a** soil and **b** flow EEM spectra

**Table 1** Major fluorescent components in soil leachate and LSBB groundwater samples

| Peak | Fluorophore Name (Parlanti et al. 2000) | Fluorophore Name (Coble 1996) | $\lambda_{ex}$ (nm) | $\lambda_{em}$ (nm) | Component type |
|---|---|---|---|---|---|
| 1 | – | – | 365–410 | 425–470 | Humic-like with high molecular weight |
| 2 | $\alpha$ | C | 320–350 | 390–460 | Humic-like |
| 3 | $\alpha'$ | A | 230–260 | 390–440 | Humic-like |
| 4 | $\delta$ | T | 275–290 | 329–345 | Protein-like |

**Fluorophore Evolution in the LSBB "Hydrosystem".** To observe the evolution of $\alpha'$-peak fluorophores, emission spectra scanned with an excitation wavelength of 254 nm, corresponding to mean excitation wavelength to obtain maximum of fluorescence intensity for $\alpha'$-peak fluorophores (Zsolnay et al. 1999) have been used. Figure 3 already shows a shifting of fluorescence maxima to lower emission wavelengths between soil leachate and groundwater samples. Recent studies have determined the mean transit time of different LSBB flows (Garry 2007). By plotting the mean emission wavelength of their fluorescence maxima, a same blue-shift according to mean transit time can be observed. Senesi et al. (1991) have shown that for a given excitation wavelength, compounds that emit at lower emission wavelengths are smaller in size than compounds emitting at higher wavelengths. This blue-shift is therefore due to a decreasing of fluorophore molecular size according to transit time.

HIX can be interpreted as the ratio between higher and lower molecular weight fluorophores. Previous works have given mean transit times of LSBB flows for 2006–2007 and 2007–2008 hydrologic cycles (Garry 2007). By plotting (i) mean HIX according to these mean transit times, and (ii) HIX values of LSBB flow and soil leachate samples with known transit time, a logarithmic relationship between HIX and transit time can be found (Fig. 4). Finally, a 85% confidence interval shows that the higher calculated transit times, the less accurate their estimations. This is due to fluorophores which evolve to small and very stable molecules.

To validate this relationship, the assumption has been tested on two different springs of the 'Vaucuse' plateau during different hydrologic conditions. These two hydrosystems have the same climatic and morphologic characteristics as LSBB flows. 'St Trinit' HIX range between 4.73 and 5.67 corresponding by calculation to transit times which range between 107 and 160 days. This transit time range is in

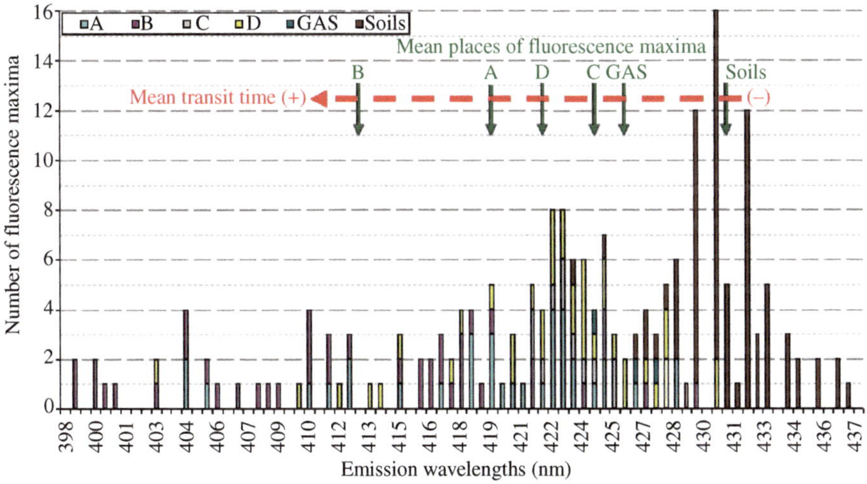


**Fig. 3** Emission wavelength distribution of fluorescence maxima for 254 nm excitation

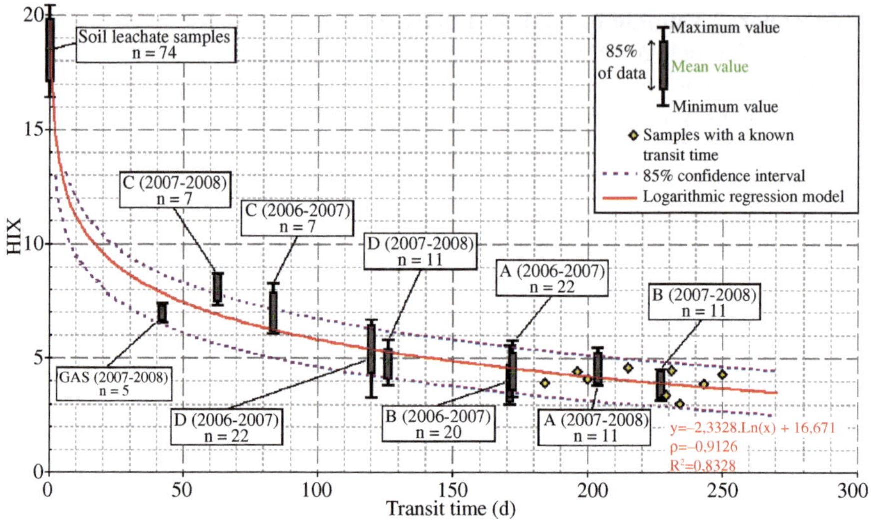


**Fig. 4** HIX value variations according to transit time

accordance with previous studies of Batiot (2002). 'Fontaine de Vaucluse' HIX values (from 4.83 to 8.40) correspond to transit time ranging between 50 and 140 days. This time seems to be low, although sampling was done after several low recharge years. These estimation difficulties can be due to important groundwater resources of the FV system, in which there is mixture between pluriannual (with altered organic matter fingerprint) and fast infiltrated (with fresh organic matter fingerprint) waters.

## 5 Conclusions

Although phenomena responsible for fluorophore evolution are not yet well understood, especially the respective role of infiltration, biological activities and variation of organic compound structure during soil humidification, experimental data provide evidence that DOM fluorescence proprieties can be used as transit time tracer in Mediterranean karstic system. HIX allows monitoring of fluorophore evolution in hydrosystems, and calibration of a logarithmic regression between evolution and transit times. However, these estimations have been validated only for hydrosystems which do not present mixture between recent and pluriannual waters. Moreover, this result corresponds to specific site and climatic conditions. Additional studies are in progress on the same site and on other karstic systems.

# References

Baker A, Genty D (1999) Fluorescence wavelength and intensity variations of cave waters. J. Hydrol. 217: 19–34

Batiot C (2002) Etude expérimentale du cycle du carbone en régions karstiques. Apport du carbone organique et du carbone minéral à la connaissance hydrogéologique des systèmes. PhD Thesis, Avignon: 247 pp

Blondel T (2008) Traçage spatial et temporel des eaux souterraines dans les hydrosystèmes karstiques par les matières organiques dissoutes. PhD Thesis, Avignon: 190 pp

Coble P (1996) Characterization of marine and terrestrial DOM in seawater using excitation-emission matrix spectroscopy. Mar Chem 51: 325–346

Garry B (2007) Etude des processus d'écoulement de la zone non saturée pour la modélisation des aquifères karstiques. PhD Thesis, Avignon: 216 pp

Garry B, Blondel T, Emblanch C et al (2008) Contribution of artificial galleries to knowledge of karstic system behaviour in addition to natural cavern data. Int J Spel 37(1): 75–82

Hudson N, Baker A, Reynolds D (2007) Fluorescence analysis of dissolved organic matter in natural, waste and polluted waters – a review. River Res Appl 23: 631–649

Lapworth D J, Gooddy A S, Butcher B L et al (2008) Tracing groundwater flow and sources of organic carbon in sandstone aquifers using fluorescence properties of dissolved organic matter (DOM). Appl Geochem 23(12): 3384–3390

Masse J P (1968) L'Urgonien de Sault (Vaucluse). Bulletin de la Société Géologique de France 9(4): 495–596

Ohno T (2002) Fluorescence inner-filtering correction for determining the humification index of dissolved organic matter. Environ Sci Technol 36: 742–746

Parlanti E, Wörz K, Geoffroy L et al (2000) Dissolved organic matter fluorescence spectroscopy as a tool to estimate biological activity in a coastal zone submitted to anthropogenic inputs. Org Geochem 31(12): 1765–1781

Senesi N, Miano T M, Provenzano M R et al (1991) Characterization, differentiation and classification of humic substances by fluorescence spectroscopy. Soil Sci 152: 259–271

Stedmon C A, Markager S, Bro R (2003) Tracing dissolved organic matter in aquatic environments using a new approach to fluorescence spectroscopy. Mar Chem 82: 239–254

Zsolnay A, Baigar E, Jimenez M et al (1999) Differentiating with fluorescence spectroscopy the sources of dissolved organic matter in soils subjected to drying. Chemosphere 38: 45–50

# Use of Dissolved Organic Carbon to Characterize Infiltration in a Small Karst System in the French Jura Mountains (Fertans, France)

J.B. Charlier, J. Mudry, and C. Bertrand

**Abstract** Due to the high input of organic soil substances in infiltrated water, dissolved organic carbon (DOC) is an interesting tracer to determine water origin in hydrosystems. The aim of this study was to use continuous measurements of DOC natural fluorescence in spring water to characterize infiltration in a karst system at the event time scale. Hydrological data were recorded at the outlet of a small perennial spring within a cliff at Fertans in the French Jura mountains since July 2009. Results showed that the response of the system was fast (response time of 3 hours) and was produced by two combined flow types. First, discharge and turbidity peaks, and electrical conductivity troughs were produced by a piston-type flow enabling an unclogging of the fractures. Second, this fast infiltration process was followed by a slower one enriched in DOC, produced by a more diffuse-type flow through the matrix compartment.

## 1 Introduction

Fissured and karst aquifers are known to be particularly heterogeneous, with transmissive fractures embedded in low permeability rock matrix volumes. Due to the high input in organic soil substances in infiltrated water, many studies have shown that DOC fluorescence and total organic carbon (TOC) were interesting tracers to determine water origin in karst systems [1–6]. TOC is a non-conservative tracer whose mineralization over time occurs at the scale of a few weeks; laboratory experiments having shown an exponential decay of TOC [7], meaning that the higher the TOC content, the faster the transit time in the system. TOC is thus adapted to study fast infiltration in karst systems.

Measuring TOC is technically difficult due to a continuous mineralization in sampling bottles and due to a difficult automation of such devices on the field. In order to overcome these drawbacks, the use of a field fluorometer, developed for ar-

J.B. Charlier, J. Mudry, C. Bertrand
Chrono-Environment, University of Franche-Comté, 16 route de Gray, 25000 Besançon, France,
e-mail: jean-baptiste.charlier@univ-fcomte.fr

B. Andreo et al. (Eds.), *Advances in Research in Karst Media*
DOI 10.1007/978-3-642-12486-0, © Springer 2010

tificial dye tracers [8] gives continuous measurements of water natural fluorescence (NF). Moreover, knowing that strong relationships exist between DOC fluorescence and TOC in fresh water [9], a hypothesis was developed that continuous NF measurements could be used as an excellent infiltration tracer.

This study aims to characterize infiltration flows through a small karst system. The original contribution of this work is the use of continuous measurements of DOC natural fluorescence in spring water to monitor infiltration processes at the event time scale. The experimental site is located at Fertans in the French Jura mountains. On the basis of the results, a conceptual model of infiltration flows during flood events is presented.

## 2 Study Site and Measurements

The study zone is located on the Amancey plateau in the French Jura mountains (Fig. 1a and b). The experimental site is located at the base of a cliff in the north of

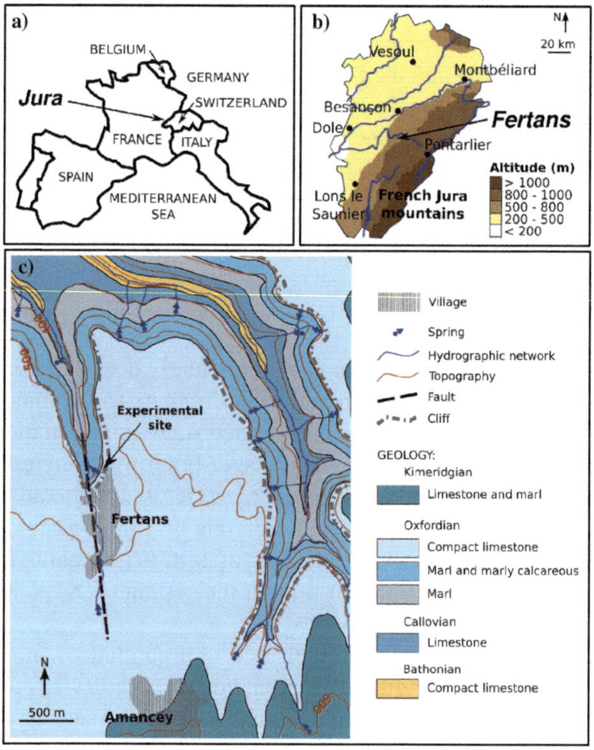


**Fig. 1** Location of the Jura in France (**a**) and of Fertans in the French Jura mountains (**b**). Geological map of the study area (**c**) modified from BRGM [11], showing the location of the experimental site

Fertans (Fig. 1c). The geological formations are composed of compact limestones ('Rauracian facies') on marl and marly-calcareous ('Argovian facies') of Oxfordian age. The geological structure is tabular, and the deformation is characterized by subvertical fractures mainly oriented N 40°, N 20°, N 0°, and N 120°. The monitoring site is the outlet of a spring (500 m ASL) draining a fracture oriented N 20° at the interface of the marly-calcareous and the compact limestone formations. The unsaturated zone of the system is thus composed of a 25 m-depth fractured compact limestone of 'Rauracien facies' – the site is considered to be a 'window' in the karst at the interface between the unsaturated and the saturated zone. The catchment area is estimated to be less than 0.5 $km^2$.

Measurements were conducted during 50 days, from 22 July to 10 September 2009. Hourly rainfall was recorded at the station of Coulans-sur-Lison at Eternoz [10], located 9 km south-west of the experimental site of Fertans. DOC natural fluorescence and turbidity were measured using a field fluorometer, GGUN-FL30 (Albillia, Neuchâtel, Switzerland), initially designed for tracer tests to measure dye tracer concentrations in the field (Schnegg 2003). Water level, electrical conductivity and water temperature were carried out every 15 min using automatic field equipment (OTT, Orpheus, Kempten, Germany).

## 3 Results and Discussion

Figure 2 shows physical measurements of spring water for three selected high-flow events during the monitoring low water period in summer 2009. The rainfall intensities (P) were moderate with a maximum of 14.4 mm/h. Excluding rainy periods, base flow was less than 0.2 L/s and water temperature (Temp) was around 11.5 °C. The baselines of electrical conductivity (EC) and DOC natural fluorescence (NF) fluctuated between 720 and 860 µS/cm, and between 3 and 6 ppb, respectively. During flood events, the response time of the system was 3 h (delay between P and flood peaks). Temp and EC troughs were observed, as were a NF smooth peak associated or not with a trough, and a turbidity (Tu) peak.

To better characterize the relationships between these parameters, a principle component analysis (PCA) was conducted on hourly data (Fig. 3). The results showed a main component on which were aligned discharge (Q) and Tu in contrast with EC and Temp. This axis, drawn partly by high Q values, represents the system dynamic in phase with the flood peak. The fact that the EC troughs and Tu peaks are associated confirms that flood peaks were generated by the arrival of less mineralized rainwater with a large flux of suspended matter (unclogging of fractures). Given the response time of 3 h of the system (given a "vertical" velocity of the transit in the unsaturated zone of about $2.10^{-3}$ m/s), infiltration is fast through a transmissive fractures network. The second component of the PCA, on which NF is aligned, showed that the DOC transport is not associated with the hydrodynamic of the first component. This pattern enabled identification of another flowpath, with a residence time longer than the fractures networks, likely a less permeable matrix compartment.

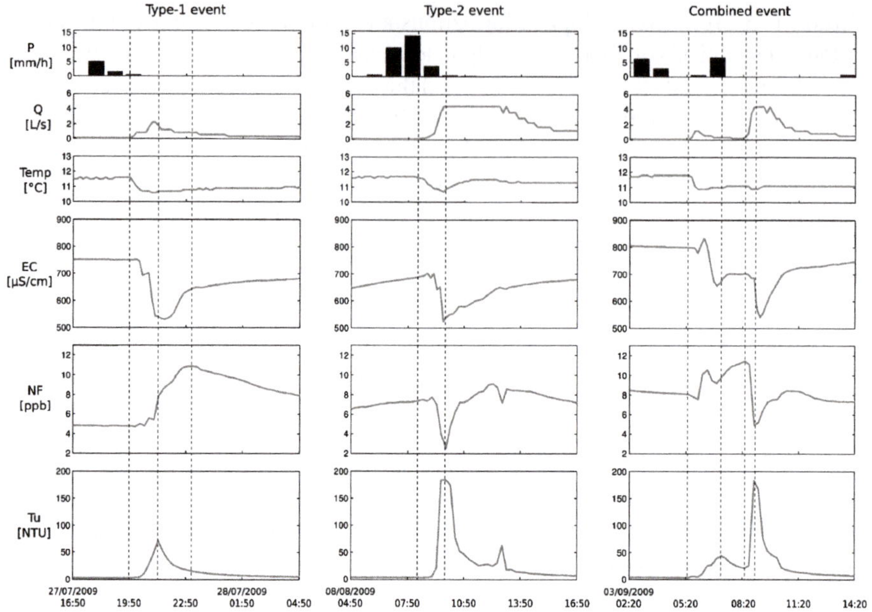


**Fig. 2** Hydrological time series monitored at the outlet of the spring for single flood events of Type 1 (27 July 2009) and of Type 2 (08 August 2009), and for a flood event combining both types (03 September 2009); with rainfall (P), spring discharge (Q), temperature (Temp), electrical conductivity (EC), DOC natural fluorescence (NF), and turbidity (Tu)

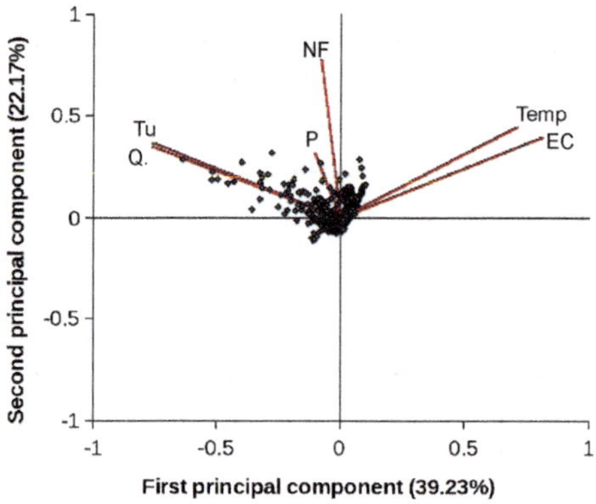


**Fig. 3** Plot based on principle components analysis of hydrological time series at a hourly time step (first and second component: 61,41%); with rainfall (P), spring discharge (Q), electrical conductivity (EC), DOC natural fluorescence (NF), temperature (Temp), and turbidity (Tu)

Time series of rainfall and discharge were divided into 11 flood events generated by 5 to 30 mm rainfall depth. A detailed analysis of the signals of EC, NF and Tu allowed the distinguishing of two types of events related to their response to rainfall (Fig. 2). For Type-1 event, two responses were observed. The first response was characterized by a EC trough and a Tu peak in phase with peakflow (response time of 3 h). It was followed by a second response characterized by a smooth and delayed NF peak (response time of 5 h). This pattern fits well with observations made by Pronk et al. [5, 6]. These authors showed that an autochtonous or pulse-through turbidity component mobilises sediments from fractures inside the system before the arrival of freshly infiltrated water from soil zone. This last component is an allochtonous or flow-through turbidity component enriched in DOC. For Type-2 event, on the contrary to Type-1, the NF response was a trough in phase with a EC trough and a high Tu peak (response time of about 3 h). Surprisingly, pre-event DOC content was diluted by infiltrated water. This pattern showed, in this case, infiltration at the soil surface was mainly localized in zones without soil cover, as karren zones. The observation of these two event types in the 'Combined event' in Fig. 2 confirmed for Type-1 event a lowest Tu peak associated with a delayed buffered NF peak. These responses suggested that i) in the first case infiltration is mainly produced by a more diffuse-type flow, through soil cover and matrix compartment, and that ii) in the second case infiltration is mainly produced by piston-type flow in transmissive flowpaths through fracture network.

The behavioral scheme of infiltration designed in this study is consistent with the concept of double permeability observed in karst and fractured media: fast processes of a piston-type flow in the fracture network, and buffered and slow processes of a diffuse-type in the less permeable matrix. Current work at Fertans experimental site will validate this model with hydrological monitoring of a piezometer drilled in the matrix. Finally, the study of flow dynamics at the event time scale shows that the occurrence of these two processes remains complex. In fact, observations at the outlet of the spring allows identification of flow paths and residence time, but not characterization of factors of such variability. The development of conceptual models calls for further studies to better understand the role of soil compartment, and of fracture/matrix relationships on karst functioning.

## 4 Conclusion

The aim of this study was to use continuous measurements of DOC natural fluorescence at the outlet of a small spring to characterize infiltration in a karst system. Results showed first that, from a methodological point of view, a continuous NF signal appears to be a relevant tracer of infiltration that improves the study of flow processes at the event time scale. Second, from a hydrogeological point of view, a study at the event time scale showed that flood and turbidity peaks were generated by fast infiltration with piston type flows enabling an unclogging of the fractures, and that they were followed by a slower infiltration process enriched in DOC.

**Acknowledgements** Many thanks to Bruno Régent for his active contribution on the field.

# References


1. Albéric P, and M. Lepiller (1998) Oxidation of organic matter in a karstic hydrologic unit supplied through stream sinks (Loiret, France). Water Res 32(7), 2051–2064
2. Emblanch, C., B. Blavoux, J.M. Puig, and J. Mudry (1998), Dissolved organic carbon of infiltration within the autogenic karst hydrosystem, Geophys. Res. Lett., 25(9), 1459– 1462
3. Batiot, C., C. Liñán, B. Andreo, C. Emblanch, F. Carrasco, and B. Blavoux (2003), Use of Total Organic Carbon (TOC) as tracer of diffuse infiltration in a dolomitic karstic system: The Nerja Cave (Andalusia, southern Spain), Geophys. Res. Lett., 30(22), 2179, doi:10.1029/2003GL018546
4. Celle-Jeanton H., C. Emblanch, J. Mudry, and A. Charmoille (2003), Contribution of time tracers ($Mg^{2+}$, TOC, δ13CTDIC, $NO_{3-}$) to understand the role of the unsaturated zone: A case study–Karst aquifers in the Doubs valley, eastern France, Geophys. Res. Lett., 30(6), 1322, doi:10.1029/2002GL016781
5. Pronk M., N. Goldscheider, and J. Zopfi (2006), Dynamics and interaction of organic carbon, turbidity and bacteria in a karst aquifer system, Hydrogeology Journal, 14, 473–484
6. Pronk M., N. Goldscheider, J. Zopfi, and F. Zwahlen (2009), Percolation and particle transport in the unsaturated zone of a karst aquifer, Ground Water, 47(3), 361–369
7. Batiot, C. (2002), Étude expérimentale du cycle du carbone en régions karstiques. Apport du carbone organique et du carbone minéral à la connaissance hydrogéologique des systèmes. PhD thesis, University of Avignon, France
8. Schnegg P.-A. (2003), A new field fluorometer for multi-tracer tests and turbidity measurement applied to hydrogeological problems. Proceedings of the Eight International Congress of the Brazilian Geophysical Society, Rio de Janeiro
9. Cumberland S.A., and A. Baker (2007), The freshwater dissolved organic matter fluorescence–total organic carbon relationship, Hydrol. Process., 2093–2099
10. METEO France (2009), hourly rainfall data, 2009, Besançon, France
11. BRGM (1975), Geological map of Quingey at 1:50 000 scale, 529, BRGM (ed), Orléans, France

# Investigation of Groundwater Dynamics in a Mediterranean Karst System by Using Multiple Hydrogeochemical Tracers

C.C. Bicalho, C. Batiot-Guilhe, J.L. Seidel, S.Van-Exter, and H. Jourde

**Abstract** The Lez karst aquifer, located in southern France, supplies the Montpellier metropolitan area with potable water and has a maximum exploitation capacity of about 1.700 L/s. The objective of this work is to improve the comprehension of groundwater dynamics in a particular Mediterranean karst system, using hydrogeochemical tracers. The different types of groundwater with similar chemical characteristics (major and trace elements) are identified. During rainy season periods, this multitracer characterization shows that deep mineralized waters emerge at the springs just before rapid infiltration waters. This phenomenon shows that hydrodynamic conditions imply water circuits, with participation of specific compartments or creation of short cuts for superficial percolating waters. The data also highlight the vulnerability of the system to natural or anthropogenic contamination. Key words: karst, hydrochemistry, natural tracing, hydrodynamics.

## 1 Introduction

The Lez spring has supplied the metropolitan region of Montpellier (France) with water since the XIXth century. Since 1981, the active pumping has been performed directly from an excavated gallery intersecting the main conduit, 48 m below the Lez spring level, with a maximum pumping rate of 1.700 L/s. This discharge should guarantee water supply throughout the year by drawing upon the water reserves in the aquifer at a larger rate than the natural water discharge during dry seasons.

This present work aims to better investigate the dynamics of the Lez KS groundwater by monitoring chemical and physical characteristics, and analyzing their variation over the hydrological cycle. The method is based on the analysis of physical

C.C. Bicalho, C. Batiot-Guilhe, J.L. Seidel, S.Van-Exter, H. Jourde
Université Montpellier 2, HydroSciences Montpellier, UMR CNRS, IRD, Université Montpellier 1, Université Montpellier 2, 34095 Montpellier cedex 5, France,
e-mail: bicalho@msem.univ-montp2.fr

B. Andreo et al. (Eds.), *Advances in Research in Karst Media*
DOI 10.1007/978-3-642-12486-0, © Springer 2010

parameters and geochemical data, such as major and trace elements, in order to asses the origin of the various water-types that contribute to the flow at the Lez spring.

Multivariate statistical techniques have several advantages such as reduction of data dimension quantities, consideration of different scales of chemical concentration with equal emphasis, and description of groups of variables with common chemical features. Jiang et al. [1] used Principal Component Analysis (PCA) to discriminate between natural and anthropogenic origins of the different factors that affect groundwater quality of the Nandong KS, China. Moore et al. [2] also used PCA for evaluating how multiple sources of water and variations in aquifer flow paths affect a spring draining the Eocene portion of the Upper Floridan aquifer.

## 2 Geological and Hydrogeological Settings

The Lez spring is one of the main karst springs in France and is located 15 km north of Montpellier (France). The Lez KS discharges also at several seasonal outlets: Lirou, Restinclières, Fleurettes and Gour Noir. The Lez spring is the main perennial outlet of a hydrogeological catchment area of around 380 km$^2$, located between the Hérault and Vidourle rivers. As a large part of the hydrogeological catchment is impermeable (Fig. 1), most of the recharge occurs in areas where limestones outcrop,

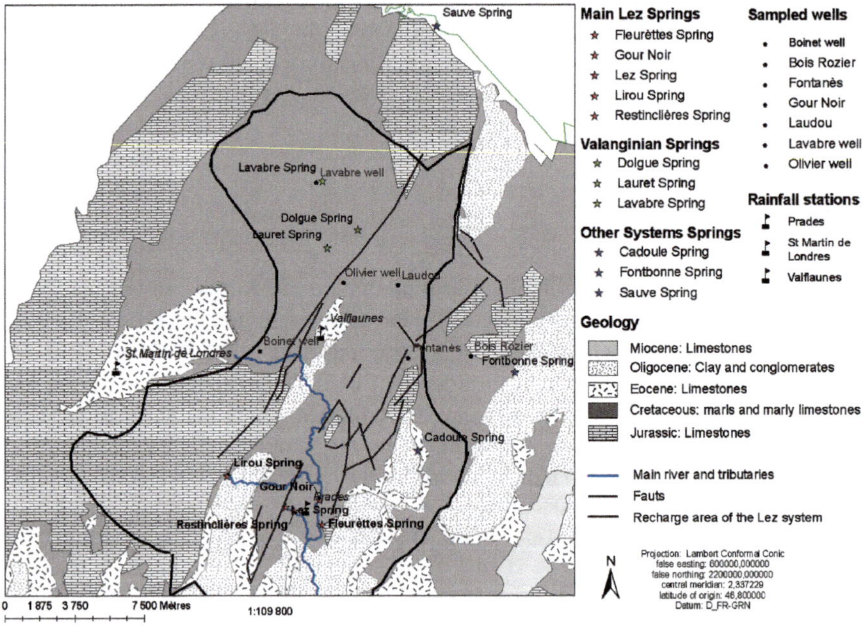


**Fig. 1** Hydrogeological map of the Lez karst system, showing the monitoring and sampling network: sampled springs, wells and rainfall stations location

which corresponds to a surface area of approximately 150 $km^2$. Lez KS works as a dammed system. The aquifer, composed by Jurassic rocks, is barred by Lower Cretaceous marls and marly-limestones.

# 3 Monitoring and Sampling

Temperature (T), Turbidity, Electrical Conductivity (EC) and groundwater (GW) level are recorded hourly in the observation well of Veolia pumping station. Hydro-Sciences Montpellier (HSM) performs T, EC, and GW level measurements hourly at the Lirou spring and at the Lez spillway. Three raingauges of Météo France (Valflaunès, Saint-Martin-de-Londres, and Prades) were considered.

Samplings were carried out twice a month and daily during floods, since March 2006. Temperature, pH and EC were measured in the field. Total alkalinity was measured within a day, by acid titration with HCl. Chemical and bacteriological analyses were performed at HSM water chemistry laboratory. Major elements ($Br^-$, $Cl^-$, $NO_3^-$, $SO_4^{2-}$, $Ca^{2+}$, $Mg^{2+}$, $Na^+$, and $K^+$) were analyzed by ion chromatography. Trace elements (Li, B, Al, V, Cr, Mn, Co, Ni, Cu, Zn, As, Rb, Sr, Mo, Cd, Ba, La, Pb and U) were analyzed with Q-ICPMS at AETE technical platform of Montpellier 2 University.

# 4 Results and Interpretation

## *4.1 Characterization of the Different Types of Groundwater at the Lez Spring*

For a preliminary verification of a multiple GW sources hypothesis, the frequency distribution of EC (CFD) of each water-year from 2005/2006 to 2007/2008 was achieved. CFD represents the sum of the populations of conductivities associated with water-types flowing through a system, assuming that a homogeneous water-type has EC normally distributed. Different modes of CFD in karst springs reflect the movement of chemically distinct water-types through the karst, and the average EC for an individual water-type depends of its origin and residence time [3].

Figure 2 presents the EC frequency distribution for 2007/2008 year with the corresponding Probability Density Function in the same plane, which was decomposed into its normally distributed component populations using a Peak-fitting software. The observed water-types are in reality a mixture of different end-members that flow together in the aquifer and emerge on the spring.

It was observed that rainier years exhibit more water-types. Apparently, stronger rainfall events provoke the activation of water flow from specific compartments, soliciting the circulation of rare water-types. Consequently, a smaller number of

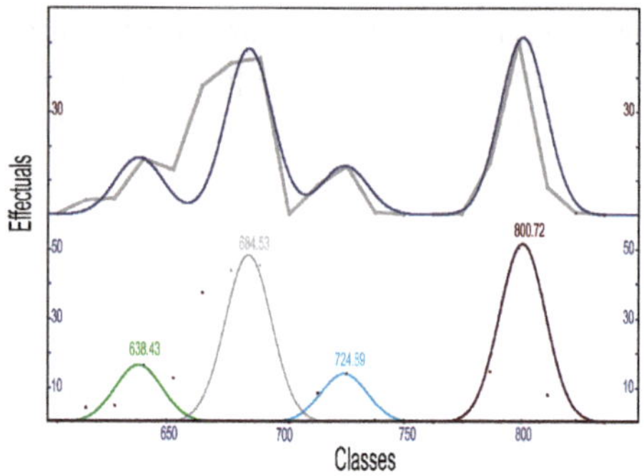


**Fig. 2** 2007/2008 water year. *Top*: EC frequency distribution with the corresponding Probability Density Function. *Bottom*: normally distributed component populations using Peak-fitting software

populations were observed for relatively dry years, like the years 2005 to 2007, with three water-types. During the 2007/2008 water-year, which was a rainy year, four water-types were identified (Fig. 2).

## *4.2 Discriminant Factorial Analysis – Lez Spring*

Once CFD showed the existence of distinct water-types composing Lez spring flowing waters, Discriminant Factorial Analysis (DFA) was applied on centred-reduced variables in order to decipher the Lez water-types chemical characteristics and obtain information about Lez KS water circulation.

DFA is a multivariate statistical technique that identifies individual groups of samples and tests the membership of each sample in a defined group, as a function of the values for a number of considered variables. It minimizes intra-group variance and maximizes intergroup variance.

The explicative variables used are EC, major ions ($Cl^-$, $Mg^{2+}$, $K^+$, $SO_4^{2-}$, $HCO_3^-$, $Na^+$ and $Ca^{2+}$), trace elements (Li, B, Rb, Ba and Sr), $pCO_2$ and calcite SI (Fig. 3).

This statistical procedure identified five water-types defined as (1) "*High waters*", discharging during the rainy season and associated with high GW level and large EC oscillations; (2) "*Low waters*", discharging during dry seasons and associated with low GW level and stable, high values of EC; (3) "*Dropping waters*", discharging during transitions between wet and dry seasons, associated with decreasing GW level and increasing EC. Waters with extremely high or low EC values were discriminated as (4) "*Flush-flow waters*" (EC $>$ 780 µS/cm) and (5) "*Dilution waters*" (EC $<$ 600 µS/cm). Both these types discharge at the beginning of the wet season, during the first floods of autumn, when highly mineralized "*Flush-flow*" waters are followed by low-mineralized "*Dilution waters.*".

The DFA shows a good discrimination between the five water-types, despite a partial superposition between "*High waters*" and "*Dropping waters*". Trace elements allow better characterisation of "*Flush-flow waters*", whereas $pCO_2$ and $HCO_3$ help to discriminate "*Low waters*".

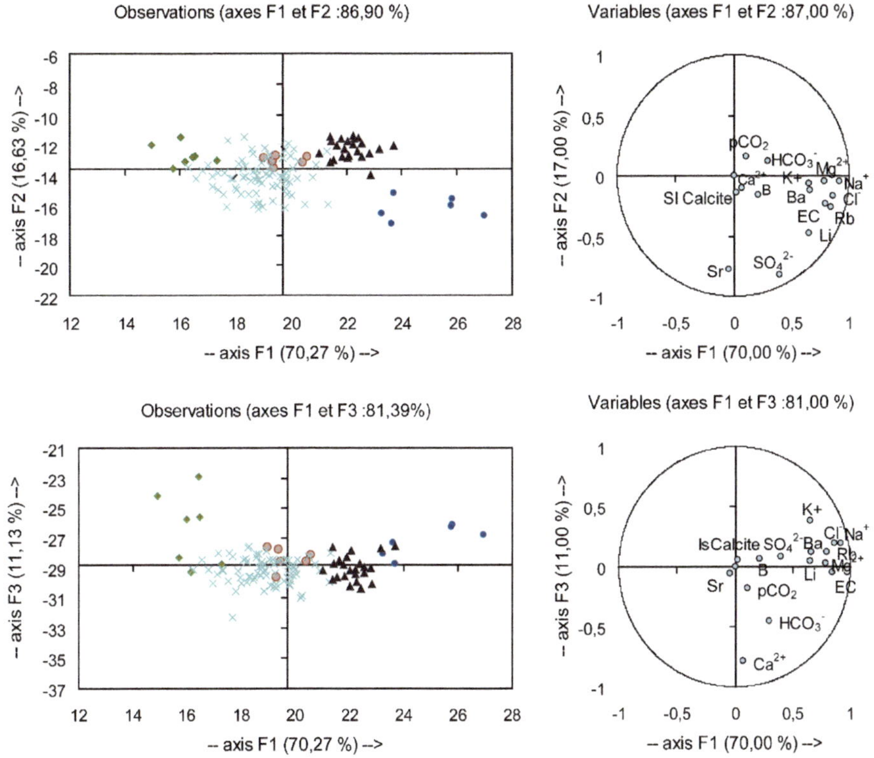


**Fig. 3** DFA of the Lez spring data using major and trace elements. *Left*: sample space; *right*: variable space

The F1 axis is positively defined by EC, Cl, Na, Mg, Ba, K, Rb and Li. The F2 axis is positively defined by $pCO_2$ and negatively by Sr and $SO_4$. F1 and F2 together represent 87% of the total variance.

The water types can be chemically characterised as follows:

- "Flush-flow waters" are highly mineralized and enriched in Cl, Mg, $SO_4$, Na, K, Li, Rb, Sr, Ba.
- "Dilution waters" are the less mineralized waters for all analysed elements.
- "High waters" are characterized by the typical presence of the same elements observed on "Flush-flow waters" but with smaller concentrations. The high data dispersion is agreed with the noticeable temporal variability observed for this water-type.
- "Low waters" are distinguished from "Flush-flow waters" by a less marked mineralization in "Cl-Na" and trace elements. They are characterized by stronger $pCO_2$.
- "Dropping waters" flow during the transition period between humid and dry season. Their mineralization properties are intermediate between "High waters" and

"Low waters", showing the evolution of the aquifer towards the dry season chemistry.

## 5 Conclusion and Perspectives

The distinct chemical characteristics, observed for each water type, suggest the probable occurrence of multiple origins and different lithologies influence on the GW mixing. Highly mineralized GW and low mineralized GW seem to be respectively related to deep GW rising and to superficial infiltration waters. The hydrological conditions induce different proportions of the end-members participation on GW mixing that emerge from the Lez spring.

Further investigations would better define the Lez spring GW origins, in order to better constrain mixing phenomenons related to the hydrodynamical functioning of the Lez karst system.

**Acknowledgements** This work was financed by CNPq, National Council of Technological and Scientific Development, of the Science and Technology Ministry of Brazil, HydroSciences Montpellier, IFR ILEE, Météo France, Montpellier city and Veolia.

## References

1. Jiang Y, Wu Y, Groves C, Yuan D, Kambesis P (2009) "Natural and anthropogenic factors affecting the groundwater quality in the Nandong karst underground system in Yunan, China." J. Contaminant Hydrol. 109, 49–61
2. Moore PJ, Martin JB, Screaton EJ (2009) "Geochemical and statistical evidence of recharge, mixing, and controls on spring discharge in an eugenic karst aquifer." J. Hydrol. 376, 443–455
3. Massei N, Mahler BJ, Bakalowicz M, Fournier M, Dupont JP (2007) "Quantitative Interpretation of Specif Conductance Frequency Distribution in karst". Ground Water. 45, 3, 288–293

# Hydrochemical Heterogeneity in the Discharge Zone of a Karstic Aquifer

M. Mudarra, B. Andreo, and J. Mudry

**Abstract** The hydrochemical response at springs in the drainage area of the Sierra del Rey – Los Tajos carbonate aquifer (province of Málaga, southern Spain) was monitored to determine the hydrogeological functioning of this aquifer. Analysis of the most important chemical parameters, using various methodologies, such as the time variation of electrical conductivity (EC), Principal Components Analysis (PCA) and Discriminant Factorial Analysis (DFA), revealed that the high level of hydrochemical heterogeneity existing in the zone of discharge, in addition to particular spatial and temporal factors, is responsible for the mineralization of the spring water. This is an example of how karstic heterogeneity may be transmitted to the discharge area of the aquifer, a fact that should be taken into account when investigating karst systems.

## 1 Introduction

Karstic aquifers present a high level of heterogeneity, from which hydrogeochemical changes in their spring water may be deduced. Such variations, or other natural responses, have long been used in karstic hydrogeology to determine the hydrogeological functioning of this type of aquifer. In many cases, studies have focused on analyzing the discharge from a single spring considered to be representative of the system as a whole. However, aquifers may exist that are not drained at a single point, by a single spring, but rather the process takes place at numerous points that

M. Mudarra, B. Andreo
Centre of Hydrogeology of University of Malaga, and Department of Geology, Faculty of Science, University of Malaga, Malaga 29071, Spain,
e-mail: mmudarra@uma.es; andreo@uma.es

J. Mudry
Laboratoire Chrono-environnement CNRS: UMR6249 – Univ. Franche-Comté, France

B. Andreo et al. (Eds.), *Advances in Research in Karst Media*
DOI 10.1007/978-3-642-12486-0, © Springer 2010

define a discharge area; it may then be difficult to define the representativity of the response measured at any given point.

Accordingly, the use of hydrochemical parameters (for example, electrical conductivity or the $Mg^{2+}$ content) and natural hydrogeochemical tracers (TOC, $NO_3^-$, natural fluorescence) may help identify the hydrogeochemical processes responsible for the distribution and hierarchization of the water flow from discharge areas: infiltration, flow velocity, degree of participation in the saturated and unsaturated zones in the functioning of the system, etc. All these processes are related to the spatial and temporal position of the springs and influence the variations in chemical composition reflected in the spring water. These factors should be borne in mind in order to correctly monitor and sample these systems.

The aim of the present study, therefore, is to illustrate the heterogeneity of hydrochemical response in the drainage zone of the carbonate aquifer of Sierra del Rey – Los Tajos, located in the province of Malaga, southern Spain (Fig. 1). To do so, a chemical characterization was performed of the water from all the springs in the discharge area. This aquifer forms part of a groundwater body known as Sierra de Enmedio – Los Tajos (20 km. The relief is rugged, with altitudes ranging from 600–1250 m above sea level. The climate in the zone is mild Mediterranean, with mean annual temperature and rainfall of 18 °C and 850 mm, respectively. Prior studies have described the general hydrogeological characteristics of this aquifer. The present study is intended to achieve an analysis of the spatial and temporal factors that affect the hydrogeochemical variability of discharge points and to establish the relation between these factors and the outflow from the saturated and unsaturated zones of the aquifer.

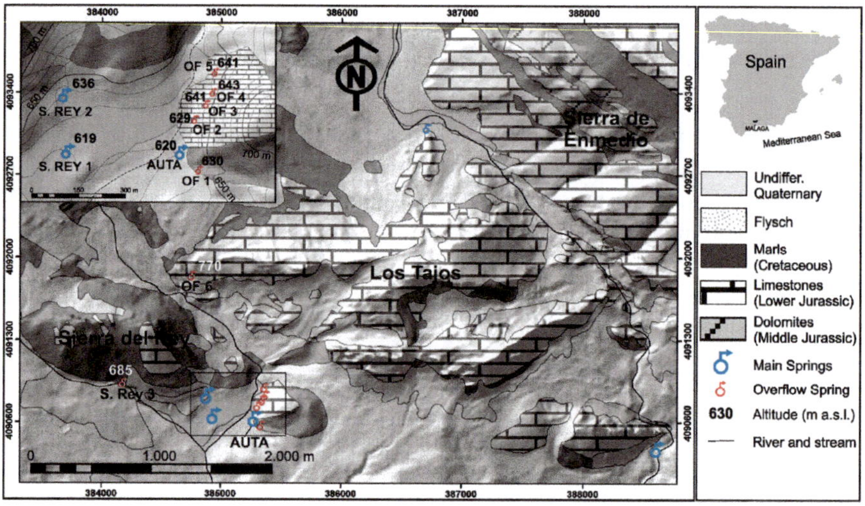


**Fig. 1** Geographic location and geological-hydrogeological sketch of Sierra del Rey – Los Tajos aquifer, showing, in detail, the situation of the Auta discharge zone

## 2 Methodology

Between November 2008 and March 2009, thermal and hydrochemical parameters were monitored at 10 different points of discharge: Auta, S. Rey 1 and S. Rey 2 and 7 overflow springs (OF 1, OF 2, OF 3, OF 4, OF 5, OF 6 and S. Rey 3), normally weekly, but occasionally daily, during high water periods. The number of samples taken (149) was not the same at every point. At the permanent springs, the monitoring was fairly continuous, while samples were taken at the overflow springs when they were active. The major components were analyzed at the Hydrogeology Laboratory of the University of Malaga, the cations – anions by ionic chromatography, and Total Organic Carbon (TOC) using a carbon analyzer. Also partial pressure of $CO_2$ ($pCO_2$) and saturation index of calcite (SIc) were determined. These parameters inform on the hydrogeological functioning of karst system (Bakalowicz 1979).

The hydrochemical data were processed by various approaches. Analysis of the time variations of the chemical components of the water for an entire hydrologic cycle or for a given high-water period provided information on the lag between rainfall (input signal) and hydrochemical response (output signal). The magnitude of the dilutions recorded during the time variations enabled deduction of the degree of functional karstification at each sector of the aquifer (Bakalowicz 1979; Mudry 1987). Multivariate statistical methods, such as Principal Components Analysis (PCA) and Discriminant Factorial Analysis (DFA), were used for the quantitative interpretation, to group variables and water types, and to identify the hydrogeological processes that may be associated with given hydrodynamic conditions or a common origin (Bakalowicz 1979).

## 3 Results and Discussion

The Auta spring is located at 620 m a.s.l. at it is permanent, with discharge values ranging between 20 and 100 L/s. Auta spring water displays the highest and less variable values of EC (from 625 μS/cm to 590 μS/cm) within a general downward trend throughout the study period. These data suggest that Auta spring is predominantly connected with the drainage from the saturated zone of the aquifer, particularly during low water conditions. However, overflow springs are active only during period of high water conditions, their altitude ranges between 629 and 643 m a.s.l., and the total flow rate of these springs is several hundred of L/s (up to 1000 L/s). Each important rainfall event is followed by a considerable fall in the EC of the overflow spring waters and also at S. Rey 1, from values of around 500 μS/cm to 325 μS/cm. Water drained by overflow springs shows the low values and the greater variations of EC. The overflow springs water comes mainly from rapid infiltration of rainwater trough unsaturated zone in high water conditions.

PCA provides a quantitative view of the main factors responsible for total variance (Figs. 3a and b). The two main axes of the PCA account for 71% of the sampling variance. Axis I (59%) is determined by EC, temperature (T), $Ca^{+2}$, $Mg^{+2}$, $Na^{+}$, $K^{+}$, TAC, $SO_4^{-2}$, $Cl^{-}$, $F^{-}$, rMg/Ca and $pCO_2$ (Fig. 3a), these variables being considered tracers of water mineralization or indicators of long residence time within the aquifer. In contrast, axis II (13%) is mainly represented, in its positive section, by the $NO_3^-$ ion, TOC and the natural fluorescence of the organic compounds dissolved in the water, which are typical tracers of water leaching a soil during a rainfall and consequently tracers of infiltration (Batiot et al. 2003).

In factorial plane I–II of the statistical units space (Fig. 3B), five water groups may be discerned. Groups 1 (G1) and 4 (G4) present a lower degree of spatial dispersion than do Groups 2 (G2), 3 (G3) and 5 (G5). The first two groups are comprised of samples with high levels of mineralization, especially G1, with samples from the Auta spring. G2 is characterized, moreover, by high levels of $NO_3^-$ corresponding to S. Rey 2 spring. The water analyzed in G3 has moderate levels of mineralization and corresponds to low-water conditions at the Auta overflow springs, and to low water conditions at the S. Rey 1 spring. This water is mineralized following the conclusion of the recharge event. G4 and G5 constitute water samples taken under high-water or very high-water conditions, with different soil tracers content.

The samples in G5 were taken after the important pluviometric event of November 2008, when the organic and $NO_3^-$ tracers obtained from the soil recorded con-

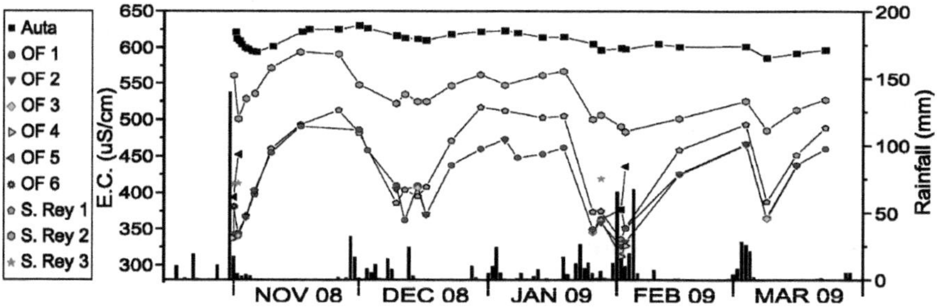


**Fig. 2** Time variations of electrical conductivity of the water from the springs, with respect to precipitation events

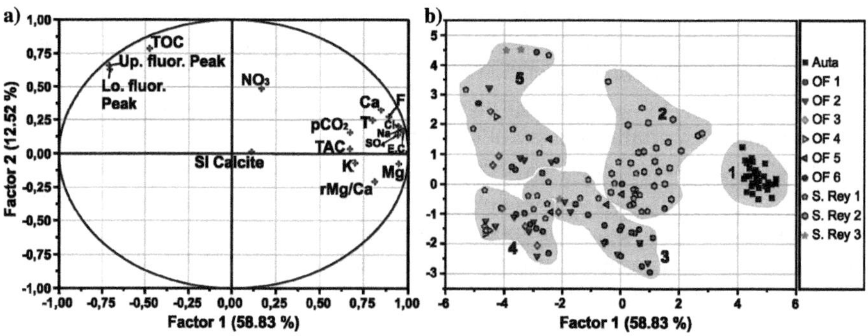


**Fig. 3** Principal Components Analysis with hydrochemical data. Plot of the variables (**a**) and of the statistical units (**b**)

siderable increases (Fig. 3). On the contrary, the samples of G4 correspond to the other autumn and winter high-water events, during which no important variations in these chemical parameters were observed. This was mainly due to the washing effect of the soil and the epikarst by the first heavy rainfall of the autumn. For this reason this group of samples can be distinguished from G3.

A Discriminant Factorial Analysis was also done to test how the hydrochemical variability and heterogeneity of the water samples was related to either of the two classification hypotheses made in this study: classification by drainage point or system (Fig. 4a), or by hydroclimatic conditions (Fig. 4b). In the first hypothesis, the system classification, the water samples were assigned to nine different groups, one per spring, except the overflow spring OF 6 which was a single sample and could not be assigned to a group; instead, it was classified as an additional observation (Fig. 4a). The position of the gravity centres is also represented. In accordance with this criterion, it was observed that 91.2% of the individual samples were well classified. The chosen qualitative variable is discriminant and it can explain the deciphering in several groups. Most of those wrongly classified *a priori* belonged to the OF 1 and OF 2 springs, and were subsequently reclassified as groups OF 2 and OF 1, respectively. Therefore, it may be assumed that the two groups represent the same type of water. The additional observation OF 6 was attributed to the S. Rey 3

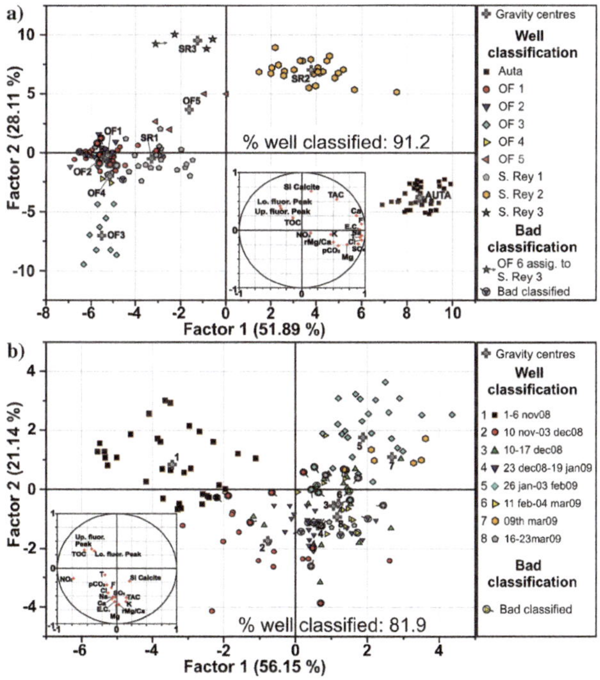


**Fig. 4** Discriminant Factorial Analysis with hydrochemical data. Plot of classification by systems (**a**) and by hydroclimatic conditions (**b**)

group, due to the chemical similarity between the water from OF 6 and that from S. Rey 3.

Under the second hypothesis, the samples were classified *a priori* into 8 different groups (Fig. 4b). Each group has its centre of gravity and it represents a different climatic or hydrodynamic situation occurring between 1 November 2008 and 23 March 2009 (high and low water levels). Note that, according to this criterion, 81.9% of the samples were correctly classified. Those that were not mainly corresponded to groups representing low water level conditions (10 November–3 December, 23 December–19 January, 11 February–4 March and 16–23 March).

Therefore hypothesis of classification by system provides greater discriminant power with the available samples. The chemical characteristics of the spring water depend, to a large extent, on the particular features of the aquifer (its lithology, functional karstification, hydrogeological functioning, etc.) rather than on the hydroclimatic conditions prevailing at any given moment.

## 4 Conclusions

The analysis and statistical treatment of the hydrogeochemical response of the water drained by the springs in the Auta discharge area (southern Spain) reveal the high heterogeneity in the Los Tajos – Sierra del Rey karst system. Functional karstification seems especially developed in the unsaturated zone, which favours rapid infiltration and draining via the overflow springs, leading in turn to a rapid decrease in the electrical conductivity of the water. On the contrary, the Auta spring, with a low degree of functional karstification, is more closely related to drainage of the saturated zone, which influences its functioning and gives rise to slow, lagged responses to rainfall events.

This hydrochemical complexity is determined more by the geological and karstic heterogeneity of the aquifer than by the hydroclimatic circumstances recorded at any given moment. All these aspects should be taken into account in future hydrogeological investigations on this type of aquifers.

**Acknowledgements** This work is a contribution to the projects CGL2005-05427 and CGL2008-06158 BTE of DGICYT, P06-RNM 2161 of Junta de Andalucía and IGCP 513 of UNESCO and to the Research Group RNM-308 of Junta de Andalucía.

## References

Bakalowicz M (1979) Contribution de la géochimie des eaux à la connaissance de l'aquifère karstique et de la karstification. Thèse Doct. Sci. Nat., Univ. P. et M. Curie, Paris-VI, Géol. Dyn.; 269 pp

Goldscheider N, Drew D (2007) Methods in Karst Hydrogeology. Taylor and Francis, London

Mudry J (1991) L'analyse discriminante, un puissant moyen de validation des hypotheses hydrogéologiques. Revue des Sciences de l'Eau 4:19–37

# Seasonal Variations of Hydrochemistry and Carbonate Precipitation Rate in a Travertine-Depositing Canal at Baishuitai, Yunnan, SW China: Implications for the Formation of Biannual Laminae in Travertine and Climatic Reconstruction

H. Sun and Z. Liu

## 1 Introduction

Travertine or tufa is a chemical sedimentary deposit that is precipitated from spring-fed stream water supersaturated with calcite due to CO2-degassing (Liu et al. 1995; Pentecost 1995; Ford and Pedley 1996). These deposits have biannual laminations which could provide valuable paleoclimatic and paleoenvironmental information with high resolution (Liu et al. 2006; Kano et al. 2007; Brasier et al. 2009; Lojen et al. 2009).

However, the relationship between hydrochemistry and carbonate precipitation rate must be determined before tufas or travertine can be confidently used as paleoclimatic tools. The purpose of this study is to examine seasonal variations of hydrochemistry, carbonate precipitation rates and formation of biannual lamination in a travertine-depositing canal at Baishuitai, Yunnan, SW China, and to understand the processes and seasonal pattern of biannual lamination in travertine and the climatic implications of these laminae of the endogenic travertine (Liu et al. 2003, 2006).

## 2 General Settings of the Study Area

Baishuitai is located in $\sim$ 100 km south of the Shangri-La Town, Yunnan Province, SW China. The elevation ranges from 2380 m to 3800 m ASL. The area is charac-

H. Sun, Z. Liu
State Key Laboratory of Environmental Geochemistry, Institute of Geochemistry, Chinese Academy of Sciences, 46 Guanshui road, Guiyang 550002, China,
e-mail: hailong-sun@hotmail.com

B. Andreo et al. (Eds.), *Advances in Research in Karst Media*
DOI 10.1007/978-3-642-12486-0, 

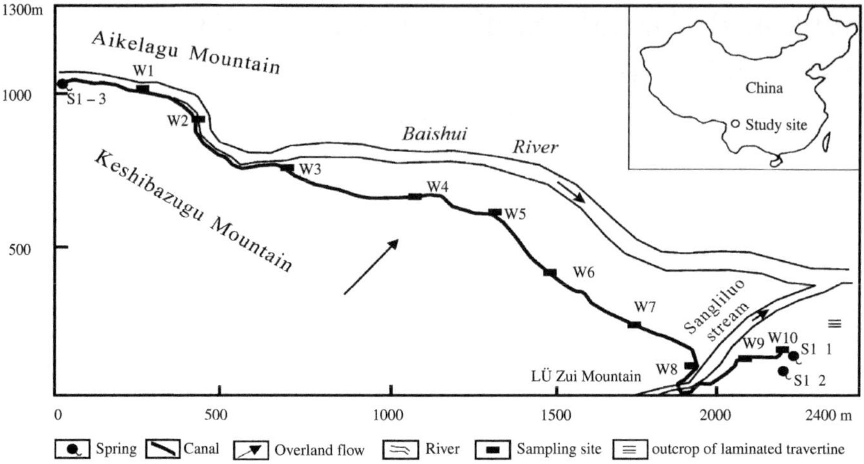


**Fig. 1** Distribution of the sampling sites in the travertine-depositing canal of Baishuitai, Yunnan, SW China

terized by a subtropical monsoon climate, with >75% of the annual precipitation (~750 mm) occurring during the rainy season from May to October and annual mean air temperature of 8 °C (Liu et al. 2003).

The endogenic travertine for this study is deposited in a canal which was supplied mainly by the spring S1-3 (Fig. 1). The canal headwaters were changed after a heavy flood on July 7, 2006. They were supplied by both the spring S1-3 and the surface river-Baishui River (with lower concentration of $Ca^{2+}$ and $HCO_3^-$) before July 7, 2006. However, on July 7, 2006, the connecting path between the canal and Baishui River was destroyed by the flood, and from this time on the canal was only supplied by the spring S1-3, which had high values of specific conductivity, [$Ca^{2+}$], [$HCO_3^-$] and $pCO_2$ (Liu et al. 2003).

## 3 Methods

In order to obtain modern travertine precipitation rates, the plexiglass substrates (5 cm × 5 cm × 0.5 cm) for depositing travertine were placed in the flowing water. The substrates were substituted semimonthly and the travertine deposited on the substrates were collected and weighted to obtain the carbonate precipitation rates.

Water temperature, pH and specific conductivity of water at each sampling site were measured in-situ daily with hand-held water quality data logger (WTW 350i), with resolutions of 0.01 pH, 0.1 °C and 1 μs/cm, respectively. In situ titrating was used to measure the [$Ca^{2+}$] and [$HCO_3^-$] of water semimonthly with Aquamerck Alkalinity Test and Hardness Test.

# 4 Results and Explanations

## *4.1 Seasonal Variations of Canal Water Chemistry*

Figure 2 shows the hydrochemical variations at sampling sites W4 and W7 from April 25, 2006 to April 20, 2007. All the chemical properties of both W4 and W7 show seasonal changes. [$HCO_3^-$], [$Ca^{2+}$], specific conductivity (spc) and $pCO_2$ de-

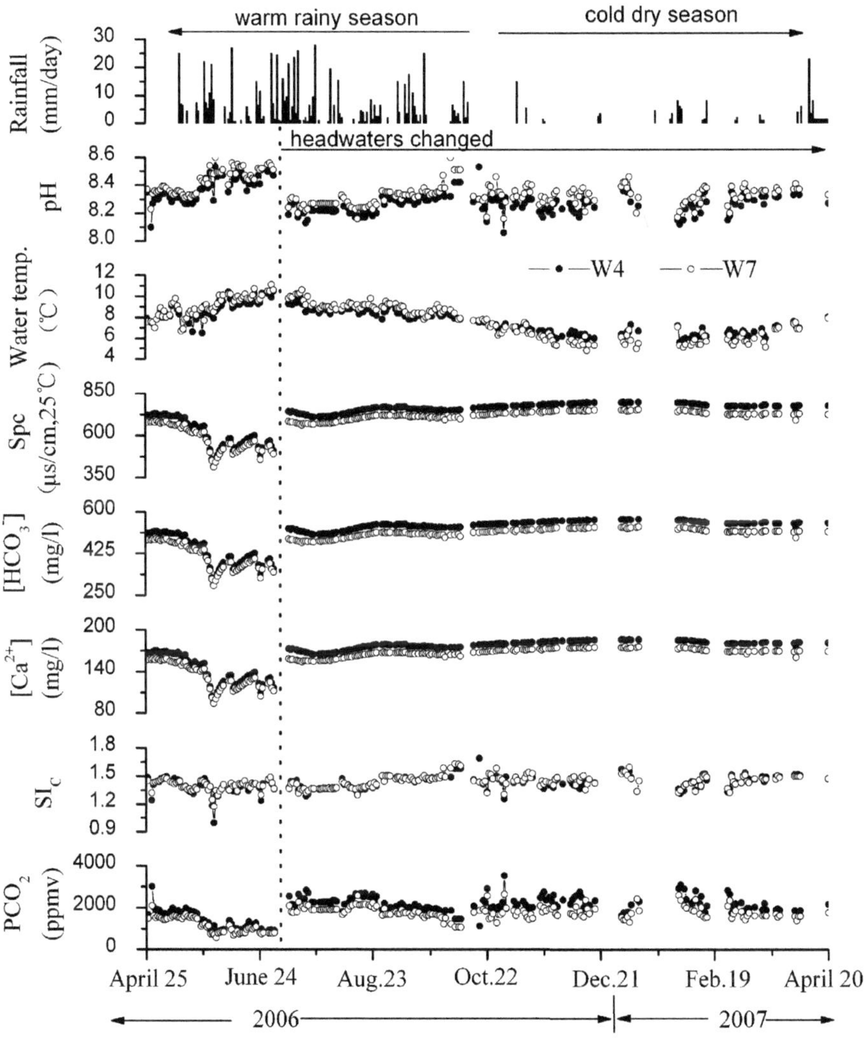


**Fig. 2** Hydrochemical variations at sample sites W4 and W7 along the Baishuitai travertine-depositing canal during the hydrological year

creased gradually in the rainy period from May 13, 2006 to May 31, 2006 (Fig. 3). The period belonged to beginning of the warm rainy season at Baishuitai and the rainfall amount increased gradually, so the synchronous decrease in spc, $[HCO_3^-]$, $[Ca^{2+}]$ and $pCO_2$ were mainly caused by the dilution effect of overland flow in the rainy season.

Figure 2 shows that the canal water temperature also had very clear seasonal variations. Temperatures were high in the warm rainy season and low in the cold dry season. However, the lower values of $pCO_2$, spc and concentrations of $Ca^{2+}$ and $HCO_3^-$ in the warm rainy season were not caused by the temperature-induced carbonate precipitation, but reflected the effect of dilution by the Baishui River water and the overland flow in the warm rainy season, because the carbonate precipitation rate was low in the warm rainy season due to the dilution effect as showed later.

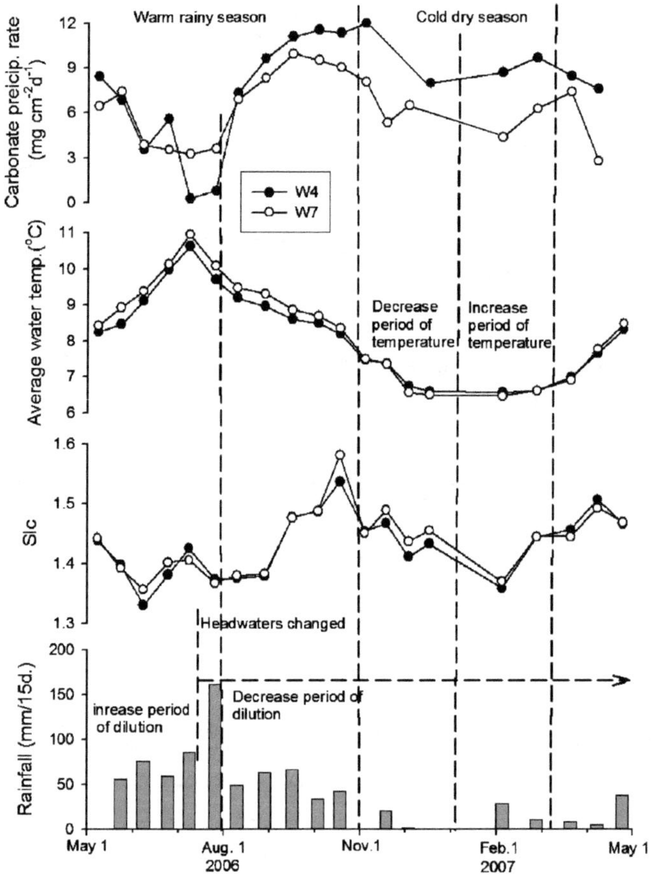


**Fig. 3** The seasonal variation in carbonate precipitation rates on the plexiglass substrates at sampling sites W4 and W7 of Baishuitai and its relation with rainfall and water temperature

### *4.2 Seasonal Variations in the Carbonate Precipitation Rates and Formation of the Biannual Laminae in Travertine*

Figure 3 shows the carbonate precipitation rates on the plexiglass substrates, the coeval rainfall amount and the coeval average water temperature at the two sampling sites against the time series. It is seen that four time periods can be defined: (1) the increase period of dilution and (2) the decrease period of dilution in warm rainy season, and (3) the decrease period of temperature and (4) the increase period of temperature in cold dry season.

In the warm rainy season, carbonate precipitation rates at the two sites all showed control of rainfall, i.e., decreased in the increase period of dilution, and increased in the decrease period of dilution (Fig. 3). The decrease of the carbonate precipitation rates was related to the decrease of [$HCO_3^-$] and [$Ca^{2+}$] in canal water caused by the dilution effect of the overland flow after rainfall as discussed in Sect. 4.1 (Fig. 2).

However, in cold dry season, carbonate precipitation rates at the two sites all showed control of temperature, i.e., decreased in the decrease period of temperature, and increased in the increase period of temperature (Fig. 3).

In order to understand the seasonal timing of the layering in the endogenic travertine at Baishuitai, plexiglass substrates were placed in the flowing waters at the sampling site W9 for one year from April 23, 2006 to April 23, 2007. Figure 4 shows the travertine collected at the sampling site W9. It is seen that the travertine exhibits one annual lamina (couplet) consists of a thin dark porous layer and a thick light dense layer.

The travertine grew from April 23, 2006, and stopped on April 23, 2007. Therefore, the thin dark porous layer (6mm thick) should form in the warm rainy season when the carbonate precipitation rate was decreased by the dilution effect of over-

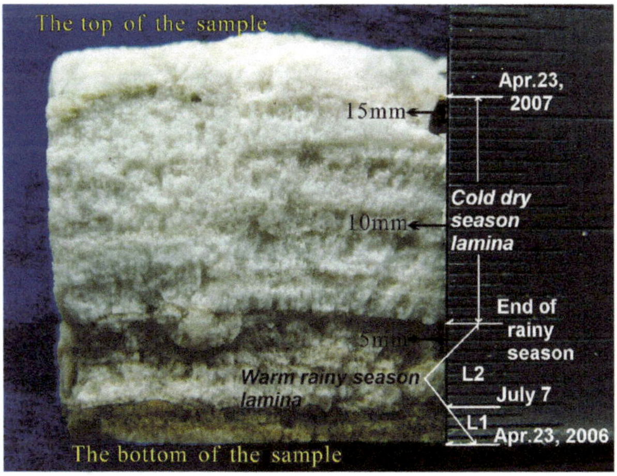


**Fig. 4** The modern biannual laminated travertine deposited on the plexiglass substrate at the sampling site W9 from April 23, 2006 to April 23, 2007 in the Baishuitai canal

land flow as stated before, and the thick light dense layer (10 mm thick) formed in the cold dry season when the carbonate precipitation rate is higher because of the lack of dilution effect. The dark-color of the warm-rainy-reason lamina was possibly related to the soil-derived clay and organic matter introduced by the overland flow in the warm rainy season. On the other hand, there was no or little contamination of the clay and organic matter in dry season. Therefore, the cold-dry-season lamina is relatively pure and appears lighter (white) color.

## 5 Conclusions

By examining the seasonal variations in hydrochemistry and carbonate precipitation rates in a travertine-depositing canal at Baishuitai, Yunnan, SW China, it was found that the hydrochemistry and carbonate precipitation rates in the travertine-depositing canal show clear seasonal patterns, which were similarly low in warm rainy seasons and high in cold dry seasons. The lower hydrochemical compositions and thus lower carbonate precipitation rates in warm and rainy seasons are mainly related to the dilution effect of the overland flow after rainfall, showing the control of rainfall on the hydrochemistry and thus the carbonate precipitation rate.

It was also found that the endogenic travertine had much clearer biannual lamination structure. The biannual lamination in the travertine was primarily controlled by changes in the dilution-dominated rate of carbonate precipitation, and its darker brown color of laminae was possibly related to the soil-derived clay and organic matter introduced by the overland flow in the warm rainy season. This study demonstrates that carbonate precipitation rates and the formation of lamination in the travertine at Baishuitai were mainly controlled by climate (e.g., rainfall).

## References

Brasier AT, Andrews, JE, Marca-Bell AD et al (2009) Depositional continuity of seasonally laminated tufas: Implications for $\delta^{18}O$ based palaeotemperatures. Global and Planetary Change (in press)

Ford TD, Pedley HM (1996) A review of tufa and travertine deposits of the world. Earth-Science Reviews 41, 117–175

Kano A, Hagiwara R, Kawai T et al (2007) Climatic conditions and hydrological change recorded in a high-resolution stable-isotope profile of a recent laminated tufa on a subtropical island, southern Japan. Journal of Sedimentary Research 77, 59–67

Liu Z, Svensson U, Dreybrodt W et al (1995) Hydrodynamic control of inorganic calcite precipitation in Huanglong Ravine, China: Field measurements and theoretical prediction of deposition rates. Geochimica et Cosmochimica Acta 59, 3087–3097

Liu Z, Zhang M, Li Q et al (2003) Hydrochemical and isotope characteristic of spring water and travertine in the Baishuitai area (SW China) and their meaning for paleoenvirnmental reconstruction. Environmental Geology 44, 698–704

Liu Z, Li H, You C et al (2006). Thickness and stable isotopic characteristics of modern seasonal climate-controlled sub-annual travertine laminas in a travertine-depositing stream at Baishuitai, SW China: implications for paleoclimate reconstruction. Environmental Geology 51, 257–265

Lojen S, Trkov A, Ščančar J et al (2009) Continuous 60-year stable isotopic and earth-alkali element records in a modern laminated tufa (Jaruga, river Krka, Croatia): Implications for climate reconstruction. Chemical Geology, 258, 242–250

Pentecost A (1995) The Quaternary travertine deposits of Europe and Asia Minor. Quaternary Science Reviews 14, 1005–1028

# Interpreting Source of Lingshui Spring by Hydrogeological, Chemical and Isotopic Methods

G. Jiang and F. Guo

**Abstract** Lingshui spring produced excellent drinking water. However there was no protection area for the spring. The goal of the study was to determine source of the spring and give suggestions for protection. Limestone with total thickness of more than 1 km composes the karst aquifer. There was siliceous rock as aquiclude in upper Permian. Lingshui spring was in a syncline. The syncline was covered by mudstone at the center, making the karst aquifer partially confined. Rainfall was the source of surface rivers and groundwater. Groundwater usually recharges to rivers. There were no obvious watersheds between Lingshui and other springs. It is important to know the relation among springs. The hydrochemical results showed that springs in front of mountains were recharged by allogenic rivers. Anthropogenic related contaminants such as $K^+$, $Na^+$, $Cl^-$, $SO_4^{2-}$ and $NO_3^-$ were higher in shallow aquifer than deep aquifer. This helped to determine that Lingshui spring came from the deep aquifer and had no allogenic source. The hydrogen and oxygen isotopes of springs, wells and rivers were all near the local meteoric water line. Karst springs including Linshui had similar concentrations of D and $^{18}O$ and exhibited much less isotopic variability than others. They came from the same karst aquifer.

## 1 Introduction

Lingshui spring was the drinking water supply of Wuming city. However, there was no protection plan for it. What had been done for protection of water resources were that (1) drilling for pumping was forbidden in urban area, (2) water samples were analyzed weekly for ions and bacteria, (3) discharge was measured by small frequency. These decisions were not enough for such large and fragile karst spring.

G. Jiang, F. Guo
Key Laboratory of Karst Dynamics, Institute of Karst Geology, MLR, Guilin 541004, People's Republic of China, e-mail: bmnxz@126.com

B. Andreo et al. (Eds.), *Advances in Research in Karst Media* 
DOI 10.1007/978-3-642-12486-0, © Springer 2010

Discharge of the spring in dry season was 2400 l/s in 2009 (the smallest number among several results), which was less than discharge at dry season in 1970s. Nitrate values of the spring were 5.26 mg/l in February 2009 which was larger than concentration in the same month in 1977. More measurements were necessary before any conclusion was made about changes of water quantity and quality.

**Geology.** Deposition in the basin from Paleozoic to Mesozoic mostly was carbonate, and it became mudstone and sandstone from middle Triassic (Fig. 1). Hiatus between Paleozoic and Mesozoic as well as Triassic and Cretaceous may be important for hydrogeology because of fossil karst. Thin siliceous and coal deposition in fossil karstified surface was not good aquiclude. So the carbonate was considered as one karst aquifer. Younger depositions (middle Triassic and Cretaceous) as aquicludes were not the divider of karst groundwater because they were upgradient of the karst aquifer (report of hydrogeological survey, 1979, unpublished).

Lingshui spring was located in north plunge of the Najia syncline. This position indicated that the syncline probably controlled groundwater flow. Surface rivers flowed along the syncline, where it was confined by the overlying aquiclude.

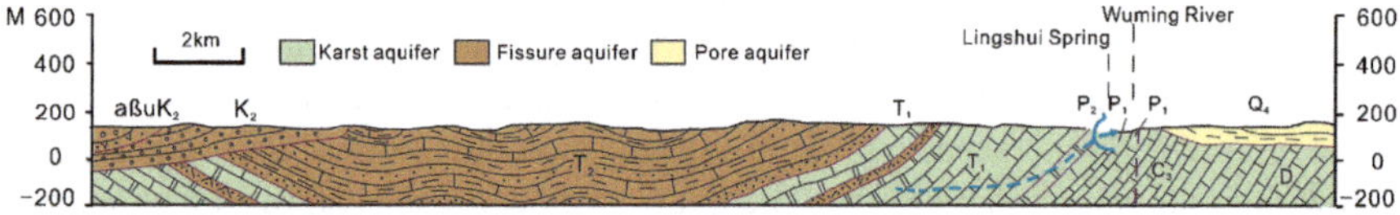


**Fig. 1** Hydrogeological section of Najia syncline with Lingshui spring. *Blue dash line* and *arrow* were conduit and Lingshui spring

**Topography.** Wuming fluvial karst basin where Lingshui spring was located was charactered by peak forest. The basin was covered by sandstone of Mesozoic, red soil and fluvial deposition. An anticline made of old stratum in the north of the basin formed mountains with highest altitude of 1700 m, which was source of rivers and the divider of groundwater. Elevation of hydrologic base level in the basin was nearly 100 m (Fig. 2).

## 2 Methods

Geological survey was done in a wide range. Discharge, pH, temperature specific electric conductivity, calcium and bicarbonate were measured in the field. Samples were taken from rivers, reservoirs, springs and wells. In laboratory, K, Na, Mg, Cl, $SO_4$, $NO_3$, F, $PO_4$, $\delta^2H$, $\delta^{18}O$ were measured. Greenspan CTDP 300 was equipped at the Lingshui spring for monitoring of discharge, temperature, pH and special conductivity.

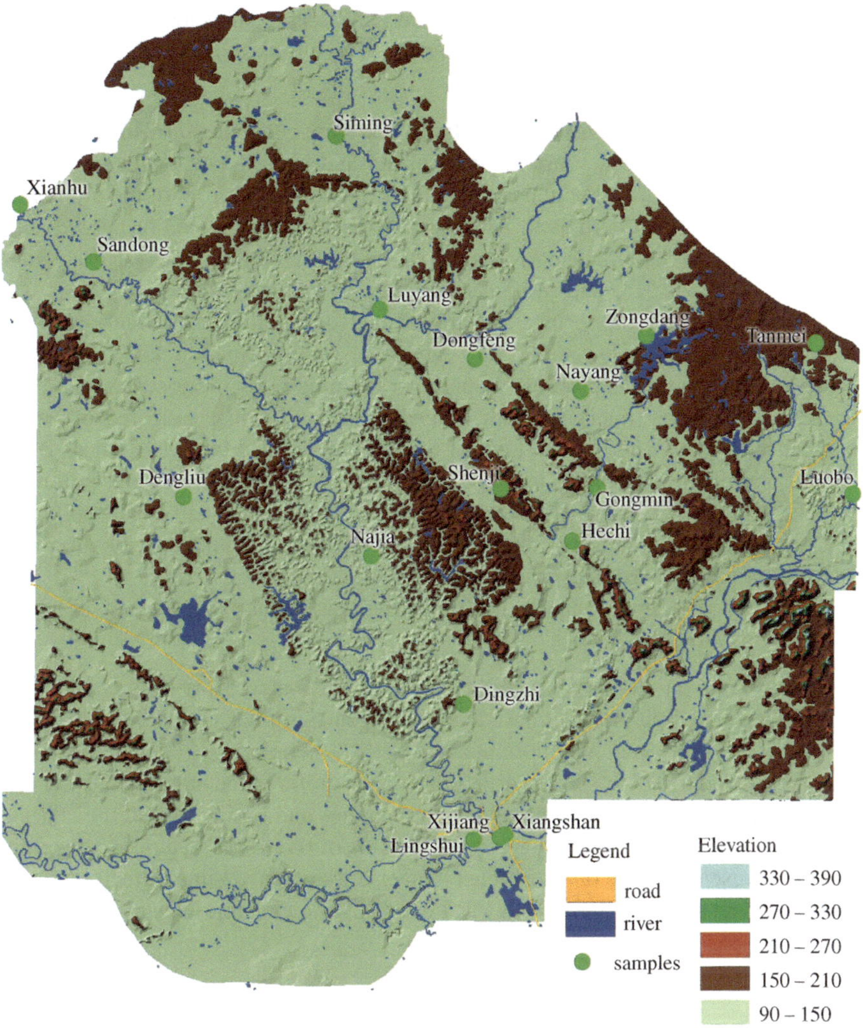


**Fig. 2** Morphological map of the study area. Sample sites are marked, except Qingjiang and Qin springs situated out of the map

## 3 Results

The karst aquifer recharged rivers in various concentrations and diffusion manners in the basin. Lingshui spring was the biggest underground tributary of the rivers. Karst aquifer is recharged by rainfall. There are few sinkholes in the karst area, so concentrated recharge was not important for the aquifer.

Springs including Lingshui in the basin were coming from the same aquifer, but they may represent different karst water systems. The water systems belonged to botj shallow and deep groundwater; however, they were connected probably. Lingshui spring came from deep karst aquifer for it was in axes of the syncline (Goldscheider 2007).

**Hydrochemistry.** Springs near mountains area were charactered by lower specific electric conductivity, calcium plus magnesium and bicarbonate. A fluvial deposition with sand and gravel which covered the karst aquifer in front of the mountains had high infiltration coefficient. Allogenic rivers from mountains recharged karst aquifer easily through the overlay depositions. Mixture of allogenic water and karst groundwater made SEC and carbonate rock related parameters lower.

It was known that the springs probably came from shallow karst water or a deep karst aquifer by $K^+$, $Na^+$, $Cl^-$, $SO_4^{2-}$ and $NO_3^-$ values. Shallow karst water was more fragile than the deep water. Cropland in Wuming basin increased dramatically in the past decades (interview with official). Large hillslopes were reclaimed for sugarcane, cassava and longan. Less of the forest remained in part of peaks. Agriculture and machining plants as well as domestic pollution led to water pollution. $K^+$, $Na^+$, $Cl^-$, $SO_4^{2-}$ and $NO_3^-$ were related to fertilization and living pollution. They entered karst aquifer by rainfall recharge. These contaminates were more rich in wells, which showed the sensitivity of shallow karst water.

It could be determined that Lingshui spring was neither impacted by allogenic water nor shallow karst water.

**Isotopes.** Concentrations of D and $^{18}O$ in meteoric water in the basin varied over time. Winter precipitation had less D and $^{18}O$ than summer precipitation (Fig. 3). Rivers in mountain area had more D and $^{18}O$ than their down streams in the basin. Rivers received tributaries from surface and karst aquifer in the basin with fewer isotopes. The mixture of end numbers from mountains and basin reduced the concentrations. The hydrogen and oxygen isotope ratios of rivers, reservoirs, karst springs and wells resembled the ratios of local meteoric waters in summer. Precipitation in winter as well as in dry season was invalid for flow water. Lingshui spring and other

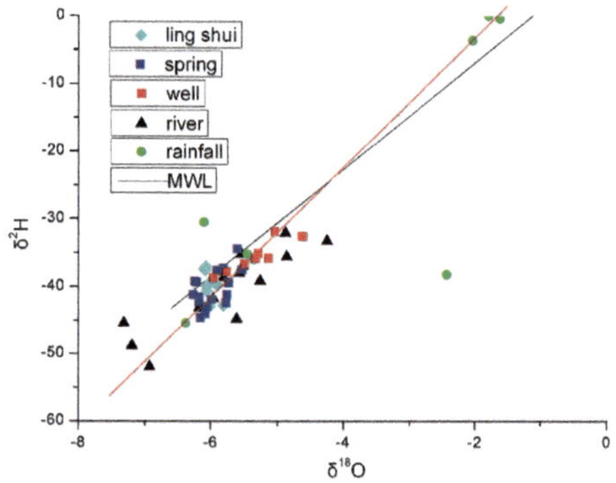


**Fig. 3** Isotopes of hydrogen and oxygen (*vs* SMOW). The *black line* was the Chinese meteoric water line and the *red* was local one

springs similarly discharge near the local meteoric water line and exhibited much less isotopic variability than others. They came from the same karst aquifer (Clark and Fritz 1997).

## 4 Discussion and Conclusion

Lingshui and other springs nearby shared the same karst aquifer according to results of geological, geomorphological, hydrochemical and isotopic methods. No phenomenon or evidence showed that there were boundaries between them. However, rivers act as discharge boundaries which divide the karst aquifer into several sub-catchments, because all springs discharged to nearby rivers. Only Lingshui spring crossed a river boundary by passing under the river bed.

Several boreholes met conduits in the recharge area of Lingshui spring. But the conduits were in the shallow aquifer. The deep aquifer was absolutely unknown. A new packinghouse was built in the catchment, and a new landfill was being planed. Local officials did not know how to protect the karst aquifer.

There were buried karst and covered karst in the drainage area of Lingshui spring. Rainfall recharged to the karst aquifer by diffusion manner. The karst aquifer recharged to the rivers, while there was small possibility for the rivers to recharge the karst aquifer in a significant wayy. Lingshui spring showed stable discharge and clean quality with little impact of rainfall. People thought that pollution impact would be difficult for so large spring. In fact, the uppermost soil laye with a thickness of about three meters had limited protection. The covered area of karst aquifer was fragile, but somewhat protected from pollution. Two suggestions were given. Firstly, a protection zone should be set up primarily according to river boundary, and classing the zone by geological conditions. Secondly, Lingshui spring, as well as other springs, should be monitored at monthly intervals.

## References

Clark Ian D, Fritz Peter (1997) Environmental Isotopes in Hydrogeology. Baker & Taylor Books

Hydrogeological group in Guangxi (1979) Hydrogeological survey report in Nanning (1:200,000). Unpublished

Hydrogeological group in Guangxi (1979) Hydrogeological survey report in Shanglin (1:200,000). Unpublished

Hydrogeological group in Guangxi (1979) Method in karst hydrogeology for water supply. Geological Publishing House, Beijing

Nico Goldscheider, David Drew (2007) Method in karst hydrogeology. Taylor & Francis, London

# Groundwater Tracing Using Stable Isotope in the Western Mediterranean (Case of Rif Chain in the North of Morocco)

M. Qurtobi, H. Marah, A. El Mahboul, and C. Emblanch

**Abstract** Nine meteorological stations, distributed between 3 and 996 m of altitude, have been used as basis of an isotopic study of rainwater carried out in the Rif chain (Northern part of Morocco) during the hydrological years 2002 to 2004. The study allowed examining the isotopic variations during the years and defining the relationship between the stable isotopes of water (oxygen-18 and deuterium). The altitudinal gradient established for oxygen-18 is $-0.27\delta‰ + 4.87$ per 100 meters. The relationship between altitude emergence points of some studied springs and their Oxygen-18 contents were also used to calculate the regional oxygen gradient. The selected springs emerge at altitudes which are very close to the highest summits of neighbouring mountains. The oxygen-18 content variation in the springs corresponds to an altitudinal gradient of $-0.27\delta‰ + 4.25$ per 100 meters. It is remarkable to note that this gradient is identical to that of the rainwater. Even if this isotopic study refers to a short time period, it seems to be sustainable to establish a correct regional meteoric line at least as to the slope. The constant difference related to the regression (4.25 for groundwater and 4.87 for rainwater), can be due to the evaporation of the rains event before their infiltration and thereafter a relative enrichment in oxygen-18 with the altitude.

M. Qurtobi, H. Marah
Unité Eau et Climat, Division des Applications aux sciences de la Terre et de l'Environnement, CNESTEN, BP. 1382, Rabat, Morocco, e-mail: qurtobi@cnesten.org.ma

A. El Mahboul
Agence du Bassin Hydraulique du Loukous, ABHL, BP. 616, Tétouan, Morocco

C. Emblanch
UMR 1114 INRA-UAPV "EMMAH", UAPV, Faculté des Sciences Exactes et de la Nature, 33 rue Louis Pasteur, 84000 Avignon, France

B. Andreo et al. (Eds.), *Advances in Research in Karst Media* 
DOI 10.1007/978-3-642-12486-0, 

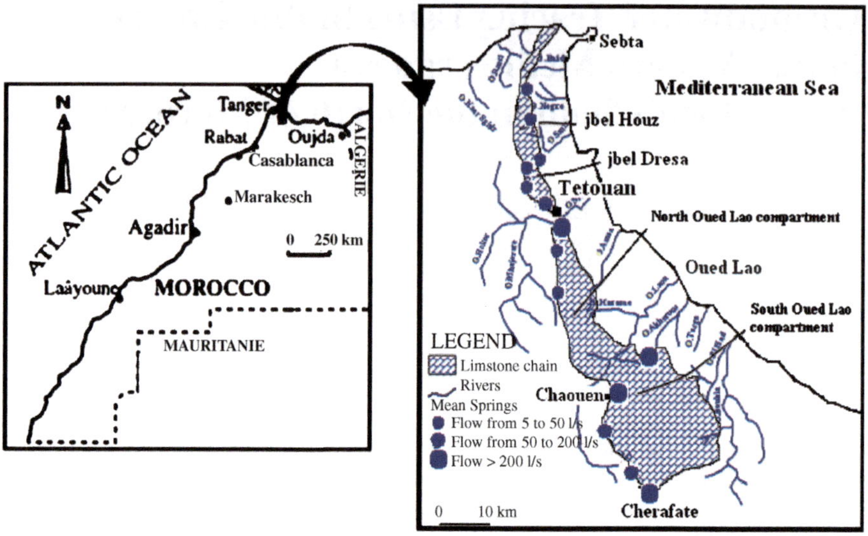


**Fig. 1** Map of Limestone chain and situation of sampling springs

## 1 Introduction and Site Description

The Rif chain is located in the north part of Morocco. In this region, the climate is influenced by two coasts: Atlantic Ocean and the Mediterranean Sea. Average precipitations range from 800 to 1400 mm, and reach certainly 2000 mm (including snow) on the highest summits (Thauvin 1971).

The Rif chain is subdivided into three domains (Fig. 1): The internal domain, the external domain and the marls domain (Martin Algarra 1987).

In the internal domain, if the Paleozoic schist is hydrogeologically impermeable, the chain of limestones bearing karst massif (1090 $km^2$) consists mainly of Triassic limestone and dolomite, with a thickness of up 1000 m, and underlying impermeable layers. The karst water is used to supply the Tetouan and Chefchaouen cities (Amraoui 1988).

## 2 Objective and Methods

This study aims to establish an altitudinal gradient of oxygen-18, which even if has to be verified by further measurements, can be used functionally to delimit the recharge areas. This has usefulness in the management and the protection of quality of numerous springs captured by local population in the Rif chain.

Seventy springs were selected for sampling during a summer period (August 2002). The sampling was primarily targeted to the main aquifers of the region: the Haouz chain (mean altitude 400 m) and Jebel Dersa (mean altitude 350 m) in the north part of Tetouan city, and the limestone chain in the south part of Tetouan city, with a north compartment of Oued Lao (mean altitude 800 m) and south compartment of Oued Lao. The last is separated by jbel Lakrâa in west compartment of jbel Lakrâa (mean altitude 1300 m) and east compartment of jbel Lakrâa (mean altitude 1100 m).

The altitudes of emersion of selected springs are very similar to those characterizing the neighbouring summits. Analysis concerned the stables isotopes of groundwater ($^{2}H$ and $^{18}O$).

## 3 Results and Interpretation

### *3.1 Origin of Water and Trajectory of the Precipitating Air*

The isotopic characterization of groundwater was carried out on emerging springs of karst systems located along a north-south direction, from jbel Dersa and chain Haouz in north of Tetouan to jbel Tissouka and jbel Bousliman to Southern Tetouan (Fig. 1). These springs are located at altitudes ranging from 100 to 1000 meters, thus their isotopic study should allow establishing the altitude effect.

The results obtained from the groundwater analysis, show a variation of $^{18}O$ contents range from −5.34 to −7.4 per mil vs. V-SMOW with an average about −6.21

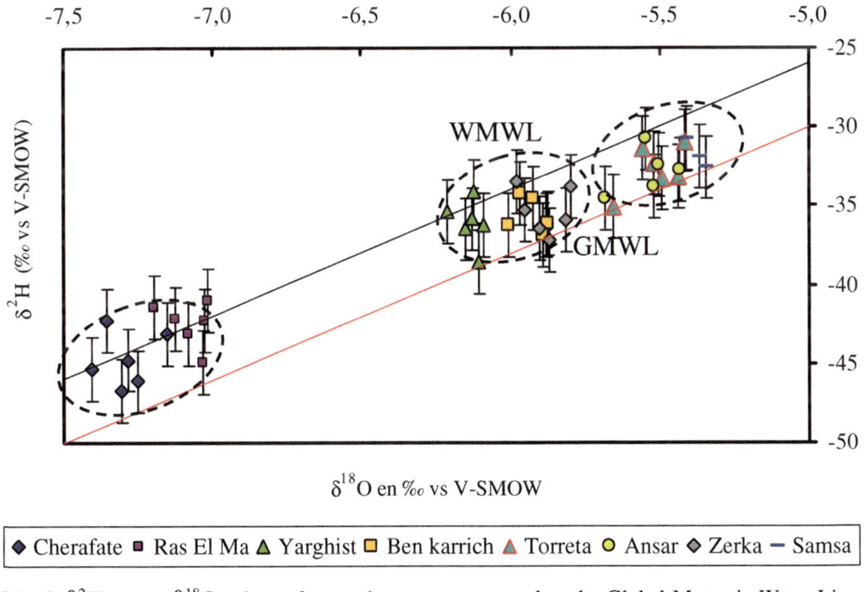


**Fig. 2** $\delta^{2}H$ versus $\delta^{18}O$ values of groundwater as compared to the Global Meteoric Water Line. Samples are sorted according by geographical distributions in Fig. 1

per mil and a variation in $^2H$ contents from −28.6 to −41.3‰ vs. SMOW with an average about −32.2‰.

The $\delta^2H$ vs. $\delta^{18}O$ diagram shows that all samples representing the springs are aligned between the Global Meteoric Water Line (GMWL), and Western Mediterranean Water Line (WMWL) (Fig. 2), and reflects mixed air masses (Mediterranean and Atlantic) generating rains that recharge aquifer karst systems of these springs (Jouzel et al. 1984). Indeed, the oceanic influence characterise springs in northern limestone chain such Ansar, Samsa and Torreta. On the other hand, the Mediterranean influence characterise the springs Ras el Ma and Cherafate located in the south part of the limestone chain. Zerka, yarghist and Ben Karrich springs show the influence of both Mediterranean and Atlantic air masses.

The first group, characterized by the samples located around the Global Meteoric Water Line, is associated with air which comes from Atlantic Ocean and enters directly in Morocco by its Atlantic coast. The second group, gathering the samples situated close to Western Mediterranean Water Line, is associated with air coming from East Mediterranean Sea before reaching Morocco by its Mediterranean coast (Celle-Jeanton et al. 2001), (Fig. 3).

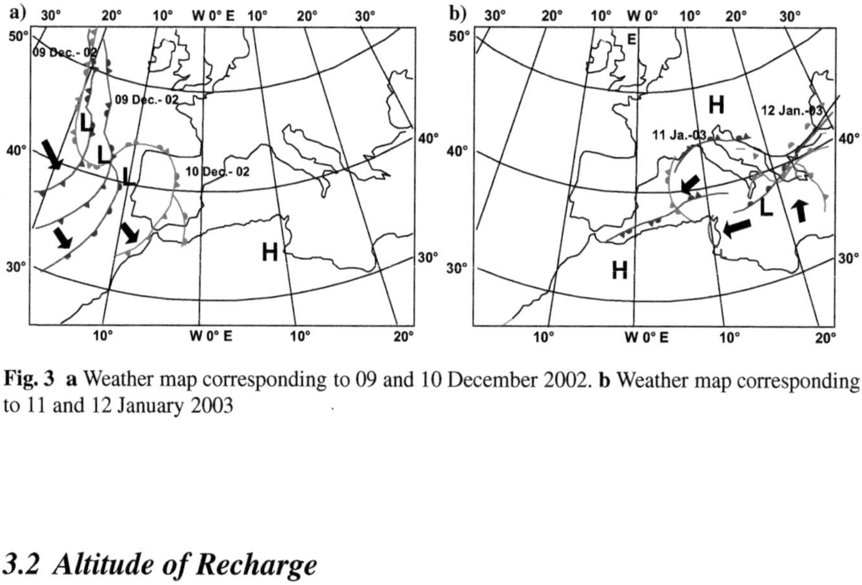


**Fig. 3** **a** Weather map corresponding to 09 and 10 December 2002. **b** Weather map corresponding to 11 and 12 January 2003

## 3.2 *Altitude of Recharge*

By using the isotope related to the numerous springs selected in the study area, an altitude gradient can be established. This allows approaching the mean altitude of recharge area (Lastennet 1994). An altitude of emergence close to that of the neighbouring summits was an important criterion of selection; the difference between the mean topographic altitude of the neighbouring summits and those of the springs selected is not more than 100 meters. Consequently, for selected springs there should be little difference between the mean altitude of recharge area and that of the emergence (maximum 50 meters). This occurrence allows using the emergence elevation as a reference.

The $^{18}O$ variation according to the spring emergence altitude corresponds to an altitudinal gradient of $-0.27\ \delta^{18}O$ by 100 m. This gradient is almost equivalent, to the one established by (Marcé 1975) and (Louvat et al. 1990), by the survey of the rainwater in Morocco to different altitudes of the Atlas chain. These authors found an altitudinal gradient of –0.3‰ in $\delta^{18}O$ per 100 m. Adopting the altitude gradient calculated (–0.27 per 100 m), the mean altitude of recharge areas for different springs sampled during this study could be determined. In the north compartment of oued Lao, the springs are supplied by recharge area with mean altitude ranging from 400 to 800 m. Springs of the south compartment of Oued Lao (Ras El Ma, Cherafate, Ain Soyah and Onsar Asaki) are supplied by recharge area having altitude ranging from 1000 and 1200 m. This result is in conformity with the observations of the relief's jbel Tissouka, jbel Bou Slimanes and jbel Lakrâas.

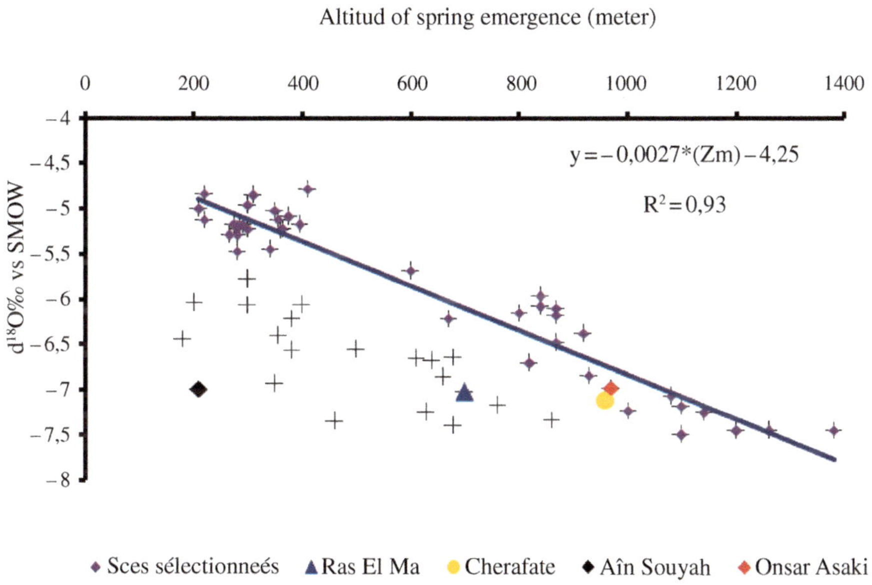


**Fig. 4** Relationship between $\delta^{18}O$ and altitude of emergence

## 4 Conclusion

Based on an instantaneous sampling of spring water at the summer season, it was possible to establish an altitudinal gradient of oxygen-18, that even it if has to be verified by further measurements, can be used functionally to delimit the recharge areas, and therefore to better manage and protect of the water quality of numerous springs used by the local population in the Rif chain.

A first practical use of this study is a better definition of the altitude recharge for good management for two important springs (Ras El Ma and Cherafate), which are the primary supplies of water for Chefchaouen city (36,600 inhabitants) and Cherafate village (1500).

Finally, this study is an interesting example of oxygen-18 analysis of an instantaneous sampling of spring at the summer season, for the establishment of altitudinal gradients and of their use in karstic aquifers management. From a reduced number of analyses on a zone deprived by previous data, results are immediately applicable in practical hydrogeology.

## References

Amraoui A (1988) Hydrogéologie de la dorsale calcaire du Rif (Maroc septentrional). Ph D Thesis, Univ. Josef Fourier-Grenoble I

Celle-Jeanton H, Travi Y, Blavoux B (2001) Isotopic typology of the precipitation in the Western Mediterranean region at three different scales. Geophysical Research Letters 28, 1215–1218

Craig H (1961) Isotopic variations in meteoric waters, Science 133 (1961) 1702–1703

Jouzel J, Merlivat L (1984) Deuterium and oxygen-18 in precipitation: modelling of the isotope effects during snow formation. Journal of Geophysical Research 89, 11749–11757

Lastennet R (1994) Rôle de la zone non saturée dans le fonctionnement des aquifères karstiques, Approche par l'étude physico-chimique et isotopique du signal d'entrée et des exutoires du massif du Ventoux (Vaucluse), Thèse Université d'Avignon et des Pays de Vaucluse, 222 p.

Louvat D, Bichara S (1990) Etude de plusieurs systèmes aquifères du Maroc, à l'aide des isotopes du milieu. Rapport A.I.E.A.., Vienne, 30 p.

Marcé A (1975) Contribution des méthodes isotopiques à l'étude des modalités d'alimentation et de renouvellement des réserves souterraines du Maroc. Rap. 75SGN447 LAB, BRGM, Orléans, 131 p.

Martín Algarra A (1987) Evolucion geologica alpine del contacto entre las zonas internas y las zonas externas de la Cordillera Bética. Tesis doctoral, Grenade, 2 vol., 1171 p.

Thauvin (1971) Ressource en eau du Maroc: Domaines du Rif et du Maroc oriental. Notes et mémoires du service géologique N° 231

# Duality of Functioning in a Karst System Under Mediterranean Climate Conditions, Deduced from Hydrochemical Characterization

J.A. Barberá and B. Andreo

**Abstract** Over a period of two hydrological years (one dry and the other rainy), physical and chemical parameters were monitored at the Cañamero spring, the main discharge point of the carbonate aquifer beneath Sierra Colorado and Sierra Merinos, in the province of Málaga, southern Spain, in order to determine its hydrogeological functioning. Analysis of the chemical data, using data processing techniques such as the distribution of electrical conductivity frequency curves, temporal evolution and principal component analysis, revealed a duality of functioning in the hydrochemical response recorded during 2007/08 (589 mm precipitation) and 2008/09 (730 mm precipitation). It is concluded that the precipitation, distribution and intensity of rainfall are factors determining the variability of hydrogeological functioning of karstic aquifers, with respect to climatology, in the Mediterranean region.

## 1 Introduction

Karstic aquifers present a marked degree of heterogeneity in their drainage systems, due to the internal hierarchization of karstic elements (channels, fractures, fissures, etc.), developed via dissolution processes within carbonate rocks. Hydrogeological studies in this respect have focused on the response, under a natural regime, of karstic springs. Analysis of hydrodynamic conditions provides information on the volumes of water transferred from the recharge zone to the discharge zone, while sampling of the chemical composition enables characterization of aquifer functioning, which is of particular interest during periods of recharge.

The aim of the present study is to highlight variations in behaviour in a karstic aquifer during hydrological years with different rainfall patterns and, thus, different

J.A. Barberá, B. Andreo
Centre of Hydrogeology of University of Malaga (CEHIUMA), and Department of Geology, Faculty of Science, University of Malaga, Malaga 29071, Spain, e-mail: jabarbera@uma.es; andreo@uma.es

B. Andreo et al. (Eds.), *Advances in Research in Karst Media* 
DOI 10.1007/978-3-642-12486-0, © Springer 2010

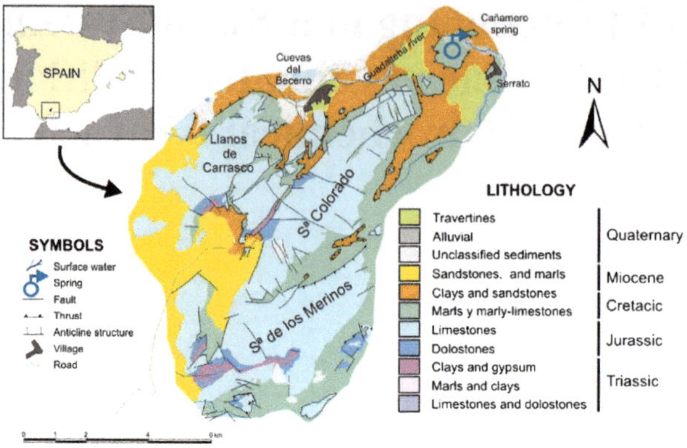


**Fig. 1** Geographical location and geology at the study area (modified from Fernández and Jiménez 2007)

types of recharge. The first year, 2007/08, was relatively dry (589 mm) and rainfall was evenly distributed in time. The second year, 2008/09, was wetter (730 mm), and precipitation took place mainly during the autumn and winter.

The aquifer beneath Sierra Colorado and Sierra Merinos has an area of 23 $km^2$ and it is located to the NE of the province of Málaga, in southern Spain. Figure 1 shows its situation, and the different lithologies present in the study area, revealing outcrops of Jurassic limestones of great thickness, which constitute the principal aquifer. The structure is made up of coffer shape folds lying NE-SW, dipping towards the NE, and has been affected by fractures in the direction N50-70E and N150E (Martín Algarra 1987; Fernández 1980).

From July 2007 to July 2009, hydrodynamic, hydrothermal and hydrochemical controls were performed at the Cañamero spring. A total of 132 water samples were taken, and the groundwater flow, electrical conductivity (EC), temperature and pH were recorded weekly under normal conditions, daily during rainfall events and fortnightly during dry periods. Moreover, continuous (hourly) record data were obtained of EC and temperature. All the majority components were analyzed at the Laboratory of Hydrogeology of the University of Málaga: cations and anions by ionic chromatography, and Total Organic Carbon (TOC), using a carbon analyzer.

## 2 Results and Discussion

Figure 2 shows the frequency curve distribution (FCD) of the electrical conductivity of the water at the Cañamero spring, revealing a high degree of variability (coefficient of variation CV = 26%) in the mineralization of the water circulating through the aquifer beneath Sierra Colorado and Sierra Merinos, mainly through a system of well developed karstic conduits. The multimodal morphology of the frequency curves reflects considerable development of functional karstification (Bakalowicz

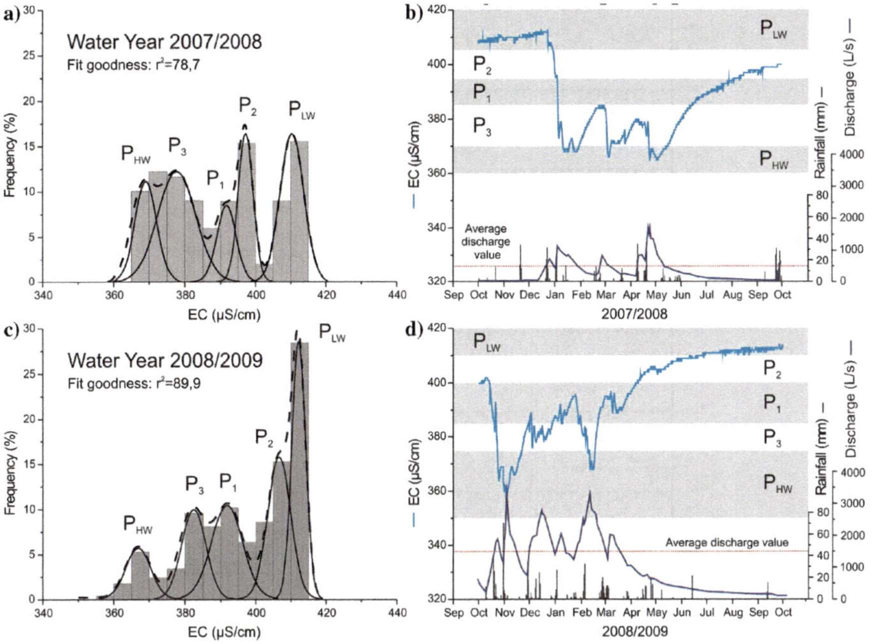


**Fig. 2** Electrical conductivity frequency curve distributions (FCDs) of the Cañamero spring waters for the two years studied (**a**, **c**). Time evolution of EC data, discharge values and rainfall distribution for the same period (**b**, **d**, respectively). The EC record shows the different water types (*grey rectangles*) provided by FCD analysis

1977; Mudry 1987). In accordance with the method proposed by Massei et al. (2007), five different water types were identified (Plw, Phw, P1, P2 and P3) that contributed to karstic flow during the sampling period.

Figures 2a and c show that, despite the difference in morphology between the two histograms, the FCD maintained the same number of modes during the two years, and thus the system can be characterized. Between 2007/08 and 2008/09, the value of the mean and the coefficient of variability varied from 390 to 397 μS/cm and from 26 to 27%, respectively, and therefore there was a change in the proportion of the FCD represented by each mode. During the hydrological year 2008/09 (the wetter year), there was a greater heterogeneity of frequencies than in the previous year, and so new classes of EC appeared, presenting low levels of mineralization (350–360 μS/cm) for high groundwater-flow levels. This is due to the fact that rainfall in 2008/09 was more intense and concentrated in short periods of time (Fig. 2d, November rains). Furthermore, the autumn rains arrived early, while the spring rains did not appear (on the contrary to 2007/08), which accounts for the high frequencies in the modes corresponding to low-water periods.

The hydrodynamic monitoring of the spring revealed the large variability in discharge rates (Figs. 2b,d and 3), with yields ranging from 47–3295 L/s. The series

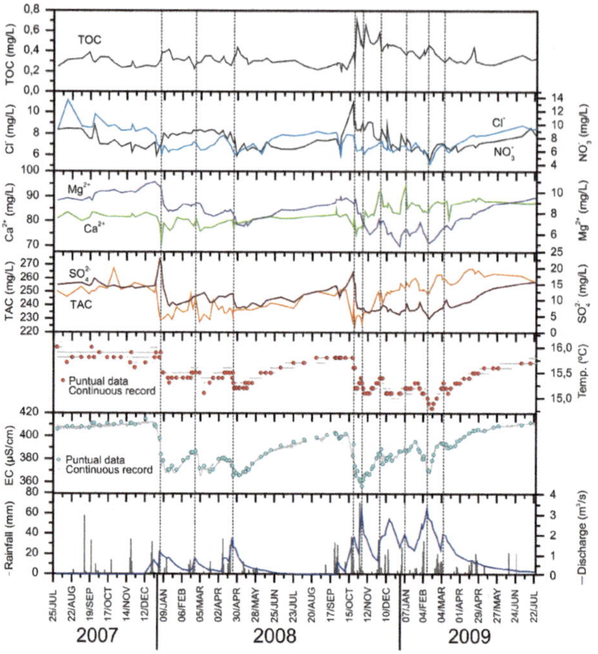


**Fig. 3** Hydrodynamic and hydrochemical time series (2007 to 2009) recorded at Cañamero karst spring

of yields recorded confirms the presence of a typically karst behaviour pattern, with sudden, rapid changes in response when rainfall events occur.

The hydrochemical evolution (Fig. 3) clearly shows the existence of different responses during the two hydrological years. During 2008/09, there were more significant increases in groundwater flow, larger falls in EC and temperature and greater dilutions in almost all the hydrochemical parameters. In general terms, each precipitation event was associated with an increase in groundwater flow and greater TAC, $Ca^{2+}$, TOC and $NO_3^-$ together with decreased EC, $Mg^{+2}$, $SO_4^{-2}$, $Cl^-$ and temperature. These variations reflect the mobilization of volumes of rapidly infiltrating water from the soil, the epikarst and the unsaturated zone as a whole – water that is more aggressive, colder, and richer in TOC and $NO_3^-$, especially during the first rains of the autumn. The $Mg^{+2}$ content, which reflects the residence time of the water (Batiot et al. 2003), together with the $SO_4^{-2}$ value, an indicator of the aquifer's evaporitic substrate, characterizes the contribution of the saturated zone to the karstic flow, which takes place mainly during low water conditions and, on occasion, when there is piston flow (see $SO_4^{-2}$ graph in Fig. 3) associated with the onset of a high-water period.

The hydroclimatic conditions and the degree of functional karstification influence the behaviour of the aquifer. In this respect, the rainfall in November 2008 was exceptional (with 70 mm being recorded on 30 October 2008), and so, following

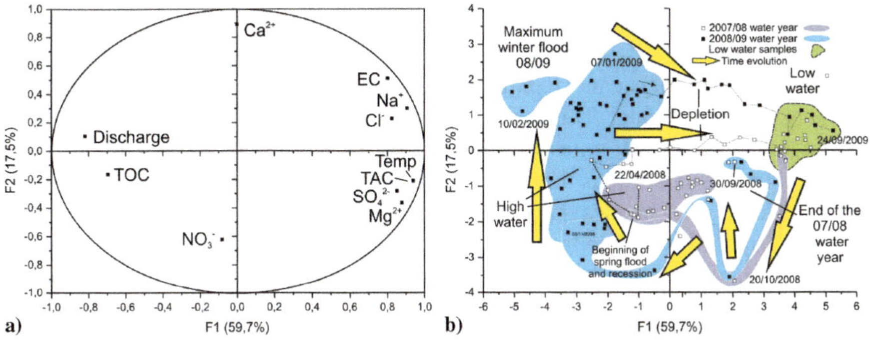


**Fig. 4** PCA on Cañamero karst spring data (from October 2007 to October 2009). Scattergram of variables (**a**) and of hydrochemical samples (**b**). *Yellow arrows* in **b** show the time evolution and hydrodynamic conditions during hydrological years

almost seven months of drought, there occurred an increase in groundwater flow to 3.3 $m^3/s$ and, associated with this, a peak in $NO_3^-$ values exceeding 13 mg/l, caused by leaching from the soil, and a progressive increase in the $Ca^{2+}$ content from the dissolution of limestone rocks, mainly in the unsaturated zone.

An analysis was made of the principal components, using 132 samples and 11 variables (cations, anions, EC, temperature, TOC and yield). Figure 4 shows the results obtained for the factorial plane of the variables (A) and for the sample space (B).

Factor 1 (59.7%) accounts for the factor of residence time in the aquifer, with more highly mineralized water, with a longer residence time (right extreme), versus rainy-period water with a higher TOC. Factor 2 (17.3%) represents the soil leaching ($NO_3^-$) process and limestone dissolution in the unsaturated zone ($Ca^{2+}$).

Figure 4b shows sample groups with similar statistical correlations, which characterize the hydrochemical response to different hydrological events. The temporal evolution of the samples for the year 2008/09 represents a cycle of hydrochemical variability. The most highly mineralized samples (from the dry period) give way, towards the negative part of both axes (during the first annual rainfall events), to samples with lower levels of mineralization and a higher TOC and $NO_3^-$ content. The next samples were obtained during high-water periods (including a very exceptional one on 10 February 2009), and the cycle then returns to the initial levels of the low-water period. This cycle was of lesser magnitude in the year 2007/08, the samples for which were distributed in a more limited space, with less variation in the chemical components.

## 3 Conclusions

An integrated analysis of hydrodynamic and hydrochemical data enabled definition of the hydrogeological functioning of the karstic aquifer drained by the Cañamero spring. There was observed to be a dual pattern in the aquifer behaviour, during two different hydrological years, 2007/08 and 2008/09. The latter was wetter, with more intense rainfall events (70 mm on 30 October 2008) which were concentrated in the autumn and winter. This aquifer system presents a considerable development of functional karstification, as may be deduced from the diversity of water types involved in flow conditions through the aquifer, and from the variability of hydrodynamic and hydrochemical parameters. There was found to be a high level of participation by the unsaturated zone during periods of high water (dilutions of almost all the chemical components and significant increases in TOC, $NO^{3-}$ and $Ca^{2+}$), and also occasionally in the saturated zone (piston effects), although the latter was observed more commonly during the low water period.

In climate regions with important variations in the quantity and distribution of rainfall events, as is the case in the Mediterranean, karstic systems may present different responses or hydrogeological behaviour patterns depending on climatic conditions, as has been found in the Colorado-Los Merinos system.

**Acknowledgements** This work is a contribution to Projects P06-RNM 2161 of Junta de Andalucía, CGL2005-05427 and CGL2008-06158 BTE of DGICYT and IGCP 513 of UNESCO, and to Research Group RNM-308 of Junta de Andalucía.

## References

Bakalowicz M (1979) Contribution de la géochimie des eaux à la connaissance de l'aquifère karstique et de la karstification. Thèse Doct. Sci. Nat., Univ. P. et M. Curie, París-VI, Géol. Dyn.; 269 pp

Batiot C, Emblanch C, Blavoux B (2003) Total Organic Carbon (TOC) and magnesium ($Mg^{2+}$): two complementary tracers of residence time in karstic systems. Comptes Rendus Geoscience 335, pp 205–214

Fernández R (1980) Investigaciones hidrogeológicas al Norte de Ronda (Málaga). Tesis de Licenciatura Univ. de Granada, 214 p

Fernández R, Jiménez P (2007) Sierras Hidalga-Merinos-Blanquilla (M.A.S. 060.043). Atlas hidrogeológico de la provincia de Málaga, 2: 49–58. Diputación de Málaga-IGME-UMA

Martín Algarra A (1987) Evolución geológica alpina del contacto entre las Zonas Internas y las Zonas Externas de la Cordillera Bética (Sector Occidental). Tesis Doctoral Univ. de Granada, 1171 p

Massei N, Mahler BJ, Bakalowicz M, Fournier M, Dupont JP (2007) Quantitative interpretation of specific conductance frequency distributions in karst. Ground Water, Vol. 45, No. 3, pp 288–293

Mudry J (1987) Apport du traçage physico-chimique naturel à la connaissance hydrocinématique des aquifères carbonatés. Thèse Sciences Naturelles, Université de Franche-Comté, Besançon, 378 pp

# Water Origin of the Kokkino Stefani Spring (W Greece) Based on Hydrogeochemical Data

K. Katsanou, G. Siavalas, and N. Lambrakis

**Abstract** The Kokkino Stefani spring is located in western Greece on the NE shore of Trichonis Lake, which is the largest lake of the country. The climate is semi-arid with water excess and deficiency during winter and summer, respectively. The area belongs to the Olonos-Pindos isopic zone consisting of a pre-orogenic sequence of deep-sea sediments successed by syn-orogenic flysch deposits. The contact of karstified Upper Cretaceous limestones with cherts gives rise to most of the springs in Trichonis catchment area with Myrtia spring as a typical example. According to the geological setting, a similar mechanism could also be applied for Kokkino Stefani. However, the discharge values are not consistent with this scenario. Hydrochemical analyses performed on samples collected on a 20-day basis showed that the Kokkino Stefani spring is classified to the Na–Ca–$HCO_3$ type, displays higher temperature, negative Eh values and is enriched in $H_2S$, F, Na, B and Li; thus being totally different from the Myrtia spring and Trichonis Lake. $H_2S$ is of biogenic origin, whereas high F concentrations are due to Ca precipitation. High B and Li values enabled the suggestion of an alternative function mechanism, which involves the existence of a deep aquifer with restricted size and the development of the spring at its current site due to isostatic pressure.

## 1 Introduction

The Kokkino Stefani spring is among the most famous mineral springs in western Greece due to its healing properties (rheumatism, dermatological diseases). It is located, on the NE shore of Trichonis Lake, at an altitude of +17 m (a.s.l.) (Fig. 1a). The major part of the lake's water originates from surrounding karstic aquifers. The hydrological balance in the region was estimated according to the method of Thornthwaite and Mather (1955). It is calculated that approximately 53% of the mean

K. Katsanou, G. Siavalas, N. Lambrakis
Department of Geology, University of Patras, Rio-Patras, GR 26 504, Greece, e-mail: katsanou@upatras.gr

B. Andreo et al. (Eds.), *Advances in Research in Karst Media*
DOI 10.1007/978-3-642-12486-0, © Springer 2010

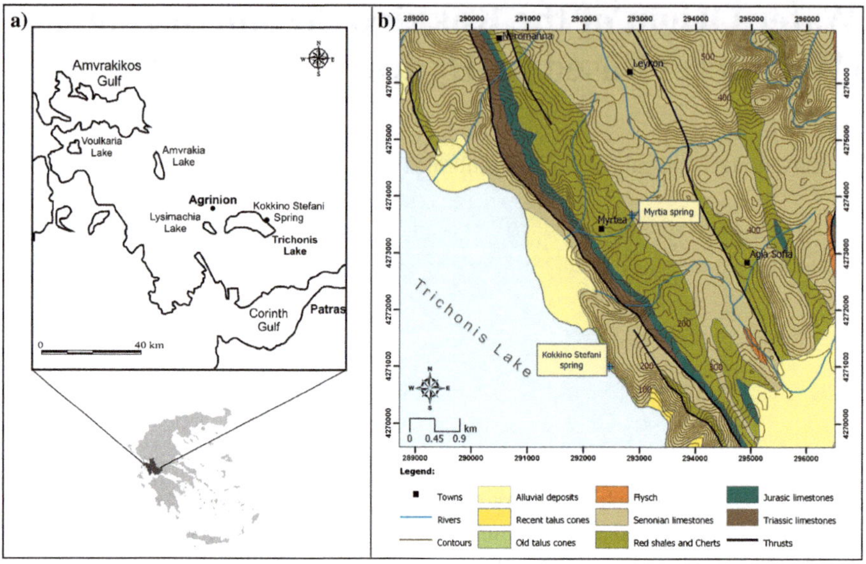


**Fig. 1** Situation (**a**) and geological map (**b**) of the study area (modified from Loftus et al. 1977)

annual precipitation in the area (929 mm) evaporates and 47% supplies surface and groundwater runoff. It is estimated that the Upper Cretaceous limestones of the area display high infiltration coefficients and 35% of the rainfall enters the groundwater (Zacharias et al. 2003). The remaining 12% is discharged through the surface runoff. According to mean monthly meteorological data and the coordinates of the spring's position, the area is classified to the Csa (Köppen and Geiger 1936) climate type and characterized as arid to semi-arid with water deficiency during summer months. Although some previous studies (Dimitropoulos 1986) suggest that the lake has intruded in the spring's aquifer, the spring's water is totally different to the other neighboring springs and also to the lake's water displaying diverse chemical composition.

The present study aims to investigate the origin of the Kokkino Stefani spring water with respect to local geology, as well as to hydrochemical data time series.

## 2 Geological and Hydrogeological Setting

Geologically, the area belongs to the Olonos-Pindos isopic zone. It is composed of deep-sea sedimentary rocks deposited from the Triassic until the Palaeocene, when syn-orogenic turbiditic sedimentation commenced lasting until the Upper Eocene (Skourlis and Doutsos 2003 and references therein). Upper Cretaceous (Senonian) limestones, which outcrop at the Kokkino Stefani site (Fig. 1b), are the most important unit of the Zone regarding hydrogeological properties, as they constitute the most important aquifers due to low degree of karstification expressed by exokastic features and tectonic deformation, that have increased their permeability. The Senonian limestones are often either in conformable or in tectonic contact with the Jurassic chert unit, resulting in the development of several springs in Trichonis' catchment. One of the most important is the Myrtia spring (Fig. 1b) with a mean discharge of approximately 41 L/s. A similar mechanism could also be expected to explain the development of the Kokkino Stefani spring. In this case, due to geological factors, the extent of the Senonian limestones which would supply the spring is estimated to 10 $km^2$.

However, the measured mean discharge of this spring is about 3 L/s or 87,600 $m^3$/y, postulating an infiltration coefficient of only 1%. This value is extraordinarily lower than that (35%) estimated for the same formation according to the data from other springs. Thus, the scenario of a karstic aquifer surrounded by impermeable rocks should be reconsidered.

## 3 Groundwater Classification and Chemical Variations Through Time

Water samples were collected according to US EPA (1976) procedures every 20 days from October 2008 to September 2009. In addition, the Myrtia (M) karstic spring was sampled on the same dates in order to be used as a reference basis. Finally, samples from Trichonis Lake, sea and rain were collected in March 2009.

According to the Piper diagram, the hydrochemical type of the KS spring is Na–Ca–$HCO_3$. On the contrary, the fresh-water spring (M) and the Trichonis Lake display Ca–$HCO_3$ and Na–Mg–$HCO_3$ types, respectively (Fig. 2a). It is evident that the KS samples display a different type to the other, being consistent with its different origin. The KS spring is classified as cold with $T_w$ values ranging between 17.8 and 19.6 °C, while $T_{air}$ ranges between 8.3 and 33.63 °C. Eh displayed negative values throughout the year indicating reductive conditions and the pH values range between 7.4 and 8.5. Both parameters display a slight shift towards lower values during winter. M spring displays significantly lower $T_w$ and positive Eh.

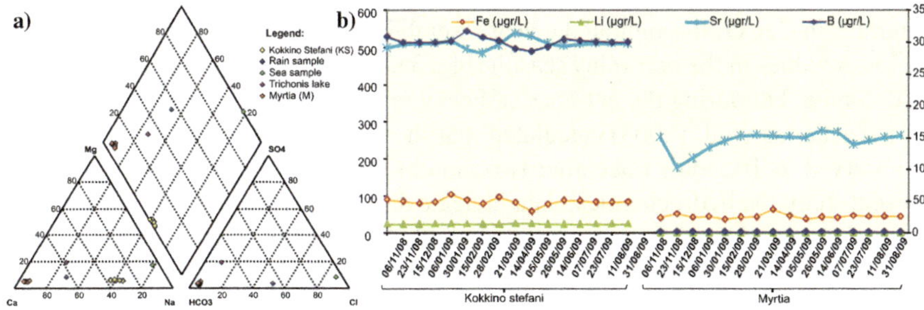


Fig. 2 **a** Piper-diagram. **b** Trace element variation in KS and M springs

The concentration of Na is one order of magnitude higher in the KS spring compared to the other two types of water, Mg and K concentrations are slightly higher, whereas Ca displays half concentration. The KS spring also proved to be rich in F (3.32 mg/L, on average) compared to the M spring (0.21 mg/L) and the lake's water (0.26 mg/L). Na may exhibit a positive correlation with F in many water types; especially those having low concentrations of Ca. High concentrations of Na will increase the solubility of $CaF_2$ in waters. This correlation has been also observed in other waters of Greece belonging to Na–Ca–$HCO_3$ type (Katsanou 2007). $H_2S$ concentration for KS ranges between 915 and 1715 μg/L with an increase during winter months but it was not detected in the M spring. The reductive conditions, which prevail, allow the maintenance of $H_2S$, which is of biogenic origin, deriving from the decomposition of sedimentary organic matter.

Concerning trace elements, no significant variations were observed throughout the year. However, the KS spring is significantly enriched in B (3.01 mg/L) compared to the M spring (0.03 mg/L) and the lake's water (0.15 mg/L), whereas all types of water display similar Sr concentration, probably associated with the hosting limestones (Fig. 2b). Li also displays elevated concentrations in the KS spring. Fidelibus and Tulipano (1990) report that Li concentration in water depends on the contact time between water and rock. Thus, Li can be used as an indicator for the residence time of water in the aquifer (Edmunds and Smedley 2000). Both B and Li concentrations can be attributed to the presence of deep waters rising along the thrust fault planes or through other underground routes. The KS samples were also proved to be enriched in Fe and Mn attributed to the dissolution of oxides from the chert unit under anoxic conditions.

## 4 Discussion

In order to better understand the groundwater chemistry variation with respect to the aquifer's chemical composition saturation indices for $CaCO_3$, dolomite and fluorite were calculated using PHREEQC. These were plotted together with pH, Eh, Ec, $T_w$, Mg/Ca ratio for KS spring (Fig. 3a–e). Additionally, precipitation data for the same period of time (Oct 2008–Sept 2009) from a local weather station (National Observatory of Athens, Greece) was also plotted (Fig. 3f). KS samples appear to be saturated in $CaCO_3$ throughout the year resembling the M samples. The index takes its lower values in the non-rainy season (Fig. 3a,b). Both springs are undersaturated in dolomite, but during the entrance of heavy rain in the aquifer they reach saturation. Zacharias et al. (2003) calculated that the response of the heavy rain period is observed in Trichonis Lake after two months through submerged springs. In the present study the hydrochemical data indicate that this period is valid for M spring, but in the case of KS this time is approximately four months later. At the same time this inflow of cold fresh water lowers Ec and $T_w$ (Fig. 3e). All samples revealed a low hardness, generally observed in waters with F exceeding 1.5 mg/L due to precipitation of Ca under common ion effects.

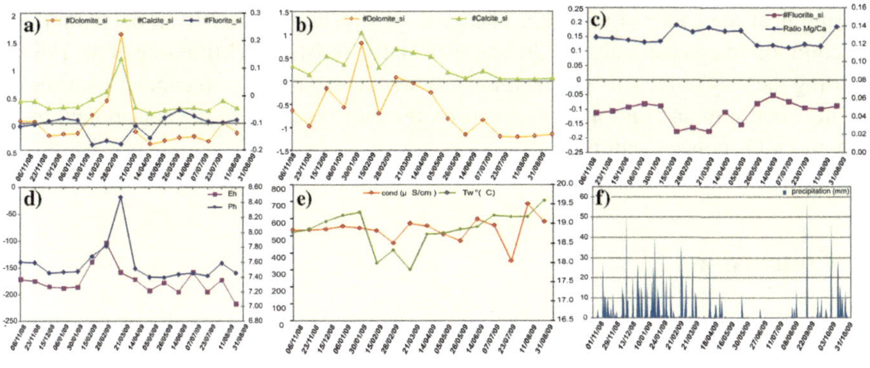


**Fig. 3 a,b** Variation of calcite, dolomite and fluoride saturation indices, **c** Variation of fluoride index and Mg/Ca ratio in KS spring, **d,e** Variation of pH and Eh, Ec and Tw for KS spring respectively, **f** Daily precipitation of Gavalou station (mm) that correspond to KS and M spring

Voroshelov (1966) indicated a strong negative correlation between Ca and F in groundwaters that contain Ca in excess of that required for the solubility of $CaF_2$. Moreover, the groundwater samples with high F values show an increase in Na/Ca ratio because of depression of Ca activity. The highest values of F correlate with residual alkalinity ($Ca^{2+} < HCO_3$) and saturation in $CaF_2$. The positive correlation between F and $HCO_3$ is explained by considering the mass law equation relationship for $CaCO_3$ and $CaF_2$ when both are in contact with water (Kundu et al. 2001). Alkalinity of the circulating water depleted in Ca is mainly responsible for the high concentration of $F^-$. The KS spring is undersaturated in $CaF_2$ (Fig. 3a,c).

$$CaCO_{3(s)} + H^+ + F \Longleftrightarrow CaF_{2(s)} + HCO_3^-$$

Voroshelov (1966) has shown that Ca in excess of that contributed by $CaF_2$ seriously affects the concentration of F at levels of 0.5–3.0 mg/L.

The amount of F released by the dissolution of $CaF_2$ in water with low ionic strength is in the order of 8–10 mg/L, causing the undersaturation in $CaF_2$ in the majority of groundwaters (Boyle 1976). However, the concentrations of Ca, Na, OH, and certain complexing ions, such as Fe, Al, B, Si, Mg, and H, can alter this concentration range (Apambire et al. 1997). F in KS samples shows a strong correlation to $SiO_2$ and Mg, indicating the offer of F through dissolution of minerals contained in the chert unit and the accompanying shales and siltstones. According to the hydrochemical data, the KS spring displays distinct features of a deep aquifer, such as the higher $T_w$ than the M spring and Trichonis Lake, as well as the enrichment in B and Li. All these, along with the spring's low discharge, point to a laterally limited, deep aquifer, supplied by rainwater from the broader area, through infiltration at greater depths following tectonic pathways or even by groundwater already stored in the Upper Cretaceous limestones. The residence time of water in the aquifer is rather long considering the Li concentration and Mg/Ca ratios (Edmunds and Smedley, 2000). It is assumed that due to isostatic pressure the water upwells to form the KS spring near the shore of the lake. The isostatic upwell of water from deep aquifers is common in coastal areas of Greece, especially on islands (Minissale et al. 1989). During the way to the surface the relatively warm water comes in contact with rocks of the chert unit and dissolves various components resulting in the enrichment in F, Na and $SiO_2$ concentrations.

# 5 Conclusions

The KS cold mineral spring displays a different chemical composition than the M spring and Trichonis Lake water. $H_2S$ originates from the decomposition of organic matter being present in the lake's sediments, whereas Ca precipitation results in high F concentrations due to common ion effects. High B and Li concentrations indicate deep aquifer and long residence time, respectively. Classical hydrological and hydrogeological methods do not explain in a satisfactory way the existence of KS spring in its specific place. Although it is hosted in Upper Cretaceous limestones like the M spring it displays significantly lower quantities of water than those expected for a typical karstic aquifer of its extent.

The hydrochemical data suggest a more reliable scenario of a deep, limited in size aquifer with relatively warm water, which upwells to the surface under the effect of isostatic pressure. The temperature was calculated with the application of Na/K geothermometer to be higher than 50 °C and is responsible for the dissolution of various elements either *in situ* or during the rise of the water in the upper layers.

# References

Apambire WB, Boyle DR, Michel FA (1997) Geochemistry, genesis, and health implications of fluoriferous groundwater in the upper regions of Chana. Environ. Geol. 33, 13–24

Boyle DR (1976) The geochemistry of fluorine and its application in mineral exploration. PhD Thesis. University of London, Imperial College Science and Technology

Boyle DR (1992) Effects of base exchange softening on fluoride uptake in groundwaters of the Moncton Sub-basin, New Brunkswick, Canada. In: Kharaka, Y.K., Matest A.S. (eds.) Water-rock interaction, pp. 771–774. Proc. 7th Int. Symp Water-rock interaction. A.A. Balkema, Rotterdam (1992)

Dimitropoulos D (1986) Hydrogeological investigation of the Kokkino Stefani spring, Aitoloakarnania prefecture. Inst. Geol and Miner. Explor. Athens (in Greek) (1986)

Edmunds WM, Smedley PL (2000) Residence time indicators in groundwater: the East Midlands Triassic sandstone aquifer. Appl. Geochem. 15, 737–752

Fidelibus MD, Tulipano L (1990) Major and minor ions as natural tracers in mixing phenomena in coastal carbonate aquifers of the Apulia. In: Kozerski, B., Sadurski, A. (eds.) Proc. of the 11th Salt Water Intrusion Meeting pp. 283–293. Gandsk

Katsanou K (2007) Environmental and hydrogeological study of the hydrological basins in the broader area of Aigion region by the use of hydrochemical methods. Master thesis. University of Patras in (Greek)

Köppen W, Geiger R (1936) Handbuch der Klimatologie. Berlin

Kundu N, Panigrahi MK, Tripathy S, Munshi S, Powell MA, Hart BR (2001) Geochemical appraisal of fluoride contamination of groundwater in Nayagargh District, Orissa, India using geochemical and resistivity studies. Environ. Geol. 41, 451–460

Loftus DL, Filippakis N, Mavridis A (1977) Geological Map of Greece, scale 1:50,000, Thermon Sheet. Inst Geol and Miner. Explor., Athens

Minissale A., Duchi V, Kolios N, Totaro G (1989) Geochemical characteristics of Greek thermal springs. J. Volcan. Geoth. Res. 39, 1–16

Skourlis K, Doutsos T (2003) The Pindos Fold-and-thrust belt (Greece): inversion kinematics of a passive continental margin. Int. J. Earth Sci. 92, 891–903

Thornthwaite CW, Mather JR (1955) The water balance. Climatology 8, 1–37

U.S. Environmental Protection Agency (1976) Quality criteria for water. Washington, DC

Voroshelov YI (1966) Geochemical behaviour of fluorine in the groundwaters of the Moscow region. Geochem. Int. 2, 261

Zacharias I, Dimitriou E, Koussouris T (2003) Estimating groundwater discharge into a lake through underwater springs by using GIS technologies. Environ. Geol. 44, 843–851

# Convective Thermal Field Reconstruction by Ordinary Kriging in Karstic Aquifers (Puglia, Italy): Geostatistical Analysis of Anisotropy

M.D. Fidelibus, L. Tulipano, and P. D'Amelio

**Abstract** The flow system in fractured aquifers is complex to define. A reliable outline of flow system in such a type of aquifers can be obtained by using natural tracers: among them, temperature demonstrates very effective usuage. Convective thermal fields, resulting from convective disturbance on conductive thermal fields, can be reasonably interpreted like analogues of flow systems. A geostatistical approach by Ordinary Kriging has been applied to groundwater temperature data obtained from temperature logs performed in two fractured karstic and coastal aquifers (Murge and Salento) of Southern Italy. The variographic analysis shows the potential of the geostatistical approach in outlining regional and local scale anisotropies. Moreover, vertical and horizontal Ordinary Kriging temperature estimations at different depths allow recognizing main recharge areas and preferential flow-pathways.

## 1 Introduction

Temperature is a powerful groundwater tracer. Starting from the 1960s, many researchers dealt with applications to hydrological and hydrogeological problems ([1] and references therein). Earliest recognition of groundwater flow role in perturbing natural conductive thermal field due to forced advection was accomplished by geophysicists studying conductive heat flow: convective thermal fields are shaped by a time-dependent exchange of heat between groundwater and surrounding rocks. The solution of the energy equation for simultaneous transport of water by hydraulic gradients and heat by forced convection [3] established that the highest perturba-

M.D. Fidelibus, P. D'Amelio
Civil and Environmental Engineering Dept., Technical University of Bari, Via Orabona 4, 70125 Bari, Italy, e-mail: d.fidelibus@poliba.it

L. Tulipano
Hydraulics, Transportation and Roads Dept., University Sapienza, Via Eudossiana 20, 00100 Rome, Italy, e-mail: luigi.tulipano@uniroma1.it

B. Andreo et al. (Eds.), *Advances in Research in Karst Media*
DOI 10.1007/978-3-642-12486-0, © Springer 2010

tion of conductive thermal field occurs where fluid flow streamlines are normal or nearly normal to heat conduction isotherms, while the slowest alteration takes place where streamlines are parallel to isotherms. Consequently, convective heat losses and geothermal gradient increasing with depth occur in groundwater recharge areas, balanced by convective heat gains and decrease of geothermal gradients with depth in discharge areas. The perturbation extent depends on topography, climate, basin geometry, water table pattern, magnitude, spatial distribution and anisotropy of formation permeability, and flow system depth [6, 9, 14].

Permeability of fractured and karstic carbonate aquifers is due to a complex structure defined by the net of fissures, fractures and karstic channels: the aquifers normally show high anisotropy of hydraulic conductivity. Conductive thermal fields of fractured and karstic aquifers can be highly perturbed, both at local and regional scale. As an example, at recharge areas, where infiltration waters move rapidly towards the saturated zone through a net of main vertical fractures/joints and swallowholes, vertical temperature gradients of saturated zone are often close to zero; vertical gradients are even negative in the presence of preferential flow pathways as karst channels [2, 7]. In suitable conditions of temperature contrast, the convective thermal field pattern can clearly outline recharge and discharge areas, preferential flow pathways [5, 11] and, in coastal aquifers, extent of seawater intrusion [13].

The interpretation of convective thermal field configuration for hydrogeological aims must be based on reliable estimations of the same fields: they can be achieved by careful geostatistical approach using the Ordinary Kriging (OK) algorithm. OK incorporates, through the variogram, the scale length, the data repeatability, the anisotropy characteristics of data, and the physical processes driving the distributions. A geostatistical study on groundwater temperature data was developed, according to the above basics, for the carbonate karstic and coastal aquifers of the Murge and the Salento (Fig. 1a, Apulia, Southern Italy).

## 2 Geological and Hydrogeological Framework of Study Areas

Murge and Salento belong to the extensive Apulia foreland (part of the Adriatic plate). The oldest deposits outcropping in the two regions are made up by a Cretaceous carbonate succession (Fig. 1a). Clays, sands and calcarenites, ranging from Miocene to Quaternary, outcrop in Salento in transgression on Cretaceous formations, while thin and discontinuous late Pliocene–Quaternary transgressive deposits of the sedimentary cycle of the Bradanic trough outcrop in the Murge [8]. Fault systems of age from Mesozoic to Pleistocene dissect the rigid Mesozoic carbonate substratum in blocks: a large deformation zone mainly oriented E-W separates Murge from Salento. In the Murge, large structural faults trending NW-SE to the N curve progressively to WNW-ESE and E-W to the S [4]. Salento is characterized by structural highs and lows separated by faults with prevailing NNW-SSE, and subordinate NE-SW and E-W directions [10]. The Murge aquifer shows variable permeability mainly in relation to the spatial distribution of carbonate facies hav-

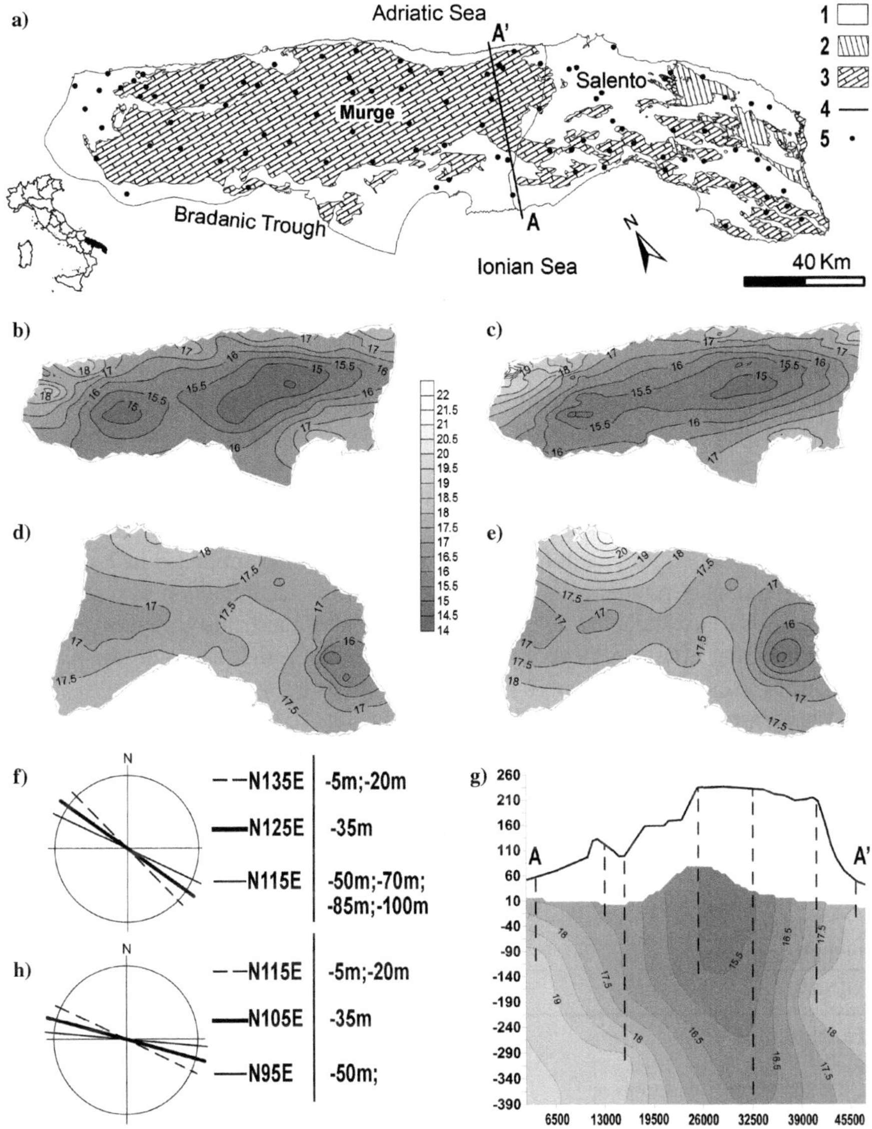


**Fig. 1** **a** Simplified geological map of Murge and Salento: 1 – calcarenites, sands and clays (Pliocene-Quaternary); 2 – calcarenites and marly calcarenites (Miocene); 3 – Limestones and dolomitic limestones (Cretaceous); 4 – Trace of thermal section; 5 – Wells of RMN; **b** and **c** Murge: OK estimation of $T$ at $-5$ and $-100$ m AMSL; **d** and **e** Salento: OK estimation of $T$ at $-5$ and $-50$ m AMSL; **f** and **h**: direction of main anisotropy axes respectively for Murge and Salento; **g** OK estimation of $T$ along the vertical section $AA'$ (a-4); *dashed lines* stand for well traces

ing different karstification aptitude; in the Salento the horizontal levels are highly karstified, connected by fissures, resulting in a higher permeability degree, quite homogeneous at regional scale. The recharge rate is about 47 m$^3$/s for Murge and

28 m$^3$/s for Salento [13]. Fresh groundwater, according to the hydraulic heads and thickness of the aquifers, floats on saltwater of marine origin, discharging into the sea through coastal springs, sub-aerial/submarine, concentrated/diffuse, under mean hydraulic gradients ranging from 0.2‰ (Murge) to 0.02‰ (Salento).

## 3 Geostatistical Study of Groundwater Temperature

The temperature data-set was built with data selected from temperature ($T$, °C) and electrical conductivity (EC, μS/cm) profiles carried out in winter 1995 in no. 107 wells of the Regional Monitoring Net (RMN) for the control of groundwater status (location in Fig. 1a). The temperatures measured respectively at −5, −20, −30, −50, −70, −85, and −100 m AMSL, constitute no. 7 sub-sets. The basic statistical analysis of each sub-set showed frequency distributions of bimodal type of the variable $T$, indicating the presence of at least two distinct groups. The cluster analysis, based on EC and $T$, allowed recognizing two main data populations corresponding to the two contiguous aquifers. The two data-bases are split each in 7 sub-sets. The subsequent geostatistical analysis focused on the identification of anisotropy. Variogram surfaces enabled visual identification of a main anisotropy direction at each elevation; afterward, directional variograms along the main and the normal anisotropy axes allowed classifying the anisotropy type.

Variographic analysis of the Murgia data sub-sets recognized at each elevation a large scale (regional) anisotropy, with maximum continuity along a direction that rotates CCW (CounterClockWise) with depth: the direction varies from N135E (NW-SE) at −5 m AMSL to N115E (roughly WNW–ESE) at −100 m AMSL (Table 1, Fig. 1f). For all the sub-sets the anisotropy is of geometric type, each var-

**Table 1** Synthesis of variographic analysis for Murge and Salento data-sets

| MURGE | | | | | | | |
|---|---|---|---|---|---|---|---|
| Elevation (m AMSL) | −5 | −20 | −35 | −50 | −70 | −85 | −100 |
| Model | Sph | Sph | Sph | Sph | Sph | Sph | Sph |
| Direction of main anisotropy axis | N135E | N135E | N125E | N115E | N115E | N115E | N115E |
| Range (main anisotropy axis) (km) | 48 | 50 | 50 | 50 | 70 | 70 | 70 |
| Range (orthogonal direction) (km) | 15 | 20 | 15 | 15 | 10 | 10 | 15 |

| SALENTO | | | | | | | |
|---|---|---|---|---|---|---|---|
| Elevation (m AMSL) | −5 | −20 | −35 | −50 | −70 | −85 | −100 |
| Model | Sph | Sph | Sph | Sph | Sph | Sph | Sph |
| Direction of main anisotropy axis | N115E | N115E | N105E | N95E | Isotropic | Isotropic | Isotropic |
| Range (main anisotropy axis) (km) | 35 | 50 | 50 | 45 | 45 | 50 | 50 |
| Range (orthogonal direction) (km) | 5 | 5 | 5 | – | – | – | – |

* Sph stands for Spherical.

iogram model being characterized by the same value of the sill in all directions, while showing different values of the range along the direction of main anisotropy axis and of the normal to the main axis. The range is higher (higher spatial correlation/continuity) along the main anisotropy axis (48–70 km) than along the normal axis (10–15 km).

Also for the Salento the variographic analysis allowed recognizing at each elevation a large scale anisotropy (with a range from 35 to 50 km) with maximum continuity along directions that rotate CCW with depth from N115E (WNW-ESE) at −5 m AMSL to N95E (roughly W-E) at −50 m AMSL. At −70, −85 and −100 m ASLM the anisotropy disappears, but this result could be false, given that the statistical significance of related sub-sets is low. Where recognized, the anisotropy continues to be of geometric type. For all elevations in both aquifers, the anisotropy at local scale shows a lower range with respect to that at the regional scale: in the Murge the range varies with depth, from 10 to 20 km, while in Salento it assumes values of 5 km, being not identifiable under the −35 m AMSL.

## 4 Discussion and Conclusions

For the Murge the directions of the main anisotropy axes, rotating CCW with depth from NW-SE to WNW-ESE, match the azimuthal trend of the main structural lineaments deduced from mesostructural analyses. The CCW rotation with depth of the main anisotropy axis could be related to the arched trend of regional fault lineaments, which, to the S, curve towards E-W. It has to be noted that the range along the regional scale anisotropy jumps from 48 to 70 km under the −50 m AMSL, indicating that spatial continuity increases at depth. Meanwhile, the range of the local scale anisotropy decreases, indicating an increase of anisotropy.

After the variographic analyses, the best variogram models were chosen as those with the best distribution of the residuals around the average, with the smaller RMSE (Root Mean Square prediction Error) index, and MSDR (Mean Squared Deviation Ratio) index closer to unity. Afterward, the horizontal distributions of temperature were estimated by OK. The horizontal OK estimations of $T$ at −5 and −100 m AMSL for Murge (Fig. 1b, c) allow recognizing the main recharge areas and the influence of seawater intrusion: the main groundwater flow-pathways develop according to the outlined geostatistical anisotropy. The thermal vertical section (Fig. 1h), which was obtained by declustering of data from profiles and horizontal OK grids, indicates a main vertical anisotropy axis with an inclination of 85°. This suggests that groundwater flows horizontally through rather vertical structures, which allow hydraulic connection between Murge and Salento: water balance studies indicate that Murge recharges laterally Salento with about 10 $m^3/s$ [13].

The directions of the main anisotropy in the Salento, which rotate CCW with depth from WNW-ESE to E-W, do not follow the main structural trend (NNW-SSE) recognized by mesostructural studies; in addition, they show a rotation of 20° CCW with respect to the main anisotropy axes of the Murge at the same depths. The

Salento aquifer shows an increasing tendency to isotropy with depth. This pattern might indicate a progressive prevailing influence on flow direction of the karstic and fissure net with respect to main structural elements. Results of geostatistical and OK estimations are surely relevant for the reconstruction of the functioning conceptual model of the aquifers and for the aims of flow modeling. In addition, the geostatistical approach on temperature data, allowing outlining the main anisotropy axes, gives a significant contribution to the knowledge of the complex tectonics of the regions. The method can deal with data both at local or regional scale, provided that $T$ data satisfy the geostatistical requirements.

## References


1. Anderson MP (2005) Heat as groundwater tracer. Groundwater 43(6):951–968
2. Cotecchia V, Tadolini T, Tulipano L (1978) Groundwater temperature in the Murgia karst aquifer ((Puglia – Southern Italy). Int Symp on Karst Hydrol, Budapest
3. Domenico PA, Palciauskas VV (1973) Theoretical analysis of forced convective heat transfer in regional ground-water flow. Geol Soc Am Bull 84:3803–3813
4. Festa V (2003) Cretaceous structural features of the Murge area (Apulian Foreland, Southern Italy). Eclogae Geol Helv 96:11–22
5. Fidelibus MD, Tulipano L (2005) Groundwater temperature as environmental tracer. In: Stournaras G et al (Eds) 7th Hell. Hydrogeol Conf, Athens, ISBN 960-88816-2-5, II:211–218
6. Forster C, Smith L (1988) Groundwater flow systems in mountanainous terrain: 2. Controlling factors. Water Resour Res 24:1011–1023
7. Ravnik D, Rajver D (1998) The use of inverse geotherms for determining underground water flow at the Ombla karst spring near Dubrovnik, Croatia. J Appl Geophys 39:177–190
8. Ciaranfi N, Pieri P, Ricchetti G (1988) Note alla carta geologica delle Murge e del Salento. Mem Soc Geol It 41:449–460
9. Smith L, Chapman, D S (1983) On the Thermal Effects of Groundwater Flow. 1. Regional Scale Systems. J Geophys Res 88(B1):593–608
10. Tozzi M (1993) Assetto tettonico dell'avampaese apulo meridionale (Murge Meridionali – Salento) sulla base di dati strutturali. Geol Romana 29:95–11
11. Tulipano L (1988) Temperature logs interpretation for the identification of preferential flow pathways in the coastal Carbonatic and Karstic Aquifer of the Salento Peninsula (Southern Italy). In: Karst Hydrogeology and Karst Environ. Protection, XXI(2):955–961, Geol Publication House, Bejing, China
12. Tulipano L, Fidelibus MD (1989) Temperature of ground waters in coastal aquifers: some aspects concerning saltwater intrusion. Natuur Tijdschr 70:308–316
13. Tulipano L, Fidelibus MD (1995) Metodologie per la valutazione degli effetti del rilascio di reflui urbani sulla distribuzione dei nitrati nelle acque sotterranee delle unità della Murgia e del Salento (Italia Meridionale). Quad. di Tec di Protez Amb 49:167–179
14. Woodbury AD, Smith L (1985) On the thermal effect of three dimensional groundwater flow. J Geophys Res 90 (B1):759–767

# Identification of Thermal Anomalies in the Carbonate Aquifer of the Lower Andarax (SE Spain) by Means of Temperature Cross-Sections

F. Sánchez-Martos, A. Pulido-Bosch, L. Tulipano, M.D. Fidelibus, and L. Molina-Sánchez

**Abstract** A significant thermal anomaly exists in the aquifer of the Lower Andarax (Almería, SE Spain), which is evidenced by the presence of several thermal springs as well as by elevated water temperatures (30–52 °C) recorded at various wells. Twelve vertical temperature logs were made to analyse these anomalies, and these yielded two-dimensional information about changes in temperature through the saturated zone. By studying these logs and drawing a series of cross-sections of temperature along the edge of the Sierra of Gádor, the variability of the thermal anomalies in the Carbonate Aquifer of the Lower Andarax is demonstrated. Variations in the morphology of the temperature logs exhibit well the variability in the thermal anomalies of the groundwaters of the Lower Andarax and their relationship with the geology and tectonics of the area. The most significant thermal anomalies occur in the zone of convergence of greatest neotectonic and seismic activity.

## 1 Introduction

The lower catchment of the river Andarax lies in the extreme southeast of Spain and corresponds to a valley of the the Sierra of Gádor and the Sierra Alhamilla (Fig. 1). Here, there is a heat flow anomaly and intense neotectonic activity (Fernández et al. 1998). These two factors are conducive to the development of varied

F. Sánchez-Martos, A. Pulido-Bosch, L. Molina-Sánchez
Department of Hidrogeology, University of Almería, Campus Universitario, Almería, Spain, e-mail: fmartos@ual.es

L. Tulipano
Department of Hydraulics, Transportation and Roads, Sapienza University of Rome, Rome, Italy

M.D. Fidelibus
Civil and Environmental Engineering Dept., Technical University of Bari, Via Orabona 4, 70125 Bari, Italy

B. Andreo et al. (Eds.), *Advances in Research in Karst Media*
DOI 10.1007/978-3-642-12486-0, 

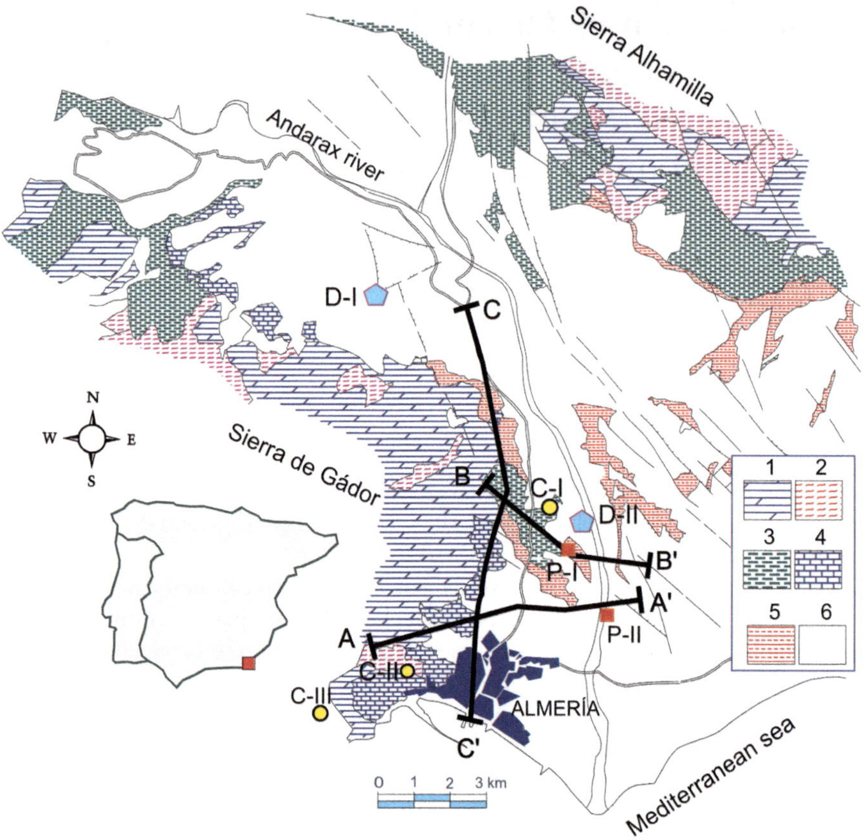


**Fig. 1** Geological escheme of the Eastern margin of the lower Andarax valley. 1: Limestones and dolomites; 2: Phyllites and calcoschists; 3: Marls ; 4: Limestones and calcarenites; 5: Silts and clays. 6: Conglomerates, sands, silts. Temperature distribution in the vertical logs recorded in the Detritic Aquifer (D-I and D-II), Carbonate Aquifer (C-I, C-II and C-III) and Deep Aquifer (P-I, P-II). A-A′, B-B′y C-C′ vertical temperature cross-section in Fig. 3

and abundant geothermal features (boreholes and springs with resurgence temperatures of between 23 °C and 52 °C, and steep geothermal gradients) over a relatively small surface area. Temperature analysis using isotherms traced at different elevations, provides a very useful means of tracking changes in water temperature and detecting the orientation of water flow at depth. Nevertheless, there are a number of factors that complicate the thermal regime of the groundwater, such as fractures, sedimentary overburden, thermal influence of deep aquifers, and karstic or even seawater flow.

This article analyses the thermal anomaly that exists along the eastern margin of the Carbonate Aquifer of the Sierra of Gádor, by means of a study of groundwater temperature. The objective is to make a qualitative approximation of the flows of thermal water, and to interpret the thermal anomalies affecting the contact between the carbonate and the detritic aquifers.

## 2 Material and Methods

The lower catchment of the River Andarax is a valley within the Sierra of Gádor and Sierra Alhamilla. Three aquifer units have been defined (Sánchez Martos 1997). The Detritic Aquifer extends along the whole of the central part of the valley and includes the quaternary detritic, alluvial and delta deposits, together with the deltaic Pliocene arenaceous and silty conglomerates. The Carbonate Aquifer basically consists of limestone and dolomite with local Miocene calcarenites that extend along the entire edge of the sierra. Its geometry is very complex and it is compartmentalized into blocks. The Deep Aquifer was detected in the centre of the valley; it is a confined aquifer, highly compartmentalized into blocks. Its lithology is limestone-dolomite with some quartzite levels (Fig. 1).

The base data are a series of temperature logs recorded from twelve aquifer points. Each log was analysed on a one-dimensional level to detect point anomalies. In two dimensions, a series of sections was drawn and isotherms plotted to show the lateral correlation between the anomalies detected in the individual logs. These profiles were drawn using data obtained from temperature logs and from temperature data taken from hydrochemical samples of pumped groundwater abstractions. The latter samples also considered the characteristics of the borehole, its static and dynamic levels and the location of the pump within the borehole. All temperature data were then integrated into an isotherm cross-section showing water temperature. The line of the isotherms was controlled using an iterative graphical technique, such that the position of each isotherm within each profile was consistent with all the others.

## 3 Results and Discussion

The vertical logs show the thermal characteristics of various aquifer points. Within the Detritic Aquifer two types of logs are identified (Fig. 2). Type D-I exhibits a gradient of 4 °C/100 m and its evolution is slightly increasing at depth, a feature characteristic of recharge areas (Domenico and Schwartz 1990). Type D-II corresponds to a borehole situated in the riverbed. Its exhibits a slight decrease in temperature related to some preferential flowpath favouring a more rapid circulation of cooler water. Within the Carbonate Aquifer, three types of log were identified. Type C-I includes the highest water temperatures and exhibits very constant values. It is situated in a tectonically-complex area where two regionally-important fracture systems of the Betic Cordillera converge, giving rise to numerous thermal manifestations (Sanz de Galdeano et al. 1995). Type C-II shows some decreases in temperature, which

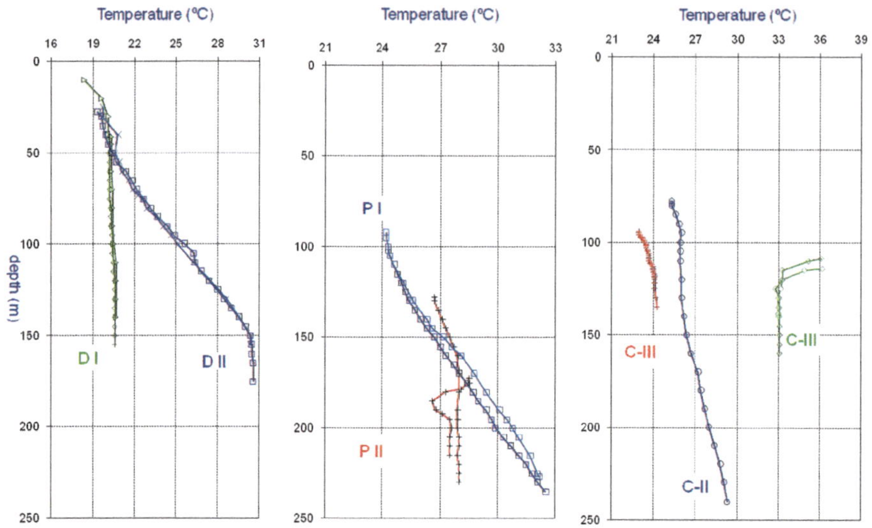


**Fig. 2** Logs temperature recorded in the Detritic Aquifer (D-I and D-II), Carbonate Aquifer (C-I, C-II and C-III) and Deep Aquifer (P-I and P-II)

suggests the existence of a rapid water flow. In the lower section, below 130 m depth, there is a constant thermal gradient. Type C-III is situated in the southern part of the Carbonate Aquifer, some 500 m from the sea. Its temperature oscillates between 22.5 and 25 °C, values that are relatively low given the marked thermal anomaly of the area. This narrow range of variation is directly related to the marine influence. In stretches of the borehole where there is a more direct seawater-aquifer relationship, there are marked jumps in salinity, while the water temperature falls slowly or is constant. Two types of log are defined in the Deep Aquifer. Log type P-I is quite homogeneous and shows a continuous increase in temperature with depth, reaching 33 °C at 240 m depth. Type P-II exhibits greater thermal variability and one of the sampling surveys even showed a drop of 2 °C in temperature with depth. The circulation of cooler, more saline water completely changes the thermal regime of the borehole, reducing water temperature and increasing electrical conductivity (5200 μS/cm).

Three vertical cross-sections were constructed along the eastern margin of the Sierra of Gádor. This orientation provided a sufficiently-dense grid to undertake a spatial analysis of the various anomalies (Fig. 1). Cross-section A shows the stepped structure of the carbonate substrate along the eastern edge of the Sierra of Gádor. The distribution of the isotherms shows that from 22 °C the gradient increases, coinciding with the contact between the carbonate deposits of the Sierra of Gádor and the detritic rocks of the depression. Cross-section B indicates a small horst structure, which reaches its maximum gradient in the highest part of the carbonate substratum. Laterally, the isotherms lay closer to the horizontal, since the zone of highest temperature is limited to the horst and to the zone of intersecting

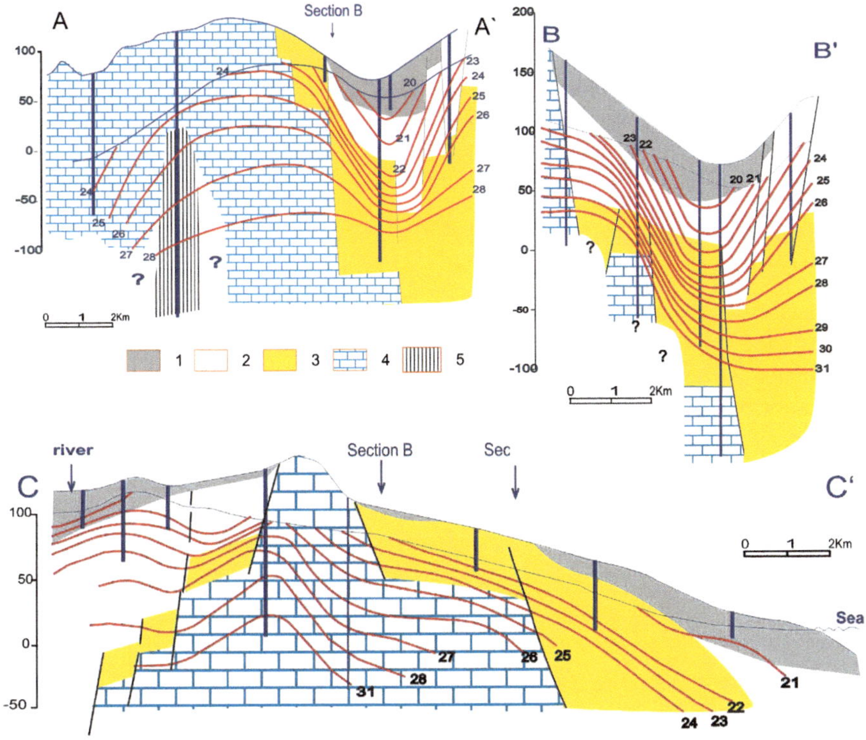


**Fig. 3** Vertical temperature profiles. The vertical scale in these cross sections is exaggerated in order to show the temperature clearly. The location of the profile is given in Fig. 2. 1: Quaternary, 2: Pliocene 3: Miocene 4: Dolomite 5: Phyllites and calcoschists

fractures. The distribution of the isotherms in profile B indicates a stretch where temperatures are less than 22 °C and with a shallower gradient, in a similar manner to that described for profile A. This is particularly clear on the edge of the Sierra of Gádor, where there is a horst-like stepped structure. The elevation of the isotherms towards the ground surface are characteristic of a preferential heat flow across the more transmissive fracture zones (Vasseur and Demongodin 1995). Water temperature rises along the edge of the Sierra, especially in areas crossed by several fracture systems of the carbonate deposits. These fractures indicate intense neotectonic activity and are closely related to the active seismicity of the region (Sanz de Galdeano et al. 1995). Laterally, the isotherms lay closer to the horizontal, since the zone of highest temperature is limited to the horst and to the zone of intersecting fractures.

Profile C reflects the block structure of the Carbonate Aquifer, which lies beneath the Miocene and Pliocene deposits of the depression. The isotherms adopt a cone shape, which is associated with a more localised groundwater flow. The sector with highest temperatures correlated to the horst structure detected in profile C. This thermal zone is situated along the margin of the Carbonate Aquifer and, as in other areas of geothermal anomalies in the Betic Cordillera, at the intersection of two regionally-important fracture systems, which are responsible for the actual location of the Sierra Alhamilla block with respect to the alienation of the Sierra Nevada-Sierra of los Filabres (Sanz de Galdeano 1985).

# 4 Final Considerations

The analysis of temperature enables the variety of thermal situations of the Carbonate Aquifer of the Lower Andarax to be represented. The various thermal situations are related both to the disposition of strata and to the tectonics of the area, which determine the thermal flows. Higher temperatures are recorded in sectors where the fractures with the greatest neotectonic and seismic activity converge; this is a consequence of the thermal flows that tend to be associated with this kind of structure.

**Acknowledgements** This work was undertaken within the framework of project PO6-RNM-01696, financed by the Junta of Andalucía

# References

Domenico PA, Schwartz F (1990) Physical and Chemical Hydrogeology. John Wiley and Sons Inc., New York

Fernández M, Marzan I, Correia A, Ramalho E (1998) Heat flow, heat production, and lithospheric thermal regime in the Iberian Peninsula. Tectonophysics 291:29–53

Sánchez Martos F (1997) Estudio hidrogeoquímico del Bajo Andarax (Almería). Tesis doctoral. Universidad de Granada

Sanz de Galdeano C (1985) Estructura del borde Oriental de la Sierra de Gádor (zona Alpujárride, Cordilleras Béticas). Acta Geológica Hispánica. 20:145–154

Sanz de Galdeano C, López Casado C, Delgado J, Peinado MA (1995) Shallow seismicity and active faults in the betic cordillera. A preliminary approach to seismic sources associated with specific faults. Tectonophysics 248:293–302

Vassseur G, Demongodin L (1995) Convective and Conductive heat transfer in sedimentary basins. Basin Research 781:67–79

# Flash Floods Forecasting in a Karstic Basin Using Neural Networks: the Case of the Lez Basin (South of France)

L. Kong A Siou, A. Johannet, S. Pistre, and V. Borrell

**Abstract** The present study focuses on the modeling of the Lez karstic system (France) using artificial neural networks. Two methods of variable selection were compared: cross-correlation and cross-validation. In both cases, the artificial neural network forecasts closely matched the measured discharge, giving Nash criteria higher than 0.8, which can thus provide satisfactory 2-day forecasts.

## 1 Introduction

Flash flood forecasting in populated areas is a difficult and crucial task. This paper addresses the case of the Lez aquifer, near the city of Montpellier (South of France). The first section shows that several effects interact non-linearly, leading to a very difficult modelling problem. The second section therefore presents neural networks, their limitations and how to use them in black-box modelling. The Lez outflow prediction is derived, and assessed on a test set, independent of the learning database, thereby obtaining a good forecast quality, which would enable two-day ahead warning for the population.

## 2 Problem Statement

**Flash Flooding of the Lez.** The Lez is a 25.8 km long coastal river in south of France (Fig. 1). It is sustained by a 380 km$^2$ karstic aquifer (Thiery et al. 1983)

L. Kong A Siou, A. Johannet
Ecole des Mines d'Alès, CMGD, 6 avenue de Clavières, 30319 Alès Cedex, France,
e-mail: Line.Kong-A-Siou@ema.fr

L. Kong A Siou, S. Pistre, V. Borrell
Université Montpellier II, Hydrosciences Montpellier, Place E. Bataillon, 34095 Cedex 5, France

B. Andreo et al. (Eds.), *Advances in Research in Karst Media* 
DOI 10.1007/978-3-642-12486-0, 

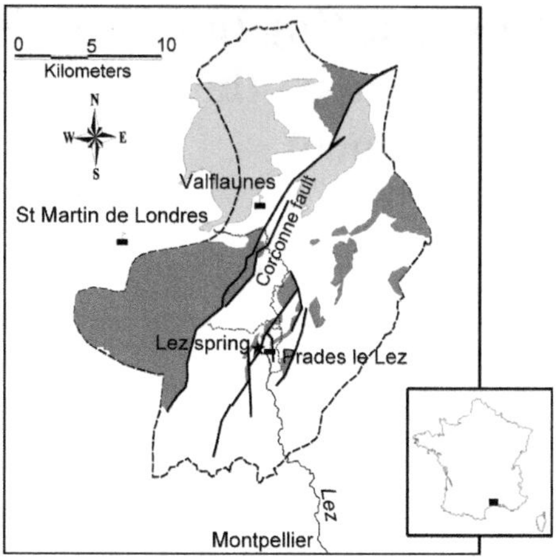


**Fig. 1** Watershed of the Lez spring and available rain gauges. *Dotted line* is the limit of the supposed hydrogeologic basin. *Dark gray areas* correspond to the main karstic aquifer outcrops, *light gray areas* to other karstic formation outcrop, *white areas* are impermeable areas

and rises 10 km north of Montpellier (400,000 inhabitants). The aquifer contains two main compartments separated by the Corconne fault, whose influence is poorly understood. This fault brings the main karstic formation (West) into contacts with the impermeable formations (East) (Fig. 1).

The climate is Mediterranean, subject to intense rainfalls during the autumn, delivering huge floods, called épisodes cévenols. These events can have catastrophic consequences (during the 2002 event, 25 fatalities and 2 billion Euros of damage occurred in the Languedoc-Roussillon Region), worsened by the proximity of the city of Montpellier and the complexity of the surface and underground water interactions (Roesch and Jourde 2006).

Three daily rainfall time series are available (rain gauges): Prades-le-Lez, near the spring, Valflaunès and Saint-Martin-de-Londres at the north of the basin. Moreover, in addition to the complexity of the heterogeneous aquifer, to the surface and underground water interactions, and to the nonlinear flash-flood behavior; the spring is exploited for providing drinkable water to Montpellier.

**Active Water Management.** Since 1982 the Lez spring is exploited with an "active water management" in order to satisfy the water demand of the city of Montpellier: instead of pumping water inside the spring pool, drillings were carried out and enable pumping inside the main drain of the spring, often drying up the spring and emptying a part of the aquifer. Nevertheless, to maintain a minimal outflow at the spring, a Public Utility Declaration imposes to release 160 l/s in the river when the spring is dry. The discharge is therefore, at least for a part, artificial.

In order to build a model able to simulate "natural" outflows at the spring, (Dörfliger et al. 2008) have taken into account the relations between: (i) water level in the pool, (ii) overflow discharge, (iii) water table in the boreholes and (iv) pumped discharge. Thus, a "natural" discharge chronicle has been generated since 1970 until 2005. Based on this chronicle and to characterize the behavior of the Lez aquifer, Fleury et al. (2008) have designed a reservoir-based model to simulate spring discharge and water level in the main drain. Its inputs were a combination of rainfall from the three pluviometers cited above. Nash criterion of 0.80 was obtained for the discharge simulation.

## 3 Neural Networks for Flash Flood Forecasting

**Nonlinear Modeling.** As pointed out above, there are multiple and complex phenomena which interact to produce outflows at the spring. The rainfall-runoff relation is thus nonlinear, depending on variables that are difficult to measure and/or estimate. Therefore, given the insufficient knowledge about physical processes, machine learning appears suitable for elaborating a nonlinear model of the Lez spring flash flooding.

Neural networks are statistical models that calculate their parameters from a calibration dataset; their generalization abilities are evaluated using an independent test set of data. Neural networks have seldom been used to simulate or forecast flash flooding. Several authors note that neural networks cannot generalize learned behavior to the test set (Piotrowski 2006). However, Toukourou et al. (2009) showed that cévenol flash floods can be forecasted accurately, without rainfall forecasts, provided overfitting is avoided throughout the modeling process by introducing regularization methods: early stopping (Coulibaly et al. 2000), model selection and complexity control by cross-validation.

**Model Design.** The proposed model is intended, at discrete time $kT$, $(k \in N+)$, to forecast or simulate the outflow at time $(k + f)T$, where $f$ is the forecasting horizon $(f \in N+)$.

Denoting the predicted value of the quantity of interest by $q(k)$, the observed outflow by $q^p(k)$, the rainfall vector by $\boldsymbol{r}(k)$, and the nonlinear function implemented by a feedforward neural network by $g_{\mathrm{NN}}$, the input-output "neural" model was designed based on (Nerrand et al. 1993):

$$q(k) = g_{\mathrm{NN}}\big(q^p(k-1), q^p(k-2), \ldots, q^p(k-n), r(k), r(k-1), \ldots, \\ r(k-w+1)\big) .$$

Rainfall information was conveyed to the network as a sliding window of width $w$, whose optimal value was chosen as described in Sect. 4. The information about past outflows was provided by the outflow for the previous day $q^p(k-1)$.

The feedforward neural network used was a multilayer perceptron with one hidden layer of $N_c$ sigmoid neurons and a linear output neuron. This model was chosen because of its properties of universal approximation and parsimony (Dreyfus 2005).

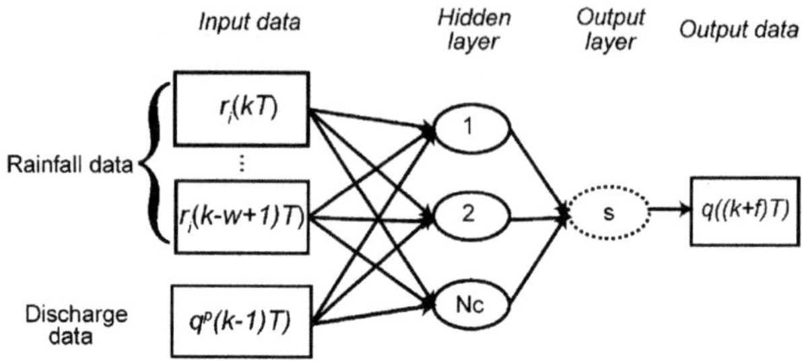


**Fig. 2** The model is fed by rainfall measurements over a temporal window of width $w$. The output is the forecast outflow, $f$ sampling periods ahead

The usual squared error cost function was minimized by the Levenberg-Marquardt algorithm (Dreyfus 2005) during calibration. Regularization was performed using early stopping (Coulibaly et al. 2000; Toukourou et al. 2009).

**Model Selection.**

*Model selection by cross validation.* Using the database of $Y = 16$ years, one year was selected as the test set, and two years as the stop set (early stopping). On the $K = 13$ remaining years, one year, each one at its turn, was set apart as the validation set; then learning was performed $K$ times on $K - 1 = 12$ years (Dreyfus 2005) and the mean squared error was calculated each time on the validation set. The generalization ability of the model was assessed by a Nash criterion cross-validation score:

$$S = \sqrt{\frac{1}{K} \cdot \sum_{i=1}^{K} \text{Nash}_i}\,, \quad \text{Nash}_i = 1 - \frac{\text{MSE}_i}{\sigma_i^2}$$

where $\text{Nash}_i$ is the Nash criterion of the forecast outflow, for year $i$ of the validation set (one complete year, $(i = 1 \ldots 13)$), $\sigma_i^2$ is the variance of the outflows observed at year $i$ and $\text{MSE}_i$ is the mean squared error calculated on the same year $i$. The model complexity was selected in picking up the configuration for which the cross-validation score is the best. The above procedure was used to select the appropriate complexity (i.e. the appropriate values of rainfall width $w$, and the number of hidden neurons $N_C$). For each model, 50 different parameter initializations were performed. After selecting, a final model was trained for each horizon $f$, from 13 sequences: every year except the test year and the early stopping years. Its performance was assessed on the test sequence.

*Rainfall Width Selection by Cross-Correlation.* Another way to select the width $w$ is to use cross-correlation between rainfall and discharge. According to (Mangin 1975), when the correlation is lower than 0.2, both variables can be considered as independent. Cross-correlations were calculated event by event, the time observed to reach the 0.2 correlation was recorded for each event: it indicates the period during which rainfall is correlated to discharge, thus determining the required width $w$.

## 4 Results and Discussion

Due to the exploitation of the spring since 1982, discharge time series for the Lez spring are available from 1988 to 2004, distributed as follows: a 13-year learning set (01/09/1988–31/08/1993 and 01/09/1995–31/08/2003), a 2-year stopping set (1993–1994 and 1994–1995) and a 1-year test set (01/08/2003–31/08/2004).

Variables were selected by cross-correlation or cross-validation. Equivalent results were obtained with both methods. Nevertheless, cross-correlation gave slightly more parsimonious models than cross-validation.

**Rainfall Width Selection Using Cross-Correlation.** Rainfall-runoff cross-correlations with the Lez spring discharge were calculated for each of the 3 rain gauges. Their maximal cross-correlation values were the same (0.6). One model was then designed for each gauge. Table 1 presents the Nash criteria for simulation ($f = 0$) and forecasts at different horizons ($f = 1, 2, 3$).

**Table 1** Nash criteria for simulation and forecasting

| Forecasting horizon $f$ | 0 | 1 day | 2 days | 3 days | Selected architecture |
|---|---|---|---|---|---|
| Valflaunès | 0.95 | 0.90 | 0.84 | 0.73 | 8 inputs, 5 hidden neurons |
| Prades | 0.94 | 0.90 | 0.84 | 0.73 | 9 inputs, 3 hidden neurons |
| St Martin | 0.94 | 0.90 | 0.87 | 0.75 | 9 inputs, 3 hidden neurons |

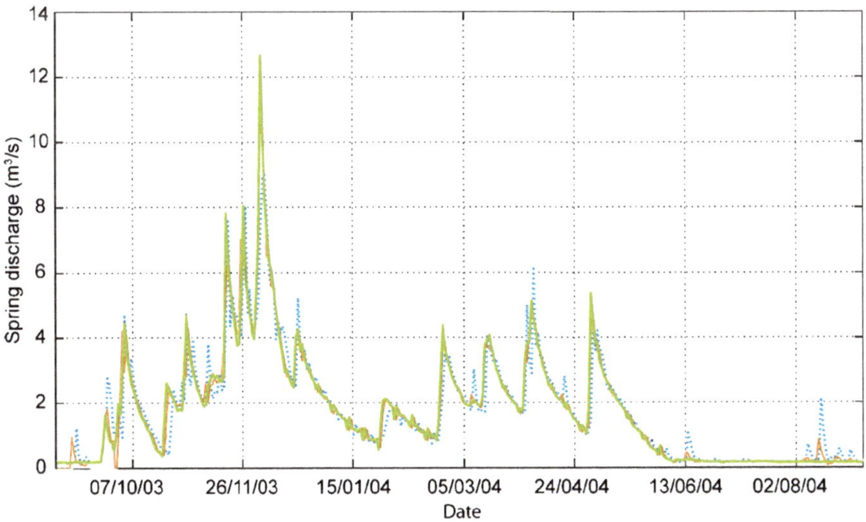


**Fig. 3** Simulation and forecasting at a 2-day horizon using *Valflaunès* data: *green bold line*: observed discharge, *orange thin line*: simulation, *blue dotted line*: two-day forecasting

The Nash criterion was about 0.95 for simulation and 0.84 for two-day forecasting (Fig. 3). Concerning forecasting, peaks remained adequately synchronized whichever the rain gauge used. Both simulation and forecasting proved satisfactory and provided reliable two-day outflow forecasts, respectively.

## 5 Conclusions

Understanding karst behavior during floods is a major issue for improving flood forecasting. The Lez karstic system is complex and affected by pumping. Neural networks therefore appeared appropriate for designing a nonlinear model of its behavior. In this study, feedforward neural networks are calibrated for each of the three rain gauges available, using two methods of input selection. Simulation is satisfactory using both methods and all three rain gauges, leading to Nash criteria higher than 0.90. In addition, the selected model is able to correctly predict hydrographs, with a forecasting horizon of two days. Given this encouraging result, a challenging aim is to capitalize on these good modeling performances to obtain information on the physical behavior of the aquifer.

**Acknowledgements** The authors acknowledge METEO-France for providing rainfall datasets, and Gilles LeGac, from the DIREN Montpellier for providing discharge data. We are very grateful to Nathalie Dörfliger, Perrine Fleury, Daniel Diep, Pierre Roussel-Ragot, Bernard Vayssade and Marc Vinches for helpful discussion. We also thank Dominique Bertin for his highly effective collaboration in the design and implementation of the Neural Network simulation tool: RnfPro.

## References

Coulibaly P, Anctil F, Bobée B (2000) Daily reservoir inflow forecasting using ANN with stopped training approach. Journal of Hydrology, 230:244–257

Dörfliger N, Jourde H, Ladouche B, Fleury P, Lachassagne P, Conroux Y, Pistre S, Vestier A (2008) Active water management resources of karstic water catchment: the example of Le Lez spring (Montpellier, France). In World Water Congress. Montpellier

Dreyfus G (2005) Neural networks, methodology and applications. Springer

Johannet A, Vayssade B, Bertin D (2008) Neural Networks: From Black Box towards Transparent Box Application to Evapotranspiration Modeling. Int. Journal of Comp. Int., 4(3)

Jourde H, Roesch A, Guinot V, Bailly-Comte V (2007) Dynamics and contribution of karst groundwater to surface flow during Mediterranean flood. Environ. Geol., 51(5), 725–730

Mangin A (1975) Contribution à l'étude hydrodynamique des aquifères karstiques. Thèse Université de Dijon. Laboratoire Souterrain du Centre National de la Recherche Scientifique

Nerrand O, Roussel-Ragot P, Personnaz L, Dreyfus G, Marcos S (1993) Neural networks and nonlinear adaptive filtering: Unifying concepts and new algorithms. Neural Computation, 5(2), 165–199

Piotrowski A, Napiorkowski JJ, Rowinski PM (2006) Flash Flood Forecasting by Means of Neural Networks and Nearest Neighbour Approach – A Comparative Study. Nonlin. Processes Geophys. 13: 443–448

Roesch A, Jourde H (2006) Incidence d'une gestion active de la ressource en eau en milieu karstique sur le risque hydrologique. Exemple du fleuve Lez (Montpellier, France). In GIRE3D Conference, Marrakech

Thiery D, Bérard P, Camus A (1983) Captage de la source du Lez. Etude de la relation entre la source et son réservoir aquifère. Rapport public BRGM/83-SGN-167-LRO. BRGM

Toukourou MS, Johannet A, Dreyfus G (2009) Flash Flood Forecasting by Statistical Learning in the Absence of Rainfall Forecast: A Case Study. In European Artificial Neural Networks Conference. pp. 98–107

# Hydrological Modeling of an Alpine Dolomite Karst System

A. Hartmann, M. Kralik, F. Humer, J. Lange, and M. Weiler

**Abstract** In most cases dolomite karst systems show less karstification than limestone karst systems, especially in terms of the formation of a conduit system. This study site is a dolomite karst system located in the northern part of the Kalkalpen national park, Austria. Preceding water balance investigations and information gained by artificial and environmental tracers provided the base for the development of a conceptual model, which represents the dominant processes of this system. Storages and fluxes are described by a series of reservoirs each representing a particular element of the system, i.e., soil, interflow, preferential flow and flow through the fissures. First, the model predicts only runoff. Then, to check the representation of subsystems and the choice of parameters, it simultaneously simulates $\delta^{18}O$ composition and runoff. Regional sensitivity analysis shows that including $\delta^{18}O$ information changes parameter identifiability but at the same time reduces model performance.

## 1 Introduction

In karst hydrology modeling of the rainfall-runoff relationship can be made from different points of view: physical (hydrogeological) models, conceptual (reservoir) models, and black-box models (Le Moine et al. 2008). However, in most cases, due to the complexity of hydrologic and hydrogeological characteristics and unavailability or limitation of groundwater monitoring data, conceptual or black-box models are applied (Jukic and Denic-Jukic 2009).

In this study a conceptual model is applied on a dolomite karst system in the Austrian Alps. The particular challenge is that this system is drained by several springs

A. Hartmann, J. Lange, M. Weiler
Institute of Hydrology, Freiburg University, Fahnenbergplatz, 79098 Freiburg, Germany,
e-mail: andreas.hartmann@hydrology.uni-freiburg.de

M. Kralik, F. Humer
Umweltbundesamt – Environment Agency Austria, Spittelauer Lände 5, 1090 Vienna, Austria

B. Andreo et al. (Eds.), *Advances in Research in Karst Media*
DOI 10.1007/978-3-642-12486-0, © Springer 2010

with strongly varying hydrological behavior and by diffuse flow to the surrounding streams. A model is Developer to simulate stream flow at two stream sections simultaneously. Additionally, mixing equations, which are commonly used to characterize karst systems (e.g. Lee and Krothe 2001), are coupled with the runoff simulations to simulate $\delta^{18}O$ variations at the stream sections and at three karst springs without runoff observations.

## 2 Study Site

The study site (Fig. 1) is located in the northern part of the Kalkalpen national park, the largest protected forest area in Austria. Its altitude ranges from 550 m to 956 m a.s.l. with Norian dolomite (Hauptdolomit) as the main rock type, which is partly overlain by Platten-Limestone and Jurassic/Cretaceous limestone and marl (Kralik et al. 2009). Due to the dominating dolomite, the catchment is not as heavily karstified as limestone systems, but shows typical karst features such as conduits and sink holes (Jost et al. 2009). Tracer investigations with artificial tracers, $\delta^{18}O$, $^{3}H$, $^{3}He$ and CFCs reported by Kralik et al. (2009) indicate that there are three different flow components: a fast (~ hours), a delayed (~ days) and also a very slow component (~ years).

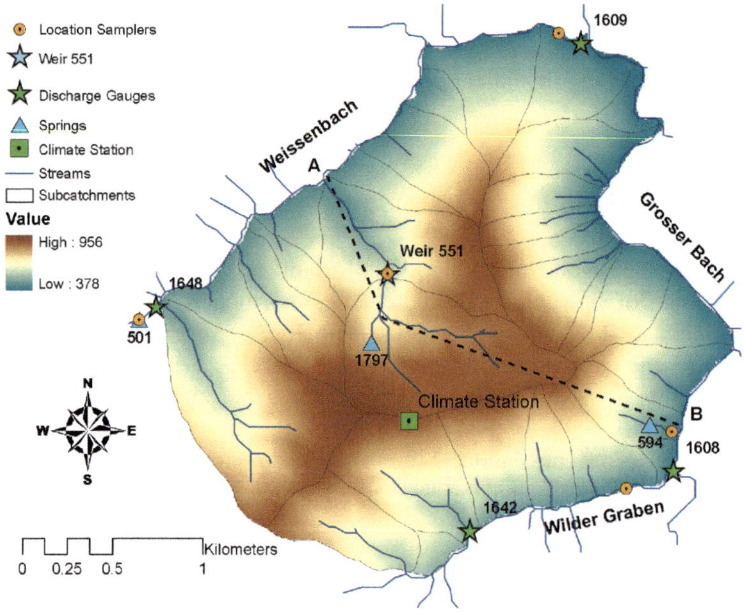


**Fig. 1** Map of the study site Zoebelboden (14°27′E, 46°50′N, Austria); the *dashed line* indicates the cross section in Fig. 2

## 3 Methodology

The integrated information given by the catchment's intrinsic properties (soils, geology, altitude difference, etc.) and the observation and interpretation of stream flow and environmental/artificial tracers yielded a conceptual description of the system's functioning presented in Fig. 2. This resulted in a model structure consisting of an overflow reservoir connected to three parallel linear reservoirs. The overflow reservoir represents the soil water storage of the system. It is controlled by two parameters, its maximum storage $OF_{max}$ and its initial overflow deficit $OF_{def}$. The linear reservoirs below represent the slow flow, interflow and preferential flow components. Each is controlled by two parameters: initial flow rate $Q_{0,slow}$, $Q_{0,interflow}$ and $Q_{0,fast}$ and storage constant $K_{slow}$, $K_{interflow}$ and $K_{fast}$. The fraction contributing to each of the reservoirs is controlled by three parameters $frac_{slow}$, $frac_{interflow}$ and $frac_{fast}$ representing their contributing volume implying that there is no interaction among the three reservoirs. These fractions may vary for each observation point but it is assumed that all other parameters stay the same since they represent intrinsic system properties. In order to additionally simulate $\delta^{18}O$, mixing equations were added to all reservoirs. Evapotranspiration was assumed to be not significant in the scale of single events.

With this structure the model was applied to an observed rainfall-runoff event in August 2009. During this event two streams passing the study site (Weissenbach and Wilder Graben, Fig. 1) were continuously monitored before (gauges 1648 and 1642) and after (gauges 1609 and 1608) passing the study site. Rating curves were used to transform the observed water levels into discharge and because of the low precision of the water level loggers ($\pm 1$ cm), a moving average was calculated. The difference between inflow and outflow was regarded as the contributions from the study site and neighboring catchments. Additionally, automatic water samplers were installed close to the stream gauges 1609 and 1608 and near to three different springs (501, 1797 and 594) to monitor $\delta^{18}O$ dynamics. Precipitation and its $\delta^{18}O$ composition were monitored at a climate station in the center of the catchment (Fig. 1).

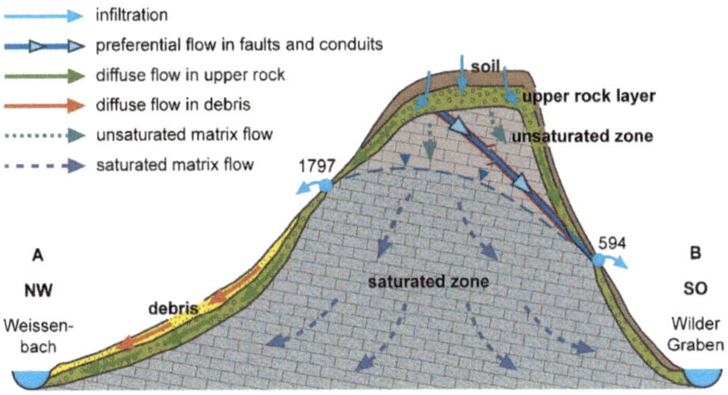


**Fig. 2** Schematic cross section of the study site from northwest to southeast intersecting the weir and springs 1797 and 594 (indicated in Fig. 1)

The model was calibrated using the Shuffled Complex Evolution (SCE) algorithm (Duan et al. 1993) in two steps: first, only runoff observations were used to find the optimal parameter set and, second, both runoff and $\delta^{18}O$ observations were considered. As objective function the Nash-Sutcliffe (NS) efficiency was used. After the calibration a modified version of the extended Regional Sensitivity Analysis proposed by Wagener et al. (2001) was applied to investigate the identifiability of the parameters with and without $\delta^{18}O$ data. In this approach a cumulative normalized distribution of the best performing parameter sets found by Monte Carlo Sampling is used. If a parameter is not sensitive, its cumulative distribution will plot on a straight line, which represents an uniform distribution. The larger the deviation from this line the higher is the identifiability of a parameter.

# 4 Results

Calibration with SCE yielded total contributing areas ($frac_{\text{slow}} + frac_{\text{interflow}} + frac_{\text{fast}}$) of 0.5 km$^2$ and 0.7 km$^2$ for stations 1608 and 1609, respectively, whereby interflow was only abundant at 1609 ($frac_{\text{interflow}} = 0.25\,\text{km}^2$). The other optimized parameters are $OF_{\max} = 119.7/102.6\,\text{mm}$, $OF_{\text{def}} = 19.4/19.6\,\text{mm}$, $K_{\text{slow}} = 15.0/23.0\,\text{a}$, $K_{\text{interflow}} = 3.9/4.1\,\text{d}$ and $K_{\text{fast}} = 18.4/16.6\,\text{h}$, and yield NS efficiencies for the runoff of 0.87 and 0.86 for calibration by flow only/calibration by flow and $\delta^{18}O$, respectively. NS efficiency for flow and $\delta^{18}O$ in the second calibration step is 0.71. Some parameters change after adding $\delta^{18}O$ (e.g. $OF_{\max}$ or $K_{\text{slow}}$) and some stay almost the same (e.g. $OF_{\text{def}}$). $Q_{0,\text{slow}}$, was set to represent pre-event discharge, while both $Q_{0,\text{interflow}}$ and $Q_{0,\text{fast}}$ were assumed to be zero.

The results of the calibration are shown in Fig. 3. For both gauging stations simulated discharge generally follows the observations. However, deviations can be observed before the raising limb and at the falling limb of the hydrograph. There are no strong differences between runoff simulations using only runoff for calibration

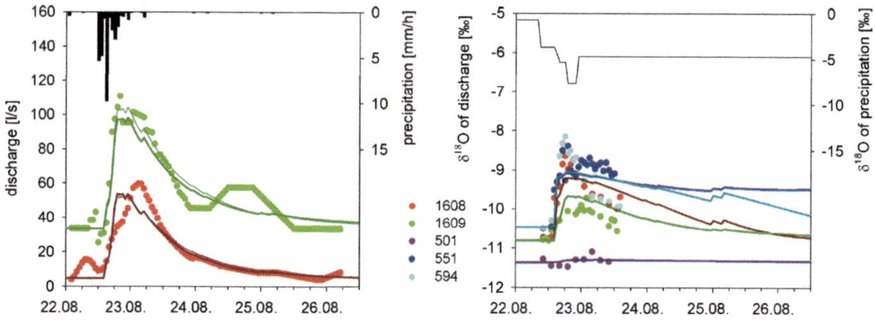


**Fig. 3** *On the left*: observed runoff (*dots*), runoff modelled by calibrating the runoff only (*thin lines*) and runoff modelled by calibrating the runoff and $\delta^{18}O$ (*thick lines*); *on the right*: observed (*dots*) and modeled (*lines*) $\delta^{18}O$ compositions

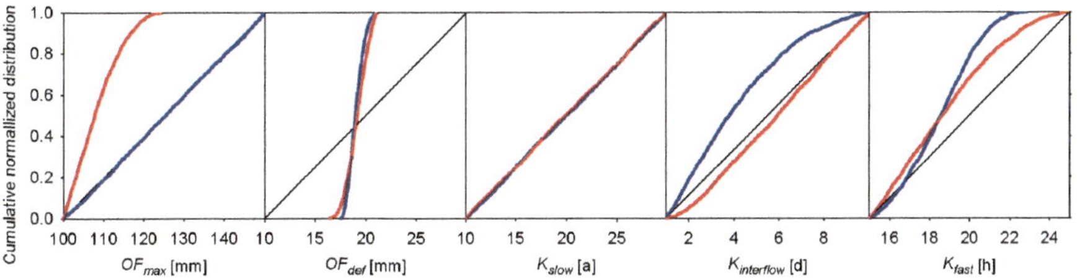


**Fig. 4** Cumulative normalized distributions of the parameters using only runoff (*blue lines*) and using both runoff and $\delta^{18}$O (*red lines*) for Monte Carlo analysis; a *straight line* (*grey*) indicates a non-sensitive parameter

and those using both, runoff and $\delta^{18}$O. $\delta^{18}$O predictions follow the overall variation of the observations (precision ±0.2‰) but almost all peaks, except for 1609, are underestimated. The simulated shape varies according to the contribution of the different reservoirs.

Monte Carlo runs totaled 50,000, from which the best 2000 were used for the regional sensitivity analysis. This was done using only runoff and using both runoff and $\delta^{18}$O. The results in Fig. 4 indicate that the identifiability of $OF_{\mathrm{max}}$ increases after including $\delta^{18}$O observations. However, the identifiability of $OF_{\mathrm{def}}$, $K_{\mathrm{slow}}$ and $K_{\mathrm{fast}}$ stays almost the same and reduces for $K_{\mathrm{interflow}}$.

# 5 Discussion

The runoff predictions obtained by only using runoff for calibration yield a good agreement with the observations (NS efficiency >0.85). Adding the $\delta^{18}$O observations to the calibration did not change runoff predictions significantly indicating an appropriate choice of the model structure; NS efficiency for runoff decreased from 0.87 to 0.86. The results of the sensitivity analysis corroborate this observation since identifiability of $OF_{\mathrm{max}}$ increases.

The simulation of $\delta^{18}$O reduced the overall model performance (NS efficiency ~0.7) and regional sensitivity analysis showed that $K_{\mathrm{interflow}}$ becomes less identifiable. This might result from simplifications of the mixing equations. It is assumed that inflow to a reservoir, with its respective concentration, is instantaneously and complete mixed with the volume already stored. Especially in the case of preferential flow in fractures or karst conduits, mixing volumes can be much smaller than the whole storage volume.

The obtained contributing areas of the two gauging stations (1.2 km$^2$ in total) seem to be quite small compared to the whole area of the system (5.7 km$^2$). Moreover, there are also flow contributions from neighboring catchments. However, considering that large parts of the runoff may leave the system towards a third stream (Grosser Bach) in the Northeast, which was not measured, and owing to large fractions of the water that might percolate through the unsaturated rock to reach the slow groundwater storage (Humer and Kralik 2008), this value is quite realistic.

Also the other parameters show good agreement with field observations. $OF_{max}$ is within the range of observed field capacities and $OF_{def}$ below their respective available water capacity. $K_{fast}$ is in the order of hours and $K_{interflow}$ within days, representing the fast and delayed flow component. Although $K_{slow}$ is in a realistic range ($\sim$ 20 a), the regional sensitivity analysis reveals that it is not identifiable using the event based dataset. But it could a priori be determined independently by age dating techniques regardless of the calibration and sensitivity analysis.

## 6 Conclusions

In this study a simple model was applied to a dolomite karst system with different flow components draining diffuse and concentrated reservoirs at different locations. A single set of parameters representing the system's intrinsic parameters could be identified using runoff, additional $\delta^{18}O$ observations and knowledge about water ages. Observed variations among different observation points could be explained by different contributions of the flow components. However, mainly due to missing observations at one stream section and due to the event based simulation period, the karst water balance could not be calculated. Additionally, deviations between observation and simulations of $\delta^{18}O$ and regional sensitivity analysis suggested that mixing equations might not be appropriate for modeling solute transport in preferential flow paths. Future work will concentrate on these aspects.

**Acknowledgements** Thanks to Maria-Theresia Grabner, Thomas Dirnböck, Florian Wenter and Johannes Kobler from the Environment Agency Austria for providing their data and valuable advice, to the measurement team, Manuela Nied, Nicole Jackisch, Matthias Ritter, Benjamin Gralher and Julien Farlin, for their brave encouragement in taking samples and measuring runoff during the rainfall-runoff event and to Irene Kohn and Jürgen Strub from the Institute of Hydrology for reviewing the paper and designing the figures.

## References

Duan QY, Gupta HV, Sorooshian S (1993) Shuffled Complex Evolution Approach for Effective and Efficient Global Minimization. Journal of Optimization Theory and Applications, 76:501–521

Humer F, Kralik M (2008) Integrated Monitoring Zöbelboden: Hydrologische und hydrochemische Untersuchungen. Unpubl. Rep. Environment Agency, Vienna, 34

Jost G, Dirnböck T, Grabner MT, Mirtl M (2009) Long term trends and inter-annual variability of nitrogen deposition and nitrogen leaching in a karst watershed. in preparation

Jukic D, Denic-Jukic D (2009) Groundwater balance estimation in karst by using a conceptual rainfall–runoff model. Journal of Hydrology, 373:302–315

Kralik M et al. (2009) Karstwater-ages in an alpine dolomite catchment, Austria: δ18O, 3H, 3H/3He, CFC and dye tracer – investigations. European Geosciences Union, General Assembly, 19.-24. April 2009 Vienna, 11(09180)

Le Moine N, Andréassian V, Mathevet T (2008) Confronting surface- and groundwater balances on the La Rochefoucauld-Touvre karstic system (Charente, France). Water Resources Research, 44(W03403)

Lee ES, Krothe NC (2001) A four-component mixing model for water in a karst terrain in south-central Indiana, USA. Using solute concentration and stable isotopes as tracers. Chemical Geology, 179

Wagener T et al. (2001) A framework for development and application of hydrological models. Hydrology and Earth System Sciences, 5(1):13–26

# Fuente de Piedra Lagoon (Málaga, Spain): a Deep Karstic Flow Discharge Point of a Regional Hydrogeological System

J. Heredia, A. García de Domingo, J.M. Ruiz, and L. Araguás

**Abstract** The Fuente de Piedra lagoon (Málaga, Spain) has a high ecological value. It has a complex hydrogeological system due to the density variation of the surface aquifer fresh water, the saline water and the deep and shallow brines. The conceptual model of this system was the classic model found in semiarid climate closed basins, where brines are located in the discharge area. Both the discovery of brines close to the watershed divide and the hydrochemical and isotopic studies of the brines from the basin and the karst springs from the Guadalhorce river identified: 1, the brines from the water bodies and from the water springs are the same type; 2, these brines are a different type than the shallow brine below the lagoon and 3, the deep water under the lagoon is a mixture between the shallow and the brine linked to a karst system. The potentiometric map of the karst brine shows a decreasing gradient towards the Mediterranean watershed and an increasing gradient under the lagoon. A new conceptual model from the hydrogeological system is thus proposed: a regional karst system with a local discharge point to the lagoon.

## 1 Introduction

The Fuente de Piedra lagoon has a surface of 13.5 $km^2$ and an endorheic basin of 150 $km^2$. It is located in the north of Málaga, between the hydrographic division of the Guadalquivir River (Atlantic watershed) and the Guadalhorce River (Mediterranean watershed) (Fig. 1). Between 1995 and 2008 precipitation, potential evapotranspiration (Thornthwaite) and average annual free laminar evaporation registered values of 467 mm, 830 mm and 1467 mm, respectively (IGME 2009). The lagoon's

J. Heredia, A. García de Domingo, J.M. Ruiz
Instituto Geológico y Minero de España (IGME), C/ Ríos Rosas, 23, 28003 Madrid, Spain,
e-mail: j.heredia@igme.es; a.garcía@igme.es; jm.ruiz.@igme.es

L. Araguás
International Atomic Energy Agency (IAEA), Vienna, Austria, e-mail: l.araguas@iaea.org

B. Andreo et al. (Eds.), *Advances in Research in Karst Media* 
DOI 10.1007/978-3-642-12486-0, 

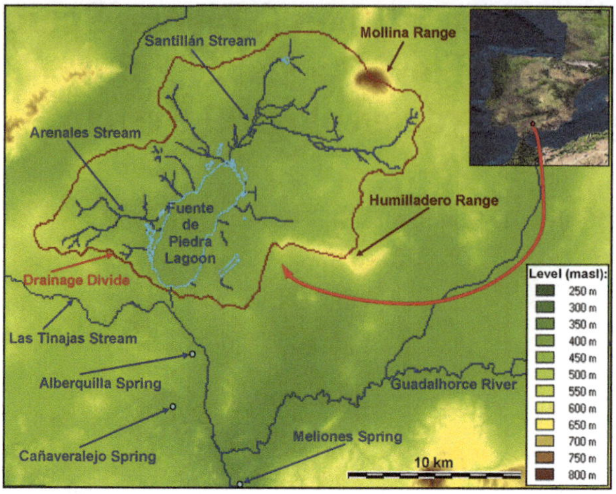


**Fig. 1** Location and geographic features of the Fuente de Piedra lagoon basin

regime is seasonal, drying up in summer. The Santillán and the Arenales streams are the main courses in the hydrological network, with 55 $km^2$ and 20 $km^2$ basins, respectively (Linares et al. 2001).

## 2 Geological Framework

The Fuente de Piedra basin is framed in the Chaotic Subbetic Complex (CSC), External Zone of the Betic Cordilleras, setting up a large olistostrome dated as medium Miocene. It is composed of a clayely, marly and salty matrix which comes from a resedimentation of upper Triassic materials and from a collection of blocks of different sizes, limestones, dolostones, marly limestones and marls dating from Jurassic to Oligocene (Vera and Martín Algarra 2004). A layer of saline gypsum materials was localised 50 m under the lagoon. It could correspond to a sedimentary concentration of these lithologies in the CSC or to the CSC para-authoctonous substrate. Upper Miocene deposits composed by biocalcarenites, sands, marls and conglomerate intercalations outcroping above the CSC. The sedimentation is discordant to the CSC and it onlaps a paleorelief that causes important thickness variations. Quaternary deposits lie above these materials.

The regional tectonic pattern is determined by CSC saline-gypsum sediment distribution. Tectonic displacement processes, deformations due to gypsum-anhydrite transformations, dissolution of evaporitic materials and halokynetic movements are the dominant structural types. Karstification is the most relevant morphologic process. It develops over the CSC gypsum-saline materials and is responsible for the lagoon's genesis.

## 3 Hydrogeology of the Basin

Two aquifers are defined inside the basin: an upper one and a lower one. At the same time, three flow systems are characterised: one inside the upper aquifer and two inside the lower one.

**Upper Aquifer.** It is formed by Upper Miocene calcarenite, sand and conglomerate deposits, Quaternary detritic sediments and Jurassic limestone and dolostone blocks. This aquifer is the upper flow system: unconfined, intergranular and locally karstic.

**Lower Aquifer.** The CSC represents a heterogeneous, mainly karstic, aquifer, developed over a clay-gypsum-saline matrix and carbonated and gypsum-saline blocks. It is a confined aquifer and it possibly extends beyond the basin. Two flow systems are identified in the CSC: the upper one, which represents a low permeability medium, mainly developed over clays and marly-gypsum layers, and the lower one, being a karst system, developed over a massive accumulation of gypsum and salts. The geometry of the karst system is difficult to define, as is typical of karst morphology (gypsum and salts).

## 4 Geophysics

Three campaigns of electric tomography and Time-Domain Electromagnetic Exploration were developed. Several brine bodies were identified, not only around the lagoon, but also in the medium basin and close to the watershed divide. Mechanical drillings confirmed the geophysical interpretation and the existence of brine associated to karstic layers about 50 m beneath the lagoon (Ruiz et al. 2007).

## 5 Hydrochemistry and Isotopes

**Hydrochemical Characterization of the Brines.** The presence of brines is the main feature of this system. Before their chemical characterization, they were classified, according to their spatial distribution and their degree of interconnection to the karst CSC lower flow system. Different types of brines were identified and sampled: a shallow brine around the Fuente de Piedra lagoon (< 40 m deep), and a deeper brine close to the. The brine bodies in different parts of the basin are considered representative of the CCS lower system. The brine found in the Meliones and Cañaveralejo hyper-saline karst springs in the Guadalhorce River basin is considered as the reference brine for this study.

The shallow brines around the lagoon are of the *sodium chloride* type (TDS < 200 g/L) and contain high contents of *magnesium*. By contrast, brines from the hyper-saline bodies and springs in the Guadalhorce basin are exclusively *sodium chloride*, originating from the dissolution of evaporitic material. The increase in

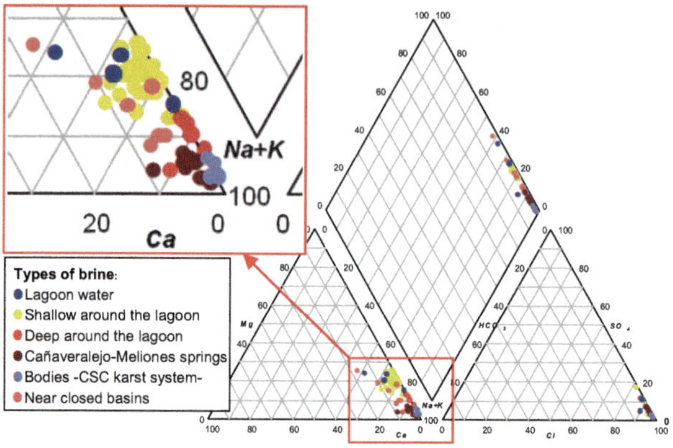


**Fig. 2** Groundwater brine (TDS > 50 g/L) Piper diagram detail

*magnesium* ions in the shallow brines is explained through evaporitic concentration, dissolution and precipitation – *calcite* precipitation and *gypsum* dissolution (dedolomitization). Deep brines around the lagoon are a mixture of *sodium chloride* brines and shallow brines in the surroundings of the lagoon (Fig. 2).

**Isotopic Characterization of the Brines.** The $\delta^{18}O$ data (Fig. 3) show that the brine bodies linked to the CSC lower system as well as the brines of the Guadalhorce basin hyper-saline springs are not evaporated. These brines have the most negative values of the $\delta^{18}O$. The brines in the surroundings of the lagoon show different origins: the shallow brine defines a mixing line from the most evaporated (north and east of the lagoon) to the least evaporated (west and south), while the deep brine is located next to the most evaporated shallow brine.

If TDS is considered (Fig. 3), the shallow and deep brines in the surroundings of the lagoon are different from the other brines in the basin and the Guadalhorce basin hyper-saline springs. The variable dynamics of the karst systems from the Guadalhorce basin hyper-saline springs could explain the wide range of salinities for the same isotopic signal. There is not isotopic evidence that the brines with the highest TDS values have suffered evaporation. Its isotopic signal is similar to that the hyper-saline springs, and its salinity seems to be due to halite dissolution.

The most saline shallow brine (TDS > 150 g/L) is located in the North and East of the lagoon. Its TDS and $\delta^{18}O$ follow a parallel evolution. In the study of the TSD/$\delta^{18}O$ relationship, the shallow brine and surface waters show low dispersion and clear tendencies, but different slopes. This could be due to a differing impact of the evaporation-concentration process in the genesis of both brines due to the existence of other inputs. The deep brine around the lagoon is more saline than the shallow brine, even though evaporation was of the same importance for both of them. At the same time, the salinity of the deep brine around the lagoon is lower than the bodies' salinity. All this could mean that the deep brine around the lagoon

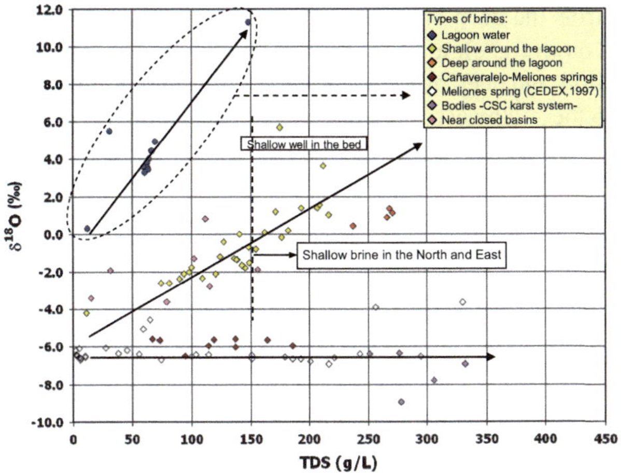


**Fig. 3** $\delta^{18}$O-TDS (g/L) relationships

is a mixture of the bodies' brine and the shallow brine. This assumption had already been suggested by hydrochemical studies.

This analysis would indicate the existence of deep brines only associated with the presence and dissolution of evaporitic materials and with no obvious connection to evaporitic concentration process taking place in the lagoon area. It would also suggest the discharge of deep brines derived from the lower CSC karst flow system of the lagoon.

## 6 Potentiometric and Hydrodynamics

The potentiometric study was carried out using equivalent heads. This approach allows for a logical and joint treatment of different water densities. In order to obtain equivalent heads, the relative density $\rho$ was obtained from the EC profiles. The relative density of the drainage basin ranges from 1 to 1.25.

The potentiometric map of the upper aquifer defines a convergent flow towards the lagoon. Aquifer discharge in the north of the lagoon has been drastically reduced, due to continuous pumping over the last 15 years. Significant discharges are only noticed in the west and southwest sides of the lagoon. The EC levels and profiles from all brines are stable both seasonally and annually. At the bodies, the brine presents a net interface, around 330 m a.s.l.. In equivalent fresh water head terms, the average brine piezometric head linked to CSC deep flows decrease from 422 m a.s.l. at 5 km north of the lagoon, to 420.25 m a.s.l. in the north of the lagoon, to 419 m a.s.l. in the south of the basin. The average heads suggest a low regional gradient decreasing towards the south (0.4 0/00), towards the Guadalhorce basin.

The shallow and deep brine potentiometric study in the surroundings of the wetland was performed through two pairs of drillings, located in the north and southeast of the lagoon. One of the drills of each pair controls the shallow brine, while the other drill controls the deep brine. The equivalent head pairs denote an increasing gradient (12% on average), a 8.8 m head difference in the north and 6.7 m in the southeast. Shallow and deep brine heads, evaporation and salinity all decrease from the northeast to the southwest, in the vicinity of the lagoon.

Descending deep flows were identified with single well flow tests in the deepest depocenter of the Mio-Quaternary aquifer. These flows denote a recharge from the upper aquifer towards the marly matrix. Deep ascending flows were identified in Jurassic blocks (CSC) in the lagoon discharge area.

## 7 Conclusions: Conceptual Model

The lagoon genesis is related to the karstification and dissolution of the CSC gypsum-saline material (Durán et al. 2002). The hydrological conceptual model considers: 1. a shallow flow system defined by the upper Miocene-Quaternary and CSC Jurassic carbonated blocks. It is recharged by precipitation and irrigation return. Discharge occurs in the lagoon and by means of pumping. 2. An intermediate flow system beneath the shallow one. It develops in the low permeability materials from the marly-gypsum matrix and in CSC blocks. Recharge occurs through infiltration from the shallow system with contribution from the underlying karstified system. Discharge takes place in the lagoon. 3. A deep karst system, developed in regional CSC evaporitic materials. Recharge and discharge take place in a regional framework and locally in the lagoon. Upper aquifer sectorized discharge and CSC deep karstic system hypersaline flow discharge through preferential pathways generate a convective flow circulation due to density gradients. This circulation dominates over the convection/diffusion circulation from evaporative process.

**Acknowledgements** We would like to thank Manuel Redón, director of the Natural Reserve Fuente de Piedra lagoon, and to all his staff for their help.

## References

Durán JJ, Molino JA (1986) Karst en yesos del Trías de Antequera (Cordilleras Bélicas). Karstologie Memoires 1. Ass. Franc. de Karstologie, 37–46

IGME (2009) Caracterización hidrogeológica y modelo conceptual de un sistema de flujo con densidad variable: Sistema Hidrogeológico de la laguna de Fuente de Piedra (Málaga). IGME (inédito)

Linares L, Andreo B, Carrasco F, Fernández G, Vadillo I (2001) La escorrentía superficial en la cuenca de la laguna de Fuente de Piedra (Málaga) durante un año húmedo (1997-98). In: Fdez.-Uría J (ed) VII SHRH. AEH

Ruiz JM, Heredia J, G. de Domingo A (2007) Characterization of subsurface brines in Fuente de Piedra lake hydrogelogical system (Málaga, Span) using hydrogeochemical and geophysical techniques. XXXV IAH Congress

Vera JA, Martín Algarra A (2004) Cordillera Bética. In: Geología de España. IGME-Soc. Geol. de España

# Numerical Modelling of Coastal Aquifer Karst Processes by Means of Coupled Simulations of Density-Driven Flow and Reactive Solute Transport Phenomena

Á.M. Sáinz, J.J. Molinero, and M.W. Saaltink

**Abstract** A number of coastal karst features worldwide are supposed to be generated by the coupling between the infiltration of under-saturated rain water together with the mixing produced in the fresh-saltwater interface. This work presents a coupled density-driven flow and reactive solute transport numerical model which is able to reproduce such a morphogenetic conceptual model.

## 1 Introduction

The interaction between fresh and saline water leads to a wide variety of karstic features such as cave formation (Hanshaw and Van Driel 1984; James et al. 1898; Smart et al. 1988). In particular, the Migjorn karstic area in the south and south-east of the Mallorca Island has more than 300 coastal karst morphologies (Encinas 1997), some of them of great tourist relevance as the Covas del Drac in Portocristo. Ginés (2001) studied the karst processes in that area and developed an evolution model of the coastal karst caves (Fig. 1). The main processes responsible for the genesis of the coastal caves are postulated to be the dissolution by infiltration of under-saturated meteoric water and the dissolution produced in the fresh-saltwater mixing zone of the aquifer (Fig. 1). On one hand, the infiltration of rain water (under-saturated in calcite) leads to dissolution of carbonates in the upper part of the aquifer (near the water table). On the other hand, the mixing of two waters of different salinities, even if both waters are in equilibrium with a solid phase, tends to be under-saturated, causing the dissolution of the mineral (Plummer 1975; Sanford and Konikow 1989; Rezaei et al. 2005).

Á.M. Sáinz, J.J. Molinero, M.W. Saaltink
Dept. of Geotechnical Engineering and Geosciences, Technical University of Catalonia (UPC), Barcelona, Spain

J.J. Molinero
Amphos XXI consulting S.L., Passeig de García i Faria, 49–51, 08019 Barcelona, Spain,
e-mail: jorge.molinero@amphos21.com

B. Andreo et al. (Eds.), *Advances in Research in Karst Media*
DOI 10.1007/978-3-642-12486-0, © Springer 2010

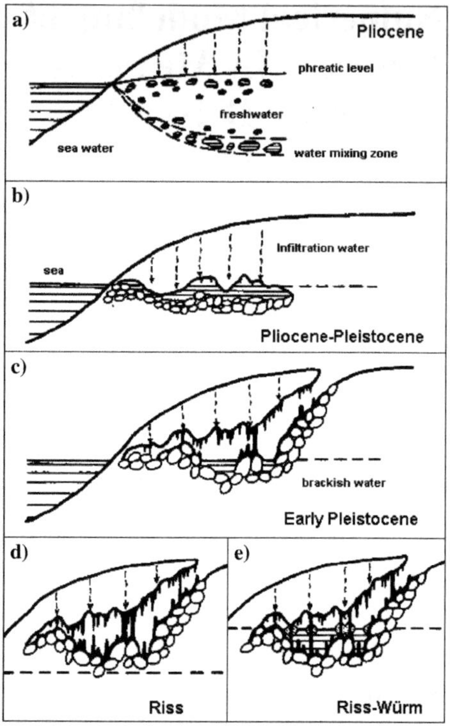


**Fig. 1** Evolution sequence model of morphogenesis of the Migjorn's caves in Mallorca. **a** Initial porosity development due to phreatic fresh-saltwater mixing dissolution. **b** Collapse of the rock that contributes to the cave genesis. **c** Cover of the boulders and stalagmitation. **d** and **e** Speleothem developing affected by the variation of the sea level. (Modified from Ginés 2001)

Reactive transport models are usually treated with some simplifications. One of the most common is to assume constant density of the water. However, in the case of coastal aquifers the influence of the high concentration of solutes on fluid density is no longer negligible when modelling groundwater flow and solute transport. A model is presented here for the coastal karst hydrogeochemical processes based on coupled simulations of density-driven fluid flow and reactive solute transport. The numerical approach has been programmed in a reactive transport code RETRASO (Saaltink et al. 2004) that has been coupled to a multiphase flow code CODE-BRIGHT (Olivella et al. 1996). The main objective of this work is to test numerically the above mentioned conceptual model for the dissolution processes of the initial stage of the genesis of coastal karst caves.

## 2 Model Description

To simulate the karst generation in coastal karst aquifers, geometry is used based on the problem described by Sanford and Konikow (1989) and modified by Rezaei et al. (2005) to model calcite dissolution in saltwater/freshwater mixing zones (Fig. 2).

The problem represents a vertical cross-section perpendicular to the coastline in which the seawater intrudes an initially fresh water filled aquifer. The higher salt

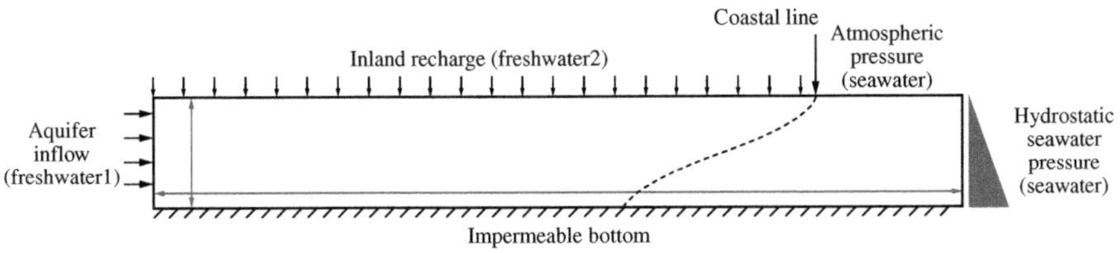


**Fig. 2** Domain and boundary conditions of the model. The *curve* displays the position of the 50% seawater mixing ratio for the stationary-state (the position of the curve is approximate). Vertical scale is exaggerated 2 times

water density induced a lateral flow towards the aquifer. As the denser fluid moves forward, the fresh water influx becomes more important, inverting the velocity direction. Therefore, the fresh water tends to flow upwards when approaching the coast, while seawater intrudes into the lower portion of the aquifer. In this case the wedge of the saline water at the bottom of the aquifer reaches about 450 meters from the coast line. The dotted line in Fig. 2 shows the position of the 50% seawater mixing ratio for the stationary-state.

The model domain is spatially discretized into a triangular finite element grid of 5511 nodes and 10,420 elements. The grid is refined in the area near the coast line, where element sizes are about three meters. To simulate such a large-scale phenomenon, a simulation of 10,000 years was performed with a time step of three days. The inflow values and other parameters of the model are described in Table 1. Table 2 shows the chemical composition assumed for the boundary waters of the problem.

Initial and aquifer inflow water (freshwater1) is a $Ca^{+2}$ and $HCO_3^-$ fresh groundwater in equilibrium with $P_{CO_2} = 0.01$ atm. The recharge water (freshwater2 in Table 2) is assumed to represent fresh rain water recently infiltrated in the carbonate aquifer and therefore under-saturated with respect to calcite. The saline water has a typical seawater composition with a relatively low pH and $P_{CO_2}$. All the boundary waters have been extracted from Appelo and Postma (1994).

**Table 1** Flow parameters

| Parameter | Value | |
|---|---|---|
| $L$ | 1500 m | Domain length |
| $D$ | 100 m | Domain thickness |
| $\Phi$ | 0.30 | Porosity |
| $K$ | $7.4 \cdot 10^{-10}$ m$^2$ | Intrinsic permeability |
| $\alpha_l$ | 12 m | Longitudinal dispersivity |
| $\alpha_d$ | 6 m | Transverse dispersivity |
| $Q$ | 250 m year$^{-1}$ | Aquifer inflow |
| $R$ | 0.5 m year$^{-1}$ | Inland recharge |
| $\rho_f$ | 1000 kg m$^{-3}$ | Freshwater density |
| $\rho_s$ | 1025 kg m$^{-3}$ | Seawater density |
| $M$ | $1.0 \cdot 10^{-3}$ Pa s | Fluid viscosity |

**Table 2** Chemical composition of boundary waters

| Primary species | Seaside boundary saline water (log mol/L) | Initial and aquifer inflow freshwater1 (log mol/L) | Inland recharge freshwater2 (log mol/L) |
|---|---|---|---|
| $H^+$ | −8.091 | −7.292 | −5.252 |
| $Na^+$ | −0.320 | −7.000 | −4.886 |
| $Ca^{+2}$ | −2.022 | −2.770 | −4.796 |
| $Mg^{+2}$ | −1.323 | −7.000 | −5.301 |
| $Cl^-$ | −0.247 | −6.700 | −4.959 |
| $HCO_3^-$ | −3.444 | −2.509 | −4.678 |

**Table 3** Hydrogeochemical processes considered

| Aqueous complexation processes | log *K* |
|---|---|
| $CaCO_{3(aq)} + H^+ \leftrightarrow Ca^{+2} + HCO_3^-$ | 7.0017 |
| $CO_{2(aq)} \leftrightarrow HCO_3^- + H^+$ | −6.3447 |
| $CO_{3(aq)}^{-2} + H^+ \leftrightarrow HCO_3^-$ | 10.3288 |
| $OH^- + H^+ \leftrightarrow H_2O$ | 13.9951 |

| Mineral processes | $\log r_m$ |
|---|---|
| $CaCO_{3(s)} + H^+ \leftrightarrow Ca^{+2} + HCO_3^-$ | $-5.866\ (\Omega - 1)$ |

| Cation exchange processes | $\log K_{Na/I}$ |
|---|---|
| $X - Na + \frac{1}{2}Ca^{+2} \leftrightarrow \frac{1}{2}X - Ca + Na^+$ | 0.4 |

The considered chemical reactions in the model are dissociation of water, aqueous carbonate complexation, cation exchange and precipitation/dissolution of calcite. The complete set of hydrogeochemical processes involved in the model are described in Table 3.

All the reactions were assumed in local equilibrium, except the precipitation-dissolution of calcite, which is treated kinetically. The quantity of calcite in the aquifer is high enough not to restrict its dissolution, and the cation exchange capacity of the aquifer is 0.6 mol/l. The thermodynamic data were taken from the EQ3/NR database (Wolery 1992).

## 3 Results and Discussion

The formation of a stable fresh-seawater mixing zone occurs and chemical reactions can require decades to reach a pseudo steady-state. However, the continuous inflow of rain water under-saturated in calcite and fresh-saltwater mixing prevent the system from reaching equilibrium with respect to calcite. This unstable situation with

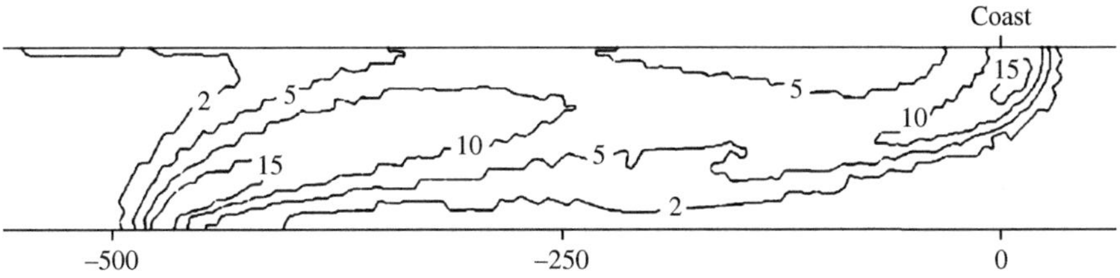


**Fig. 3** Porosity changes after 10,000 years expressed in %

respect to the carbonates system provides the driving force for the long-term development of carbonate dissolution in the aquifer and the subsequent generation of the karst structures.

Figure 3 shows the porosity changes after 10,000 years of simulation. As mentioned, the development of the karstic features is due to the coupling of two main processes which coalesce near the coastal area: 1) dissolution by mixing and, 2) dissolution by meteoric water infiltration.

The mixing dissolution takes place along the whole mixing zone, although the maximum dissolution rates are concentrated at the toe of the saltwater wedge and in the coastal discharge area, where convective flows due to the water mixing are greater. The meteoric water dissolution is only concentrated in the upper part of the aquifer. The under-saturated fresh rain water penetrates into the aquifer and dissolves calcite, increasing its aqueous calcium content. This process occurs until the inflow water becomes saturated in calcite. In the simulated case presented here, the dissolution front due to infiltration of meteoric water reaches a depth of about 6 m.

The porosity increase due to phreatic dissolution is lower than the one generated by water mixing. The total amount of water inflow due to the inland recharge is 40 times lower than the aquifer inflow, and therefore its effect on the system is lower.

As can be observed, the development of porosity by dissolution is quite important and may generate mechanical instabilities on the rock structure, which could finally cause collapse and the formation of the typical karst cave system. It is worth mentioning that the coupling of changes in permeability due to the increase of porosity has not been implemented in the current model. This coupling would enhance the flux velocity and, therefore, the real mixing and dissolution could be even faster in a real system. Future improvements of the model would require considering such a porosity permeability dependency.

## References

Appelo C, Postma D (1994) Geochemistry, groundwater and pollution. A.A.Balkema, Amterdam

Encinas JA (1997) Inventari espeleològic de les Illes Balears. Endins. 21 103–128, Palma de Mallorca

Ginés J (2001) El karst litoral en el levante de Mallorca: una aproximación al conocimiento de su morfologénesis y cronología. Endins 24

Hanshaw BB, Van Driel JN (1984) Role of groundwater in shaping the eastern coastline of the Yucatan Peninsula, Mexico. Groundwater as a geomorphic agent. Allen & Unwin Inc. 281–293, Boston

James JM, Rogers P, Spate AP (1989) The role of missing corrosion in the genesis of the caves of the Nullarbor plain, Australia. Proc. 10th Int. Congress Speleol. 1. 263–265, Budapest

Olivella S, Gens A, Carrera J, Alonso J (1996) Numerical formulation for a simulator (CODE_BRIGHT) for the coupled analysis of saline media. Engineering computations 13(7), 87–112

Plummer L (1975) Mixing of sea water with calcium carbonate ground water. Memoir, 219

Rezaei M, Sanz E, Raeisi E, Ayora C, Vázquez-Suñé E, Carrera J (2005) Reactive transport modeling of calcite dissolution in the fresh-salt water mixing zone. Journal of Hydrology 311(1–4), 282–298

Saaltink M, Batlle F, Ayora C, Carrera J, Olivella S (2004) RETRASO, a code for modeling reactive transport in saturated and unsaturated porous media. Geologica acta 2(3), 235–251

Sanford W, Konikow L (1989) Simulation of calcite dissolution and porosity changes in saltwater mixing zones in coastal aquifers. Water Resources Research 25(4), 655–667

Smart PL, Dawans JM, Whitaker FF (1988) Carbonate dissolution in a modern mixing zone, South Andros, Bahamas. Nature 335, 811–813

Wolery T (1992) EQ3/6, a software package for geochemical modeling of aqueous systems: Package overview and installation guide (Version 7.0). Technical report, UCRL-MA-110662-Pt. 1, Lawrence Livermore National Lab., CA (United States)

# Characteristic Curve of a Groundwater Reservoir. Calculation Models

J.A. López-Geta and J.A. Navarro Iáñez

**Abstract** The sustainable management of an aquifer requires a knowledge of the characteristic curve of the groundwater reservoir, which links the reservoir's topographic elevation to its storage capacity. It is a static concept (invariable in time) constructed on the basis of the available geological and hydraulic data, and corresponds to a polynomial mathematical function by which the storage volume of a groundwater reservoir may be calculated according to its topographic elevation. In the calculation models proposed in this paper, only effective porosity was included as a storage parameter for the characteristic curve.

## 1 Definition of a Groundwater Reservoir

A *groundwater reservoir* is defined on the basis of an aquiferous geological formation that constitutes a system with differentiated hydrogeological behaviour, inasmuch as it is possible to establish hydrogeological limits (associated with structural or hydraulic constraints), a specific conceptual behaviour model (defined according to a certain piezometric regime and hydrogeological balance with clearly defined terms) and distinct hydrochemical characteristics. A groundwater reservoir constitutes a dynamic natural system as a result of the variations imposed by its own natural behaviour patterns, and so the concept of hydrogeological reserve is necessarily linked to the time variable (Honnungar and Uddmari 2005; Rubin et al. 2006).

J.A. López-Geta
Instituto Geológico y Minero de España (IGME), C/ Ríos Rosas, 23, 28003 Madrid, Spain, e-mail: lopez.geta@igme.es

J.A. Navarro Iáñez
University College of Mining Engineering, Madrid Polytechnic University (UPM), C/ Ríos Rosas 21, 28003 Madrid, Spain, e-mail: juanantonio.navarro@upm.es

B. Andreo et al. (Eds.), *Advances in Research in Karst Media*
DOI 10.1007/978-3-642-12486-0, © Springer 2010

## 2 Definition of the Characteristic Curve

The characteristic curve links the groundwater reservoir's topographic elevation to its storage capacity. Todefine it requires knowing the reservoir's geometry and its storage parameters. It constitutes a static concept (invariable in time) of underground water storage and is constructed on the basis of the available geological and hydraulic data, corresponding to a polynomial mathematical function by which the storage volume in the groundwater reservoir may be calculated according to its topographic elevation. That is:

$$V = f(z) = \sum_{n=0}^{N} A_n \cdot z^n \quad (1)$$

Where: $V$, groundwater reservoir volume; $A_n$, adjustment coefficients (depending on the geometry and the storage parameter $S$); $z$, topographic elevation; $N$, degree of the adjustment polynomial.

## 3 Calculation Methods

The models proposed are defined on the basis of geometry and storage parameters (Lambán et al. 2001). According to the availability of data, they allow the characteristic curve to be calculated by models of varying complexity, as set out in Table 1.

**Table 1** Types of characteristic curve in a groundwater reservoir

| TYPES of CHARACTERISTIC CURVE | | STORAGE PARAMETERS | | |
|---|---|---|---|---|
| | | Invariant model | Two-dimensional model | Three-dimensional model |
| GEOMETRIC MODEL | 2D | 2D-Invariant | 2D-2 dimensional | – |
| | 3D simple geological structure | 3D Simple-Invariant | 3D Simple-2 dimensional | 3D Simple-3 dimensional |
| | 3D complex geological structure | 3D Complex-Invariant | 3D Complex-2 dimensional | 3D Complex-3 dimensional |

**2D-Invariant Model.** The geometry is characterized by the following elements: a roof area $z_t(i, j)$ (ground area, if the piezometric regime is free, or area of the roof of the permeable geological formation in the case of a confined aquifer), and an average thickness for the groundwater reservoir $B$. Moreover, the storage parameter $S$[1] is invariant (constant for the whole volume of rock forming the groundwater

[1] This corresponds to the parameter of gravitational storage or effective porosity $S_f$, as the aim is only to calculate the volume of cavities associated with the texture and/or structure of the permeable formation constituting the groundwater reservoir.

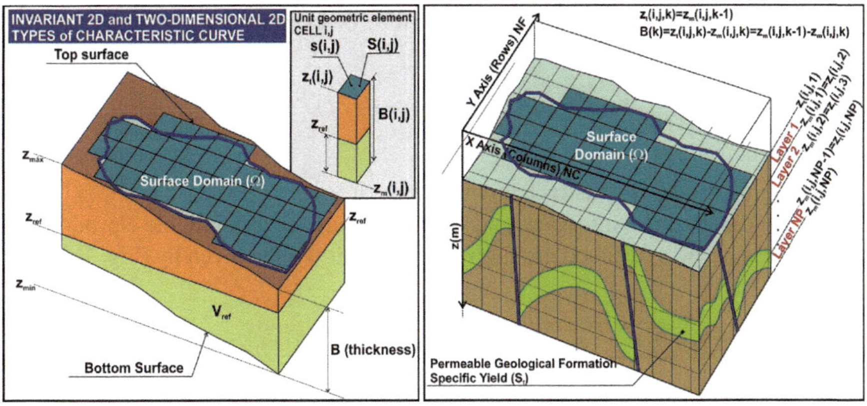


**Fig. 1** *Left*: Graphic explanation of 2D-invariant and 2D two-dimensional characteristic curves; *right*: Univariate-complex 3D geometric model

reservoir). On these premises is formulated (Fig. 1, left):

$$V = S_{\mathrm{f}} \cdot \sum_{n=0}^{N} a_n \cdot z^n \tag{2}$$

Moreover, on the basis of the 2D geometric model, it is possible to formulate the following expression:

$$V_{\mathrm{total}} = B \cdot S_{\mathrm{f}} \cdot \sum_{i=1}^{\mathrm{NC}} \sum_{j=1}^{\mathrm{NF}} s_{ij} \cdot s(i, j) \tag{3}$$

If the digital model is constructed using a regular grid, then $s(i, j) = \mathrm{cte} = \sigma$, and taking a reference topographic elevation $z_{\mathrm{ref}}$, the volume of the associated groundwater reservoir will be given by:

$$V_{\mathrm{ref}} = \sigma \cdot N_{\mathrm{s}} \cdot S_{\mathrm{f}} \cdot \sum_{i=1}^{\mathrm{NC}} \sum_{j=1}^{\mathrm{NF}} \left[ (z_{\mathrm{ref}} - z_{\mathrm{t}}(i, j)) \cdot \left(1 - \lambda_{ij}\right) - B \right] \cdot \delta_{ij} \cdot s_{ij} \tag{4}$$

Setting out from the digital model of the ceiling $z_{\mathrm{t}}(i, j)$, the average thickness value $B$, and the value assigned to the storage parameter $S_{\mathrm{f}}$, with expression 4 yields a series of values $z_{\mathrm{ref}} \rightarrow V_{\mathrm{ref}}$ allowing us, with a polynomial mathematical adjustment, to obtain the $A_n$ coefficient in Eq. 1.

**2D Two-Dimensional Model.** The geometry is characterized by a ceiling area $z_{\mathrm{t}}(i, j)$ and an average thickness for the groundwater reservoir B. A contour map for the storage parameter shows effective porosity, $S_{\mathrm{f}}(i, j)$, in two-dimensional model.

On these premises, the characteristic curve may be expressed as:

$$V_{\text{total}} = B \cdot \sum_{i=1}^{NC} \sum_{j=1}^{NF} s_{ij} \cdot s(i,j) \cdot S_{\text{f}}(i,j) \tag{5}$$

If the digital model is constructed with a regular grid, then $s(i,j) = \text{cte} = \sigma$ and setting out from this expression, and taking a reference topographic elevation $z_{\text{ref}}$, the volume of the associated groundwater reservoir will be given by (Fig. 1, left):

$$V_{\text{ref}} = \sigma \cdot N_{\text{s}} \cdot \sum_{i=1}^{NC} \sum_{j=1}^{NF} s_{ij} \cdot S(i,j) \cdot \left[ (z_{\text{ref}} - z_{\text{t}}(i,j)) \cdot \left(1 - \lambda_{ij}\right) + B \right] \cdot \delta_{ij} \tag{6}$$

**Simple-Invariant 3D Model.** The geometry is characterized by a ceiling area $z_{\text{t}}(i,j)$ and a wall area $z_{\text{m}}(i,j)$. Moreover the storage parameter (effective porosity) $S_{\text{f}}$ is invariant (average value assignable to any groundwater reservoir). On these premises, Eq. 1 may be expressed:

$$V_{\text{total}} = S_{\text{f}} \cdot \sum_{i=1}^{NC} \sum_{j=1}^{NF} [z_{\text{t}}(i,j) - z_{\text{m}}(i,j)] \cdot s_{ij} \cdot s(i,j) \tag{7}$$

If the digital model is constructed using a regular grid, then $s(i,j) = \text{cte} = \sigma$, and taking a reference topographic elevation $z_{\text{ref}}$, the volume of the associated groundwater reservoir will be given by (Fig. 1, left):

$$V_{\text{ref}} = \sigma \cdot S_{\text{f}} \cdot N_{\text{s}} \cdot \sum_{i=1}^{NC} \sum_{j=1}^{NF} \left[ \left( z_{\text{ref}} \cdot (1 - \lambda_{ij}) - z_{\text{t}}(i,j) \cdot \lambda_{ij} \right) - z_{\text{m}}(i,j) \right] \cdot \delta_{ij} \cdot s_{ij} \tag{8}$$

**Two-Dimensional Simple 3D Model.** The geometry is characterized by a ceiling area $z_{\text{t}}(i,j)$ and a wall area $z_{\text{m}}(i,j)$. There is a contour map for the storage parameter (effective porosity) $S_{\text{f}}$ (two-dimensional model). On these premises, Eq. 1 may be expressed:

$$V_{\text{total}} = \sum_{i=1}^{NC} \sum_{j=1}^{NF} S_{\text{f}}(i,j) \cdot [z_{\text{t}}(i,j) - z_{\text{m}}(i,j)] \cdot s_{ij} \cdot s(i,j) \tag{9}$$

If the digital model is constructed using a regular grid, then $s(i,j) = \text{cte} = \sigma$, and taking a reference topographic elevation $z_{\text{ref}}$, the volume of the associated groundwater reservoir will be given by:

$$V_{\text{ref}} = \sigma \cdot N_{\text{s}} \cdot \sum_{i=1}^{NC} \sum_{j=1}^{NF} S_{\text{f}}(i,j) \cdot \left[ \left( z_{\text{ref}} \cdot (1 - \lambda_{ij}) - z_{\text{t}}(i,j) \cdot \lambda_{ij} \right) - z_{\text{m}}(i,j) \right] \cdot \delta_{ij} \cdot s_{ij} \tag{10}$$

**Three-Dimensional Simple 3D Type.** The geometry is characterized by the following elements: a ceiling area $z_t(i, j)$ and a wall area $z_m(i, j)$. Moreover, there is a three-dimensional digital model for the storage parameter (effective porosity) $S_f(i, j, k)$ (distribution of the storage coefficient with depth), which may correspond to a continuous or discontinuous function. Thus it may be expressed as:

| *Continuous function* | *Discontinuous function* |
|---|---|
| $S_f(i, j, z) = \sum_{n=1}^{N} \beta_n \cdot z^n$ | $S_f(i, j, z) = \begin{cases} z \le z_1 \Rightarrow S(i, j, z) = S_1 \\ z_1 \le z \le z_2 \Rightarrow S(i, j, z) = S_2 \\ \dots \\ z_{N-1} \le z \le z_N \Rightarrow S(i, j, z) = S_N \end{cases}$ |

(11)

The 3D geometric model and the three-dimensional storage model can be expressed as:

$$V_{\text{total}} = \sum_{k=1}^{\text{NP}} \sum_{i=1}^{\text{NC}} \sum_{j=1}^{\text{NF}} s_{ijk} \cdot s(i, j) \cdot S_f(i, j, k) \cdot [z_t(i, j, k) - z_m(i, j, k)] \tag{12}$$

If the digital model for the storage parameter corresponds to a function discontinuous with depth, but takes a constant value for each layer considered, then: $S_f(i, j, k) = S_f(k)$. If the digital model is constructed using a regular grid, then $s(i, j) = \text{cte} = \sigma$, and taking a reference topographic elevation $z_{\text{ref}}$, the volume of the associated groundwater reservoir will be given by:

$$V_{\text{ref}} = \sigma \cdot \sum_{k=1}^{\text{NP}} N_s(k) \cdot \left[ \sum_{i=1}^{\text{NC}} \sum_{j=1}^{\text{NF}} s_{ijk} \cdot S_f(i, j, k) \cdot \left[ \left( z_{\text{ref}} \cdot (1 - \lambda_{ijk}) - z_t(i, j, k) \cdot \lambda_{ijk} \right) - z_m(i, j, k) \right] \cdot \delta_{ijk} \right] \tag{13}$$

| | Case 1 | Case 2 | Case 3 |
|---|---|---|---|
| Layer $k$ | $z_{\text{ref}} < z_t(i, j, k)$<br>$z_{\text{ref}} > z_m(i, j, k)$ | $z_{\text{ref}} \ge z_t(i, j, k)$<br>$z_{\text{ref}} > z_m(i, j, k)$ | $z_{\text{ref}} < z_t(i, j, k)$<br>$z_{\text{ref}} \le z_m(i, j, k)$ |
| $\lambda_{ijk}$ | 0 | 1 | 0 |
| $\delta_{ijk}$ | 1 | 1 | 0 |
| $v(i, j, k)$<br>unit volume | $s(i, j) \cdot S_f(i, j, k)$<br>$\cdot [z_{\text{ref}} - z_m(i, j, k)]$ | $s(i, j) \cdot S_f(i, j, k)$<br>$\cdot [z_t(i, j, k) - z_m(i, j, k)]$ | 0<br>0 |

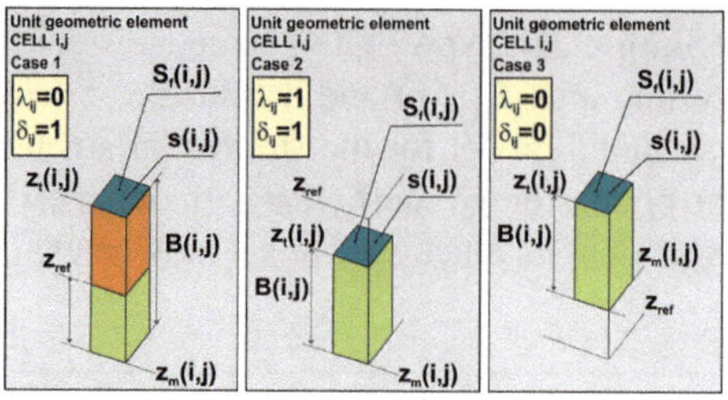


**Fig. 2** Graphical explanation of the variables $\lambda_{ijk}$ and $\delta_{ijk}$

**Complex 3D Model.** In this case the geometry is complex as a result of an intricate geological structure that does not allow defining the groundwater reservoir's geometry by ceiling and wall areas, so it is necessary to define a homogenous three-dimensional grid characterized by a series of depth layers delimited by a series of parallel areas (Fig. 1, rigth). Considering the 3D geometric model and the three-dimensional storage model, it may be expressed as:

$$V_{\text{total}} = \sum_{k=1}^{\text{NP}} \sum_{i=1}^{\text{NC}} \sum_{j=1}^{\text{NF}} s_{ijk} \cdot s(i,j) \cdot S_{\text{f}}(i,j,k) \cdot [z_{\text{t}}(i,j,k) - z_{\text{m}}(i,j,k)] \tag{14}$$

If the digital model is constructed using a regular grid, then $s(i, j) = \text{cte} = \sigma$. If the distribution function of the storage coefficient according to depth corresponds to a discontinuous function, it would yield:

$$V = \sigma \cdot \sum_{k=1}^{\text{NP}} N_{\text{s}}(k) \cdot \left[ \sum_{i=1}^{\text{NC}} \sum_{j=1}^{\text{NF}} s_{ijk} \cdot S_{\text{f}}(i,j,k) \cdot [z_{\text{t}}(i,j,k) - z_{\text{m}}(i,j,k)] \right] \tag{15}$$

Taking a reference topographic elevation $z_{\text{ref}}$, the volume of the associated groundwater reservoir will be given by:

$$V_{\text{ref}} = \sigma \cdot \sum_{k=1}^{\text{NP}} N_{\text{s}}(k) \cdot \left[ \sum_{i=1}^{\text{NC}} \sum_{j=1}^{\text{NF}} s_{ijk} \cdot S_{\text{f}}(i,j,k) \cdot \left[ \left( z_{\text{ref}} \cdot \left(1 - \lambda_{ijk}\right) - z_{\text{t}}(i,j,k) \cdot \lambda_{ijk} \right) - z_{\text{m}}(i,j,k) \right] \cdot \delta_{ijk} \right] \tag{16}$$

**Symbols:** **NC**, number of columns defined in the digital model; **NF**, number of rows defined in the digital model; **NP**, number of layers defined in the three-dimensional model; $\boldsymbol{z_{\text{t}}(i, j, k)}$, digital model of the ceiling of the $k$-th layer; $\boldsymbol{B(i, j, k)}$, average thickness of the groundwater reservoir; $\boldsymbol{z_{\text{m}}(i, j, k)}$, digital model of the wall of the $k$-th layer; $\boldsymbol{s_{ijk}}$, parameter which takes a unit value (1) if the cell belongs to domain (s) of the surface digital model corresponding to the $k$-th layer, and a zero value (0) if it does not belong to this domain:

$$s_{ijk} = \begin{cases} 1, & si \quad s(i,j,k) \in \Omega_k \\ 1, & si \quad s(i,j,k) \notin \Omega_k \end{cases} \tag{17}$$

$\boldsymbol{Ns(k)}$, corresponds to the total number of active cells for the $k$-th layer: $\Omega_k = \sigma \cdot N_s(k)$; $\boldsymbol{s(i,j,k)}$, area of the unit element (cell) for the $k$-th layer that does not vary for the various layers, though it may correspond to a regular grid $\boldsymbol{s(i,j) = \sigma}$ or an irregular one; $\boldsymbol{V_{ref}}$, volume of the groundwater reservoir for a certain topographic elevation $z_{ref}$; $\boldsymbol{z_{ref}}$, reference elevation for calculating reservoir volume according to elevation; $\boldsymbol{S_f(i,j,k)}$, three-dimensional digital model for the storage parameter (effective porosity); $\boldsymbol{\lambda_{ij}}$ and $\boldsymbol{\delta_{ij}}$ variables which takes a unit value or a zero value according to:

$$\lambda_{ij} = \begin{cases} 1 & si \quad z_{ref} > z_t(i,j) \\ 0 & si \quad z_{ref} \leq z_t(i,j) \end{cases} \tag{18}$$

$$\delta_{ij} = \begin{cases} 1 & si \quad z_{ref} > z_m(i,j) \Rightarrow z_{ref} > z_t(i,j) - B \\ 0 & si \quad z_{ref} \leq z_m(i,j) \Rightarrow z_{ref} \leq z_t(i,j) - B \end{cases} \tag{19}$$

# 4 Results and Discussion

The characteristic curve of a groundwater reservoir offers information on storage capacity according to topographic elevation, and accordingly it is a static concept, as its calculation involves parameters invariant in time (geometry and storage coefficients). Moreover, the calculation and cataloguing of hydrogeological reserves is a dynamic concept, given that the piezometric surface is inevitably involved. For the calculation and cataloguing of hydrogeological reserves it is necessary to have their characteristic curve, for from a combination of this with the piezometric surface, it is possible to obtain concrete reserve data.

In the calculation models proposed for the characteristic curve, only effective porosity was included as a storage parameter in the belief that elastic porosity (elastic storage coefficient) applies only in calculating hydrogeological reserves where the piezometric surface must be involved. However, the relative value of elastic porosity as against effective porosity is inconsequential in practical terms when calculating the characteristic curve.

# References

Honnugar V, Uddameri V (2005) Estimating Water Availability and sustainable yield in a coastal semi-arid region of South Texas. Texas University

Lambán Jiménez, LJ, Aragón Rueda R, Rodriguez Hernández L (2001) Determinación de las reservas hídricas subterráneas en el acuífero de Jijona (provincia de Alicante). 7th Hydrogeological Symposium. Murcia, Spain

Rubin H, Rubin A, Reuter C, Köngeter J (2006) Sustainable integrated water resources management (IWRM) in a semi-arid area. The International Journal of Environmental, Cultural, Economic & Social Sustainability, Vol. 2, No. 3

# How to Model Realistic 3D Karst Reservoirs Using a Pseudo-Genetic Methodology – Example of Two Case Studies

A. Borghi, P. Renard, and S. Jenni

**Abstract** This paper describes the application of a new pseudo-genetic approach for modeling of karstic aquifers. Two case studies are presented. The first is a karstic aquifer in the Swiss Jura Mountains, which is characterized by a very complex geometry due to tectonics. The second is a synthetic case of a submarine spring in a karstic area in the Mediterranean basin. For both examples, the same methodology has been applied. First, a 3D deterministic model of the regional geology is built, then based on field observations stochastically heterogeneous hydrogeological properties are modeled within the different formations. The analysis of the hydrological conditions helps in defining the type of karstic systems that is expected, the recharge area (that can be diffuse or localized), and the discharge area. Finally, karst conduits are simulated that connect the recharge area to the main discharge area using a shortest path technique based on the 3D simulated distribution of the hydrogeological properties of the geological formations. Doing so, one can obtain equiprobable conduit simulations from the different realizations of the hydrogeological properties. The results of these simulations can then be used to model flow and solute transport.

## 1 Introduction

Karst aquifers are difficult to characterize due to their extreme heterogeneity induced by the presence of sparse conduits embedded in a carbonate matrix. The presence and position of these conduits (i.e., their density and geometry) depends on two main factors. The first is the geology that controls their shape and distribution:

A. Borghi, P. Renard
CHYN – Center of Hydrogeology and geothermic, University of Neuchâtel, 11 rue Emile Argand, CP 158, 2000 Neuchâtel, Switzerland, e-mail: andrea.borghi@unine.ch; philippe.renard@unine.ch

S. Jenni
Schlumberger Water Services, Immeuble Madeleine D, 92057 Paris la Défense Cédex, France, e-mail: sjenni@la-defense.oilfield.slb.com

B. Andreo et al. (Eds.), *Advances in Research in Karst Media*
DOI 10.1007/978-3-642-12486-0, © Springer 2010

conduits are formed preferentially along bedding planes and/or initial fracturation of the carbonate formation (Filipponi 2009; Palmer 1991; Worthington 1999). The second main factor are the hydrological and hydrogeological conditions (Bakalowicz 2005) such as recharge and position of recent or paleo base levels. Therefore, to try to predict the geometry and position of the karstic conduits, it is necessary to account for these factors. One approach is to model the physics and chemistry of the speleogenesis processes (Dreybrodt et al. 2005). The difficulty with this approach when modeling an actual system is to reconstruct all the conditions that were prevailing during its formation. Consequently, these models are often extremely difficult to condition. Another approach is to build a purely statistical model. For example, Henrion (2008) proposes to model complex cave geometries by generating first a discrete fracture network and then combining it with a truncated multi-Gaussian field in order to represent the 3D geometry of the caves around the conduit network.

This study proposes an alternative methodology that aims at simulating the geometry of karstic conduits at a regional scale constrained by the regional geology, by the regional hydrological/hydrogeological knowledge and by the knowledge of the genetic factors that control the conduits genesis without solving the physical and chemical equations of the speleogenesis. The motivation is to include the speleogenesis processes in a simplified manner to ensure that the modeled geometry is realistic. For that purpose, stochastic methods are used that allow generation of multiple equiprobable realizations that can all be conditioned to field observations. The methodology is illustrated by two case studies. The first is a synthetic case of a submarine karst spring catchment ($20\,\text{km}^2$) inspired from a real site (whose location must remain confidential). The geology is simple: the formations dip monoclinally to the sea. The second site is the Noiraigue spring catchment ($80\,\text{km}^2$). It is situated in the Swiss Jura mountains, close to Neuchâtel and has been intensively studied (Atteia et al. 1998). The structural geology of this catchment is very complex (Valley et al. 2004).

## 2 Methodology

The proposed methodology consists of three main steps: modeling the 3D geology of the site, analyzing the regional hydrology/hydrogeology, and simulating the karst conduits.

### *Step 1: 3D Geological Modelling*

Because karstification processes are controlled by the 3D geology, the first step consists in building a 3D model of the area to define the geometry of the karstifiable formations and the aquitards. The model provides information about the kind of formation in each point of the domain. In the examples (Fig. 1), the models were

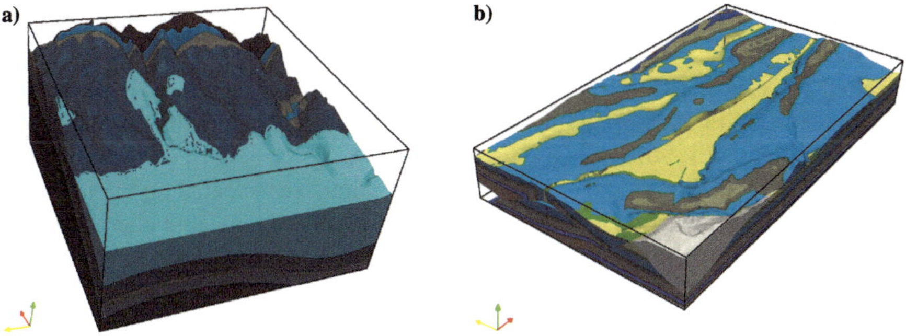


**Fig. 1** Geological model for **a** marine spring in mediterranean basin, **b** Noiraigue spring catchment basin the swiss Jura Montains. *Blue*: karst formations, *other colors*: aquitards

built using an implicit approach (Calcagno et al. 2008) that accounts for orientation data, faults and stratigraphical rules.

## *Step 2: Hydrogeological Analysis*

The hydrogeological analysis of the site aims at defining the boundaries and boundary conditions of the future flow model. One aim is to identify the recharge processes of the aquifer and the locations of possible sinkholes and dolines because the type of recharge and its location control the type of karstic system that should be modelled (Palmer 1991). In addition, the location of the springs (active and fossils) must be defined, as well as any data such as tracer tests which aid in describing the connectivity between the sinkholes and the springs to identify the possible extension of the catchment. These data will be used in the next step to constrain the modelling of the conduits.

## *Step 3: Modelling the Conduits*

In this approach, the conduits are simulated by following the shortest path computed using a fast marching approach (Sethian 1996). These computations are very fast, and can be based on the geometrical constraints that can be added to the 3D regional geological geometrical model such as initial fracturation, bedding planes, etc. In fact, all the initial heterogeneities can be added to constrain the shape of the conduits while honoring the actual knowledge about karst conduits geometry constraints and generation processes.

The modelling of the conduits involves three sub-steps that can be repeated when there are different phases of karstification.

**Step 3.1** Stochastic modelling of internal properties of the different geological formations to obtain a 3D field of maximum local velocity. This field can be controlled by fracturation, bedding, or any internal heterogeneity. This step is using standard stochastic techniques.

**Step 3.2** Computation of the minimum travel time between a given outlet (for example the Noiraigue Spring) and any point in the domain using the fast marching approach (Sethian 1996).

**Step 3.3** Modelling of the conduits. Starting points are selected on the topography both from known sinkhole locations and random but plausible alternative locations. From these points, random walkers follow the shortest path to the spring according to the 3D time map.

When a conduit is generated, its presence will influence the velocity field because it will be associated to a high speed path, i.e., the presence of a pre-existing conduit influences the shape of the next ones to be simulated. This leads to a hierarchical organization of the resulting karstic network (Fig. 2).

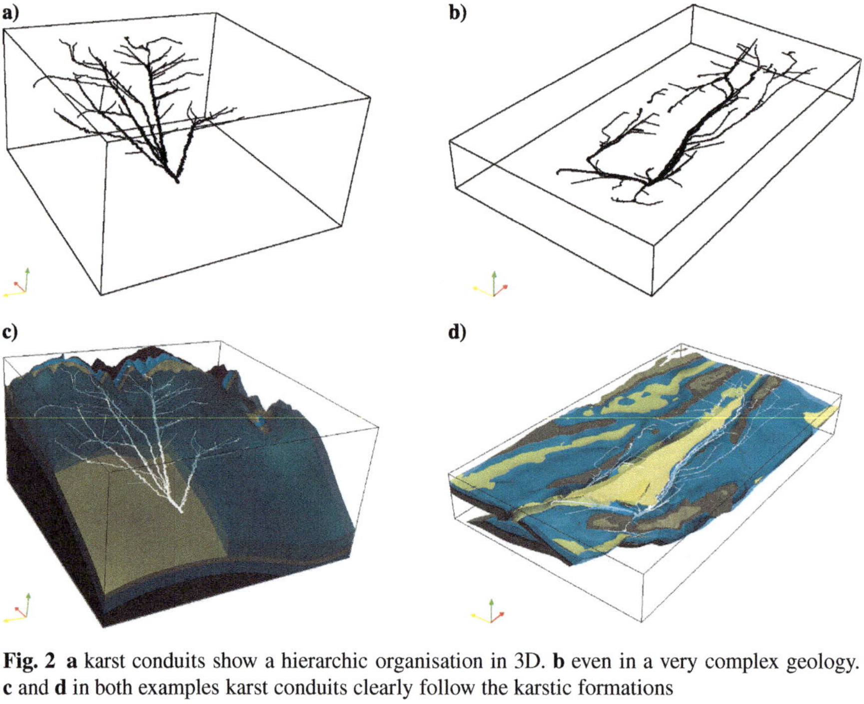


**Fig. 2** **a** karst conduits show a hierarchic organisation in 3D. **b** even in a very complex geology. **c** and **d** in both examples karst conduits clearly follow the karstic formations

## 3 Conclusions

These two examples illustrate a new method to model realistic karst aquifers that accounts for the main factors controlling the shape and distribution of karst conduits. However, the speleogenetic equations are not solved, but only try to mimic realistic geometries based on the analysis of the processes. This is why it is a "pseudo-genetic" approach. The final result of this methodology can be used to perform flow and transport simulations. The method is planned to be further improved by including a probabilistic model of potential starting points for the conduits based on a geological and hydrogeological analysis of the zone. In addition, if it is known that several karstification phases occurred, the method can be applied to create successive networks honoring the paleo-conditions. This can lead to a very rich and complex network. Finally, because the proposed algorithm is very fast, it is possible to use it in an inverse modeling framework to honor field data, such as spring hydrograms and tracer tests results, and so to better constrain the geometry and properties of the conduit network.

## References


Atteia O, Perret D, Adatte T, Kozel R, Rossi P (1998) Characterization of natural collodis from a river and spring in a karstic basin. Env Geol 34: 257–269

Bakalowicz M (2005) Karst groundwater: A challenge for new resources. Hydrogeology Journal 13 DOI 10.1007/s10040-004-0402-9

Calcagno P, Chilès JP, Courrioux G, Guillen A (2008) Geological modelling from field data and geological knowledge: Part I. Modelling method coupling 3D potential-field interpolation and geological rules. Physics of the Earth and Planetary Interiors 171: 147–157 DOI doi:10.1016/j.pepi.2008.06.013

Dreybrodt W, Gabrovsek F, Douchko R (2005) Processes of speleogenesis: a modelling approach Karst Research Institute ZRC Postojna – Ljubljana

Filipponi M (2009) Spatal analysis of karst conduit networks and determination of parameters controlling the speleogenesis along preferential lithostratigraphic horizons. PhD, EPFL – Ecole Polytechnique Fédérale de Lausanne

Henrion V, Pellerin J, Caumon G (2008) A stochastic methodology for 3D cave system modeling. In: Ortiz JM, Emery X (eds) Geostats 2008 Proceedings of the Eighth International Geostatistics Congress: 525–534

Palmer AN (1991) Origin and morphology of limestone caves. Geol Soc Am Bull 103

Sethian JA (1996) A fast marching level set method for monotonically advacing fronts. Proc Natl Acad Sci USA 93

Valley B, Burkhard M, Schnegg P-A (2004) Dépliage 3-D des anticlinaux bordant le synclinal fermé de la vallée des Ponts, Jura central, Suisse. Eclogae Geologicae Helvetiae 97: 279–291

Worthington SRH (1999) A comprehensive strategy for understanding flow in carbonate aquifers. In: Palmer AN, Palmer MV, Sasowsky ID (eds) Karst modelling

# Insights of 3D Geological Modelling in Distributed Hydrogeological Models of Karstic Carbonate Aquifers

A. Fournillon, S. Viseur, B. Arfib, and J. Borgomano

**Abstract** Flow simulation in karstic aquifers with distributed hydrogeological models may be used to numerically model the spatial distribution, the connectivity and the geometry of the karstic network. A new approach is proposed to characterize the karstic network and to reproduce it by the use of 3D geological modelling. The observed karstic network is first decomposed into sets of unidirectional elements which are analysed in 3D with variograms in term of dimension and distribution. These sets of karst directions are then stochastically simulated in 3D in a curvilinear grid that is consistent with the orientation of inception horizons. Several probability maps or 3D probability fields can be added as secondary data in order to constrain the distribution and the connectivity of the simulated karstic network.

## 1 Introduction

Flow simulation in karst aquifers has been extensively studied. Recently, distributed hydrogeological models have dealt with karstic flow behaviour. A great advance in such models was the integration of discrete conduit elements, as was done in Feflow (Diersch 2007) or Conduit Flow Process in Modflow-2005 (Shoemaker et al. 2008). These models need the location of the karstic network as a prerequisite input. However, only some karstic conduits can be observed among the whole set of existing karstic network in a given geological unit. Therefore some authors have proposed methods to simulate the unobserved karst conduits. Most of the existing methods are process-based, using: chemical dissolution reactions (e.g. Clemens et al. 1996; Kaufman 2009) or particles displacements (Jaquet et al. 2004). However, process-based methods are difficult to constrain to observed data (Koltermann & Gorelick

A. Fournillon, S. Viseur, B. Arfib, J. Borgomano
Laboratoire de Géologie des Systèmes et des Réservoirs Carbonatés, EA 4234, Université de Provence, Case 67, 3 Place Victor Hugo, 13003 Marseille, France,
e-mail: afournillon@gmail.com; sophie.viseur@univ-provence.fr; bruno.arfib@univ-provence.fr; jean.borgomano@univ-provence.fr

B. Andreo et al. (Eds.), *Advances in Research in Karst Media* 
DOI 10.1007/978-3-642-12486-0, 

1996). Other have authors proposed geostatistical methods that can fit existing data. Labourdette et al. (2007) have developed an approach based on the stochastic simulation of cavities from geometrical parameters and dissolution probability 3D fields. However, it is dedicated solely to carbonate tropical island settings. Henrion et al. (2007) have proposed a method based first on the recognition of preferential paths in a 3D stochastically generated fracture network and second on the simulation of cavities around these paths. Karstic networks are reproduced with efficiency but this approach requires detailled knowledge of the fracture network and strong assumptions on the past flow paths.

An alternative approach is proponed which focuses on the reproduction of karst network geometry and connectivity. This method consists first on analyzing statistical parameters from observed karsts or other data defining the geometry of karst features, and second on using geostatistical approaches to reproduce the complex karst geometry. After describing the proposed approach, an application of a case study is presented to illustrate its flexibility and potentiality by comparing two models of kartic networks obtained using two different assumptions.

## 2 Methodology

Karstic networks are often organized around particular horizons known as inception horizons (Lowe 2000). For this reason, the simulations of the karst network are performed in a 3D curvilinear grid whose cells are aligned according to these horizons.

In this grid, the karst occurrence is considered as an indicator variable $Ik(x)$.$Ik(x) = 1$ if $x$ belongs to a karst and 0 otherwise. Therefore, the proposed approach aims at simulating the value of $Ik(x)$ for each cell of the grid. In geostatistics, the Sequential Indicator Simulation (SIS) is dedicated to stochastically simulate such variables, conditioned to available data and to estimated statistical parameters: (1) the variogram which defines the spatial distribution of the simulated variable and (2) the proportion of each variable classes. However such a method can not be directly used to reproduce the complex karst geometry: the variograms only characterize the spatial distribution of ellipsoid-like shapes and not highly curvilinear features. To address this, it is proposed to analyse the karstic network while decomposing the discontinuities network obtained from cave surveys into unidirectional elements. These elements are classified into families according to their orientation; each family is characterized by a variogram model and a proportion. To produce a consistent karstic network with each direction connected, the following workflow is proposed: (1) Simulating the major karst (karst1) direction, given its variogram and its proportion. A probability 3D field or map of the preferential karst area, constructed from *a priori* knowledge (e.g. landscape karstic features, major joint direction, weathered rock zone), may be used as secondary data. (2) Creating a probability 3D field from the karst1 in order to yield to karst2 more occurence to be simulated in connection with karst1. (3) Simulating karst2 using the obtained probability 3D field, its variogram and its proportion. (4) Simulating similarly to karst2 the other karst families. By using different probability maps compiled from *a priori* knowledge or simulated karsts, it is possible to constrain the connectivity between the different karst elements. (5) Merging all simulated karst elements into unique bodies. The simulated karstic network is then obtained and can be exported into a distributed parameter hydrogeological model as a double continuum matrix or as a discrete pipe network after conversion of the grid cell properties.

The aim of this approach is no longer to create a precise image of the unexplored karst, but to provide several equiprobable models of the karstic network, in respect of quantitative and qualitative data from a wide range of origins. The ultimate goal is to quantify the uncertainty of flow in a carbonate karstic aquifer by realizing flow simulations on these models.

## 3 Case Study

The Morieres Massif is in Provence (SE France) (Fig. 1). It consists of three main stratigraphic units from lower to upper Jurassic ages (Fig. 1). This area has been intensively explored by cavers (Lucot & CDS 83 2005). Moreover, sub-surface data are also available from a tunnel dug through the massif in the upper and mid Jurassic (Combes 1976). The speleological surveys and the observations in the tunnel show that the main karstic network, with horizontal development, is inferred from a single 20-m-thick layer from Bathonian age.

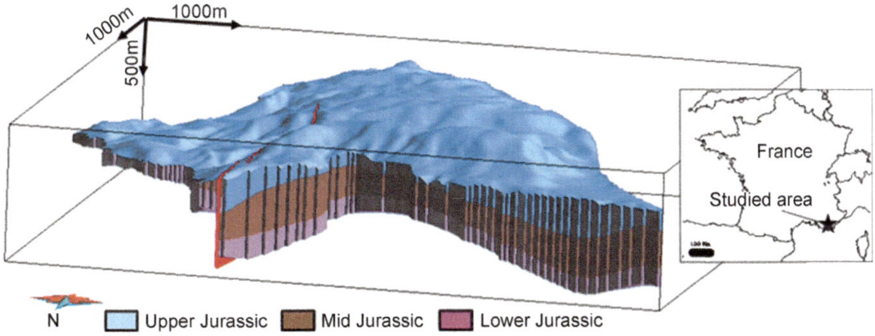


**Fig. 1** Localisation of the studied area and the 3D stratigraphic grid in which the simulations are performed. The targeted layer of the study is on the *top* of the Mid Jurassic (*brown unit*)

## 4 Analysis and Simulation of the Karstic Network

A 3D curvilinear grid has been built from structural data. It consists in a 3 layered grid (Fig. 1). Karst occurs only in the second layer and so only karst on this layer has been simulated. A cell size of 8 m×6 m has been chosen. The targeted layer has been explored in terms of fracture analysis from outcrops, tunnel operations and aerial photographs; the resulting directions are consistent at all scales (Combes 1976; Blanc and Nicod 1990). These directions are: N090°, N045°, N135° and N000°.

The same directions are found in karst conduits, and in karsts intersected by the tunnel (Combes 1976). Two different simulations have been performed: (1) the simulation #1 (Fig. 2a and b) which accounts for the relative connectivity of each karst element family and probability maps; (2) the simulation #2 (Fig. 2c) for which all karst element families have been simulated independently.

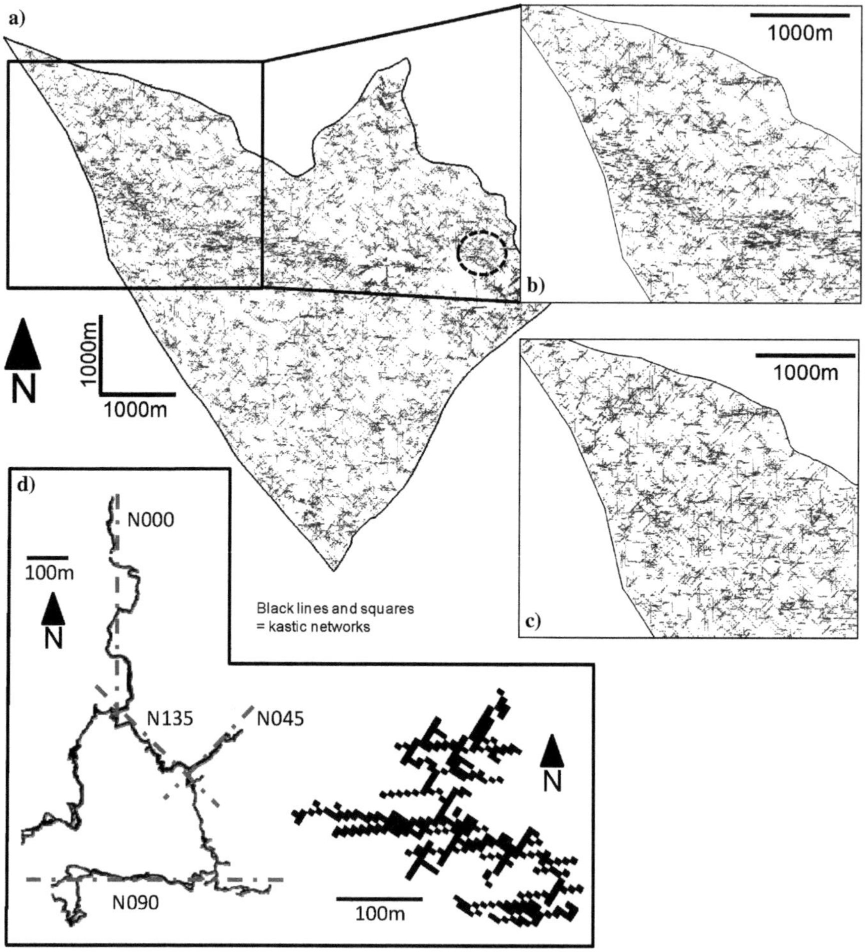


**Fig. 2** Main results of the two simulations. **a** Top view of the whole simulation #1. **b** Close up of the simulation #1. **c** Close up of the simulation #2. **d** Comparison between the Rampins Cave (main directions of cave conduits are drawn in *dashed lines*) which is developed in Bathonian near the study area and a simulated karstic network from simulation #1 (located on *dashed line circle* on Fig. 3a)

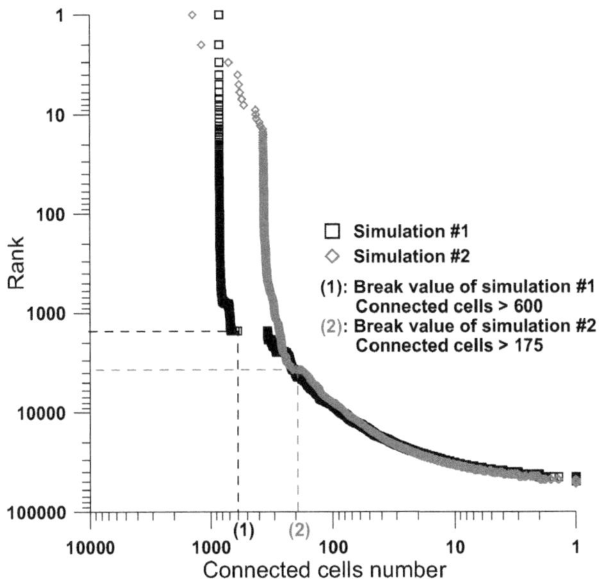


**Fig. 3** Zipf diagram of the number of connected cells of each independent karst network for the simulation #1 (*black squares*) and the simulation #2 (*gray diamonds*)

The results of both simulations are consistent in terms of geometry with the known karstic networks (Fig. 2d). It confirms that the chosen input data are then representative of the observed geometry. A connectivity analysis has been computed on each model to quantify the differences between the two simulations: the number of connected cells for each simulation has been estimated. The results of this analysis have been plotted on a Zipf diagram (Fig. 3). Prior to reaching a break point, the two simulations have similar behaviour but beyond this point the two simulations reach a sill that is higher for simulation #1 (>600) than for simulation #2 (>200). Simulation #1 shows a greater number of connected cells than simulation #2 does, except for its highest value. The presence of a body with a very high number of connected cells in simulation #2 could be simply due to randomness, as shown by the grey diamonds in the Zipf diagram (Fig. 3). It suggests that conditioning from probability fields, in simulation #1, is efficient for constraining the connectivity between each karst family and to avoid randomness in the distribution of the karstic network.

## 5 Conclusion

The use of 3D geological models provides tools for the integration of numerous data types, from outcrop to sub-surface, into a coherent framework. The proposed approach reproduces karstic networks that are consistent with observed geometry. The connectivity analyses, performed on two different kinds of simulation, have shown that both the decomposition in karst elements and the integration of an *a priori* knowledge can control the karst connectivity.

These efforts should be pursued by determining other geometrical and topological parameters (number and type of karst element connectivity, karst section area or their tortuosity) for the characterization of the geometry and the connectivity of the karstic network. These parameters could either condition the simulation or be computed *a posteriori* on the generated 3D models in order to classify them and then guide the selection of models to export to flow simulators. This defines the method to easily export the obtained karstic models into a flow simulator, while preserving the original parameters. Flow simulations obtained from these selected models will finally serve as support for quantifying the uncertainty of flow in karst terranes.

**Acknowledgements** This work is part of the KarstEAU project, funded by the "Agence de l'Eau Rhône-Méditerranée-Corse", the "Conseil Général du Var", the "Conseil Général des Bouches-du-Rhône", and by the "Région Provence-Alpes-Côte d'Azur". A special thanks to Alex Hairabian for his useful help in the field. Authors also thank an anonymous reviewer for his helpful corrections and comments.

## References

Blanc JJ, Nicod J (1990) Les surface karstiques du plateau de Montrieux (Var), (secteur de Valbelle–Morières–Siou Blanc), étude quantitative de la fracturation. Karstologia 16:17–28

Clemens T, Hückinghaus D, Sauter M, Liedl R, Teutsch G (1996) A combined continuum and discrete network reactive transport model for the simulation of karst development. In: Kovar K, van der Heijde P (eds.) Calibration and reliability in groundwater modelling. IAHS Publication 237:309–318

Combes A (1976) Essai de méthodologie en pays karstique. Etude de problèmes hydrogéologiques et géotechniques dans le Massif des Morières. PhD thesis. Université de Grenoble, 174 p

Diersch HJG (2007) Discrete feature modeling of flow, mass and heat transport processes by using FEFLOW. In: White papers volume 1. Berlin, 191–198

Henrion V, Camoin G, Vitel S, Kedzierski P (2007) Stochastic simulation of cave systems in reservoir modeling. 27th GOCAD meeting, 11 p

Jaquet O, Siegel P, Klubertanz G, Benabderrrhamane H (2004) Stochastic discrete model of karstic networks. Advances in Water Resources 27:751–760

Kaufmann G (2009) Modelling karst geomorphology on different time scales. Geomorphology 106:62–77

Koltermann C, Gorelcik S (1996) Heterogeneity in sedimentary deposits; a review of structure-imitating, process-imitating, and descriptive approaches. Water Resources Research 32(9): 2617–2658

Labourdette R, Lascu I, Mylroie J, Roth M (2007) Process-like modeling of flank-margin caves: from genesis to burial evolution. Journal of Sedimentary Research 77:965–979

Lowe D (2000) Role of stratigraphic elements in speleogenesis: the speleoinception concept. In: Klimchouk AB, Ford DC, Palmer AN, Dreybrodt W (eds.), Speleogenesis, evolution of karst aquifers. National Speleological Society, Huntsville (Alabama), pp 65–76

Lucot JP, Comité Dép. de Spéléologie 83 (2005) Fichier des cavités du Var. Spelunca 99

Shoemaker WB, Kuniansky EL, Birk S, Bauer S, Swain ED (2008) Documentation of a Conduit Flow Process (CFP) for MODFLOW-2005: U.S. Geological Survey Techniques and Methods 6-A24, 50 p

# Hydrological Modelling of a Karst System Using non Linear Hysteretic Conceptual Model

S. Tritz, H. Jourde, and V. Guinot

**Abstract** A two reservoir conceptual model with a hysteretic transfer function is proposed for hydrological modelling of a karst system. The main features of the model are (i) a simple model for evapotranspiration based on potential rates, (ii) a nonlinear, hysteretic discharge law accounting for the variations of connectivity in the soil/epikarst zone, (iii) a threshold-based discharge law accounting for the time variability of the active catchment area via losses to secondary springs. The structure of the model and the governing equations are proposed on the basis of physical considerations. The model is applied to the karst system of the Durzon spring (France). Model robustness is assessed by using only one year for calibration and four years for validation. The Nash-Sutcliffe criterion, used for performance measure, decreases from 91 to 87% between the calibration and validation phase, which demonstrates the robustness of the model. The proposed model is compared to a model of similar complexity available from the literature, previously applied to similar catchments. The comparison indicates the superiority of the hysteresis-based model in terms of robustness and overall performance.

## 1 Introduction

From a physical point of view, the hydrologic functioning of a karst system is driven by the typical hydrodynamic behaviour of each of its compartments, which can be classified according to three distinct zones: epikarst, infiltration zone and saturated/phreatic zone (Mangin 1975). In karst hydrology, different types of models are generally used to simulate the rainfall-discharge relationship. Earlier reservoir models were based on simple relationship, such as the linear law of Maillet (1906), which linearly links the discharge at the outlet of the reservoir to the water level

S. Tritz, H. Jourde, V.Guinot
Hydrosciences Montpellier UMR 5569 (CNRS, IRD, UM1, UM2) Université Montpellier 2, CC MSE, 34095 Montpellier Cedex 5, France, e-mail: herve.jourde@univ-montp2.fr

B. Andreo et al. (Eds.), *Advances in Research in Karst Media* 
DOI 10.1007/978-3-642-12486-0, © Springer 2010

in the reservoir. Other models, such as MODALP (Arikan 1988), Tridep model (Padilla-Benitez 1990), GR4J model (Perrin 2000), Vensim (Fleury 2005), Geyer model (Geyer et al. 2008), RR model (Jukic and Denic-Jukic 2009), consider non-linear discharge laws.

The purpose of the proposed hysteretic model is to better represent the influence of the connectivity of the soil/epikarst zone on the response dynamics of the karst system. Furthermore, as water does not flow necessarily towards one single outlet (Jukic and Denic-Jukic 2009), a parameter is introduced to account for the discharge outside of the karst system main outlet.

## 2 Physical Processes and Model Structure

The hysteretic model (Fig. 1) uses two reservoirs, each of which corresponds to a physical storage entity over the catchment. The upper reservoir, denoted by E hereafter, represents the epikarst/soil zone. The lower reservoir, denoted by L hereafter, represents the lower, vadose and saturated zone. In what follows, the volumes of water stored in reservoirs E and L are denoted by $E$ and $L$ respectively:

1. The epikarst reservoir E receives an incoming precipitation rate $P$. Part of this incoming flux goes back to the atmosphere via an evapotranspiration rate ET.
2. Part of the water in reservoir E leaks to the reservoir L with a discharge $Q_{\mathrm{EL}}$.
3. Part of the water contained in E flows outside of the catchment with a discharge $Q_{\mathrm{sec}}$ to secondary springs, when $E$ exceeds a given threshold $E_{\mathrm{sec}}$.
4. A third discharge path for the water is the direct connection to the outlet of the catchment with a discharge $Q_{\mathrm{EH}}$. Such a connection, allowed by the epikarst via a network of fractures and preferential flowpaths, is responsible for the fast component of the catchment response to rainfall events. A nonlinear, hysteretic transfer function is proposed to account for the influence of karst connectivity and initial state on the fast response of the catchment to sudden rainfall events.

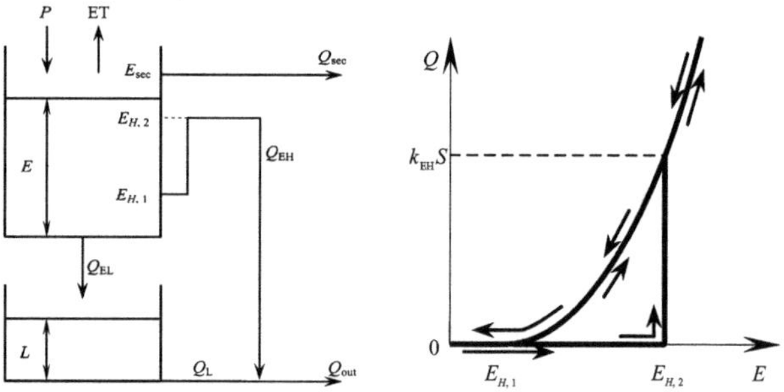


**Fig. 1** *Left*: structure, variables and parameters in the proposed model; *right*: hysteretic discharge function given by Eq. 2

5. The water in the lower reservoir L leaks to the outlet of the catchment at a discharge $Q_{\mathrm{L}}$ via a classical, linear discharge law (Maillet 1906).

In classical models, the water level (or volume stored) in the reservoir E must reach a certain upper threshold value $E_{H,2}$ before the connectivity of the karst system becomes nonzero and the rapid discharge $Q_{\mathrm{EH}}$ is initiated. However, karst connectivity may be preserved even if the water level in the reservoir E falls below the threshold level $E_{H,2}$. This behaviour is accounted for with the hysteretic transfer function (Fig. 1, right), as connectivity is destroyed only if the water level drops below a lower threshold level $E_{H,1}$.

## 3 Governing Equations

The model has two balance equations

$$\frac{\mathrm{d}E}{\mathrm{d}t} = \begin{cases} P - \mathrm{ET} - \frac{Q_{\mathrm{sec}}+Q_{\mathrm{EH}}+Q_{\mathrm{EL}}}{S} & \text{if } E > 0 \\ 0 & \text{if } E = 0 \end{cases} \tag{1a}$$

$$\frac{\mathrm{d}L}{\mathrm{d}t} = \begin{cases} \frac{Q_{\mathrm{EL}}-Q_{\mathrm{L}}}{S} & \text{if } L > 0 \\ 0 & \text{if } L = 0 \end{cases} \tag{1b}$$

where $E$ is the water level in the reservoir E, ET is the evapotranspiration rate, $L$ is the water level in the reservoir L, $P$ is the precipitation rate, $S$ is the plan view area of the catchment, $Q_{\mathrm{EH}}$ is the fast flow component through the epikarst zone, $Q_{\mathrm{EL}}$ is the infiltration rate to the lower reservoir, $Q_{\mathrm{sec}}$ is the discharge to secondary springs, and $Q_{\mathrm{L}}$ is the slow flow component (baseflow) from the lower reservoir L to the outlet of the catchment. The discharges $Q_{\mathrm{sec}} Q_{\mathrm{EL}}$ and $Q_{\mathrm{L}}$ are assumed to obey classical, linear laws. The total discharge $Q_{\mathrm{out}}$ is the sum of $Q_{\mathrm{EH}}$ and $Q_{\mathrm{L}}$.

The nonlinear, hysteretic transfer function (Fig. 1, right) takes the form

$$Q_{\mathrm{EH}} = \varepsilon \left[ \frac{\max(E - E_{H,1}, 0)}{E_{H,2} - E_{H,1}} \right]^{\alpha} k_{\mathrm{EH}} S \tag{2}$$

where $E_{H,1}$ and $E_{H,2}$ are the lower and upper threshold levels for the hysteretic discharge function, $k_{\mathrm{EH}}$ is a specific discharge coefficient, $\alpha$ is a positive exponent, and $\varepsilon$ is the connectivity indicator for the karst system. $\varepsilon$ is switched to 1 if $E$ rises above $E_{H,2}$ and is switched to 0 if $E$ falls below $E_{H,1}$.

## 4 Application Site

The karst system of Durzon is located in the Grands Causses area in the Southern Massif Central (France). It is embedded in a 400 m thick formation of middle to upper Jurassic limestones and dolomites, deposited on top of a 200 m thick upper-Liassic marl formation. This latter formation acts as an impermeable barrier which defines the lower boundary, but also the North Eastern and Southern boundaries of the karst catchment. The other boundaries are delimited by the topography. The estimated recharge area is 116.8 $km^2$ (Fleury 2005). The precipitation rate $P$ and the temperature data needed for the estimate of the potential evapotranspiration (PET) are measured at "Le Caylar" Meteo-France meteorological station, located 10.3 km S-SE of the Durzon Vauclusian spring (average outflow of 1.4 $m^3/s$ and maximum discharge of 18 $m^3/s$ over the period 2002–2008). Daily PET values are estimated from the monthly PET using a sine function-based interpolation. The choice of Thornthwaite's formula is justified by the limited quantity of data needed, its widespread use in hydrological modelling (Arikan 1988; Padilla-Benitez 1990) and its close correlation with actual evapotranspiration (Lu et al. 2005).

## 5 Comparison with a Classical Reservoir-Based Model

The response of the proposed hysteresis-based model is compared to that of the Vensim model (Fleury 2005), a three reservoir model (Fig. 2) developed specifically for modelling of karst aquifers. This model is selected as a reference because its complexity is similar to that of the proposed model. The first reservoir in the Vensim model is the soil reservoir. It manages the production function that accounts for the input (precipitation rate $R$) and outputs (potential evapotranspiration PET and infiltration $I$). As long as the depth $H$ of water stored in the reservoir is smaller than a threshold value $H_0$, the infiltration is zero. When $H$ reaches $H_0$, the net rainfall $R$ – PET is converted directly into infiltration $I$. $H_{\min}$ is a minimum threshold level below which the water cannot drop. The infiltration is routed to the so-called slow and rapid reservoirs $R_1$ and $R_2$ via a distribution coefficient. The coefficient is equal to $x$ if $R_1$ is smaller than a given threshold $R_{\text{sill}}$. If $R_1$ rises above $R_{\text{sill}}$, then the discharge coefficient switches to a new value $x' < x$, which allows stronger flow peaks to be obtained. The discharges $Q_1$ and $Q_2$, respectively the discharges from the reservoirs 1 and 2, are obtained from linear discharge laws (discharge coefficients respectively equal to $\alpha_1$ and $\alpha_2$). In the case of the Durzon catchment's modelling (Fleury 2005) $H_0$ equals 0, and the model uses only six parameters ($H_{\min}$, $x$, $x'$, $R_{\text{sill}}$, $\alpha_1$, $\alpha_2$).

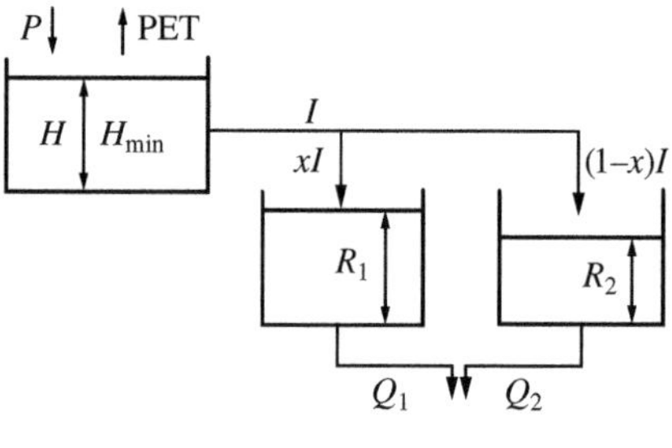


**Fig. 2** Structure, variables and parameters in the Vensim model (Fleury 2005)

# 6 Methodology and Calibration

Both models use 2002 as a warm-up year that allows the influence of wrongly defined initial conditions to be eliminated from the simulation results. Calibration is carried out using 2003 alone (Fig. 3).

The model is then validated against measured discharges at the Durzon spring over the period from 1 January 2004 to 25 May 2008 (Fig. 4). For the hysteresis-based model, the Nash-Sutcliffe (1970) criterion for the calibration phase is 91% and 87% for the validation phase. For the Vensim model, it is 92% and 59% over the calibration and validation period respectively. Overall, the Vensim model allows both a correct average baseflow and the peak values at the outlet of the catchment to be reproduced correctly over the entire calibration period (Fig. 3, right). However, both for the calibration and validation period (Fig. 4): (i) the Vensim model is less accurate than the hysteresis-based model in representing the recession part of the discharge time series, (ii) the time derivative of the simulated discharge is overesti-

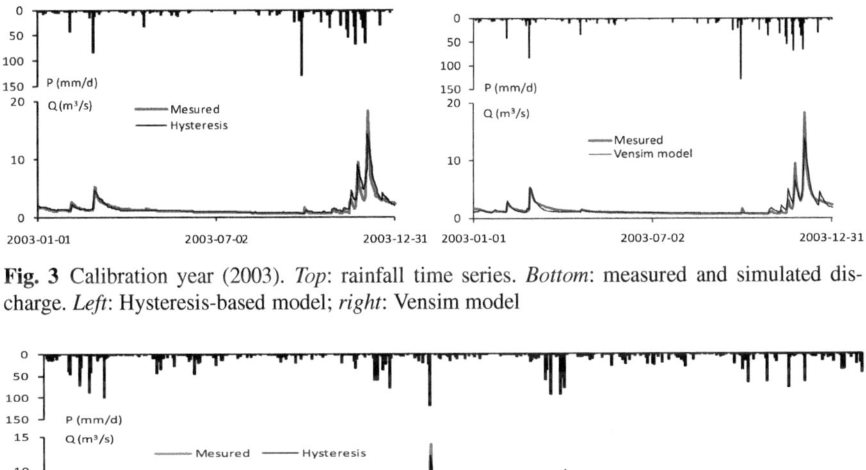


**Fig. 3** Calibration year (2003). *Top*: rainfall time series. *Bottom*: measured and simulated discharge. *Left*: Hysteresis-based model; *right*: Vensim model

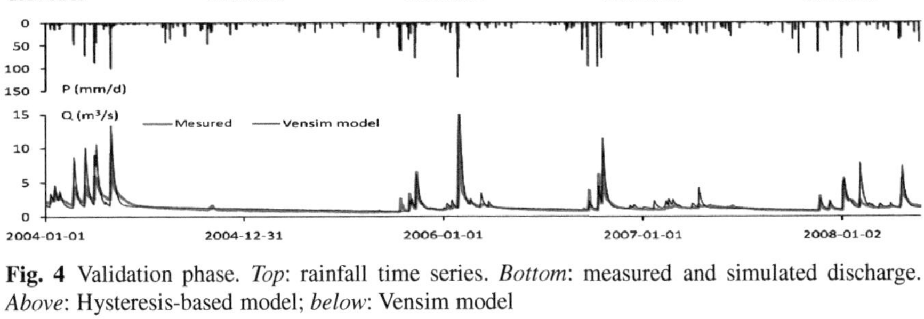


**Fig. 4** Validation phase. *Top*: rainfall time series. *Bottom*: measured and simulated discharge. *Above*: Hysteresis-based model; *below*: Vensim model

mated over the dry period; (iii) it fails to reproduce the small, isolated peak (see e.g. 4th quarter of 2003).

## 7 Conclusions

A two reservoir model has been presented for modelling response dynamics of karst systems. Its main features are (i) the division between the upper and lower reservoirs which correspond to exchange fluxes based on a physically realistic description of the hydrological processes, (ii) taking into account temporary losses to secondary springs via a threshold-based linear discharge law, (iii) the time variability in the connectivity of the soil/epikarst system represented using a nonlinear, hysteretic discharge law, (iv) the long term behaviour of the catchment during dry periods via a linear discharge law. An application to a karst system and a comparison to an existing model lead to the following conclusions: (i) the proposed model is robust, with only a few percents variation in the modified Nash criterion between the calibration period and the validation period, (ii) the nonlinear character of the hysteretic discharge law plays a key role in the hydrograph recession, (iii) the hysteretic discharge law is essential in predicting the influence of past hydrological time series on karst response dynamics. More specifically, it allows for a correct identification of the periods when the karst system is connected or disconnected, with the subsequent activation or de-activation of the fast response components of the model.

## References

Arikan A (1988) MODALP: a deterministic rainfall-runoff model for large karstic areas. Hydrological Sciences Journal 33(4):401–414

Fleury P (2005) Sources sous-marines et aquifères karstiques côtiers méditerranéens. Fonctionnement et caractérisation. PhD thesis, University Paris 6, 286 p

Geyer T, Birk S, Liedl R, Sauter M (2008) Quantification of temporal distribution of recharge in karst systems from spring hydrographs. Journal of Hydrology 348:452–463

Jukic D, Denic-Jukic V (2009) Groundwater balance estimation in karst by using a conceptual rainfall–runoff model. Journal of Hydrology 373:302–315

Lu J, Sun G, McNulty SG, Amatya DM (2005) A Comparison of Six Potential Evapotranspiration Methods for Regional Use in the Southeastern United States. JAWRA 31(6):612–633

Maillet E (1906) La vidange des systèmes de réservoirs. Annales Ponts et Chaussées, 21

Mangin A (1975) Contribution à l'étude hydrodynamique des aquifères karstique. Thèse de doctorat de l'Université des Sciences de Dijon

Nash JE, Sutcliffe JV (1970) Uncertainty in rainfall-runoff model simulations and implications for predicting the hydrologic effects of land-use change. Journal of Hydrology 192:211–232

Padilla-Benitez A (1990) Los modelos matématicos aplicados al analisis de los acuiferos karsticos. PhD thesis, University of Granada, Spain

Perrin C (2000) Vers une amélioration d'un modèle global pluie-débit au travers d'une approche comparative. Thèse de doctorat de l'INP Grenoble / Cemagref (Antony), 530 p

Rimmer A, Salingar Y (2006) Modelling precipitation-streamflow processes in karst bassin: the case of the Jordan River sources, Israel, Journal of Hydrology 331:524–542

Thornthwaite CW (1948) An approach toward a rational classification of climate. Geogr Rev 38:55–94

# Object-Oriented Modelling as a Decision-Making Tool in Agriculturally Overexploited Karstic Aquifers

J.L. Molina, J. Bromley, J.L. García-Aróstegui, M. Molina, and J. Benavente

**Abstract** Object-Oriented Bayesian Networks (OOBNs) utilise the power of Object-Oriented Programming (OOP) and offer a novel approach to the problems of integrated water management. This paper describes the building of an OOBN Decision Support System (DSS) that allows complex domains to be described in terms of inter-related objects. Thus, the DSS structure is able to represent an accurate reflection of a complex real-world water system made for an aquifer that has been used as an example of a successful application. In this research, conventional Bayesian Networks (BNs) are used to describe the probabilistic relationships between variables (objects) within each network. A network is a group of objects that can be described as a class. Different classes can possess similar sets of objects and be linked through other networks having common variables. Classes inherit commonly used states and behavior from other classes in a hierarchical way. This model of networks represents a participatory DSS for helping water managers.

## 1 Introduction

Groundwater-based systems are particularly vulnerable to the over-exploitation of the aquifers on which they depends. This problem is most common in arid and semi-

J.L. Molina, J.L. García-Aróstegui
Instituto Geológico y Minero de España, Avda. Miguel de Cervantes, 45 5° A, Edificio Expo Murcia 30009 Murcia, Spain, e-mail: jl.molina@igme.es; j.arostegui@igme.es

J. Bromley
Oxford University Centre for the Environment, South Parks Road, Oxford, OX1 3QY, United Kingdom

M. Molina
Dpto. Didáctica de la Matemática, Facultad de CC de la Educación, Univ. de Granada, 180071 Granada, Spain

J. Benavente
Instituto del Agua, Universidad de Granada, C/Ramón y Cajal 18071 Granada, Spain

B. Andreo et al. (Eds.), *Advances in Research in Karst Media*
DOI 10.1007/978-3-642-12486-0, © Springer 2010

arid regions where precipitation is low but other conditions make it favourable to human settlement and economic development. The use of groundwater in such areas to supply large extensions of irrigated land has been the key to agricultural and economic development in a large number of countries over the past few decades (Martín de Santa Olalla 2006). Groundwater for irrigation has transformed areas of limited agricultural potential into regions of high productivity leading to unprecedented economic development. This has been accompanied by increased agricultural incomes and a growing rural population.

A model of an agricultural water system, which draws its supply from four different aquifers, has been used to illustrate the principles of OOBN construction. The model is a series of linked networks that can be used by managers as a DSS to identify the most appropriate water management scenarios for the region. The example used is located in the Altiplano of Murcia in south-east Spain, a region in which groundwater provides the sole water supply and where agriculture is the main consumer. This situation is commonly encountered in arid or semiarid regions, where the need to find a balance between economic development and environmental preservation requires an integrated and holistic approach.

The goals of this paper are, firstly, to explain the structure and operation of Object Oriented Bayesian Networks (OOBNs), secondly to illustrate the increased potential of OOBNs over conventional Bayesian Networks (BNs), and finally, to demonstrate the utility of OOBNs as a Decision Support System (DSS) in the context of IWRM for modelling real complex multi-aquifer water systems.

## 2 Bayesian Networks (BNs) and Object-Oriented Bayesian Networks (OOBNs)

A BN is a type of DSS based on probability theory, which uses Bayes' rule of probability. There are many publications that describe the theory and application of BNs (Jensen 1996; Koller and Pfeffer 1997; Jensen 2001; Bromley 2005). According to Cain (2001), a BN can be defined as follows: "... some nodes (variables) that represent random variables that interact with others. These interactions are expressed like connections between variables". BNs have been used as decision support systems for many years in fields such as medicine, road safety and artificial intelligence. However, they have only recently been widely applied to environmental systems. (Henriksen et al. 2006; Molina et al. 2009). The technique offers a number of advantages over other types of DSS (Bromley 2005).

BNs are based on the concept of conditional probability, or more specifically on Bayes' Theorem (Bayes 1764). Bayes' Theorem takes the prior (a priori) probability of a variable being in a particular state and through the introduction of new knowledge (measurement, observation etc) calculates the posterior (a posteriori) probability of the same variable under the new conditions. The a posteriori probability is the probability of a variable or subset of variables, in the light of the new evidence '$e$'. This is termed the conditional probability and is written as $P(x|e)$. The analysis of stochastic processes in which uncertainty is treated via probability, characterises the BN.

A BN is made up of three main elements: (1) a set of variables (nodes) that represent the factors relevant to a particular environmental system or problem, (2) the relationships (links) between these variables, and (3) the conditional probability tables (CPTs) that lie behind each variable and which are used to calculate the state of the node. The first two elements form a Bayesian Diagram and the addition of the third forms a full network.

A conventional BN is a single system that is unable to receive or transmit information from outside the network. In contrast an OOBN represents a number of networks that can be linked together such that it is possible to transfer information from one to the other. The transfer of information is accomplished through the creation of output and input nodes in each network. These types of node are able to import and export information outside individual networks; these linking nodes are known as interface nodes. Together the interface nodes form what in Object Oriented programming terminology is known as an 'instance node', which in effect represents an 'instance' of another network. In object oriented terms each network becomes equivalent to a class.

OOBNs are a hierarchical description (or model) of real-world problems that mirror the way in which humans conceptualise complex systems. To cope with complexity humans think in terms of hierarchies of different classes. The use of instance nodes provides support for working with these different levels of abstraction in the construction of object oriented network models.

There are three main programming features that characterise Object Oriented programming and consequently OOBNs: 1) Encapsulation: The encapsulation of the internal details of a class means that some objects can be hidden; only those objects required for interfacing with other classes need to be exposed. 2) Inheritance: Inheritance allows a class to inherit the attributes and methods of another class. 3) Polymorphism: Finally, the feature of polymorphism allows objects to be of different types or nature, thus allowing an accurate representation of the real world.

## 3 Case Study

The Altiplano groundwater system comprises four independent karstic groundwater bodies, the Serral-Salinas, Ascoy-Sopalmo, Jumilla-Villena and Cingla aquifers situated in SE Spain (Fig. 1). All these aquifers have been intensively exploited, and groundwater is the unique resource for all classes of water users, resulting in extreme drawdowns. Thus, the water balances calculated from the above data are clearly deficient, which is evidenced as a notable consumption of water reserves ($-111$ million of $m^3$/year and an accumulated drawdown exceeding 3,000 million of $m^3$). The rates of drawdown of the piezometric levels have exceeded 10 m/year in some sectors of the aquifers, these figures being amongst the highest in Spain.

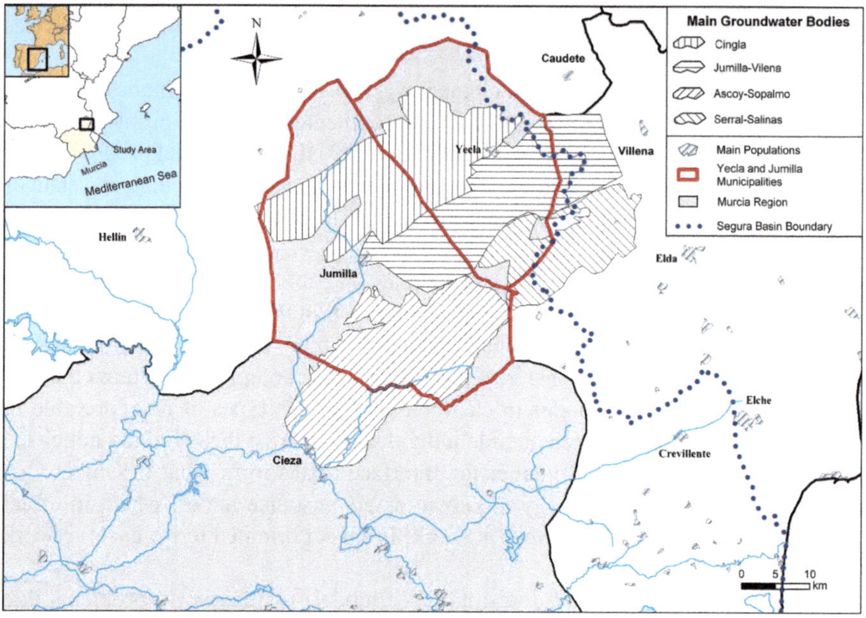


**Fig. 1** Location of the study area

Each aquifer exhibits a different hydrogeological behaviour, and each is hydraulically separated from the other. The hydrogeology of the system can thus simply be represented in terms of these four separate aquifers.

On the other hand, considering the social and economic state-of-affairs in the region, the picture is more complicated. In the Altiplano the greatest demand for water comes from a very profitable irrigated agriculture sector with rates of more than 8,000 €/ha. Water for irrigation is supplied from all four aquifers, but water from each aquifer is not restricted for use within the area of that aquifer; some is exported outside its boundaries. There is thus a complex transfer of water from one aquifer to the irrigated areas of others.

# 4 Results – DSS Structure

Social and economic conditions can be evaluated using one network to represent the whole region; however, the aquifers that supply the water are separate hydraulic units that can best be dealt with using four separate networks. To use only one network to represent this complex system would require all four networks plus the social and economic variables to be included; the resulting network would be so large and complex it would be impractical to use. Instead an OOBN approach has been used. With this approach each of the four aquifer regions are represented as

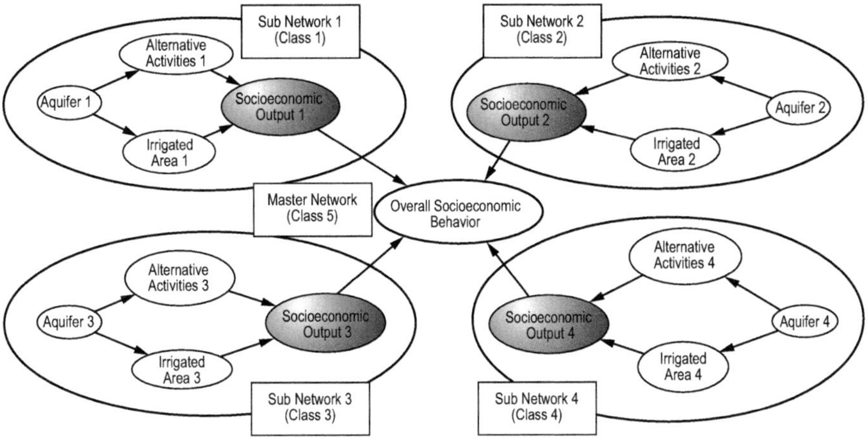


**Fig. 2** Logic scheme of the applied OOBN model

separate networks (aquifer network or sub-networks) that are all linked to a fifth (master) network, which is used to describe the social and economic conditions of the entire region, covering the area of all four aquifers (Fig. 2). In object oriented language the individual networks are classes and the variables that make up each network are objects.

The BN used to describe each aquifer is divided into two main sections; one deals with the hydrogeology, the other with social and economic conditions. The aquifers are not identical, which means that the configuration of each one is slightly different to reflect the specific circumstances of each aquifer. In the master network the socioeconomic conditions identified in each aquifer sub-network are integrated and expressed through two variables: the "Total Number of Agricultural Employments" and the "Total Income". These two social and economic variables are common to all the networks and serve to link the entire water system. The two variables appear as output nodes in each of the four sub-networks that represent the four aquifers, and as input nodes in the overarching master network that describes the regional socio-economic conditions integrated over the four aquifers. Any change that takes place in any of the sub-networks affects the state of the whole system and vice-versa.

## 5 Conclusions

This research has shown OOBNs as a great practical value for the management of multiple groundwater systems. By representing aquifer units as a hierarchy of interconnected objects, an OOBN provides an effective organizational structure with which to plan management strategies. The advantages of OOBNs over conventional BNs are as follows: 1) Using a traditional BN, the structure to represent the problem would need to be much more complex and confused; in the example presented here, all the aquifer units, plus all the nodes in the master network, would have to be represented in the same network. The subsequent network would be extremely large, difficult to run and even more difficult to understand. 2) No matter how many aquifers (sub networks or classes) are involved in a water system, the structure remains simple, clear and intelligible. It then becomes straightforward to evaluate and interpret the impact of each aquifer on the whole model; in this way the relative importance of every aquifer can be assessed. 3) OOBNs allow the logical structure of the real world to be reproduced. The hierarchical structure and the principles of inheritance permit a realistic representation of the interactions between, in this particular case, aquifer units.

In this case OOBNs have been applied to water management and specifically to integrated management of hydrogeological resources. Water systems supplied by groundwater resources are frequently supplied from different aquifers. There is a big potential for the widespread application of this type of approach in the field of groundwater resource management; this study can represent a starting point for further work.

## References


Bayes T (1764) An essay towards solving a problem in the doctrine of chances. Philosophical Transactions of the Royal Society of London 53, 37–418. (Reprinted in: Pearson ES, Kendall MG (eds), 1970, Studies in the History of Statistics and Probability. Charles Griffin, London)

Bromley J (2005) Guidelines for the Use of Bayesian Networks as a Participatory Tool for Water Resource Management, Centre for Ecology and Hydrology, Wallingford

Cain J (2001) Planning Improvements in Natural Resources Management. Centre for Ecology and Hydrology, Wallingford

Duntemann J, Marinacci C (1990) New objects for old structures. Byte 15(4):261–266

Henriksen HJ, Rasmussen P, Brandt G, von Bülow G, Jensen FV (2006) Public participation modelling using Bayesian networks in management of groundwater contamination. Environ. Model. Softw 22(8):1101–1113

Jensen FV (1996) An Introduction to Bayesian Networks. UCL Press, London

Jensen FV (2001) Bayesian Network and Decision Graphs. Department of Computer Science, Aalborg University, Aalborg, Denmark

Koller D, Pfeffer A (1997) Object-oriented Bayesian networks. In: Proceedings of the Thirteenth Annual Conference on Uncertainty in Artificial Intelligence (UAI-97), Providence, Rhode Island

Martín de Santa Olalla FJ, Domínguez A, Ortega JF, Artigao A, Fabeiro C (2006) Bayesian networks in planning a large aquifer in Eastern Mancha, Spain. Environ. Model. Softw 22:1089–1100

Molina JL, García-Aróstegui JL, Benavente J, Varela C, De la Hera A, López-Geta JA (2009) Aquifers overexploitation in SE Spain: A proposal for the integrated analysis of water management. Water Resour. Manage. 23:2737–2760

# Toxicity and Ecotoxicity of By-Products Resulting from Degradation of Fluorescent Tracers Used in the Karstic Chalk of Normandy (France)

P. Gombert, D. Granier, R. De Seze, P. Pandard, F. Gondelle, F. El Koulaly, F.Z. Benkada, and R. Toussaint

**Abstract** The chalk is the most extensive French aquifer, and numerous tracer tests are executed here each year to delimit the protection zones of water supply wells. In the karstic chalk of Normandy, only 49% of the tracer tests emerge at one of the monitoring points and subsequently pass into the water distribution network where they generally undergo chlorination. The majority of the tracers remains adsorbed in the subsurface or emerges at other points, notably untapped springs, in which case the tracers reach surface streams where they can be degraded by photolysis. Here, we study the degradation in the laboratory of three of the main fluorescent tracers, submitted to a chlorine flow or a strong luminosity. The tests were carried with concentrations greatly higher than the maximum ever recorded during tracer tests in chalk. In spite of an important degradation, the tests carried out on the by-products of degradation show an absence of acute toxicity and a moderate ecotoxicity, except for naphthionate (to relativize due to the high tested concentrations).

## 1 Introduction

The chalk has a specific hydrogeological behaviour, resulting from thick overburden of clay and loam, high matrix porosity, small hydraulic gradients and dry valleys with sinkholes related to the springs of humid valleys. The chalk outcrops cover 70,000 $km^2$ in the Paris basin, and contains the most important French aquifer. Tapped by thousands of boreholes and springs, it provides most of the drinking water of the administrative region of Haute-Normandie (Fig. 1) and almost half of the water consumed in Paris.

To delimit the protection zones of groundwater catchment area, numerous tracer tests are carried out by consulting agencies. The main tracers used are fluorescent

P. Gombert, D. Granier, R. De Seze, P. Pandard, F. Gondelle, F. El Koulaly, F.Z. Benkada, R. Toussaint
INERIS, BP 2, 60550 Verneuil-en-Halatte, France, e-mail: philippe.gombert@ineris.fr

B. Andreo et al. (Eds.), *Advances in Research in Karst Media*
DOI 10.1007/978-3-642-12486-0, © Springer 2010

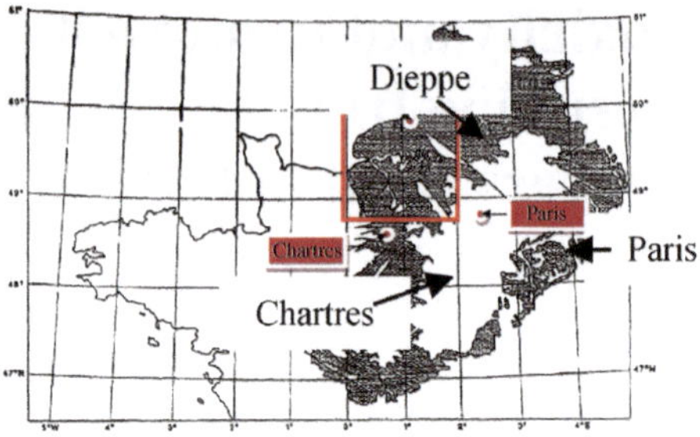


**Fig. 1** Outcrop of the chalk of the Paris basin and studied zone

(uranine, tinopal, naphtionate, sulforhodamine B) or ionic (iodide, bromide, lithium, zinc).

In the karstic chalk of Haute-Normandy, less than half of the injected tracers emerge at one of the monitoring points, which generally is a catchment work for drinking water supply. That means one half of the tracers enter the distribution network and show up at the consumer's tap, while the fate of the other half injection in the aquifer is unknown; they can remain absorbed in the subsurface or emerge in surface water via springs. In the last case, they may react with light or air and form by-products of which we know neither the nature nor the ecotoxicity.

Concerning the tracers that enter the water supply network, they are subjected to the action of purification agents, most of which are strong oxidizers (chlorine, chlorine dioxide, ozone, bromine, ultraviolet light): these tracers can degrade to by-products of which we do not know the nature, nor the toxicity in humans.

Upon the request of the Regional Direction of the Environment (DREAL) of Haute-Normandie, INERIS has realized a study concerning the environmental fate of the main tracers employed in the chalk and the risks these tracers and their degradation by-products have for human health or the environment.

## 2 Results of the Tracer Tests

The results presented originate from three studies, covering approximately the same geographical zone from 1960 to 2007. The success rate of the 1,893 tracer tests varies from 44–49%, according to the period taken into account. The interpretation by de Bechillon (2009) of the database managed by the French Geological Survey (BRGM) indicates that in the period 1970–2007, a total tracer mass of about 6 tons has been injected into the chalk of Haute-Normandie, about 76% of which was uranine. Of about 210 tracer concentration values, the maximum peak of the breakthrough curves is 8000 μg/l. A more detailed statistical analysis allows to show the specificity of the different "tracer test systems", i.e., the sections of the karstic network connecting an injection point (sinkhole, borehole, surface point) to a breakthrough point (spring or borehole).

Thus Gombert (2009) has shown that the "sinkhole-spring" systems perform best, probably because of the continuity of the karstic network at the entry and the exit of the system: the success rate reaches 62% and the average transfer speed is 144 m/h. At the bottom is the “surface-borehole” system, where, because of the absence of karstification, the success rate is only 22%, at an average transfer speed of 12 m/h. It is likely that phenomena as adsorption of the tracers while traversing the overburden or the unsaturated zone affect the breakthrough of the tracers. Between these two extremes are the system’s sinkhole-borehole and borehole-spring, which display together intermediary results: a success rate of 40–55% and a transfer rate of 101 to 142 m/h.

## 3 Description of Tracers Used

**Presentation of the Tracers.** The tracers selected for this study, are three of the main fluorescent tracers used in the chalk in Normandy: uranine, tinopal and naphtionate. Uranine (C20H10O5Na2) is the most frequently used tracer in the karst, because of its low adsorption, low detection limits and low procurement and analysis costs. Naphtionate (C10H 8NSO3Na) is a colorless tracer that emits a violet-blue fluorescence under ultraviolet radiation. Tinopal (C28H22O6S2Na2) is an optical brightener, colorless after dilution, that emits also a violet-blue fluorescence under UV radiation.

**Toxicological and Eco-Toxicological Parameters.** Studied by Behrens et al. (2001), the toxicology of fluorescent tracers shows no effect except for rhodamine (Wt, B & 6G) upon the genotoxicity and for sulforhodamine-B and rhodamine (B & 6G) upon the ecotoxicity. The USEPA has evaluated the toxicity of fluorescent molecules whose uranine and “bleaching agent 351” (which is tinopal): the results show a low toxicity and a moderate ecotoxicity. Carré et al. (2007) confirm no significant risk to the health of consumer at an exposure level corresponding to the concentrations likely to be found in water for fluorescent tracers. No carcinogenic effect has been demonstrated in animals.

*Note:* concerning ionic tracers, Behrens et al. (2001) indicate a potential ecotoxicity of lithium and a risk of toxicity with bromide in case of chlorination.

## 4 Degradation of Fluorescent Tracers

**Degradation Protocol.** Two types of degradation have been reproduced in the laboratory: a sudden degradation with chlorine (simulating the treatment which the water undergoes when entering the distribution network) and a slow degradation under the combined effects of light and oxygen in the air (conditions prevailing in a water course). Microbiological activity out of the aquifer but in the sampling bottles, as observed by Goldscheider et al. (2001) for naphtionate, has not been taken into account.

The tracer solutions have been prepared at a concentration of 1 g/L, an intermediate value between the tracer concentration at the injection point (about 100 g/l) and at the restitution point (maximum 8 mg/L). In the first case, 1 l of tracer solution was subjected to bubbles of gaseous chlorine, resulting from the introduction of sulphuric acid into household bleach. The mixture of tracer solution and chlorine was then left in contact for half an hour. In the second case, 2.5 l of tracer solution was poured in a flat tray and subjected during 24 hours to:

- Permanent agitation produced by magnetic bars rotating at a slow speed,
- Strong illumination (20,000 lux) of white light at 4000°K, UV-free,
- Ultraviolet radiation through a 400 W bulb.

**Protocol of the Toxicological Study on Rats.** The acute toxicity of the degradation products of tracers after chlorination was studied by oral administration to rats. The substance was tested according to a protocol derived from the test method in the OECD guidelines for testing of chemical products n° 423 "Acute Oral Toxicity – Acute Toxic Class Method". Batches of 3 female rats were formed and the animals received the test substance, in a concentration of 10 mg/kg in a single dose the first day. The animals had been fasted overnight and access to food was restored four hours after treatment. The tested dose belongs to group II of the packing groups described in the "abstract ADR-2005 – toxic materials", corresponding to moderately toxic substances. A control batch has received the medium alone in the same conditions as the animals tested. All animals were observed during 14 days during which their weight and clinical signs were recorded. They have then been sacrificed and a full autopsy was conducted by detailed macroscopic examination.

**Protocol of the Ecotoxicity Tests.** This approach on the ecotoxicity of the degradation products due to light and air has measured their effects on daphnia and unicellular green algae.

The first test, carried out according to the ISO 6341 standard NF is based on the measurement of the mobility of daphnia (Daphnia magna), exposed to various dilutions of the initial solution for 48 hours. The second test, executed according to the ISO 8692 standard NF, aimed to determine the inhibition of the growth of unicellular green algae (Pseudokirchneriella subcapitata) during a period of 72 hours. The inhibition corresponds to the reduction of the rate of growth of specific algae cultures, exposed to various dilutions of the initial solution, compared to the growth of control cultures incubated under the same conditions. The results obtained are used to calculate values of $EC_{50}$ i.e. the concentration (expressed as a percentage of the initial solution) having an effect on half of the tested population.

## 5 First Results Obtained

**Degradation in Chlorine.** It is known that strong oxidizing agents used in the purification of water, like chlorine, destroy uranine (Käss 1998). In the course of the experiments, the effect of the chlorine on the tracers resulted in a decoloration of the uranine and in a gray coloring of the naphtionate. The solutions of degraded tracers are currently analysed by mass spectometry and by HPLC chromatography in order to determine the by-products obtained.

**Degradation in Air and Light.** The tracers were decomposed in solution under the influence of an intense white light flow (10,000 lux), without UV, during 24h with continuous recording of their fluorescence. The tracer most influenced by light is uranine, which looses 82% of its fluorescence. Next come tinopal and naphtionate, with respectively 37% and 9% losses.

**Toxicity and Ecotoxicity of Degradation Products.** The tracers degraded by chlorination did not exhibit toxicity in rats (no decease nor clinical suffering) and their autopsy did not show macroscopic anomalies in the digestive tract.

The results obtained during the ecotoxicity tests are presented in Table 1. They demonstrate a moderate toxicity of uranine and tinopal CBS-X on the two species considered. In contrast, an inhibition effect is displayed for naphtionate for P. subcapitata ($EC_{50}$ : 4.0%), but it must be relativized by high concentrations tested.

Additional experiments are currently underway to connect the effects observed on these decanted and/or filtered solutions with concentrations of the initial compounds and compounds resulting from degradation in air and in light.

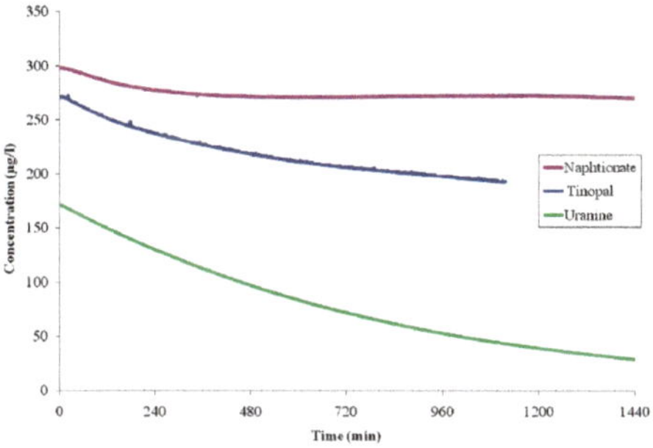


**Fig. 2** Evolution of the fluorescence of tracers under the influence of light

**Table 1** Ecotoxicity tests with tracers degraded in air and in light

| Tracer | Test | | Results |
|---|---|---|---|
| Uranine | Daphnia | $EC_{50}$ 48h | ND* (40% inhibition in the undiluted solution) |
| | Alga | $EC_{50}$ 72h | 77.8% (67.9–90.2) |
| Tinopal CBS-X | Daphnia | $EC_{50}$–48h | ND* (20% inhibition in the undiluted solution) |
| | Alga | $EC_{50}$ 72h | ND* (41% inhibition in the undiluted solution) |
| Naphtionate | Daphnia | $EC_{50}$–48h | 56.4% (52.1–60.3) |
| | Alga | $EC_{50}$ 72h | 4.0% (3.8–4.3) |

* ND = Not determinable

# 6 Conclusions

Uranine, tinopal and naphtionate are three of the main tracers used in the karstic chalk in Normandy. The literature shows that they have a moderate to zero toxicity and ecotoxicity. However, they may be subjected to degradations when they emerge from the aquifer in which they were injected. They may undergo chlorination upon arrival in a drinking water catchment network, or photolysis if they emerge to the open air via a spring. A study has therefore been carried out to determine, the toxicity and ecotoxicity of their degradation by-products. Solutions were prepared of each of the three tracers at a concentration of 1 g/L, intermediate between the concentrations observed at the injection and breakthrough points: these solutions are 125 times more concentrated than the maximum breakthroughs observed during 1893 recorded tracer tests. One part of these solutions was subjected to sparging by gaseous chlorine. The degradation by-products have been orally administered to rats, and did not show toxicity. The other part of the solutions has been subjected during 24 h to an intense flow of white and ultraviolet light as well as to mechanical agitation. A photolysis of these tracers was observed, but the ecotoxicity of the degradation products on daphnia and microalgae was weak, except for naphtionate. In conclusion, the structure of these fluorescent tracers is altered when they are submitted to sparging by chlorine or an intense light flow. However, the degradation products apparently do not present significant toxic properties at the usual levels, except for naphtionate on the growth of green microalgae. The precise characterization of these products by mass spectometry and by HPLC is currently underway.

**Acknowledgements** We would like to thank Mrs. Feeny as well as MM. Ferlin and Appéré of the DREAL of Haute-Normandie, for the financing of this study.

# References

Bechillon M de (2009) Inventaire régional Haute-Normandie des bétoires, itinéraire souterrauns des eaux (traçages) et des exutoires: Etude préalable à l'élaboration d'un guide de bonnes pratiques des opérations de traçages hydrogéologiques en Haute-Normandie, Mém. de stage réalisé au BRGM de mars à août 2009, Master 2, Univ de Perpignan – BRGM

Behrens H, Beims U, Dieter U, Dietze G, Eikmann T, Grummt T, Hanisch H, Henseling H, Käß W, Kerndorff H, Leibundgut C, Müller-Wegener U, Rönnefahrt I et al. (2001) Toxicological and ecotoxicological assessment of water tracers, Hydrogeology Journal (2001) 9:321–325

Carre J, Joyeux M, Montiel A (2007) Risques sanitaires associés aux traceurs fluorescents utilisés en hydrologie, Environnem, Risques & Santé 6/6:443–452

Goldscheider N, Hötzl H, Kottke K (2001) Microbiological decay of naphthionate in water samples as a source of misinterpretation of tracer tests, Proc 31th IAH Congr Munich, 77–81

Gombert P (2009) Synthèse des traçages réalisés dans la craie karstique de Haute-Normandie et proposition de normalisation, Géologues 159:13–30

Käss W (1998) Tracing technique in Geohydrology. Balkema, Rotterdam, NL

# Distribution of Trace Elements in a Karst Environment

B. Trček and R. Šajn

**Abstract** Monitoring of trace elements was established in main springs and swallow holes of the Ljubljanica karst river basin, in soils and river sediments to study their distribution and to analyze the contamination risk of the study area. Elements were divided into groups that are presented with statistical factors reflecting their geogenic or anthropogenic origin. The environment load due to metal industry is stressed. Results indicated significant impacts on sampled waters, soils and sediments. The heavy metal transport strongly depends on hydrodynamic conditions and a drainage system development. The study pointed out significant underground connections, which is important for protection of this karst water body.

## 1 Introduction

The unique attributes of the Ljubljanica karst river basin (Fig. 1) represented great challenges for scientists and researchers in the past (Gospodarič and Habič 1976). This area is an important drinking water reservoir, but very vulnerable (Trček 2006). Metal industry, agriculture and urbanization should be set out as the most hazardous activities. Their impacts on a karst groundwater body were assessed with geochemical researches. The distribution of trace elements in soil, water and river sediments was studied, as well as migration processes. The impacts resulting from the metal industry are considered in this study. Ferrous metal processing industry has had a historical tradition in the karst poljes A and B (Fig. 1). Consequently, galvanization activities were developed in karst poljes A, B and D.

B. Trček
University of Maribor, Faculty of Civil Engineering, Smetanova 17, SI-2000 Maribor, Slovenia,
e-mail: branka.trcek@uni-mb.si

B. Trček, R. Šajn
Geological Survey of Slovenia, Dimičeva, 14, SI-1000 Ljubljana, Slovenia,
e-mail: robert.sajn@geo-zs.si

B. Andreo et al. (Eds.), *Advances in Research in Karst Media*
DOI 10.1007/978-3-642-12486-0, © Springer 2010

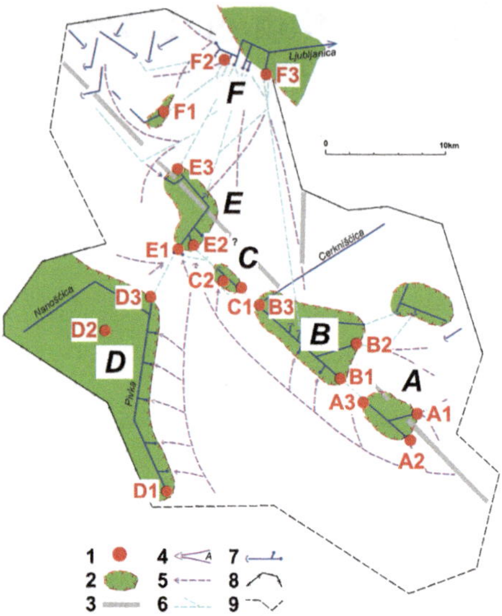


**Fig. 1** Review of superficial and underground waters in the Ljubljanica karst river basin: 1 – sampling places, 2 – karst polje with sediments, 3 – central hydrogeologic relative barrier, 4 – catchment area of permanent karst springs, 5 – periodic high-water discharge of karst water, 6 – underground water connections, 7 – superficial stream with springs and ponors, 8 – superficial watershed, 9 – supposed karst watershed (Modified after Habič 1976)

## 2 Description of the Study Area

Triassic, Jurassic, and Cretaceous rocks dominate in this area, while Quaternary sediments cover only karst poljes (Habič 1976). The altering of permeable and impermeable rocks and the tectonic structure influence both the surface runoff and groundwater flow (Fig. 1). The majority of karst groundwaters discharge from higher to lower basins, via a series of sinks and karst poljes. Only a small part discharges across the karstified dolomitic barrier directly to the Ljubljanica springs.

## 3 Methods and Techniques

The research methodology based on monitoring of trace elements in water, soils and river sediments. For two years water was sampled at all important springs, streams and swallowholes of the Ljubljanica karst river basin:

- the Loško polje area (A in Fig. 1) – the Mali and Veliki Obrh springs (A1 and A2) , the Loški Obrh stream near the Golobina swallowhole (A3);
- the Cerkniško polje area (B in Fig. 1) – the Cemun and Obrh springs (B1 and B2), the Cerkniščica stream before its outflow to the Velika Karlovica cave (B3);
- the Rakov Škocjan area (C in Fig. 1) – the Rak underground stream in the Zelške jame cave, the Kotliči spring (C1 and C2);
- the Postojna basin (D in Fig. 1) – the Pivka and Korentan springs (D1 and D2), the Pivka stream before its outflow to the Postojna jama cave (D3);

- the Planinsko polje area (E in Fig. 1) – the Unica spring in the Planinska jama cave (E1), the Malenščica spring (E2), the Unica stream before the Škofji lom swallowhole (E3);
- the Logaško polje and Ljubljanica spring area (F in Fig. 1) – the Jačka stream before the swallowhole (F1), the Ljubljanica and Bistra springs (F2 and F3).

The in-situ measurements of specific electroconductivity (SEC), temperature and pH were performed concurrently. Water samples were analyzed by the enhanced ICP-ES/ICP-MS methods for 72 elements. Data were processed with multivariate R-mode factor analyses to reveal the associations of chemical elements in a sample media (Reimann et al. 2002). The analysis was performed on variables standardized to zero mean and unit standard deviation. As a measure of similarity between variables, the product-moment correlation coefficient was applied. For orthogonal rotation, the varimax method was used.

## 4 Results

In the final multivariate factor solution of water data only 28 statistically significant elements retained. They were divided into six groups that are presented with statistical factors indicating their geogenic or anthropogenic origin:

- the statistical factor 1 associates B, Br, Cl, Na, P and S – the very soluble elements resulting from agriculture and housekeeping waste water;
- the statistical factor 2 represents a natural distribution of trace elements that are residues of the calcareous cambisol (Al, Ce, La, Nd, Y and Zr);
- the statistical factor 3 incorporates Ca, Mo and U elements and reflects the natural distribution of the Dinaric karst linked to active fault zones;
- the statistical factor 4 represents the geochemical association Ba, Co, K, Li, Mn, Ni, Rb and Si that results from a weathering process of flysch sediments;
- the statistical factor 5 incorporates Cu, Pb and Zn elements and reflects the anthropogenic impacts on water due to a non-ferrous metal processing;
- the statistical factor 6 associating Sn and W reflects the anthropogenic impacts on sampled water due to a ferrous metal processing.

The presented geochemical associations are comparable with those referring to soil and sediments. Statistical characteristics of data indicate the anthropogenic impacts of metal industry that are presented with factors 5 and 6. The impacts of non-ferrous metals are evidenced at all sampling places (Fig. 2a). The variation of factor 5 in A and C poljes and F2 coincides with oscillations of Zn concentrations (Fig. 3). The highest Zn concentrations were registered in A and C poljes in May 2006 during a relatively dry season. On the other hand increased Pb and Cu concentrations registered after a long rainy period in October 2005 (Fig. 3) are responsible for high significance of the factor 5 in B, E and D poljes. Pb critical concentration for drinking water (10 ppb) was even exceeded in B2g, where high Cu concentrations were also measured, as in B3 and E1. The Cu concentration increased again in B3 during

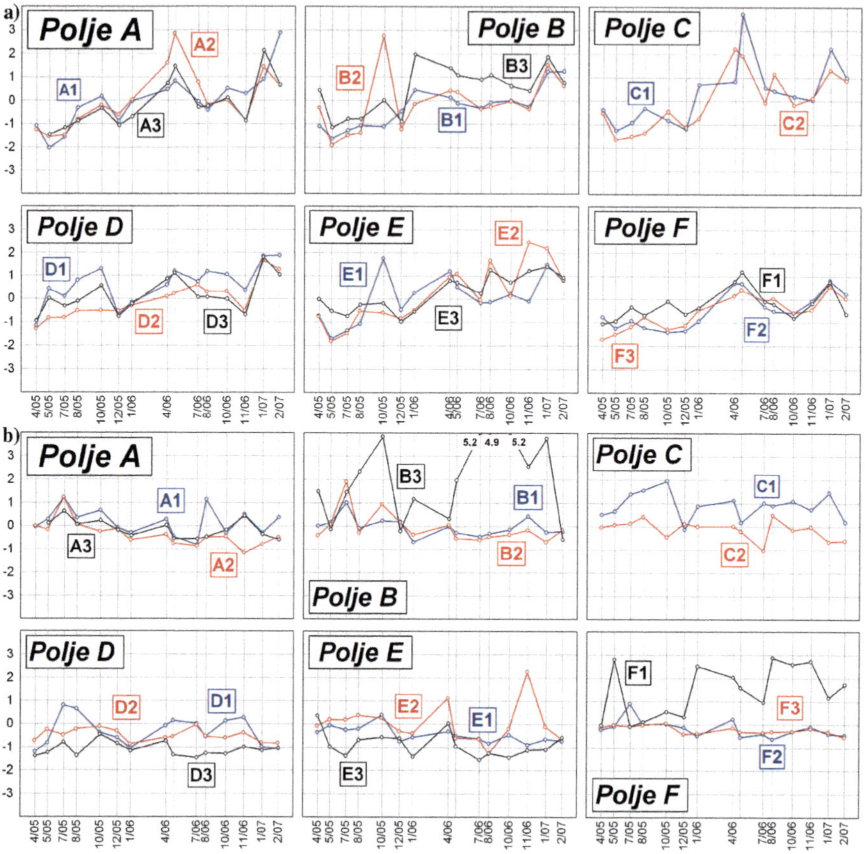


**Fig. 2** Distribution of **a** Factor 5 (Cu, Pb and Zn) and **b** Factor 6 (Sn and W) regarding date and location of water sampling

a wet summer 2006, which should be reflected in the hydraulically connected E2 (Fig. 3).

The ferrous metal impacts are particularly significant in B3 (Fig. 2b), which should result from both A and B poljes. Due to underground hydraulic connections Sn and W were transported also to C1 and occasionally to E2. Furthermore, it is evident that the ferrous metal processing is present also in the F1 area (Fig. 2b). Critical concentrations of Sn and W for drinking water were not exceeded in water samples during the monitoring period.

## 5 Discussion and Conclusions

The metal industry has a significant impact on sampled waters, soils and sediments in the study area. It the heavy metal content (As, Cd, Hg, Pb and Zn) exceeded the

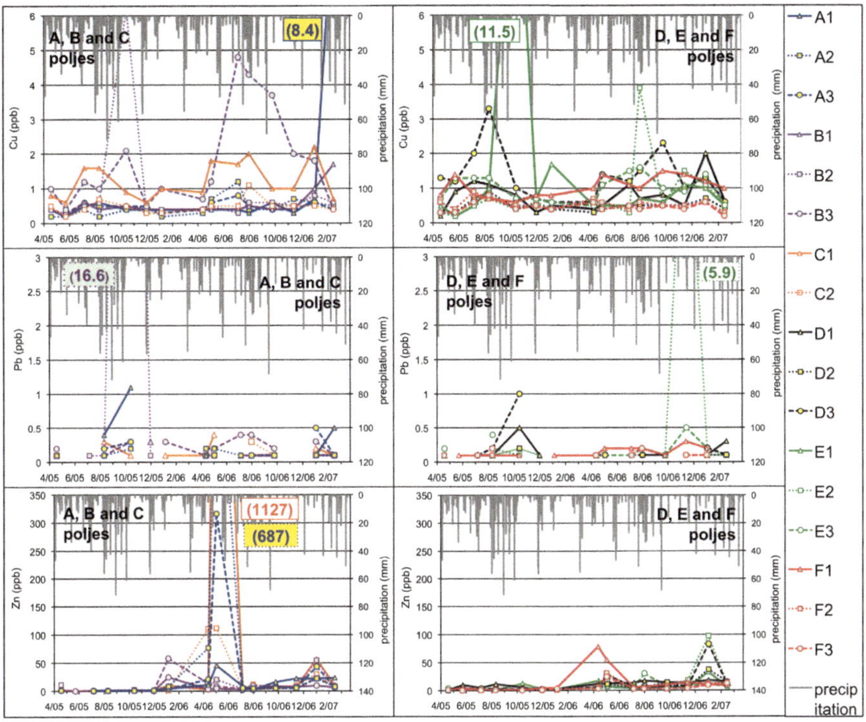


**Fig. 3** Time-trend plot of Cu, Pb and Zn in sampled water

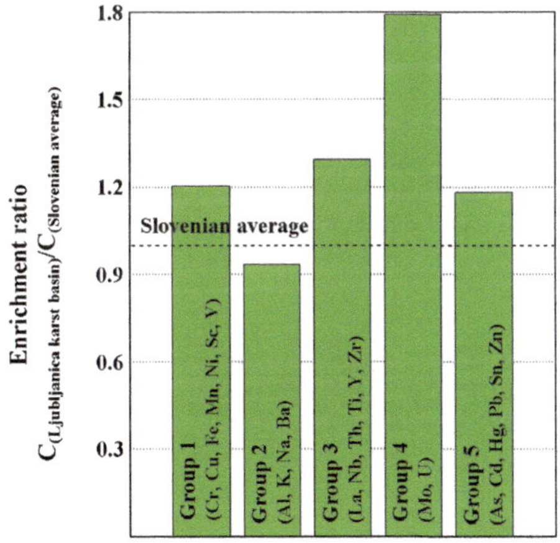


**Fig. 4** Average enrichment ratios regard to element groups in soil and river sediments

Slovene average in soil and river sediments (Šajn 2003) for about 20% (Fig. 4). The geochemical associations presented with factors 5 and 6 should result from both sources, the metal processing industry and natural erosion. The concentration of heavy metals decreased with the soil and sediment depth, which is typical for an anthropogenic origin. High concentrations of discussed metals were not linked to high Al and Fe concentrations. SEC values were slightly increased, while pH did not change significantly. It should be stressed that the factor 5 distribution in water, referring to non-ferrous metals, has a positive trend in all monitoring locations, while the distribution of factor 6, referring to ferrous metals, has a positive trend only in B3 and F1. The wash-off of heavy metals strongly depends on hydrodynamic conditions of the karst aquifer system. The underground hydraulic connections were indicated. The interactions between B3, C1 and E2 are particularly evidenced in Fig. 2b. These underground connections were proved also with tracing tests (Habič 1976; Kogovšek et al. 1999, 2008). The presented results contribute to better understanding of the complex karst environment and solute transport, which is important for improvement of management strategies of the study area and for protection of drinking water resources. E2 is an invaluable drinking water resource that supplies more than 20,000 inhabitants.

## References


Gospodarič R, Habič P (1976) Underground water tracing – Investigations in Slovenia, 1972–1975. Institute for Karst Research SAZU, Postojna

Habič P (1976) Goemorphologic and hydrographic characteristics. In: Gospodarič R, Habič P (eds) Underground water tracing-Investigations in Slovenia, 1972–1975. Institute for Karst Research SAZU, Postojna

Kogovšek J, Knez M, Mihevc A, Petrič M, Slabe T, Šebela (1999) Military training area in Kras (Slovenia). Environmental Geology 38/1: 69–76

Kogovšek J, Prelovšek M, M Petrič (2008) Undergroundwater flow between Bloke plateau and Cerknica polje and hydrologic function of Križna jama. Acta carsologica 37/ 2–3: 213–225

Reimann C, Filzmoser P, Garrett RG (2002) Factor analysis applied to regional geochemical data: problems and possibilities. Applied Geochemistry 17, 185–206

Šajn R (2003) Distribution of chemical elements in attic dust and soil as reflection of lithology and anthropogenic influence in Slovenia. Journal de Physique 107: 1173–1176

Trček B (2006) Heavy metals in the Ljubljanica catchment area. Geologija 49/2: 267–277

# Scenarios of Groundwater Pollution in a Karst Watershed: a Case Study in the Pinar del Rio Province at Cuba

H. Farfán, C. Dias, M. Parise, and C. Aldana

**Abstract** Groundwater pollution scenarios in the Santo Tomás watershed, a site of great importance for Cuban karst, are described in this work. The area hosts one of the most remarkable karst system of the island (Gran Caverna de Santo Tomás, over 45 km long). The Santo Tomás watershed has been an object of study since the 1950s, but in the last decades the researches have increased significantly, and were focused on different issues, including the hydrogeological and hydrodynamic characteristics of the aquifer, tracing tests, and hydro-geochemistry analysis. Within this framework, assessment of the groundwater vulnerability to pollution is an essential stage in the process addressed to safeguard the water resources. This has to be realized through a two-fold approach: i) prevention, in order to avoid contamination of karst groundwater, and ii) mitigation of the effects deriving from the existing activities. With reference to this latter point, knowledge of location and typology of the human activities potentially able to produce contamination is fundamental for land use planning programmes.

## 1 Introduction

Safeguard of the groundwater quality, as a crucial supply for the population and the economic development, requires particular attention due to the high possibil-

H. Farfán
Parque Nacional Viñales, ECOVIDA-CITMA, Pinar del Río, Cuba

C. Dias
Dept. Geology, Univ. Pinar del Río, Cuba

M. Parise
CNR-IRPI, Bari, Italy, e-mail: m.parise@ba.irpi.cnr.it

C. Aldana
Centro Nacional de Áreas Protegidas, CITMA, Cuba

B. Andreo et al. (Eds.), *Advances in Research in Karst Media* 
DOI 10.1007/978-3-642-12486-0, 

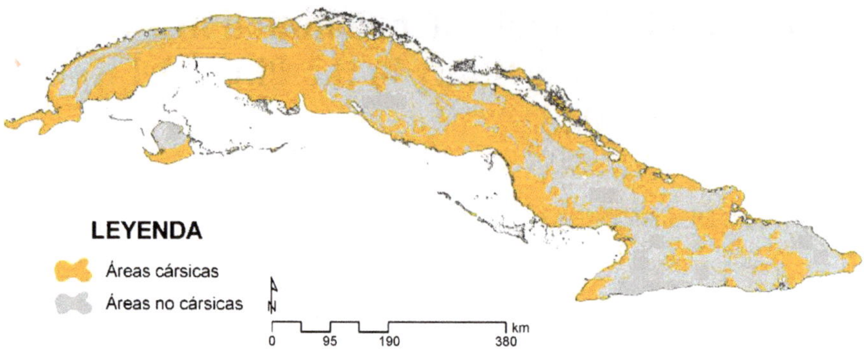


**Fig. 1** Karst areas in Cuba

ity of pollution, and cost and difficulty of recovering operations. Karst environments are not immune to such processes. Being characterized by rapid transport of water as turbulent flow through the network of karst fissures, conduits and caves, which implies very low possibilities of auto-depuration, at the same time allowing preservation of the high potential of pollution, they have to be included among the most fragile settings in the world as regards groundwater degradation (White 1988; Bakalowicz 2005; Ford and Williams 2007). Karst aquifers are extremely susceptible to contamination, with effects that are in most of the cases very difficult to forecast, and to remediate as well. This is due to the hydrological processes taking place in karst, the direct relationship between surface and subterranean drainage, and the degree of hydrological activity of karst systems (Palmer 2000).

In countries as Cuba, where over 65% of the land is affected by karst phenomena, and approximately 80% of the exploitable water resources are of karstic origin, studying karst hydrogeology and hydrology has a crucial importance, since also over these territorios most of the economic and social activities are concentrated (Molerio León and Gutiérrez Díaz 1999; Molerio León and Parise 2009).

In this work, groundwater pollution scenarios in the Santo Tomás watershed are described. The area is of great importance for Cuban karst. The Santo Tomás watershed has been at object of study since the 1950s, but in the last decades the hydrogeological researches have increased significantly. Within this framework, assessment of groundwater vulnerability to pollution is an essential stage in the process of safeguarding the water resources. The knowledge of location and typology of the human activities potentially able to produce contamination is fundamental for land use planning programmes.

## 2 Geological Setting

The Santo Tomás watershed is part of the Sierra de los Órganos in western Cuba. It is characterized by a typical tropical karst developed in Jurassic–Cretaceous carbonate rocks. The typical landscape presents low karst denudated mogote-type mountains, marginal poljes and dolines, and a diffuse development of extensive caves.

Geologically, three deposits are recognized. One terrigenous, corresponding to the oldest deposits is a Jurassic delta complex formed by sandstones, clays, schists. Unconformably above these deposits, the massive or grossly stratified limestones are present (Upper Oxfordian–Lower Tithonian). Other terrigenous with olistostromic character (Middle Eocene), produced by turbidites associated to overtrust of the complex at Cretaceous lie over Guaniguanico deposits. An Upper Pliocene – Upper Pleistocene marine transgression determined the formation of an alluvial plain, and deposition of the terrigenous material. Further alluvial and eluvial deposits of recent age complete the stratigraphic setting of the area.

From the groundwater viewpoint the area belongs to the watershed of Cuyaguateje, and consists of five sub-watersheds of the asymmetric dendritic type: Santo Tomás, Peñate, el Bolo, la Tierra, and los Cerritos. The waters from the five sub-watersheds join in the Sierra de Quemado to give origin to the Gran Caverna de Santo Tomás (46 km), the longest at Cuba, consisting of seven levels of cave galleries, and with a potential of some 70 km of development within the subterranean system of Sierra de Quemado, as demonstrated by tracer tests (Molerio León 2004). The Gran Caverna de Santo Tomás has a hydrologically active level (the lower one), two levels of seasonal, occasional or episodic functioning, and four fossil levels. Several permanent lakes, mostly fed by autochtonous water, are present in the seasonal levels of the system. The main discharge occurs with the river Santo Tomás, integrated by allogenic inputs.

## 3 Contamination Scenarios for the Groundwater of the Santo Tomás Watershed

Since the great majority of the social and economical activities in the region are concentrated in the valley of Santo Tomás, there the main likely sources of pollution for both surface and groundwaters are present (Molerio León and Parise 2009). The first step in the study of the hazard scenarios as regards the groundwater pollution was to identify, describe and map those anthropogenic activities that are potentially able to degrade the groundwater quality (Fig. 2), and are mainly distributed in the lower sector of the valley, at the margins of Sierra de Quemado.

The small village "El Moncada" occupies approximately 2 $km^2$, with a population ranging from 800 and 1,800 inhabitants, and a minimum density of about 400 inhabitants/$km^2$. These numbers are increased daily by tourists directed to the part of the Gran Caverna de Santo Tomás that is exploited as show cave, corresponding to the sixth level of cave development (Parise and Valdes Suárez 2005). The tourist visits determine a greater impact on the territory, due to the amount of vehicles and people reaching daily the site.

One of the main uses of the poljes is the agricultural one, with wide sectors devoted to farming tobacco, coffee and other products. The agricultural impact on the

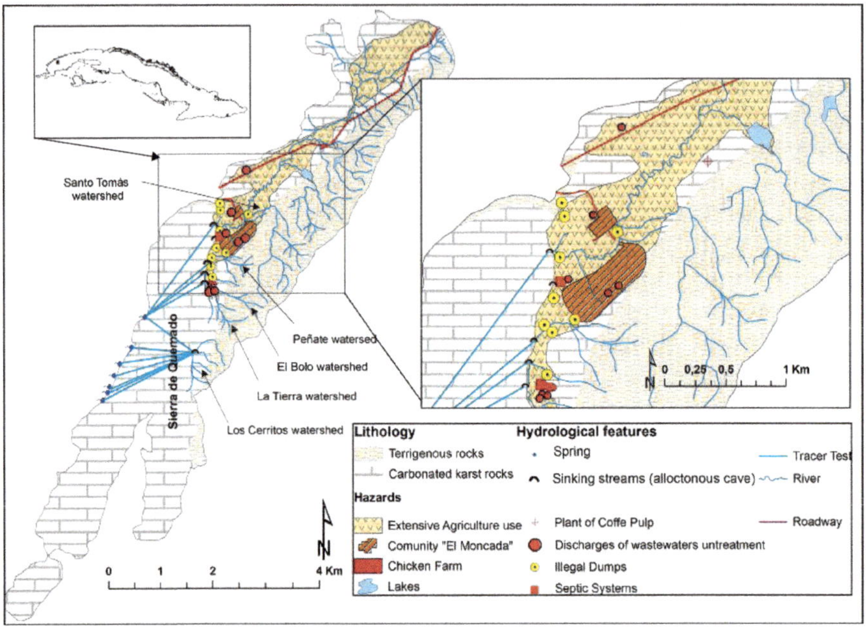


**Fig. 2** Location of the significant elements for assessing groundwater contamination in the Santo Tomás watershed

groundwater derives from the land use, the irrigation, the practices of farming and lifestock breeding, and also depends on the properties of the soil cover. The nitrate contamination of the groundwater is, for example, almost exclusively linked to the agricultural land use (Molerio León and Gutiérrez 1999). In the area, extensive and intensive agricultures exist, and approximately 1 ton/ha/year of chemical fertilizer (NPK complete formula) are used on the soil dedicated to tobacco cultivate. According with Ravbar and Goldscheider (2007), this quantity very high, nevertheless provokes a relatively low hazard. The more intensive agriculture use is associated with subsistence agriculture development by the people in the areas, using principally low quantity organic fertilizing.

The main hazard scenarios for groundwater contamination are associated to direct discharge of organic and inorganic domestic wastes and of the sewer system into the creeks and rivers flowing into the karst massif of Sierra de Quemado. Several sites of illegal discharge have been identified in the area, mainly at the banks of the permanent or seasonal creeks, as demonstrated by the finding of inorganic material transported by the flow within the cave system. The dimensions are variable, with $100\,m^2$ as average, and the biggest have $400\,m^2$. These illegal discharges have a big potential contribution to contaminated waters because the strong rainfall provokes large floods in this little allogenic watershed to the detriment of the underground environment. Traditional approaches are not capable of explaining the mechanisms of self-purification in so short distances (Días Arenas 1988). In several place of the

**Table 1** Average major constituents, nutrients and dissolved oxygen (mg/L) in selected point (Aldana et al. 2007)

| Localization | $HCO_3$ | Cl | $SO_4$ | Ca | Mg | Na | K | $NH_4$ | $NO_2$ | $NO_3$ | TDS | DO |
|---|---|---|---|---|---|---|---|---|---|---|---|---|
| Sumidero Santo Tomás | 108 | 8 | 13 | 28.2 | 2.6 | 6.5 | 1.6 | 0.35 | <0.03 | <10 | 152 | 4.89 |
| Arroyo La Tierra | 221 | 12 | 10 | 30 | 5.5 | 8.9 | 13.1 | 15.5 | <0.03 | <10 | 267 | 0 |
| Descarga Río Frío | 167 | 8 | 12 | 46 | 3.9 | 6.4 | 1.4 | 0.38 | <0.03 | <10 | 229 | 7.9 |
| Resolladero Santo Tomás | 108 | 8 | 10 | 24.9 | 3.2 | 6 | 1.3 | 0.31 | <0.03 | <10 | 146 | 7.9 |
| Lago Permanente 2do Cauce | 114 | 6 | 10 | 32 | 5.2 | 2.5 | 1.0 | <0.25 | <0.03 | 10.9 | 161 | 1.43 |
| Represas-Hoyo de fanía | 152 | 8 | 10 | 42.1 | 4.2 | 4.1 | 2.3 | <0.25 | <0.03 | <10 | 208 | 3.8 |
| Lago Charco Hondo | 36 | 7 | 10 | 5.7 | 2.0 | 1.7 | 4 | <0.25 | <.125 | <10 | 53.7 | 3.85 |

cave at places of active hydrological behavior, occasional and episodic solid waste from the surface are found.

Only six discharge of waste waters are describe, five of them are handcrafted construction, with a very low spillage and without serious problems. The most critical situations are the discharges of untreated waters in the Cueva de la Letrina, and the waste waters from Granja Avicola directly in the Arroyo de la Tierra. The principal sewer system of El Moncada is located less than 10 m from Cueva de la Letrina. This is a seasonally active gallery where the sewers discharge without any treatment or control on the occasion of intense rainstorms. Recently, monitoring of this water has started, and the first results indicate low values of dissolved oxygen, ranging between 0 and 1.3 mg/l, with mean value of 0.6 mg/L.

Arroyo de la Tierra has the highest values of mineralization of the surface waters that enter the karst system, due to the use of carbonate products in the Granja Avicola, where most of the 80,000 chickens are produced. The daily spillage of the products from the farm, as well as of the cleanliness of illegal rustic facilities to produce pork located in the margins of the river, provokes a higher rate of pollution.

In consequence, high NH4 concentrations are registered. Dissociation of ammonia represents a toxic compound that enters the system and threatens the aquatic life because of the alkaline environment it generates, thus decreasing water quality. At this site the values of dissolved oxygen present the lowest value (OD $=$ 0; Aldana et al. 2007), with mean value of 0.26 mg/L.

Recently, the speleological explorations, still on going in the Gran Caverna de Santo Tomás, have found further pollution problems related to the presence of hydrocarbons within the karst system coming from the swallet of the arroyo Santo Tomás. The presence of this pollutant may relate to the uncontrolled spillage of a workshop cars located inside the village, and in the general surroundings of the sinking stream.

# 4 Conclusions

The hazard scenarios and the consequent risks suffered by karst aquifers depend upon a number of factors. These include but are not limited to climate, land and water use, and to the local economy as well. To actually have a positive impact on groundwater protection, it is extremely important to develop in the local population an awareness of karst and of the processes acting in it. Understanding the high vulnerability of karst to contamination, and being aware that the changes it may suffer are in most of the cases unrecoverable (Zaporozec 2002), is the first step to directly involve the population.

At the same time, scientific research must proceed to collect the data necessary for a full comprehension of the hydrogeological and hydrochemical characterization of the area, as have been started in the last decades at the Santo Tomás watershed. Protection of the natural karst environment, and the resources contained therein, may be achieved only through this difficult but necessary two-fold approach, combining scientific research and direct involvement of the local communities.

## References

Aldana C, Farfán H, Molerio L, Parise M (2007) Self-purification capability of the course of underground water in tropic humid karst mountain. Geophys Res Abstracts 9:01841

Bakalowicz M (2005) Karst groundwater: a challenge for new resources. Hydrog J 13(1):148–160

Días Arenas A (1988) Something more than a Tropical Climate in the Caribbean Basin. In: Quesada V, Gutiérrez J, Landner L (eds) Water resources management and protection in tropical climates. Selected papers from the First International Symposium. 8–12 February. Havana, Cuba, 38–44

Ford D, Williams P (2007) Karst hydrogeology and geomorphology. Wiley, New York

Molerio León LF (2004) El enlace absorción-descarga de la Gran Caverna de Santo Tomás: evidencias derivadas de un ensayo con trazadores artificiales. Ing Hidr Amb 25(3):22–26

Molerio León LF, Gutiérrez Díaz J (1999) Agricultural impacts on Cuban Karstics aquifers. In: Drew D, Hötzl H (eds) Karst Hydrogeology and Human Activities. Impacts, consequences and Implications, A.A. Balkema, Rótterdam

Molerio León LF, Parise M (2009) Managing environmental problems in Cuban karstic aquifers. Environ Geol 58:275–283

Palmer AN (2000) Hydrogeological control of cave patterns. In: Klimchouk A, Ford DC, Palmer AN, Dreybrodt W (eds) Speleogenesis. Evolution of karst aquifers. Nat Spel Soc, Huntsville

Parise M, Valdes Suarez M (2005) The show cave at "Gran Caverna de Santo Tomás" (Pinar del Rio province, Cuba). Acta Carsologica 34:135–149

Ravbar N, Goldscheider N (2007) Proposed methodology of vulnerability and contamination risk mapping for the protection of karst aquifers in Slovenia. Acta Carsologica 36(3):397–411

White WB (1988) Geomorphology and hydrology of karst terrains. Oxford Univ. Press, Oxford

Zaporozec A (ed) (2002) Groundwater contamination inventory. A methodological guide. IHP-VI, Series on Groundwater 2, UNESCO

Zwahlen F (2004) Vulnerability and Risk Mapping for the Protection of Carbonate (Karstic) Aquifers. Final report COST action 620. European Commission, Brüssel

# Assessment of Pollution Risk in the Western La Mancha Carbonated Aquifer Due to Anthropogenic Degradation of the Tablas de Daimiel National Park (Spain)

H. Aguilera, L. Moreno, A. de la Losa, M.E. Jiménez, and S. Castaño

**Abstract** The inversion of the hydraulic gradient, primarily due to overexploitation and inadequate hydraulic management of the Western La Mancha carbonated aquifer, has led to the desiccation of the Tablas de Daimiel National Park (TDNP). As a consequence, a permanent unsaturated zone (UZ) has appeared beneath the former wetland area, making the carbonated aquifer extremely vulnerable to contamination. Besides, the strong human induced disturbance of the UZ (i.e. smouldering peat fires) is releasing nutrients that used to be retained in the soil matrix under natural conditions. The present research demonstrates how this degradation process might result in aquifer contamination.

## 1 Introduction

The Tablas de Daimiel National Park (TDNP) is a very important ecosystem, representative of Mediterranean wetland areas linked to groundwater dynamics. The TDNP used to be a discharge area for the 5,500 $km^2$ Western La Mancha carbonated aquifer.

During the last 40 years, the TDNP has suffered a strong process of anthropogenic degradation caused by intensive exploitation of the aquifer; the average water table drawdown for the last 30 years is 24 m. This has provoked a complete drying out of the wetland area, followed by a hydraulic gradient inversion and a development of permanent unsaturated zone (UZ). The system has turned from being an aquifer discharge zone into a recharge one.

The carbonated nature of the underlying aquifer, constituted by Miocene limestone and marl materials, implies a high vulnerability of the system, as the response rate of groundwater to disturbance is very elevated.

H. Aguilera, L. Moreno, A. de la Losa, M.E. Jiménez, S. Castaño
Spanish Geological Survey (IGME) Ríos Rosas 23, 28003 Madrid, Spain,
e-mail: h.aguilera@igme.es

B. Andreo et al. (Eds.), *Advances in Research in Karst Media*
DOI 10.1007/978-3-642-12486-0, © Springer 2010

The objective of this study is to determine whether the vulnerability to pollution of the carbonated Western La Mancha aquifer has been modified by the notable anthropogenic degradation of the natural hydrological cycle in the TDNP.

## 2 Materials and Methods

Groundwater sampling network in the TDNP area consists of 17 observation boreholes (Fig. 1), 7 of which were sampled monthly from April 2006 to October 2008. The other 10 have been monitored biannually. For this study, only the 4 observation boreholes that best represent the behaviour of the global system have been selected to be interpreted.

Surface water sampling network comprises seven permanent points complemented with 19 points in the Cigüela ditch. The ones selected for the study are: water from the Tajo-Segura transfer in August 2007 sampled in the Cigüela river gauging station, the 19 points in the Cigüela ditch located downstream of a sewage effluent discharge and the Navarro Bridge dam (Fig. 1).

To estimate the amount of labile nutrients stored in the unsatured zone (UZ), 120 samples were take in July 2006 (Eijkelkamp P1.01 auger) in 24 points located inside the TDNP to a maximum depth of 120 cm at 20 cm intervals (Fig. 1).

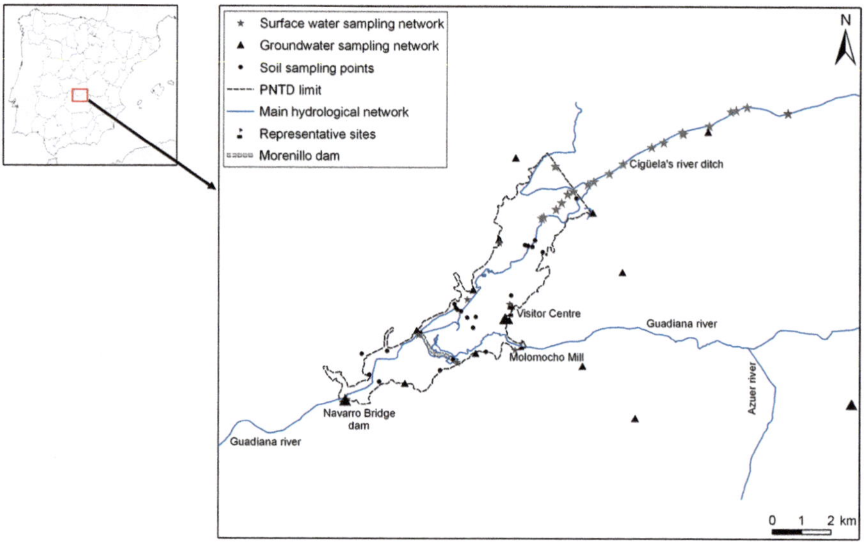


**Fig. 1** Soil and water sampling network. *Bigger icons* represent the surface and groundwater sampling points used in this study

# 3 Results and Discussion

Figure 2 shows two important characteristics of the aquifer: the general decreasing trend in groundwater levels and the rapid response capacity. Average concentration of nutrients and EC for selected water sampling points are shown in Table 1. There is an increase in salinity along the path of the water transfer through the Cigüela riverbed caused by evaporation and weathering of gypsic lithologies. Significant concentrations of TOC and $N_{org}$ in this sample are possibly due to pollution sources upstream. Inorganic compounds remain at low levels.

On the other hand, waters circulating through the Cigüela ditch downstream from the sewage effluent discharge show similar salinity values (lithological weathering), but concentration of inorganic nutrients ($NO_3^-$ and $PO_4^{3-}$) are quite high because of organic matter degradation and the existence of two agricultural farms.

The Navarro Bridge dam directly recharges a shallow perched water table level over carbonated fluvial silts. Observed concentrations point out the process of drainage and evaporation (CE > 18,000 μS/cm) and shift to anoxic euthropic conditions (TOC > 100 mg/l and presence of $NH_4^+$ produced by denitrification).

Groundwater parameters in the shallow perched level beneath the Navarro Bridge dam reflect the organic matter oxidation process that takes place in the UZ and the leaching of salts. Mean nitrate concentration doubles (from 12.6 to 25.1 mg/L) and nitrite concentration notably increases (from 0.2 to 1.4 mg/L), while EC reaches values around 7,000 μS/cm. However, water quality in the deeper aquifer level in this zone is much more acceptable, with lower nutrient and organic matter content, and

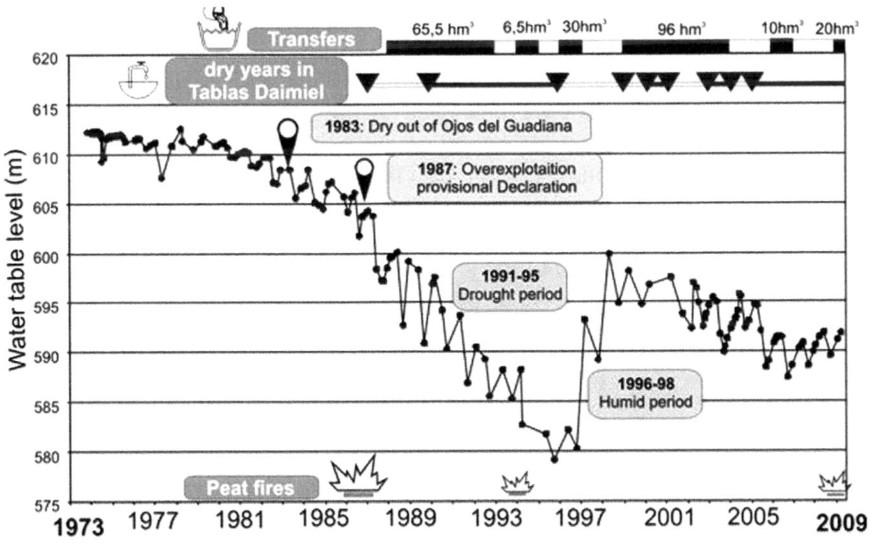


**Fig. 2** Groundwater level evolution for the 1973–2009 period in a representative observation borehole (IGME observation borehole n° 193030014) and major events occurred along this time, such as water transfers derived from the Tajo-Segura channel to the National Park

**Table 1** Mean and standard deviation (in bracket) values of nutrient concentration (mg/L) and electrical conductivity (EC) from surface and groundwater samples. TOC: total organic carbon; $N_{org}$: organic nitrogen

| | *n* | $NO_3^-$ | $K^+$ | EC(μS/cm) | $NO_2^-$ | $NH_4^+$ | $PO_4^{3-}$ | TOC | $N_{org}$ |
|---|---|---|---|---|---|---|---|---|---|
| Transfer (08/08/2007) | 1 | 1.0 | 14.0 | 3,543 | 0.9 | 0.0 | 0.7 | 9.4 | 2.1 |
| Cigüela ditch | 22 | 16.9 (14.0) | 20.0 (8.1) | 3,224 (2,575) | 0.0 (0) | 0.0 (0) | 3.7 (3.4) | 6.3 (4.2) | 5.2 (3.1) |
| Navarro Bridge dam | 27 | 12.6 (17.5) | 44.6 (58.3) | 18,691 (26,413) | 0.2 (0.6) | 1.8 (2.8) | 0.1 (0.1) | 104.3 (99.0) | 18.4 (30.3) |
| Pumping recirculation | 5 | 0.6 (0.5) | 24 (1.4) | 12,611 (536) | 0.0 (0) | 0.0 (0) | 0.1 (0.1) | 9.2 (4.0) | 3.1 (1.7) |
| Perched water table (Navarro Bridge) | 26 | 25.1 (9.4) | 20.5 (3.1) | 7,053 (332) | 1.4 (3.8) | 0.2 (0.8) | 0.0 (0.1) | 29.5 (5.1) | 7.3 (4.6) |
| Carbonated aquifer (inside TDNP) | 24 | 6.5 (4.1) | 11.2 (2.1) | 4,245 (203) | 0.1 (0.4) | 0.0 (0.1) | 0.1 (0.1) | 4.8 (1.3) | 3.6 (5.4) |
| Carbonated aquifer (outside TDNP) | 5 | 24.2 (1.6) | 2.0 (0) | 666 (33) | 0.0 (0) | 0.0 (0) | 0.0 (0.0) | 1.1 (0.8) | 1.5 (0.6) |

salinity ($\approx$ 4,000 μS/cm). Actual concentrations of dissolved chemical species in the deeper carbonated aquifer outside the TDNP are close to what would be expected naturally in this kind of waters: low EC (< 700 μS/cm) and generalized absence of nutrients, except for some agriculturally originated $NO_3^-$ (24.2 mg/L).

Groundwater quality of recirculating pumps poured onto the TDNP surface to artificially keep a minimum inundated area has been measured. The abnormally elevated EC (> 12,000 μS/cm) and organic matter content can be related to an evidence of the dissolution and leaching during infiltration.

The UZ developed in the TDNP area can reach a thickness of up to 15 m, but in many zones, due to the existence of perched aquifers, the phreatic level is found at 1–2 m depth. The high $SO_4^{2-}$ concentrations, over 1,800 mg/l, and EC values, over 3,000 μS/cm, in soil aqueous extracts, reflect lithological properties (gypsum weathering) and the active salinization process (Table 2). The predominance of oxidized forms of nitrogen and sulphur ($NO_3^-$, $SO_4^{2-}$) is also indicative of actual drainage conditions (Reddy and DeLaune 2008). The fact that the C/N ratio is lower than 10 implies a mineralization trend (Bohn et al. 2001).

Soils contain very high average concentrations of nutrients, particularly in the topsoil (Corg = 5.7%, Total N = 0.5%; $NO_3^-$ = 57.3 mg/L; available P = 27.4 mg/kg). Their release is accelerated by soil matrix aeration and wetting-drying cycles. These nutrients are mostly stored in peat and fluvial silts that fill up drained riverbeds and ditches, and also in the topsoil of edaphizated charopyte layers. Overabundance of nutrients allows classifying the TDNP as an eutrophic system.

The smouldering peat fires that began in summer 2009 increase even more the risk of lixiviation of contaminants (Rein et al. 2008). At the same time, there is an increase in hydrophobicity and a decrease in water retention capacity in soil surface layers up to 40%.

**Table 2** Summary of chemical composition of sampled soils ($n = 120$) and mean values in depth. SD: standard deviation; EC: electrical conductivity; pH ($H_2O$): pH in soil-water suspension; $C_{org}$: organic carbon; Total N: total nitrogen; Available P: available phosphorus (mg/kg); C/N: organic carbon-total nitrogen ratio

| | Depth (m) | | | | | | Mean | Median | SD |
|---|---|---|---|---|---|---|---|---|---|
| | 0–20 | 20–40 | 40–60 | 60–80 | 80–100 | 100–120 | | | |
| $Na^+$ (mg/L) | 217.5 | 160.5 | 136.6 | 153.5 | 153.4 | 76.3 | 162.3 | 125.5 | 165.2 |
| $K^+$ (mg/L) | 20.7 | 11.8 | 9.3 | 9.4 | 9.0 | 6.0 | 12.0 | 10.0 | 8.3 |
| $Ca^{2+}$ (mg/L) | 515.9 | 535.5 | 494.8 | 482.5 | 471.2 | 545.3 | 501.9 | 530.0 | 114.2 |
| $Mg^{2+}$ (mg/L) | 286.9 | 248.8 | 214.3 | 236.3 | 235.2 | 140.3 | 241.9 | 205.0 | 198.7 |
| $SO_4^{2-}$ (mg/L) | 1,995.0 | 1,965.0 | 1,824.4 | 1,862.3 | 1,830.9 | 1,780.0 | 1,894.2 | 1,870.0 | 700.3 |
| $Cl^-$ (mg/L) | 494.0 | 377.8 | 298.6 | 327.8 | 344.2 | 118.0 | 362.9 | 269.0 | 407.2 |
| $NO_3^-$ (mg/L) | 57.3 | 35.3 | 20.7 | 24.5 | 16.3 | 15.3 | 30.8 | 17.5 | 41.1 |
| $NO_2^-$ (mg/L) | 0.9 | 0.5 | 0.7 | 0.2 | 0.1 | 0.2 | 0.5 | 0.1 | 1.6 |
| $NH_4^+$ (mg/L) | 0.9 | 0.8 | 1.1 | 1.6 | 0.9 | 0.6 | 1.1 | 0.4 | 1.6 |
| $PO_4^{3-}$ (mg/L) | 0.3 | 0.2 | 0.0 | 0.0 | 0.0 | 0.0 | 0.1 | 0.0 | 0.4 |
| EC (μS/cm) | 4,020 | 3,593 | 3,346 | 3,507 | 3,417 | 2,717 | 3,559 | 3,165 | 1,672 |
| pH ($H_2O$) | 7.8 | 8.0 | 8.1 | 8.1 | 8.2 | 7.9 | 8.0 | 7.9 | 0.5 |
| $C_{org}$ (%) | 5.7 | 4.4 | 2.7 | 3.2 | 3.2 | 1.9 | 3.8 | 2.7 | 3.8 |
| Total N (%) | 0.5 | 0.4 | 0.3 | 0.3 | 0.3 | 0.2 | 0.3 | 0.2 | 0.3 |
| Available P | 27.4 | 12.9 | 8.0 | 8.6 | 8.4 | 4.5 | 13.0 | 7.0 | 18.6 |
| C/N | 6.2 | 6.1 | 5.4 | 5.1 | 5.3 | 6.0 | 5.6 | 6.0 | 2.0 |

Dispersion values of chemical parameters, both in water and soil, are very high (Tables 1 and 2), as expected for a heterogeneous, anthropic and strongly variable system like the TDNP. Typical infiltration capacities of present TDNP soils are extremely high, between 500 and 1,000 cm/d, with maximum values of 3,000 cm/d in peaty areas. There are also huge amounts of labile nutrients stored in UZ. For example, nitrate concentrations up to 270 mg/L in soil aqueous extracts and 110 mg/L in the saturated zone around the TDNP have been reported. Data from the perched water table below the Navarro Bridge dam reflect the surface water-soil-groundwater interaction. TOC and Norg in dam waters, which under normal flooding conditions would be around 80 and 10 mg/l, respectively, are retained in the UZ (Corg $\approx$ 10%; Total N $\approx$ 1%) and subsequently washed towards the saturated zone, where TOC and Norg reach concentrations of up to 40 and 16 mg/L. The high permeability and transmissivity of the saturated carbonated media ensure a fast spread of pollutants.

The electrical conductivity in soil aqueous extracts shows average values of 3,600 μS/cm (Table 2) and of 7,800 μS/cm in groundwater. The active ongoing process of salt accumulation and dissolution basically affects the northern zone of the TDNP, associated with the flooding area of the Cigüela river. Groundwater recirculation in the artificially inundated area of the Park and irrigation return flows from nearby farms accelerate this salinization.

Furthermore, the release of nutrients in organic soils is favoured by the infiltration of poor quality waters (Koerselman et al. 1993) like, for example, the sewage effluents poured onto the Cigüela ditch, or the recirculating pumps.

## 4 Conclusions

The Western La Mancha carbonated aquifer in the TDNP area was protected against contamination because it was a natural discharge area of groundwater and at the same time a nutrient accumulation zone in the peaty sediments. The inappropriate management of water resources has promoted the Park desiccation and the appearance of a UZ that reaches up to 15 m thick. This fact favours the oxidation of organic matter and, consequently, the release of labile nutrients. The TDNP area is a very permeable zone that in case of a great water inflow is able to mobilize large quantities of nutrients to the groundwater. From all of this, it can be concluded that the aquifer has become extremely vulnerable. The situation is aggravated by recent smouldering peat fires in the TDNP.

**Acknowledgements** This research has been financed by the CICYT project CGL2005-06458-C02-01 and has had material and logistic support from the Department of Edaphology of the Complutense University of Madrid.

## References

Bohn H, McNeal BL, O'Connor GA (2001) Soil chemistry. 3rd ed. John Wiley & Sons, Inc, New York

García Rodríguez M (1996) Hidrogeología de las Tablas de Daimiel y de los ojos del Guadiana. Bases hidrogeológicas para una clasificación funcional de humedales ribereños. PhD dissertation, Complutense University of Madrid

Koerselman W, Van Kerkhoven M, Verhoeven J (1993) Release of inorganic N, P and K in peat soils; effect of temperature, water chemistry and water level. Biogeochemistry 20:63–81

Martínez-Santos P, Llamas MR, Martínez-Alfaro PE (2008) Vulnerability assessment of groundwater resources: A modelling-based approach to the Western La Mancha aquifer, Spain. Environ Modell Softw 23:1145–1162

Reddy KR, DeLaune RD (2008) Biogeochemistry of Wetlands: Science and Applications. CRC Press, Taylor & Francis Group, New York

Rein G, Cleaver N, Ashton C, Pironi P, Torero JL (2008) The severity of smouldering peat fires and damage to the forest soil. Catena 74:304–309

# The Protection of Groundwaters Destined for Human Consumption in Karstic Aquifers. Advances Towards Safeguard Zones

A. Jiménez-Madrid, F. Carrasco, and C. Martínez

**Abstract** Carbonate aquifers constitute a water reserve of critical importance as a source of drinking water. For this, it is necessary to establish suitable protection measures so that groundwater bodies can achieve good status as required by the Water Framework Directive (WFD). As a general protection measure, the first advances in the development of a methodology to delimit safeguard zones are presented here. Carbonate aquifers in the south of Spain have been selected as pilot zone for the application of the three phases or stages of work proposed. Initial results obtained, following the application of the selected criteria (vulnerability, pressures, recharge and hydrogeological characteristics) show that the advance in the successive phases allows the necessary protection areas for these types of aquifers to be delimited.

## 1 Introduction

Groundwater constitutes a basic resource in Europe, which is demonstrated in countries such as Austria, Germany, Italy or Denmark, where more than 70% of the population's water supply comes from groundwater (Martínez Navarrete et al. 2008). Carbonate aquifers are especially important in the European continent from a human water supply point of view given that they represent 35% of the surface area and their resources make up 50% of the total water for supply (COST 65 1995).

With the Water Framework Directive, WFD (UE 2000) coming into effect, water protection has turned into one of the priority environmental targets of European policies. For this, it is necessary to develop specific methodologies for carbonate

A. Jiménez-Madrid, C. Martínez
Instituto Geológico y Minero de España (IGME), C/ Ríos Rosas, 23, 28003 Madrid, Spain, e-mail: a.jimenez@igme.es

F. Carrasco
Centre of Hydrogeology of University of Malaga, and Department of Geology, Faculty of Science, University of Malaga, Malaga 29071, Spain

B. Andreo et al. (Eds.), *Advances in Research in Karst Media* 299
DOI 10.1007/978-3-642-12486-0, © Springer 2010

materials that improve the protection against contamination in order to reach a good status regarding both quantity and quality of groundwater bodies used for human consumption as required by the WFD.

Safeguard zones are areas (given as an option in Article 7.3 of the WFD) whose scope focuses on measures to protect groundwaters with the aim of avoiding the deterioration of water quality and reducing the level of purification treatments required for human water consumption. They are therefore equivalent to wellhead protection areas of groundwater bodies intended for human consumption. The size of safeguard zones can be highly variable; in many cases they will be smaller than the groundwater body and several can co-exist in the same groundwater body, whilst others may extend outside of it. On the other hand, safeguard zones may correspond to the entire extension of the groundwater body or cover the protection perimeters of the existing water abstraction points (Jiménez-Madrid et al. 2008a).

The objective of the present work is to make an advance in the development of methods to delimit safeguard areas as a general measure of protection in all water bodies intended for human consumption. For this, the carbonate aquifers of the Sierra de Cañete, found in the Western Mediterranean to the south of the Iberian Peninsula, have been selected as a pilot area. Outcropping materials belong to the Subbetic and correspond to Jurassic limestones and dolomites that have been made permeable through fissures and karstification. They constitute several aquifers due to the existence of compartments caused by tectonic events (Linares and Carrasco 2007).

## 2 Methodology

To consider safeguard zones as a general measure of protection for carbonate aquifer groundwater bodies used for human consumption, and to be able to delimit them in a large number of groundwater bodies, the methodology should consider, among others parameters, water abstraction points intended for human consumption, existing pressures, an evaluation of the intrinsic vulnerability to contamination, an evaluation of the recharge and the water table elevation.

Based on the proposed methodology, the different groundwater bodies destined for human consumption are intended to be distinguished into different areas classified into one of the following groups:

A. Safeguard zone with heavy restrictions
B. Safeguard zone with moderate restrictions
C. Safeguard zone of future prevention
D. Safeguard zones not established

Safeguard zones with heavy restrictions will be similar to those contemplated in zone I of microbiological protection of wellhead protection areas to abstraction points, whilst those of moderate restrictions will be similar to areas indicated as zone II for dilution and control of wellhead protection areas in accordance with the circumstances contemplated for these protection areas to be integrated into land-use planning policies (MIMAM 2002).

To obtain the different areas, three phases or working stages are proposed. As result of each of them, a delimitation scheme is obtained that is complemented by the following phases until reaching the definitive delimitation map (Fig. 1).

A) First phase. A zoning is obtained from the evaluation of the intrinsic vulnerability using the COP method (Vías et al. 2006) and by the existence or absence of significant pressures on the groundwater body (Zones A, B, C and D, with scores from 6 to 3 of Fig. 1).

B) Second phase. Evaluation of aquifer recharge. The score obtained in the previous stage is increased according to the amount of recharge by means of establishing three intervals in the percentage of recharge as a proportion of the total precipitation (Fig. 1) to which a score between 0 and 2 is assigned. This way, for example, a B zone (with moderate restrictions) from the first phase and with high recharge values shifts to the A zone with heavy restrictions when establishing the safeguard area. For the evaluation of recharge, the use of the APLIS method (Andreo et al. 2008) developed specifically for carbonate aquifers is recommended.

C) Third phase. Delimiting of catchment areas. This phase consists of characterising the behaviour of an aquifer by carrying out an inventory of abstraction points, water table elevation, groundwater flow direction and hydrogeological watersheds, to delimitate compartments, with their corresponding zoning, in order to carry out an improved characterisation of the groundwater body and to readjust the different protection zones.

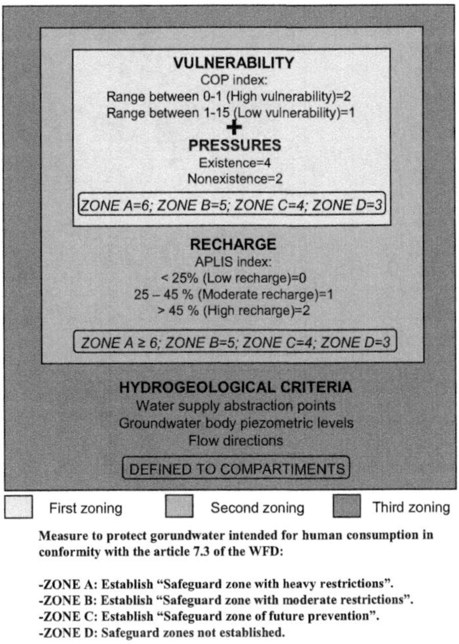


**Fig. 1** Methodological phases to delimit safeguard zones

# 3 Results

The first results obtained following the application of the methodological approach to the carbonate aquifers in Sierra de Cañete are presented in Fig. 2.

The results of the COP index (Fig. 2a, Jiménez-Madrid et al. 2009) show that the high and very high vulnerability classes are predominant (together covering more than 52% of the surface area). If the existence or absence of significant pressures is considered (Fig. 2b) and the evaluation is combined with the vulnerability (Fig. 2c), an initial delineation of safeguard zones with heavy restrictions is determined, which occupy 18% of the surface area, as well as a delineation of safeguard zones with moderate restrictions that occupy 30% of the area (zones A and B in Fig. 2c and Table 1). The existence of areas of high vulnerability without significant pressures that occupy 35% of the surface area is observed, which are proposed as safeguard zones of future prevention (zones C of Fig. 2c and Table 1), and it is also noticed that 17% of the surface area does not require protection (zone D in Fig. 2c).

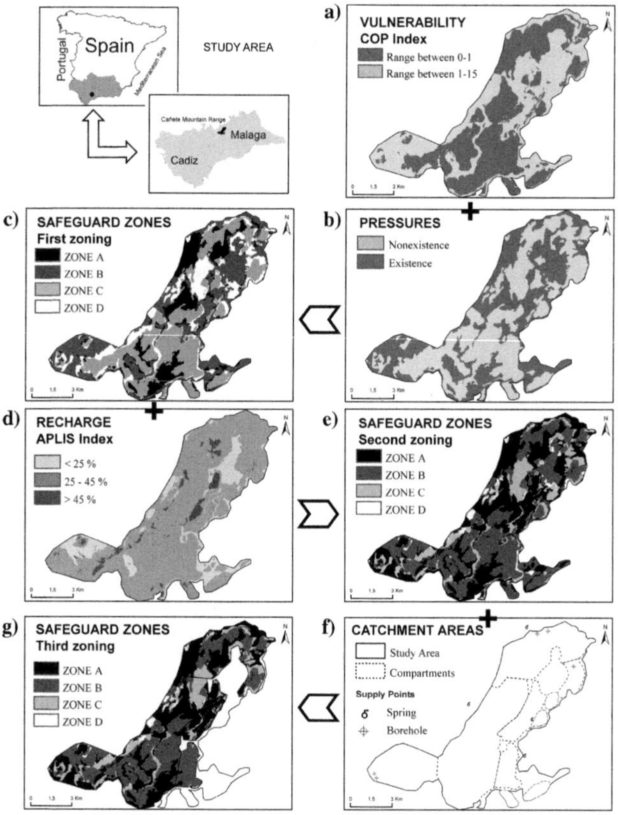


**Fig. 2** The first advances towards safeguard zones. Application in Sierra de Cañete (Málaga, Spain)

**Table 1** Percent of the territory occupied by areas delimited as safeguard zones

| | SAFEGUARD ZONES | | | |
|---|---|---|---|---|
| | A | B | C | D |
| First zoning | 18 | 30 | 35 | 17 |
| Second zoning | 38 | 45 | 14 | 3 |
| Third zoning | 33 | 35 | 12 | 20 |

The evaluation of recharge, its spatial distribution in the study area (Jiménez-Madrid et al. 2008b) carried out in the second phase of the study (Fig. 2d), and its contribution to the protection of the water resources reveals an important increase in the number of safeguard zones with both heavy and moderate restrictions (zones A and B in Fig. 2e and Table 1), with a decrease in the zone of future prevention and in those that do not require protection. Therefore, in this second phase, there is an increase in the protection of water resources principally in areas of greater recharge.

Based on third phase, ten compartments have been defined in Sierra de Cañete (Fig. 2f). Water resources from six of them are used for supplying water to eight municipalities located in the area for which it is necessary to establish protection measures. The consideration of these criteria allows the protection measure to be localised (Fig. 2g) avoids the establishment of protection areas covering the whole groundwater body, and in this way areas that do not feed to the existing abstraction points and the compartments that are not used for human water supply can be excluded. Following this criteria, 20% of Sierra de Cañete's surface area does not require the establishment of safeguard zones and a considerable decrease in safeguard zones with high and moderate restrictions is achieved (Table 1). A further advanced step using water table contour maps and mathematical models would enable a greater knowledge of hydrogeological features. In this case, areas where the groundwater flow direction is not towards abstraction points would be separated in order to reduce the extension of safeguard zones.

## 4 Conclusions

The establishment of safeguard zones as a general measure of protection for groundwater bodies intended for human consumption is recommended option as a means of complying with the requirements imposed by the WFD.

In this work, an initial methodological approach has been developed for the delineation of safeguard areas which is divided into three working phases. In the first phase the evaluation of intrinsic vulnerability to contamination and the existence of pressures are considered, in the second phase the existence of recharge areas is addressed and in the final phase to delimit catchment area criteria are applied.

The application of this methodology to the Sierra de Cañete shows that, with each working phase, the appropriate protection zones can be located within a groundwater body, while not encompassing the whole water body being studied.

In future work, it will be necessary to contemplate other factors that allow the characterisation of safeguard zones of groundwaters used for human consumption with more accuracy, such as for example, the evaluation of risk concerning pressures or the use of mathematical models dealing with the flow and transport of pollutants which allows an improved hydrogeological characterisation of groundwater bodies.

**Acknowledgements** This work is a contribution to the project CGL2008-04938 of the Ministry of Science and Innovation of Spain and to the Unit Partner IGME-GHUMA "Advanced Hydrogeological Studies".

## References

Andreo B, Vías J, Durán JJ, Jiménez P, López-Geta JA, Carrasco F (2008) Methodology for groundwater recharge assessment in carbonate aquifers: application to pilot sites in southern Spain. Hydrogeology Journal 16:911–925

COST 65 (1995) Hydrogeological aspects of groundwater protection in karstic areas, Final report (COST action 65). European Commission. Brussels, Luxembourg

Jiménez-Madrid A, Martínez Navarrete C, Carrasco Cantos F (2008a) Comparative analysis of analytical methods for wellhead protection areas implementation. Application to different types of aquifers in the south of Spain. GroPro – Groundwater Protection, ATV Jord og Grundvand, 67–74, Denmark

Jiménez-Madrid A, Martínez Navarrete C, Carrasco Cantos F (2008b) Estimation of recharge by the APLIS method in the Sierra de Cañete aquifers (Malaga). VII Simposio del Agua en Andalucía. Agua y Cultura, Tomo I: 293–303

Jiménez-Madrid A, Martínez Navarrete C, Carrasco Cantos F (2009) Groundwater pollution risk assessment. Application to different carbonate aquifers in south Spain, European Geosciencies Union, General Assembly, Abstracts book-Vienna 2009

Linares L, Carrasco A (2007) Sierra de Cañete. In: Durán JJ (ed) Atlas hidrogeológico de la provincia de Málaga, IGME-DPM, 2, 60–64

Martínez Navarrete C, Grima Olmedo J, Durán Valsero JJ, Gomez Gomez JD, Luque Espinar JA, De la Orden Gomez JA (2008) Groundwater protection in Mediterranean countries after the European water Framework directive. Environ Geology 54:537–549

MIMAM (2002) Guía para la delimitación e implantación de perímetros de protección de captaciones de aguas subterráneas para abastecimiento público

UE (2000) Directiva 2000/60/CE del Parlamento Europeo y del Consejo, de 23 de Octubre de 2000, por la que se establece un marco comunitario de actuación en el ámbito de la política de aguas. DO L 327 de 22-12-2000

Vías JM, Andreo B, Perles MJ, Carrasco F, Vadillo I, Jiménez P (2006) Proposed method for groundwater vulnerability zapping in carbonate (karstic) aquifers: the COP method. Application in two pilot sites in Southern Spain. Hydrogeology Journal 14:912–925

# Wellhead Protection Areas Delimitation in Karstic Aquifers. Aplication in Guadalquivir River Basin (Spain)

A. Jiménez-Madrid, C. Martínez, J.A. Luque, J.A. Zuazo, and P. Jiménez

**Abstract** Protection of groundwater destined for human consumption is one of the environmental priority aims since the Water Framework Directive (WFD) was put into effect. With this aim, traditionally wellhead protection areas have been delimited using analytical methods. Due to the carbonate aquifers singularities, a methodology to delimit wellhead protection areas is presented that complements the above mentioned methods with a hydrogeological characterization. This methodology has been applied in four main phases of work: firstly, water supply point features, secondly hydrogeological and hydrodynamic parameters data collection, boundaries and hydrogeological running outline; thirdly, hydrogeological surfaces meaning and zoning and finally, hydrogeological running and budget analysis at body water scale. The methodology offers more accurate delimitation and reduces the uncertainties derived from the hydrogeological features of these aquifers. Finally, the obtained results show the zoning established around each catchment areas by means of this methodology.

## 1 Introduction

Groundwater protection is a priority objective in EU environmental policy, which is reflected specifically in Directive 2000/60/EC, Water Framework Directive, which in the Article 7.1, establishes limits to designate a body of water as a Drinking Water Protection Area (UE 2000).

A wellhead protection area of human consumption water is a figure largely reflected in water legislation in different European Union countries as a protection

A. Jiménez-Madrid, C. Martínez, J.A. Luque
Instituto Geológico y Minero de España (IGME), C/ Ríos Rosas, 23, 28003 Madrid, Spain, e-mail: a.jimenez@igme.es

J.A. Zuazo, and P. Jiménez
CRN, S.A.

B. Andreo et al. (Eds.), *Advances in Research in Karst Media* 
DOI 10.1007/978-3-642-12486-0, 

against pollution (Martinez Navarrete et al. 2008). In Spain there is a specific legislation concerned with this while in countries like Germany the protection perimeters are 20% of the territory (Vorreyer 1998), in Spain its implementation is practically nonexistent (Martinez Navarrete and Garcia Garcia 2003).

The present work aims to complement the traditional analytic methods used in wellhead protection areas, delimitation of detritical aquifers with the detailed hydrogeological studies, which allow protection areas to be better defined in carbonate aquifers.

As a previous stage in the wellhead protection areas delimitation, the main points of human water supply of carbonated water bodies of the Guadalquivir River Basin were chosen. Also, a detailed analysis of supply infrastructure and data of quantity demand has been collected, including geological and hydrogeological features of the aquifer, potential contamination sites and pressure assessment, intrinsic vulnerability to the contamination movility proccesses and, finally, the groundwater pollution risk assessment (Jiménez-Madrid et al. 2008). Due to the particularities and diversity of the carbonate aquifers systems, a complementary detailed hydrogeologic analysis has been made, which includes the study of the boundaries and aquifer geometry, water table maps and flow nets and hydrogeological running (isolines, flow paths), hydrochemical parameters, a input output budget of the resources (verified by calculus) and identification of influential zones and supply zones.

Finally, it is necessary to point out a set of recommendations related to the control of the monitoring well network and regarding surveillance to monitories the wellhead protection areas efficiently which will ensure water quality protection at the supplying points.

# 2 Methodology

## *2.1 Issues Presented in Karstic Aquifers*

Due to the singularities and particularities of the hydrogeologic behaviour of karstic aquifers, the generic methods (based on bidimensional flow calculations: Wyssling, isochrones, radius of influence) employed to estimate and define the wellhead protection areas do not provide adequate and accurate results to guarantee functionality and effectiveness (Jiménez-Madrid et al. 2008). In karstic aquifers, hydrogeological specific peculiarities, spatial heterogeneity and anisotropy, outer recharge basins, punctual inflows and flow directions, and temporal variability and constraints determine the hydrogeological running of each aquifer. That is why in karstic aquifers, generic methods for protection areas have to be consider in an orientative or site-specific way, and must be supported with hydrogeological studies and criteria.

Given this limited effectiveness of the numeric analytical methods for the identification of the parameters in karstic areas, hydrogeologic analysis is especially relevant.

## 2.2 Phases of Work

Taking into account the items mentioned above, four main work phases have been defined:

1. Study of the water point: Well (depth and pumping rate) or spring (discharge rates) and main well design features.
2. Study of the karstic system characteristics
    a. Climatic and geographical conditions: Elevation, topography and drainage, soils, gradients, infiltration,
    b. Hydrogeological mapping, thickness and lateral extent of formations, structure, hydrodynamic parameters, boundary conditions, limits and impermeable substrate contacts, water table, preferential flow directions and pathways, existing wells, outer slopes, concentrated / diffuse inflow and discharge rates.
3. Identification of hydrogeological surfaces: Influence radius and catchments groundwater basins
4. Hydrogeological running and budget. Recharge, circulation and discharge. Running typology, conceptual analysis of behaviour and resources quantification in budget.

The 2nd and 3rd steps of work can be very well supported by a vulnerability background zoning for karstic ranges using classical COP, EPIK or PI methods for vulnerability and risk mapping for protection of carbonate aquifers.

This thematic information collected in the above phases can be overlapped in a Geographical Information System GIS, with the aim to correlate the information with the potential or confirmed contaminant sources and sites. The different stages

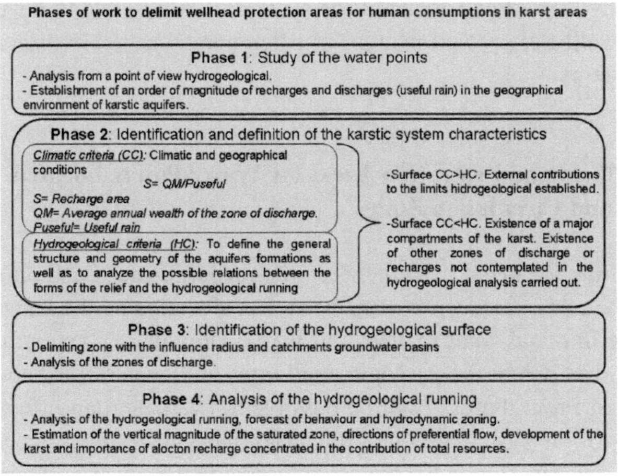


**Fig. 1** Methodological phases to delimit wellhead protection areas

in this methodology to define the wellhead protection catchment of karstic areas are described in Fig. 1.

## 2.3 Water Supply Points Location

Besides these sequence of phases, at least two possible hydrogeological situations have been differentiated in karstic systems for the water supplying points, depending on whether the points are in natural discharge sectors of the system or whether they are located in recharge and flow areas.

### Situation 1. Wellhead Protection Areas for Human Water Supply in Discharge Zones of Karst Areas

From the analysis of all these elements in discharge zones, mainly springs, the possible surfaces that can compound the catchment's water area are the following:

1. Surface of outcrops of carbonate terrains in the basin
2. External basins of no carbonates annexed terrains and formations
3. Satellite carbonate outcrops and other formations laterally connected with the system

Therefore, it is possible to establish wellhead protection areas which will not be linked exclusively to the spatial variable and that take into consideration preferential pathways with high hydraulic conductivity and other features specific to each karstic aquifer. These may determine the behaviour of a contaminant introduced into the system.

After the delimitation of the study area (karstic system) and the identification of the different hydrogeologic zones of the system; it is possible to develop a further proposal for wellhead protection areas of catchments located in discharge zones of carbonate systems.

### Situation 2: Wellhead Protection Areas for Water Points Located in Recharge and Circulation Zones

In the case of the localized catchments in the recharge zones and circulation pathways along the carbonate formations, the steps for analysing the karstic system are to identify, delimit and analyze. The aim is to contextualize the extractions within the system and to define the geologic conditions and constraints between the well and the aquifer. From there, it will be possible to evaluate their effect on the natural conditions of the system and its limitations and vulnerability according to their location in the system. Once the hydrogeological factors of the site are known, the application of the numeric analytical method is a good tool (Wyssling for example) for the wellhead protection area calculations (Wyssling 1979 in Lallemand-Barrés and Roux 1999) to define protection zones inside the system boundaries.

The results obtained by these methods are compared with the hydrogeological analysis of the system to make adjustments and for cross validation to correct deviations and inconsistencies with the configuration and running of the system.

## 3 Results

The first results after the application of the methodology described above are shown in Fig. 2. In the environment at collection points for human supply, zoning has been undertaken in each area of protection according to the following structure:

- Zone I, immediate or absolute restrictions (Transition time of 1 day)
- Zone II, near or of maximum restrictions (Transition time of 60 days)
- Zone III, remote or of moderate restrictions (Transition time of 4 years)
- Protection Zone of quantity

Wellhead protection areas are shown in the Guadalquivir river basin delimited in connection with this work, as well as existing studies both of sufficiency and mineral waters. Zone I occupies 0.4 $km^2$ while Zone II extends over 105 $km^2$. Zones III and the zone of protection of quantity have a similar extension of around 350 $km^2$. An amplification can be seen of the results of the water body of Sierra Mágina.

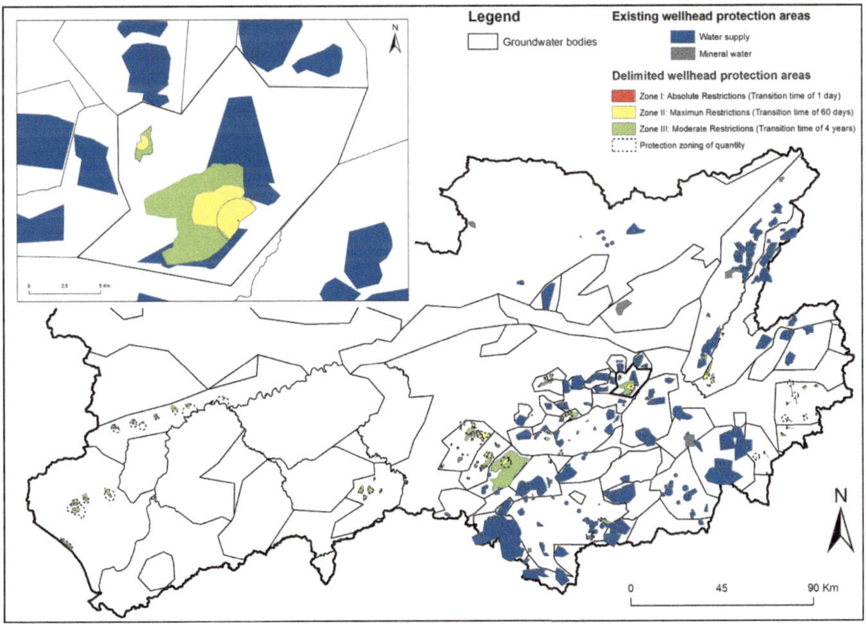


**Fig. 2** Wellhead protection areas existing and delimited in Guadalquivir River Basin (Spain)

# 4 Conclusions

Wellhead protection areas have been traditionally defined to establish appropriate measures to protect catchments suitable for human consumption. This paper presents a methodology that complements the classical analytical methods used in detritical porous aquifers by means a detailed hydrogeological analysis for application in carbonated aquifers. This provides an increased level of accuracy in the delimitation of the wellhead protection areas and reduces the uncertainties of the specific hydrogeological behaviour of these kind of aquifers.

After applying this methodology to the Guadalquivir basin bodies of water zoning, the results obtained show wellhead protection areas adapted to the carbonate aquifers features, limits and boundaries and which support the calculations of the analytical results. Additionally, the methodology has been applied with an GIS integration of different thematic layers according with four work steps for a systematic collecting and analyze data.

In future studies, recommendations will be made relating to the control and monitoring well network necessary to test the yield of the wellhead protection areas and to guarantee the protection of drinking water quality at the supply points.

# References

Jiménez-Madrid A, Martinez Navarrete C and Carrasco Cantos F (2008) Comparative analysis of analytical methods for wellhead protection areas implementation. Application to different types of aquifers in the south of Spain. GroPro – Groundwater Protection, ATV Jord og Grundvand, 67–74 Denmark

Lallemand–Barrès A et Roux JC (1999) Périmètres de protection des captages d'eau souterraine destinée a la consommation humaine. Éditions du BRGM

Martinez Navarrete C and Garcia Garcia A (2003) Wellhead protection areas in water suplí. Methodology and application. IGME

Martinez Navarrete C, Grima Olmedo J, Durán Valsero JJ, Gomez Gomez JD, Luque Espinar JA and De la Orden Gomez JA (2008) Groundwater protection in Mediterranean countries after the European water Framework directive. Environ Geology 54:537–549

UE (2000) Directiva 2000/60/CE del Parlamento Europeo y del Consejo, de 23 de Octubre de 2000, por la que se establece un marco comunitario de actuación en el ámbito de la política de aguas. DO L 327 de 22-12-2000

Vorreyer C (1998) Delineating surface source water protection areas in Germany. Source water assessment and protection 98, Dallas, Proceedings, 61–64

# Definition of Statutory Water-Supply Protection Zones in Mountainous Karstic Aquifers of the Guadalquivir Basin

A. Jiménez-Madrid, C. Martínez, J.A. Zuazo, and P. Jiménez

**Abstract** A method is proposed for identifying the basic features of karstic areas as a starting phase in the definition and outline of wellhead protection areas and safeguard zones in large regional or supra-regional ranges. This characterization is based on analysis of four topics or control items that shape the hydrogeological scenario and constrain the karst system. Due to the singularities of karst aquifers, as well as the scarce availability of specific information, it is essential to have a basic hydrogeological characterization of the karst system to outline its structure and to understand its hydrogeological processes. A better understanding of the groundwater flow and geometry of each karst system allows the delimitation of wellhead protection areas to be more accurate and effective.

## 1 Introduction

In vulnerability studies of aquifers, wellhead protection areas, groundwater management projects and land planning, karstic ranges use to have the singularity of a scarcity of specific information, which is frequently solved in other situations applying generalist criteria and parameters.

Karstic aquifers show specific features that make it more difficult to apply general criteria in delimiting the wellhead protection areas and safeguard zones (Ravbar 2007). The "safeguard zones" are areas, that can be established optionally as stated in the WFD, whose scope focuses on measures to protect groundwater with the aim of avoiding the deterioration of water quality and reducing the level of purification treatments required for human water consumption (Jiménez-Madrid et al. 2008).

A. Jiménez-Madrid, C. Martínez
Instituto Geológico y Minero de España (IGME), C/ Ríos Rosas, 23, 28003 Madrid, Spain, e-mail: a.jimenez@igme.es

J.A. Zuazo, P. Jiménez
CRN, S.A.

B. Andreo et al. (Eds.), *Advances in Research in Karst Media* 
DOI 10.1007/978-3-642-12486-0, 

They are therefore equivalent to "wellhead protection areas" of groundwater bodies intended for human consumption according to article 7.3 of the WFD (Martínez and García 2003; Martínez Navarrete et al. 2008). Nevertheless, a hydrogeology analysis under the karstology point of view provides worthy information about the configuration, behavior and flow patterns of karstic units.

To make up for this information scarcity about the singularities of the karstic units, a methodology is proponed to undertake the preliminary analysis of this area, with the aim of getting basic information about its configuration, groundwater flow and system behaviour. This information permits the integration of the specific variables of each karstic system in the vulnerability definition of the aquifer and, consequently, in the statutory water-supply protections zones.

This methodology has to be fully functional and yet simple to apply. Several factors must be considered to exceed more than the basic knowledge of the analyzed karstic systems. These factors correspond to the different hydrogeological variables that determine the groundwater flow in karstic system.

The main considered variables describe four kinds of items: lithological control that determines the hydraulic parameters and its distribution; structural control that determines the aquifer geometry with its limits and hydrodynamic relationships; geographic control that determine the outcrops morphology, baseline level, etc.; and the climatic control that supplies the water resources inflow. These controls consider the geometrical and physical basic parameters that configure a karstic system (Goldscheider and Drew 2007; Jeannin 1998; Klimchouk et al. 2000; Dreybrodt 2005). They control the conditions for groundwater flow. Such controls or information inputs deal with both qualitative and quantitative items.

From the variability of the different combinations that can provide these four controls, different hydrogeological "scenarios" are obtained in which it is possible to foresee the system vulnerability and its distribution, making possible the definition of protection perimeters more fitted to the reality of each karstic system.

This method has been applied in the study to define the wellhead protection areas and safeguard zones in a set of karstic areas of the groundwater bodies in the Guadalquivir River Basin. The results permit a better fit for protection zones than those obtained previously by analytic means and methods.

## 2 Methodology

In delimiting wellhead protection areas and safeguard zones, the starting point of the analysis begins in the well or the spring point (or a set of them), which correspond to artificial or natural outflow points. Although the hydrogeological implications of one or another kind of water-supply point can be different, they can all be considered as outflow zones of the system in their hydrogeological location.

Taking this into account, the general analysis and features of the karstic system has been developed by considering four controls or factors that determine the system flow and hydrogeological process. These four controls are explained as follows:

**Lithological Control.** This control corresponds mainly to the different lithological features of the karstic system, specifically dealing with the intrinsic hydraulic properties of the rock (porosity, transmissivity, composition, etc.) and their distribution.

Lithological control determines the karstic porosity and fissure network of pathways in the geological materials and the occurring of higher transmissivity layers and impermeable barriers due to the permeability contrasts between lithologies, etc. In short, these features are going to map the efficiency of the aquifer units in the transmission of the groundwater flows.

**Structural Control.** This control defines the structural configuration and geometry of the aquifer system, taking into account the geometry of the lithological bodies and the structural framework of the karstic system (lateral changes, contacts, folding, main fractures and joints in the rock mass).

Generally, it is possible to establish a hierarchy of the structural elements (mainly fractures) depending on their extension and the hydrogeological role that play in the system flow. It is possible that only the general flow directions or constraints determine the pathways of the flow.

Structural control influences two fundamental items: groundwater network geometry and the distribution of the water flows drains between high transmissivity structures against attached blocks, with water storage and slow flows.

**Geographical Control.** This item defines the geometry of the outcrops of the system range. It determines the location of the outflow zones, the inflow typologies, the extension of the system (massivity), vertical development (saturated and no saturated zones), and base level drainage. In short, it settles the hydrodynamic general

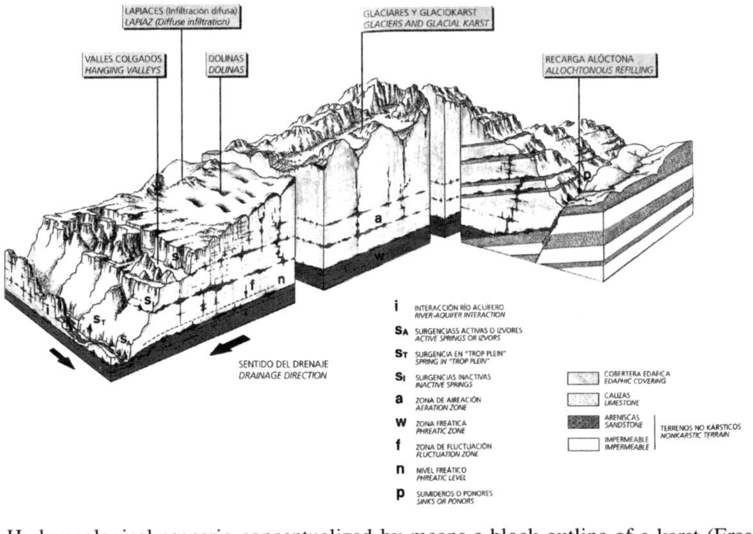


**Fig. 1** Hydrogeological scenario conceptualized by means a block outline of a karst (Eraso and Fernández 1993)

conditions of the system, mainly the gradients and general groundwater flow direction.

**Climatic and Temporal Control.** This factor defines the relationship between the quantity of the discharge and the system extent. Departing from a yearly outflow module estimation, it is possible to set the inflow system surface with enough accuracy. After that, it is possible to fit this surface to the hydrogeological limits obtained from the previous controls factors.

It is recommended to consider climatic evolution that the karst development has suffered through time due to different climates (cold and wet periods) that have could configured features in the karstic system which depart from the current conditions of groundwater flow.

These four factors or controls describe the karstic configuration (structure, geometry, hydrodynamic and boundary conditions, etc) and consequently the hydrogeological running, processes and water budget of the karst, conceptualized in Fig. 1.

## 3 Results

By applying this four controls method, an outline or hydrogeological scheme has been obtained of the karstic system and an approximation to the conceptual model of groundwater running.

**Hydrogeological Scheme Definition.** The hydrogeological scheme or scenario fits to the physical framework of the karst, its geometry of aquifers and the outcrops, hydrodynamic parameters, groundwater flows (vertical development, gradients, inflow typologies, etc) and boundary conditions.

**Approximation to the System Running Conceptual Model.** Once the physical framework of the karst system is characterized, its flow and behavior in different scenarios of inflow are analyzed, approaching the conceptual model of the system.

Based upon karstic system delimitation and the conceptual model of groundwater flow, it is possible to develop a proposal of water-supply point protection perimeter and safeguard zones, not linked exclusively with the distance variable. This proposal will include the singularities of each karstic area. The obtained zoning map is fitted with higher accurate to the reality of the system.

This methodology has been used to define the karstic ranges' safeguard zones and wellhead protection areas for the Guadalquivir Basin. Previously, the hydrogeological basic parameters were defined and analytical methods (mainly Wissling method) have been applied to get an approximation of the perimeters' magnitudes.

Sierra Mágina and Estepa karstic mountains are clear examples of extent ranges of middle-low mountains located in Jaen province, in the southern edge of the Guadalquivir Depression.

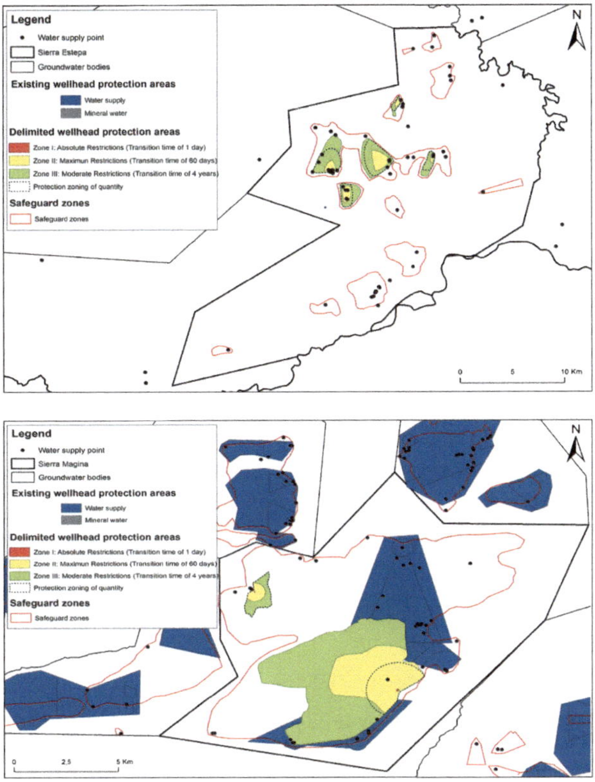


**Fig. 2** Example of a Guadalquivir Basin karstic area (Sierra de Estepa and Mágina). Hydrogeological analysis versus analytical methods

# 4 Conclusions

Karst media has been traditionally a complicated hydrogeological system to apply successfully standard hydrogeological and hydraulic parameters, due to:

- Doubly nature of groundwater flow (continuous porous and discontinuous – complex network of fractures)
- Scarce knowledge available about the singularities of each area and also the variability in time (karstic evolution).

A methodology is proposed to perform the analysis and characterization of karstic areas in extent geographical ranges (regional and interstate ranges) to settle drinking-water-supply points protection zones or safeguard zones. The proposal is based upon the analysis of four indicators or control factors that permit outlining the "hydrogeological scenario" of a determined karstic aquifer. This hydrogeological scenario has to define and delimit the physical framework of the system and to approach the conceptual model of the groundwater flow.

Based on this analysis, it is possible to delimit the drinking water protection areas and safeguard zones fitted, with higher accuracy, to the singularities of each karstic system. This permits more efficient protection to preserve in quantity and quality the water resources for drinking water supply.

The obtained results in carbonate media of the Guadalquivir Basin highlight the importance of the hydrogeological methods versus analytical methods, because analytical methods do not consider the main factors that constraint and determine flow in the karstic aquifers.

Finally, it is important to emphasize that the proposed analysis is a first step to undertake the delineation of wellhead protection areas and safeguard zones quickly and efficiently in regional and vitiated or impaired karstic ranges.

This preliminary approach is undertaken based on the standard available information. The methodology will always permit the incorporatation of new inputs of data that refine the results more and more.

As it has been also commented, this methodology has been used in the delimitation of the wellhead protection areas and safeguard zones in the karstic media of the Guadalquivir Basin in Spain.

## References


Dreybrodt W, Gabrovsek F and Romanov D (2005) Processes of Speleogenesis: A Modeling Approach. ZRC Publishing, Karst Research Institute at ZRC SAZU. Carslogica, 4

Eraso A and Fernandez Gilbert E (1993) The karst. Instituto Técnológico Goeminero de España

Goldscheider N and Drew D (2007) Methods in karst Hydrogeology. International Association of Hydrogeologists. Taylor & Francis Group, London

Jeannin P-Y (1998) Structure et comportement hydraulique des aquifères karstiques. Thése de doctorat. Université de Neuchâtel

Jiménez-Madrid A, Martínez Navarrete C and Carrasco Cantos F (2008a) Comparative analysis of analytical methods for wellhead protection areas implementation. Application to different types of aquifers in the south of Spain. GroPro – Groundwater Protection, ATV Jord og Grundvand, 67–74 Denmark

Klimchouk AB, Ford DC, Palmer AN, Dreybrodt W et al (2000) Speleogénesis. Evolution of Karst Aquifers. National Speleological Society, Inc

Martínez Navarrete C and Garcia Garcia A (2003) Wellhead protection areas in water suplí. Methodology and application. IGME

Martínez Navarrete C, Grima Olmedo J, Duran Valsero JJ, Gomez Gomez JD, Luque Espinar JA and De la Orden Gomez JA (2008) Groundwater protection in Mediterranean countries after the European water Framework directive. Environ Geology 54: 537–549

Ravbar N (2007) The protection of karst waters: a comprehensive Slovene Approach to vulnerability and contamination risk mapping. Karst Research Institute at ZRC SAZU; Ljubljana. Carsologica 6

# Delineating Source Protection Zones of Karst Springs. The Case Study of Villanueva del Rosario Spring (Southern Spain)

A.I. Marín and B. Andreo

**Abstract** As part of a research programme into karst systems in Europe located under different climatic and hydrogeological contexts, this paper presents the results obtained from applying the COP+K method to the Villanueva del Rosario spring (province of Málaga, southern Spain). This spring is the main source of water supply for the village of the same name, and presents the typical karst behavior pattern of carbonate aquifers in southern Spain. Evaluation of the source vulnerability to contamination and the delineation of protection boundaries were performed using the COP+K method, designed for carbonate aquifers. The zones of maximum protection are mainly the routes of preferential infiltration where, despite the distance separating them from the spring, transit times of about one day have been recorded. The results obtained are coherent with the data of a multitracer test and an analysis of the hydrodynamic and hydrochemical response at the spring, especially with regard to Total Organic Carbon. The results of this study constitute an instrument that may be used by the competent authorities in delineating protection perimeters for aquifers with similar characteristics.

## 1 Introduction

The methodology used to delineate protection perimeters for water supply springs needs to be adapted to the hydrogeological characteristics of the medium that is to be protected. Karst aquifers present a high degree of heterogeneity and anisotropy, which complicates the prediction and modelling of water flow patterns and velocities. In these aquifers, the distribution of protection zones differs from the traditional type, in which there is a graduation of protection levels that is directly proportional

A.I. Marín, B. Andreo
Centre of Hydrogeology of University of Malaga, and Department of Geology, Faculty of Science, University of Malaga, Malaga 29071, Spain, e-mail: aimarin@uma.es; andreo@uma.es

B. Andreo et al. (Eds.), *Advances in Research in Karst Media*
DOI 10.1007/978-3-642-12486-0, © Springer 2010

to the distance from the spring or borehole. Investigations are currently underway for the delineation of protection perimeters in karst aquifers on the basis of results obtained from pilot areas in various countries (including Spain, France, Slovenia and Portugal). Among the goals of these studies is to compare the methods currently being employed and to improve validation techniques and interpretation by means of tracer tests and the analysis of hydrochemical and hydrodynamic responses from the spring.

This paper presents some results from the above studies. Specifically, the delineation of protection perimeters for a karst aquifer are examined by applying the COP+K method (Andreo et al. 2009), validated by means of artificial and natural tracers. From a practical standpoint, delineation of protection perimeters are proposed for the Villanueva del Rosario spring, which is the water supply for nearby populations.

## 2 Description of the Test Site

The study zone is the catchment area for the Villanueva del Rosario spring (Fig. 1), one of the springs draining the Alta Cadena aquifer in the province of Málaga, southern Spain. This area of 14 $km^2$ was delineated by geological and hydrogeological characterization of the aquifer, taking into account the results obtained from a multitracer test.

A detailed description of the aquifer's geomorphologic and hydrogeological aspects is given by Mudarra et al. (2008) and Marín (2009). From the hydrogeological

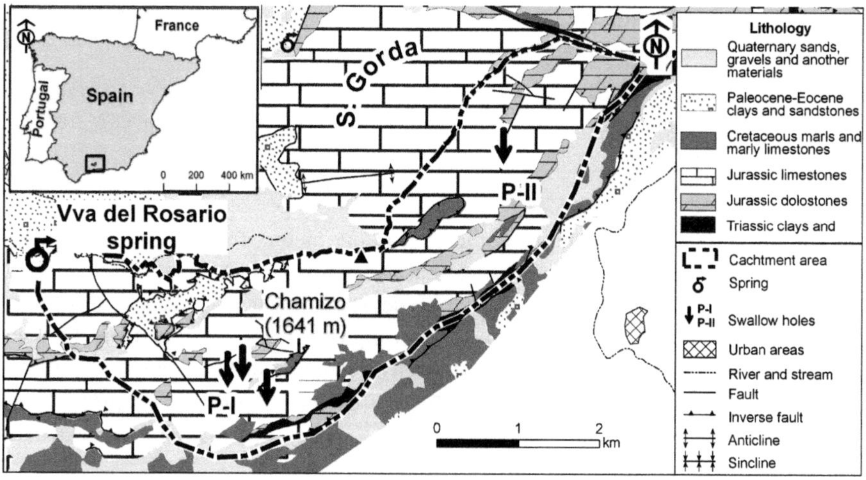


**Fig. 1** Map of catchment area of Villanueva del Rosario spring (and adjacent area) showing the main geological features

standpoint, the area is constituted of Jurassic dolostones and limestones, with a basement of clays and Triassic evaporites. Recharge to the aquifer takes place entirely by the filtration of precipitation, both in concentrated form (via swallow holes) and in diffuse form. It has been shown that a connection exists between the main karst swallow holes of Alta Cadena (Fig. 1) and the Villanueva del Rosario spring, which presents the most typically karst behaviour of all those draining the aquifer. The yield from this spring ranges from 15–2,500 L/s, and it responds to recharge events with rapid diminutions in electrical conductivity and water temperature, as well as in the hydrochemical parameters of the water. Moreover, the flow velocities measured (exceeding 200 m/h) probe the high degree of functional karstification of the area drained by this spring.

## 3 Methodology and Results

The norms and standards concerning the delineation of protection perimeters for sources of groundwater used to supply human populations vary from country to country. This disparity is reflected in the number of zones established, the requirements concerning minimum dimensions and regulations on allowable activities within each such protection zone. Normally, however, three zones are distinguished: one of maximum protection, termed the "immediate protection zone" (IMPZ), which conceptually is the "operation courtyard" in the expression of Foster et al. (2002). Zone 2 is basically an area of microbiological protection; it is also known as the "inner protection zone" (IPZ). Finally, the protection zone with fewest restrictions is the "outer protection zone" (OPZ) (WPO, 1998).

The COP+K method (Andreo et al. 2009) is an extension of the COP method (Vías et al. 2006), by which an evaluation is made of the vulnerability to contamination within karst aquifers; from this, by direct conversion, protection perimeters can be delineated. The COP method has been widely applied in aquifers throughout the world, and results have been more satisfactory than those obtained from other methods (Vías et al. 2006; Plan et al. 2009). The COP+K method incorporates the conceptual model and the recommendations of COST Action 620 (Zwahlen 2004), evaluating the protection capacity of the layers overlying the aquifer (the O factor), the recharge conditions (the C factor) and the precipitation conditions (the P factor), together with the precipitation in the saturated zone, recording water flow velocities and the degree of karstification (the K factor).

Figure 2 shows the different maps obtained from application of the COP+K method in the recharge area at the Villanueva del Rosario spring. Figure 2a shows the results obtained with the COP (resource vulnerability) method (Marín 2009). In general, the recharge area presents a "High" degree of vulnerability to contamination, and this rises to "Very High" in the zones of preferential infiltration (karrenfields and swallow holes) and at carbonate outcrops where slopes are shallow and there is a thin coverage of the unsaturated zone. Using the groundwater flow velocities estimated from the multitracer test carried out in the study area (Table 1), the

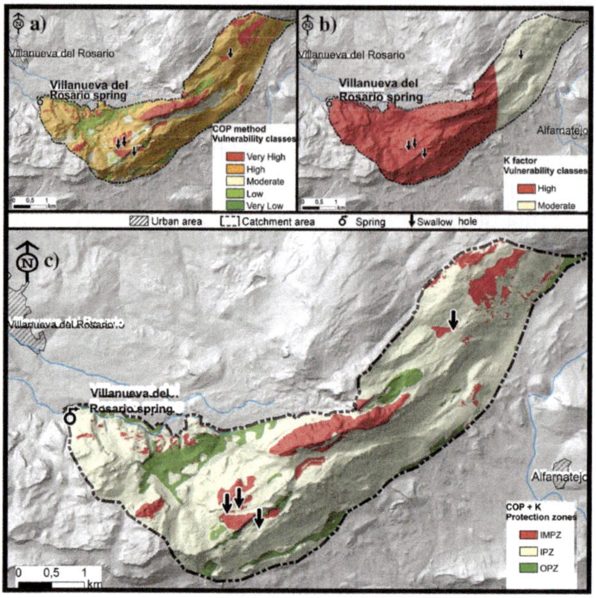


**Fig. 2** **a** Resource vulnerability estimated by COP method, **b** K factor, **c** Proposal for protection zoning in Villanueva del Rosario catchment area according to COP+K method

parameter "travel time of the K factor" is evaluated (Fig. 2b). This parameter determined the spatial distribution of the K factor (the boundary between two classes is defined by the travel time of 24 hours). From a combination of the above maps (COP and the K factor), the spatial distribution of the protection zones was obtained (Fig. 2c). According to the COP+K method, at the Villanueva del Rosario spring, the zones of maximum protection (IMPZ) are limited to the areas where the vulnerability is Very High, irrespective of the value of the K factor. The swallow holes must be protected with the maximum level of protection, even if the travel time estimated with tracer test is higher than one day. In these areas the natural protection of the aquifer is very low. For this reason, the swallow hole PII (Fig. 1) has been included as IMPZ according to the delineation determining by means of COP+K method, although the first time detection of artificial tracer was greater than one day (Table 1).

The rest of the recharge area with outcrops of carbonate rocks (limestones and dolostones) is classed as IPZ. The lowest degree of protection, OPZ, is afforded to the zones where impermeable materials overlie the aquifer. In addition to these zones, there remains the possibility of systematically incorporating an IMPZ zone around the spring to protect it from direct influences. To do this, criteria must be identified so that such a zone may be correctly defined, because delineation on the basis of a one-day transit time, as is specified in some methodological guides, would create a very large area with respect to karst springs such as Villanueva del Rosario, where flow velocities are high.

**Table 1** Main characteristics of the tracer test

| Tracer | Mass injected (kg) | Distance to spring (m) | Swallow hole* | First time to detection (h) | Maximum velocity (m/h) | Recovery rate (%) |
|---|---|---|---|---|---|---|
| Uranine | 3 | 5880 | P-II | 27 | 220 | 60 |
| Eosin | 2 | 3230 | P-I | 17 | 190 | 21 |

* See location of injection points in Fig. 1

# 4 Validation

Protection perimeters need to test the validity and coherence of the results obtained. Tracer tests are widely used for validating maps of vulnerability to contamination and are equally applicable to the validation of protection perimeters.

Table 1 shows the principal parameters of the multitracer test carried out in the Alta Cadena aquifer during a general hydrodynamic context of maximum recharge. Eosin was injected at point P-I and uranine at P-II (Fig. 1). Although the distance separating the two swallowholes into which the tracer was injected exceeded 4 km, the flow velocities estimated in each case were very similar (220 m/h for the uranine, and 190 m/h for the eosin). This finding reflects the important degree of general endokarstic development in the study area, and allows extrapolation of these flow velocities to the other zones where the vulnerability to contamination is Very High.

Total Organic Carbon (TOC) is a natural chemical component of spring water by which flow velocities may be estimated, at least qualitatively (Batiot et al. 2003). At the Villanueva del Rosario spring, in response to precipitation events, there were rapid increases in the TOC, with response times of around one day both for flow rate and hydrochemical parameters (Mudarra et al. 2008). This behaviour, together with the hydrodynamic and hydrochemical responses measured, which also reflected mean response times of one day, leads to the belief that the study area constitutes a vulnerable zone that needs to be protected in accordance with the karst features influencing recharge and the hydrogeological functioning of the aquifer.

In the case of zones where the aquifer is covered by relatively impermeable materials, it is classified as OPZ, as these materials have been shown to be efficient in slowing the advance of contamination (Perrin et al. 2004).

# 5 Conclusions

The spatial distribution of the protection perimeters for karst aquifers should not be determined by the delineation of isochrons based on the transit time. These aquifers require specific methodologies such as that applied in the present study, including hydrogeological techniques, artificial tracer tests and the evaluation of intrinsic vulnerability. The suitability of the protection perimeters suggested for the Villanueva del Rosario spring has been validated both by tracer tests and by analysis of the hydrochemical and hydrodynamic responses from the spring.

Karst features, especially dolines, karrenfields and swallow holes, are the main areas to be protected in karst aquifers, irrespective of the distance separating them from the spring. The estimation of flow velocities by means of artificial tracer tests is very important in ensuring the correct protection of karst springs, as this factor strongly influences the spatial distribution of the protected zones. Nevertheless, tracer test data are rarely available, and so the characterization of flow velocities on the basis of analyses of hydrochemical and hydrodynamic responses (especially as concerns the Total Organic Carbon content) is a very interesting alternative, and one that may be of use in decision taking and map validation.

**Acknowledgements** This work is a contribution to projects CGL2005-05427 and CGL2008-06158 BTE of DGICYT, P06-RNM 2161 of Junta de Andalucía and IGCP 513 of UNESCO and to Research Group RNM-308 of Junta de Andalucía.

## References

Andreo B, Ravbar N, Vías JM (2009) Source vulnerability mapping in carbonate (karst) aquifers by extension of the COP method: application to pilot sites. Hydrogeol J 17(3):749–758

Batiot C, Liñán C, Andreo B, Emblanch C, Carrasco F, Blavoux B (2003) Use of Total Organic Carbon (TOC) as tracer of diffuse infiltration in a dolomitic karst system: The Nerja Cave (Andalusia, southern Spain). Geophys Res Lett 30(22):2179, doi: 10.1029/2003GLO18546

Foster S, Hirata R, Gomes D, D'Elia M, Paris M (2002) Groundwater Quality Protection: a Guide for Water Utilities, Municipal Authorities and Environment Agencies. World Bank Publication. Washington, DC

Marín AI (2009) The application of GIS to evaluation of resources and vulnerability to contamination of carbonated aquifer. Test site Alta Cadena (Málaga province). Degree thesis. University of Málaga

Mudarra M, Andreo B, Marín AI (2008) Considerations about hydrogeological behaviour of carbonated aquifer Alta Cadena (Malaga province, Spain). Geogaceta 44:163–166

Perrin J, Pochon A, Jeannin PY, Zwahlen F (2004) Vulnerability assessment in karstic areas: validation by field experiments. Environ Geol 46:237–245

Plan L, Decker K, Faber R, Wagreich M, Grasemann B (2009) Karst morphology and groundwater vulnerability of high alpine karst plateaus. Environ Geol 58(2):285–297

Vías JM, Andreo B, Perles JM, Carrasco F, Vadillo I (2006) Proposed method for groundwater vulnerability mapping in carbonate (karstic) aquifers: the COP method: application in two pilot sites in southern Spain. Hydrogeol J 14(6):912–925

WPO (1998) Water Protection Ordinance of 28 October 1998 (814.201), UNECE, Geneva

Zwahlen F (ed) (2004) Vulnerability and risk mapping for the protection of carbonate (karst) aquifers. Final report of COST Action 620. European Commission, Directorate-General XII Science, Research and Development, Brussels

# PaPRIKa, the French Multicriteria Method for Mapping the Intrinsic Vulnerability of Karst Water Resource and Source – Two Examples (Pyrenees, Normandy)

V. Plagnes, K. Kavouri, F. Huneau, M. Fournier, J. Jaunat, C. Pinto-Ferreira, B. Leroy, P. Marchet, and N. Dörfliger

**Abstract** PaPRIKa is an updated intrinsic vulnerability mapping method derived from previous specialized works on karst groundwater. It assesses vulnerability by using four criteria: **P** for Protection (considering the most protective index that results from the combination of Soil cover, Unsaturated zone and Epikarst aquifer), **R** for Reservoir type, **I** for Infiltration and **Ka** for Karstification development. These criteria are chosen based on the structure and functioning of the karst aquifers. PaPRIKa distinguishes source and resource vulnerability, so it constitutes a common basis for hydrogeologists and French administration helping to delineate water catchments protection zones. Two test sites (a Pyrenean karstic system and a chalk karstified aquifer in the Paris basin) are here presented to illustrate its application.

V. Plagnes, K. Kavouri
Université Pierre et Marie Curie – Paris 6, UMR 7619 Sisyphe, 4 place Jussieu, cc 105, 75252 Paris Cedex 05, France, e-mail: valerie.plagnes@upmc.fr

F. Huneau, J. Jaunat
Université de Bordeaux, GHYMAC Géosciences Hydrosciences, B18 avenue des Facultés, 33405 Talence cedex, France

M. Fournier, C. Pinto-Ferreira
Université de Rouen, UMR 6143 MC2, Bât. IRESE A, Place Emile Blondel, 76821 Mont-Saint-Aignan Cedex, France

B. Leroy
Agence de l'Eau Seine-Normandie, Espace des Marégraphes, B.P. 1174, 76176 Rouen cedex 1, France

P. Marchet
Agence de l'Eau Adour-Garonne, 90 rue du Férétra, 31078 Toulouse Cedex 4

N. Dörfliger
BRGM, Water division, RMD Unit, 1039 rue de Pinville, 34000 Montpellier, France

B. Andreo et al. (Eds.), *Advances in Research in Karst Media*
DOI 10.1007/978-3-642-12486-0, © Springer 2010

## 1 Introduction

One of the main problems for karst groundwater abstraction and resource management is due to groundwater quality changes in the case of diffuse and/or accidental pollution in the catchment area. The absence of filtration in the carbonate reservoir and the high velocities of underground flows coupled with a concentrated recharge with fast infiltration in conduits network give to karst aquifers a high degree of vulnerability. The fact that karst groundwater requires appropriate groundwater protection schemes is generally accepted as the only way to combine economic development with land use practices and sustainable water management. Thus, the concept of groundwater vulnerability mapping constitutes the common approach shared by different European countries that have significant karst aquifers (France, Germany, Italy, Spain, Switzerland, etc.). Several methods were specifically developed for karst aquifers in recent years (EPIK, REKS RISKE, KARSTIC, COP and COP+K). In France, carbonate terrains occupy 30% of the land-surface. To reach the conformity with the guidance of the Water Framework Directive (WFD) and to preserve drinking water quality for the future, the French water agencies have to complete the definition of the protection zones outlining water catchments in karst aquifers. To standardize the strategy for resource protection zoning at the national scale, an original partnership between three French universities, the geological survey (BRGM), private companies and the three Water agencies whose territory includes a significant proportion of karst terrain (Adour-Garonne, Rhône Méditerranée-Corse and Seine-Normandie) has been created. This collaboration leads to develop a new method to map the intrinsic vulnerability of karst aquifers, the PaPRIKa method. It was developed and tested on nine different French pilot sites characterized by different geological, topographic and hydrological settings in order to evaluate its feasibility on variable conditions encountered across the country.

## 2 PaPRIKa Method

PaPRIKa is an intrinsic vulnerability mapping method for the **P**rotection of karst **a**quifers, which is based on four criteria: **P**rotection, **R**eservoir, **I**nfiltration and **Ka**rstification. This method consists of a resource and source vulnerability mapping. Two major differences exist between PaPRIKa and derived methods from COST Action 620 (Zwahlen 2003). The first one is the absence of considering the precipitation regime as an inherent and intrinsic attribute. The second one is a simple indexation in five classes of the vulnerability of each criterion as well as of the final index (0 for the most protective and 4 for the most vulnerable). The four attributes of PaPRIKa were defined in regard to the structure (R and P criteria) and the hydrogeological behavior (I and Ka) of the karst aquifer (Table 1). Details of the method are presented in the French guideline used as a common basis in France for the karst groundwater protection (Dörfliger and Plagnes 2009). An application of PaPRIKa on a large karst plateau in the south-west of France is given in Dörfliger

**Table 1** Criteria used in PaPRIKa method (for details see Dörfliger and Plagnes 2009)

| | |
|---|---|
| P map for the carbonate reservoir<br>It contains all surface (S) and subsurface elements (USZ, E) that can provide a significant delay to infiltration. The most protective value in each pixel is kept to realize the P map. | S: soil characteristics<br>(texture, structure and thickness) |
| | USZ: non saturated zone<br>(thickness, lithology and fracture degree)<br>E: epikarst aquifer |
| P map within catchments of water losses | Run-off properties of the surface (P4 for impervious formations). |
| R map<br>It characterizes the aquifer body | Lithology and the degree of fracturing of the aquifer body |
| I map<br>It distinguishes concentrated from diffuse infiltration | Topographic slope coupled with karst morphology (swallow-holes, sinkholes, karren fields...). Features allowing direct infiltration are considered as the most vulnerable (I4) |
| Ka factor<br>It characterizes the karst development considering the drainage capacity and the functioning of the karst network of both unsaturated and saturated zones. | The spatial representation of Ka requires a sufficient level of knowledge of the system's functioning (analysis of discharge flows and springs'chemical properties, tracing tests, speleological data). A distinction is made between autogenic and allogenic systems. |

et al. (2009). The weights of the four criteria are equal to 1, the total weight for the structure factors (R, P) is equal to 0.4 and the total weight for the hydrogeological functioning (I, Ka) is equal to 0.6 to give more emphasis to the special functioning of this type of aquifers.

The obtained vulnerability map is considered as a **resource vulnerability map**. To obtain a **source vulnerability map** (spring or borehole), the mapping strategy and the criteria definitions are still the same, excepted for I map which includes the underground transit time as an additional factor in order to simulate the worst sce-

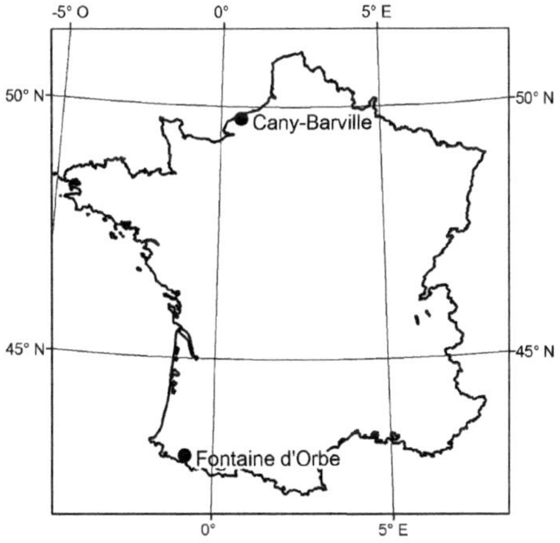


**Fig. 1** Location of the two test sites

nario through the source. The I map is modified for source vulnerability purposes; it shows isochrones defined between the source and different spatial directions using velocity data from artificial tracing tests and the available knowledge on karst conduits. The application of PaPRIKa to two French test sites (Fig. 1) are presented here.

**Cany-Barville boreholes** extract the water from the karstified chalk of the Paris basin; the chalk is overlaid by a thick layer of clays with flints (Laignel et al. 2004).

**Fontaine d'Orbe spring** outflows from fractured cretaceous limestones located between 400 and 800 m of elevation in the Pyrenees. Limestones outcrop all over the catchment area (Rey 2007).

## 3 Application on a Small Mountainous Pyrenean Karstic System

Fontaine d'Orbe test site is a small catchment (2.15 $km^2$), located in the Cretaceous limestones in the Pyrenees (Fig. 1). The karstic system and its functioning are largely described in Rey (2007). Resource and source vulnerability have been estimated (see details in Dörfliger and Plagnes, 2009). The resource vulnerability map and two source vulnerability maps are presented. Each map corresponds to the same criteria weighting: $0.4I+0.2R+0.2P+0.2Ka$. The resource vulnerability map was first realized (Fig. 2). Then, for source vulnerability map, the I map was modified keeping the higher vulnerability index only in the inner zone of two isochrones (12 and 48 h) defined from artificial tracing tests. Two source vulnerability maps were obtained illustrating the vulnerability of the spring for two transit times (12 and 48 h, Fig. 2). Both maps show very high vulnerability close to the spring. An immediate protection zone is suggested in this area. For longer transit times (48 h), the swallow holes located far from the spring should also be protected by satellite immediate protection zones in order to protect the spring from accidental pollution that could come from these swallow holes.

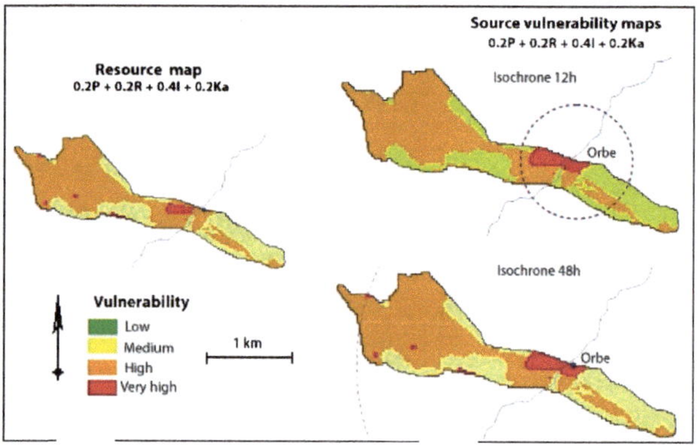


**Fig. 2** Resource and two source vulnerability maps for Fontaine d'Orbe

# 4 Application on a Covered Kartified Chalk Aquifer (Normandy)

The Cany-Barville karst system test site is located north-west of the Paris basin. The hydrologic network is typical of a karstic zone, with mostly low-order streams as most of the flow is through the subsurface. The Cretaceous chalk aquifer is the main water resource for the region (50–300 m thick), it exhibits karstic characteristics. The chalk plateaus are covered with clay-with-flints resulting from weathering of the chalk and quaternary loess (Laignel et al. 2004).

This superficial layer constitutes a semi-permeable cover from 5 to 40 m thick over the chalk substratum. It protects the aquifer but its thickness presents a strong spatial variability: it is thicker on the plateau and less on the alluvial plain. Surface runoff only infiltrates locally through sinkholes that penetrate through the impervious superficial layer into the chalk.

Groundwater is extracted from five boreholes drilled in the alluvial plain and screened in the chalk. The maps of the four criteria and the map of the intrinsic vulnerability of the resource are shown in Fig. 3.

**Concerning P criteria,** many sinkholes are aligned at the right river bank. So, the whole sinkhole catchment is defined by the highest vulnerability. Because of the presence of thick (>5 m) semi-permeable superficial layers that protect the chalk aquifer, the plateau presents a low vulnerability. The superficial layer is thinner on the alluvial plain, thus the protection criterion is defined by medium vulnerability in this area.

Then, the artificial tracing tests realized on the right river bank are used to decipher the **Ka criterion**. They showed a hydraulic connection between sinkholes and two boreholes (40 to 140 m/h). This part of the karst system appears to be more functional than the left river bank where karstic features are rare.

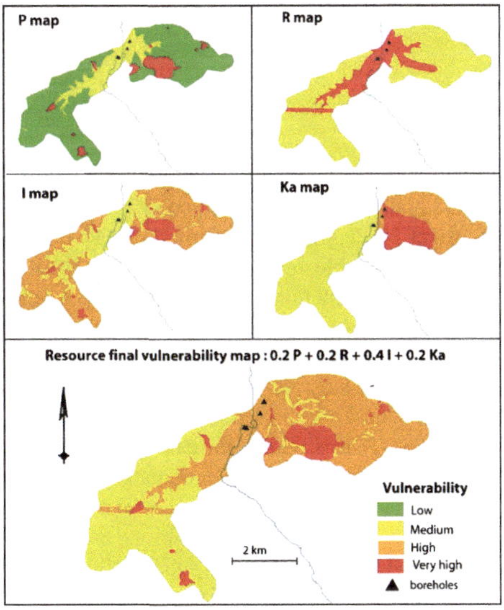


**Fig. 3** Criteria and resource vulnerability map of Cany-Barville test site

For the **R criteria**, the chalk reservoir is represented by a medium vulnerability. Locally, due to the faulting and the karst network located in the valleys because of a higher fracturing, the vulnerability increases to very strong. **For I criteria**, topographic data (division of four classes of slopes) are combined and the map of karstic features such as sinkholes and their catchment areas are mapped as highly vulnerable due to the concentrated infiltration.

The final weighting of the resource vulnerability map is 0.4I + 0.2R + 0.2P + 0.2Ka. It is consistent with field observations, notably with the high vulnerability due to hydraulic connection between sinkholes and the aquifer (Fig. 4). This final vulnerability map shows that the strong vulnerability is systematically due to sinkholes; otherwise the semi-permeable cover protects the aquifer. The appropriate protection could consist in satellite protection zones around the sinkholes to prevent fast and accidental pollution combined with specific actions to reduce the slow diffuse pollution over the rest of the catchment area.

## 5 Conclusion

The PaPRIKa method allows obtaining resource and source vulnerability mapping. Resource vulnerability maps highlight zones where actions should be taken to reduce diffuse pollution (in particular nitrate pollution). The source vulnerability mapping provides an extra tool for water catchments protection against accidental pollutions. To maximise the protection strategy for short and long term purposes, both maps should be used together. As PaPRIKa method was developed with the objective to make it as easy to operate as possible, it is now a method affordable in terms of cost and technically feasible for private consulting hydrogeologists and local administrations. With this method, the protection zoning of French water catchments in karst should be soon achieved.

## References

Dörfliger N, Plagnes V. (2009) Cartographie de la vulnérabilité des aquifères karstiques, guide méthodologique de la méthode PaPRIKa, Rapport BRGM RP-57527-FR, 100 p

Dörfliger N, Plagnes V, Kavouri K (2009) PaPRIKa a multicriteria vulnerability method as a tool for sustainable management of karst aquifers, example of application on a test site in SW France. International interdisciplinary scientific conference "Sustainability of the Karst Enviroment – Dinaric Karst and other karst regions", Croatie, 23–26 september 2009

Laignel B, Dupuis E, Rodet J, Lacroix M, Massei N (2004) An example of sedimentary filling in the chalky karst of the Western Paris Basin: Characterization, origins and hydrosedimentary behaviour. Zeitschrift für Geomorphologie N.F., 48(2):219–243

Rey F (2007) Ressources en eau souterraine dans les chaînons béarnais (Pyrénées-Atlantiques, France), Géométrie et fonctionnement hydrogéologique de quatre aquifères carbonatés, Thèse de l'Université de Bordeaux 1, 466 p

Zwahlen F (ed) (2003) Vulnerability and risk mapping for the protection of carbonate (karst) aquifers. Final report of COST Action 620. European Commission, Directorate-General XII Science, Research and Development, Brussels

# Comparative Application of Two Methods (COP and PaPRIKa) for Groundwater Vulnerability Mapping in the Lez Karst System (Montpellier, South France)

A.I. Marín, N. Dörfliger, and B. Andreo

**Abstract** A comparative test of two vulnerability mapping methods (COP+K and PaPRIKa) specifically developed for karst aquifers have been applied to Lez karst system located in the South of France. The Lez spring is captured for water supply to the town of Montpellier and the surroundings and it is managed in an active way, through pumping boreholes. The mean pumping rate is around 1300 l/s. The catchment area is about 350 $km^2$ with 110 $km^2$ of limestone outcrop. Due to the importance of resource of this Mediterranean karst system, the vulnerability mapping is part of a decision support tool to cope with water management and land use projects. The obtained results of the comparison of the two methods are discussed to identify major differences and draw some recommendations in terms of methodology and validation of the vulnerability map.

## 1 Introduction

In Europe, carbonate terrains occupy 35% of the land surface, contributing up to 50% of drinking water in some countries. Karst groundwater is thus an important water resource, which is, however, particularly sensitive to contamination, due to structure and hydrological behaviour. Karst aquifers require consequently, specific groundwater protection. Intrinsic groundwater vulnerability assessment schemes have been developed during the last decade, specifically for karst aquifers, based on the early EPIK method (Dörfliger 1996) that influenced later ones in the framework of the European COST Action 620 as well as at the national level.

A.I. Marín, B. Andreo
Centre of Hydrogeology of University of Malaga, and Department of Geology, Faculty of Science, University of Malaga, Malaga 29071, Spain, e-mail: aimarin@uma.es; andreo@uma.es

N. Dörfliger
BRGM, Water Division, RMD Unit, 1039 rue de Pinville, 34000 Montpellier, France, e-mail: n.dorfliger@brgm.fr

B. Andreo et al. (Eds.), *Advances in Research in Karst Media*
DOI 10.1007/978-3-642-12486-0, © Springer 2010

Two specific karst methods for intrinsic groundwater vulnerability have been applied in this paper: COP and PaPRIKa. Though both methods allow assessing the source vulnerability, the nonexistence of reliable data of velocity flow based on dye tracers has forced postponement of the validation of this assessment. Consequently, outlining the protection zones of Lez Spring according to a source vulnerability map will be possible only when trustworthy information related to velocity flow based on future dye tracer tests is available.

The main aim of this paper is to present the preliminary results on the defining protection schemes of a large karst catchment whose outlet (Lez spring) is used for water supply Montpellier city (Southern France). An additional aim is to compare the results between the COP method, already applied worldwide during the last five years, and PaPRIKa method, a new method based on the EPIK method.

## 2 Description of the Test Site

The Lez karst system is drained by one main outlet, the Lez spring, and several seasonal springs, namely the Lirou springs. The Lez spring is the source of the Lez River, a small Mediterranean coastal River, 26 km northern to the shoreline. The catchment of this karst spring has an area of about 380 km$^2$ located between Hérault and Vidourle rivers, in the north of Montpellier, South France (Fig. 1). The catchment area has been assessed by considering dye tracer tests realised in 1960s, and considering groundwater levels and discharge of the spring as function of recharge. From a total of 380 km$^2$, approximately 110 km$^2$ of outcropping limestone contribute to the recharge of the aquifer. The aquifer lithology is made up of massive limestones of Late Jurassic age (650 to 1100 m thick) and marls and limestones of Early Cretaceous age. Moreover, there are some perched unconfined aquifers formed by Eocene limestones. From hydrogeological point of view, the recharge of the karst aquifer occurs both by direct infiltration of rainfall over limestone outcrops and by concentrated infiltration through swallow holes of the water runoff coming from impervious layers.

The Lez spring groundwater is actively management for water supply of the town of Montpellier and the surroundings using three pumps located within boreholes intercepting the karst conduits upstream the Lez outlet. The mean pumping rate is around 1300 l/s.

## 3 Vulnerability Mapping: Results and Comparisons

The two groundwater vulnerability assessment methods applied on the Lez catchment have been designed specifically for karst aquifers. They are based on various types of information concerning physical characteristics of unsatured and saturated zones, structure of the aquifer as well as hydrological behavior.

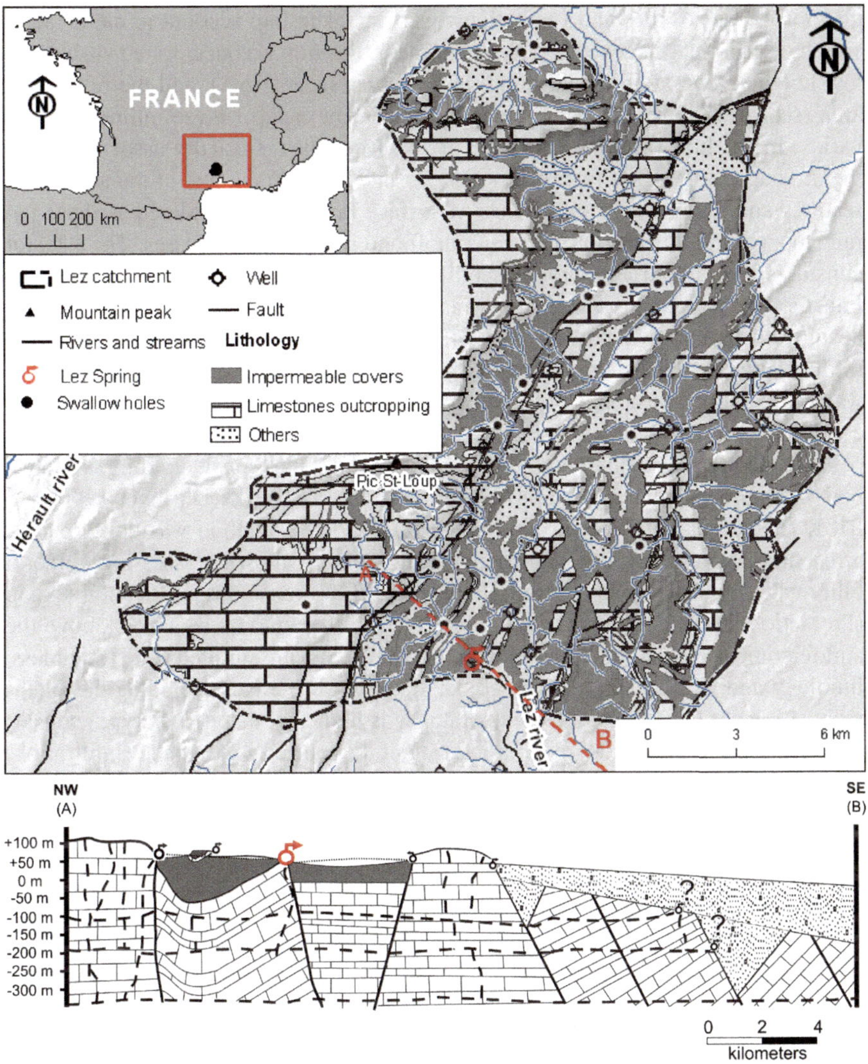


**Fig. 1** Map and cross section of catchment area of Lez spring (and adjacent area) showing the main geological features (after Conroux 2007)

The COP method (Vías et al. 2006) was developed in the Framework of the European COST Action 620 taking into account specific properties of karst aquifers. The method considers the characteristics of layers overlying the water table (O factor), the parameters which control the flow concentration (C factor) and precipitation (P factor). O factor reflects the protective capacity of the overlying layers by means of soils (texture and thickness) and lithology of unsaturated zone (fracturation degree and thickness of each layer, and the confining conditions). Karst geomorphology, slopes and vegetation cover are taken into account in the C factor which discriminates between areas where the infiltration is concentrated via a swallow hole and overlaying layer might be bypassed (scenario 1), and the rest of the area (scenario 2). Karst features may be present in each of these infiltration scenarios. In the first case, the existence of swallow holes is usually associated with the drainage of the karst features (poljes, dolines, etc). In the rest of the area, karst features and the presence or absence of surface layers are considered as these parameters control the relation between runoff and infiltration processes. The P factor considers spatial and temporal variability of precipitation that plays a role in the transfer of contaminant, specifically in large aquifer. The COP+K method has been developed recently by Andreo et al. (2009) in order to characterize the vulnerability of the source by extension of the COP method. The K factor considers the groundwater travel time, the connection and contribution to the source, as well as the active conduit network.

The PaPRIKa (Dörfliger et al. 2009) method takes into consideration criteria for both structure and functioning of the aquifer. Based on EPIK (Dörfliger 1996) and RISK (Pételet et al. 2000) resource methods, PaPRIKa method was developed as a resource and source vulnerability mapping method, allowing assessing vulnerability with four criteria: Protection, Rock type, Infiltration and Karstification. P map (Protection) considers the protection provided to the aquifers by layers above the aquifers: the soil (texture, structure and thickness), the unsaturated zone (thickness, lithology and fracture degree) and epikarst aquifer. Moreover, including the catchments of water losses where the vulnerability is higher. R map (Rock type) considers the lithology and the degree of fracturing of the satured zone. I map (Infiltration) distinguishes concentrated from diffuse infiltration. Ka map (Karstification development) assesses the drainage capacity and the organization of the karst conduits network. Additional criteria are applied such as groundwater travel time and active conduit network, on the resulting vulnerability map. In this manner adequate source protection can be assessed.

Figure 2 shows the results of COP and PaPRIKa methods in the catchment area of Lez spring. The vulnerability map obtained with COP method (Fig. 2a) is more contrasted, with all vulnerability degrees expressed, from very low to very high. Very high values concern the large area in the NW part of the catchment and the part close to swallow holes where infiltration of sinking streams exist. The very low and low vulnerabilities concern the areas where the aquifer is protected by marls (see cross section Fig. 1). The PaPRIKa method allows to choose between different combinations of weights variables and to select one according to coherency with hydrogeological knowledge. Related to Ka parameter the previous information is confusing: this variable can be either Ka = 2 or Ka = 3 (Perriquet 2006; Conroux 2007). The two options have been tested waiting for studies clarify this parameter. To sum up, four scenarios have been tested. On the maps B, C, D and E, the three vulnerability classes are present, from moderate to very high. On the map C (Fig. 2) the very high degree of vulnerability is almost absent.

Although the parameters considered by two methods are similar, the weighting system and the combination between them result in different final maps. In the final

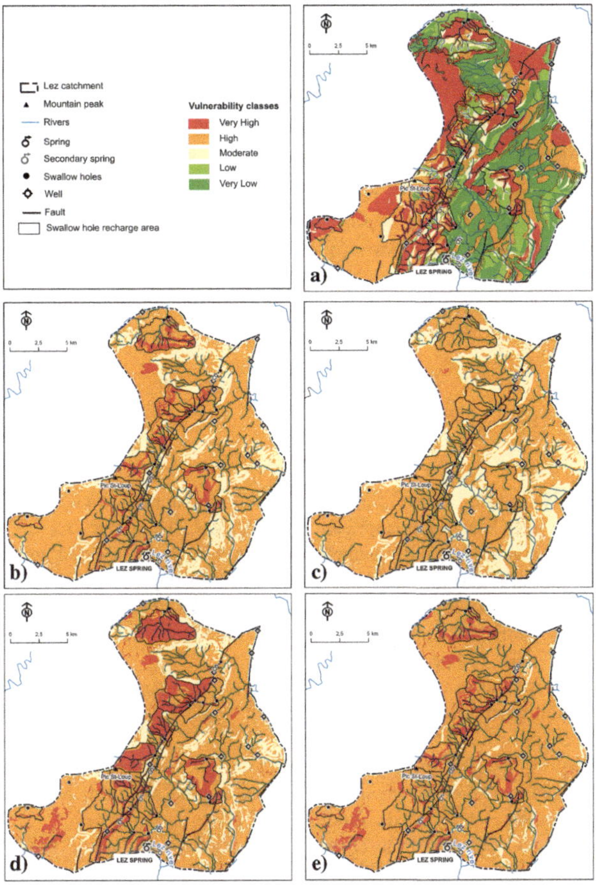


**Fig. 2** Resource vulnerability maps obtained by COP method (**a**) and PaPRIKa method with Ka = 2 and weighting parameters 0.2P + 0.2R + 0.4I + 0.2Ka (**b**), Ka = 2 and 0.2P + 0.2R + 0.3I + 0.3Ka (**c**), Ka = 3 and 0.2P+0.2R+0.4I+0.2Ka (**d**), Ka = 3 and 0.2P+0.2R+0.3I+0.3Ka (**e**)

map of vulnerability, the weight of lithology of unsaturated zone by COP method is much bigger than PaPRIKa method. The last one does not show low or very low vulnerability even where there is a great thickness of marls overlying the aquifer.

## 4 Conclusions

The methods (COP and PaPRIKa) applied for karst groundwater vulnerability mapping take into account similar factors that control the infiltration of water and contamination from the land surface towards the source. However, their application to Lez catchment area (Montpellier, France) shows different results. The major difference between the two methods is due to the way of considering the variables related to overlaying layers. COP results are more contrasted, with all the vulnerability classes distributed more congruently with the geological map.

Nevertheless, vulnerability maps need to be reliable and consequently validation should be done by chemical tracers (nitrates, pesticides) and/or artificial dye tracer tests under different hydrological conditions. As previously stated, this document shows the preliminary results of an ambitious research which aims are comparing the methodologies currently being employed for vulnerability mapping and improving validation techniques by means of tracer tests and the analysis of hydrochemical and hydrodynamic responses. This field data will be necessary to perform final source vulnerability mapping of Lez spring.

## References


Andreo B, Ravbar N, Vías JM (2009) Source vulnerability mapping in carbonate (karst) aquifers by extension of the COP method: application to pilot sites. Hydrogeol J 17 (3):749–758

Conroux, Y (2007) Caracterisation du fonctionnement hydrodynamique de l'aquifere karstique du Lez (Hérault) a l'etat naturel. Université d'Avignon et des pays de Vaucluse et BRGM, 227 pp

Dörfliger N, Plagnes V avec la collaboration de K. Kavouri, J. Gouin (2009) Cartographie de la vulnérabilité des aquifères karstiques, guide méthodologique de la méthode PaPRIKa, Rapport BRGM RP-57527-FR, 100 pp

Dörfliger N (1996) Advances in karst groundwater protection strategy using artificial tracer test analysis and multiattribute vulnerability mapping. – PhD Thesis, University of Neuchâtel, Switzerland, 292 pp

Perriquet M (2006) Caractérisation de la structure et du fonctionnement du système karstique du Lez en gestion active – Effet des pompages sur la source – Rapport de Master 2 Professionnel. Université d'Avignon et des pays de Vaucluse, 135 pp

Pételet GE, Dörfliger N, Crochet P (2000) RISKE: Méthode d'évaluation multicritère de la cartographie de la vulnérabilité des aquifères karstiques. Applications aux systèmes des Fontanilles et Cent-Fonts (Hérault, France), Hydrogéologie, Vol. 4, pp 71–88 (paru)

Vías JM, Andreo B, Perles JM, Carrasco F, Vadillo I (2006) Proposed method for groundwater vulnerability mapping in carbonate (karstic) aquifers: the COP method: application in two pilot sites in southern Spain. Hydrogeol J 14(6):912–925

# Karst Geomorphology

# Geomorphology and Structural Control in Caves: a Research in Torca Teyera (Picos de Europa, NW Spain)

D. Ballesteros, M. Jiménez-Sánchez, J. García-Sansegundo, and S. Giralt

**Abstract** Research in caves is rare and restricted to speleological explorations. Picos de Europa Massif shows a spectacular development of vertical caves deeper than 500 m. Torca Teyera is a 738 m-deep and 4 km-long vertical cave. The aim of this work is to establish a methodology to characterize the geological and geomorphological aspects of this special group of caves, as well as to identify the factors conditioning karst development. The research involves (1) the cave survey at a 1:500 scale and the construction of a 3D model, (2) the geomorphological mapping on the topographical survey, (3) the geological mapping of the cave environment at 1:5000 and (4) the comparison in stereographic projection of the obtained survey data and joint measures.

## 1 Introduction

Research in large vertical caves has always been scarce due to difficult access and methodological constraints. Scientific research in vertical cavities is conditioned by cave exploration, usually including the elaboration of associated documentation (Kambesis (2007). The speleological documentation in Picos de Europa, a mountain massif in the West Cantabrian Mountains (NW Spain), is extensive and is not systematized. Picos de Europa contains 12 out of the 90 caves deeper than 1000 m known in the world.

It is characterized by a rough and calcareous relief with peaks exceeding 2500 m a.s.l. This massif consists mainly of 1200 m of carboniferous limestone and is char-

D. Ballesteros, M. Jiménez-Sánchez, J. García-Sansegundo
Departamento de Geología, Universidad de Oviedo, C/ Arias de Velasco s/n, 33005 Oviedo, Spain,
e-mail: ballesteros@geol.uniovi.es

S. Giralt
Instituto de las Ciencias de la Tierra Jaume Almera (CSIC), C/ Lluís Solé i Sabarís s/n, 08028 Barcelona, Spain

B. Andreo et al. (Eds.), *Advances in Research in Karst Media*
DOI 10.1007/978-3-642-12486-0, 

acterized by E-W to NW-SE, South-directed imbricated system thrust (summarized in Merino-Tomé et al. 2009).

The Torca Teyera cave (Fig. 1a) is North of the Picos de Europa and was explored and surveyed by the Société Suisse de Spéléo and other groups between 1979 and 1982 (Borreguero 1986). During these explorations, a first karst research took

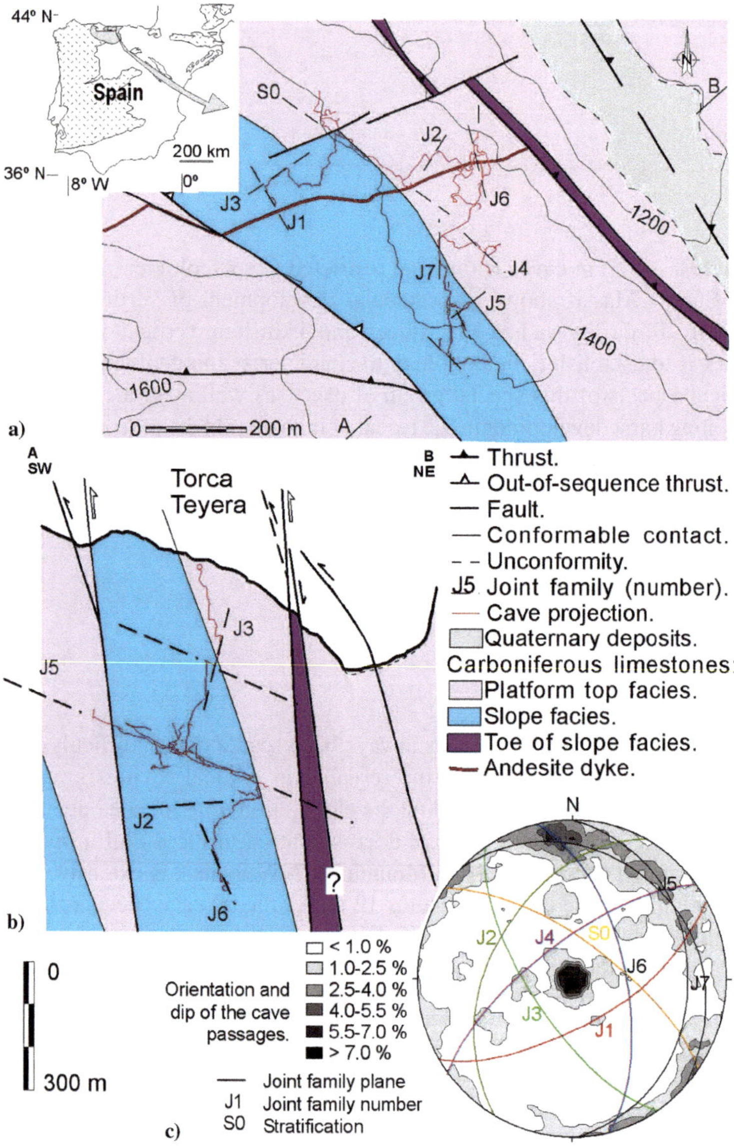


**Fig. 1** **a** Geological map and **b** cross-section of Torca Teyera area. **c** Stereographic projection of the orientation and dip of the passages and the families of joints and the stratification

place (Borreguero 1986), allowing the establishment of the first hypothesis about the cave structural control. From 2007 and 2009 new passages were discovered by the GEMA group.

In this paper, a methodological approach is proposed to establish the geological characterization of these special environments.

## 2 Methodology

The methodology includes fieldwork: the cave survey, the cave 3D model, the cave geomorphological mapping, the geological mapping of the cave surroundings and the definition of the joints families.

The cave survey includes the passages projection in a horizontal plane. The survey was mapped by using the speleological classical method at a 1:500 scale. In the passages, successive stations were defined. Distances, orientation and dip data were measured between stations using a tape measure or laser, ruler and clinometer, respectively. The topography has been determined by considering a reference level between 30 cm and 1.5 m above the cave ground. The collected data was managed using the VisualTopo.503 software program (David 2009) to define the survey line (a line connecting the stations). The VisualTopo.503 produced a 3D model approximating the passages by an octagonal conduit whose axes were the height and width of the passages.

Cave geomorphological mapping was carried out at a 1:500 scale, taking the cave survey as a topographical basis. Cave features were inventoried and classified according to genetic, morphological and sedimentary criteria (Jiménez-Sánchez et al. 2006; Ford and Williams 2007). The limits of the different features were established and projected on the survey. A geological map covering a surface of 1.1 $km^2$ and a geological cross-section were carried out at a 1:5000 scale. The cave was projected over the map and the cross-section. Joint data measures (dip and direction) were taken (124 in the surface and 33 into the cave). These data were represented in stereographic projection for further analysis. Data referring to the direction and angle of each meter of the survey were also projected, to obtain an approach of the direction and dip of the passages. Subsequently, a density analysis has been made with these data to compare it with data obtained from joints.

## 3 Results and Discussion

The cave, with 4 km of development, consists of three levels of passages (1260, 900–750 and 615 m altitude) and seven shafts that intersect them. The 1260 m level is a short passage hanged and the 900–750 m level is formed by canyon-shaped passages. The 615 m level is an active passage located 18 m over the saturated zone, inferred by the presence of a siphon located at a depth of 738 m.

The legend and selected sites of the geomorphological map of the cave are illustrated in Fig. 2. The legend is divided in three parts: issues relating to the survey, the geomorphologic features and geological aspects. The first group includes morphometrical data related to the passages: contour (using the upper and lower contour when there is overlap), scarps, pits, presence of rivers or lakes, slope of the ground and altitude and depth data.

Fluviokarst, gravity and precipitation forms have been recognized in the cave. The fluviokarst forms are classified as erosive and sedimentation features. The first include scallops, roof pendants and corrosion notches that are mostly located in ac-

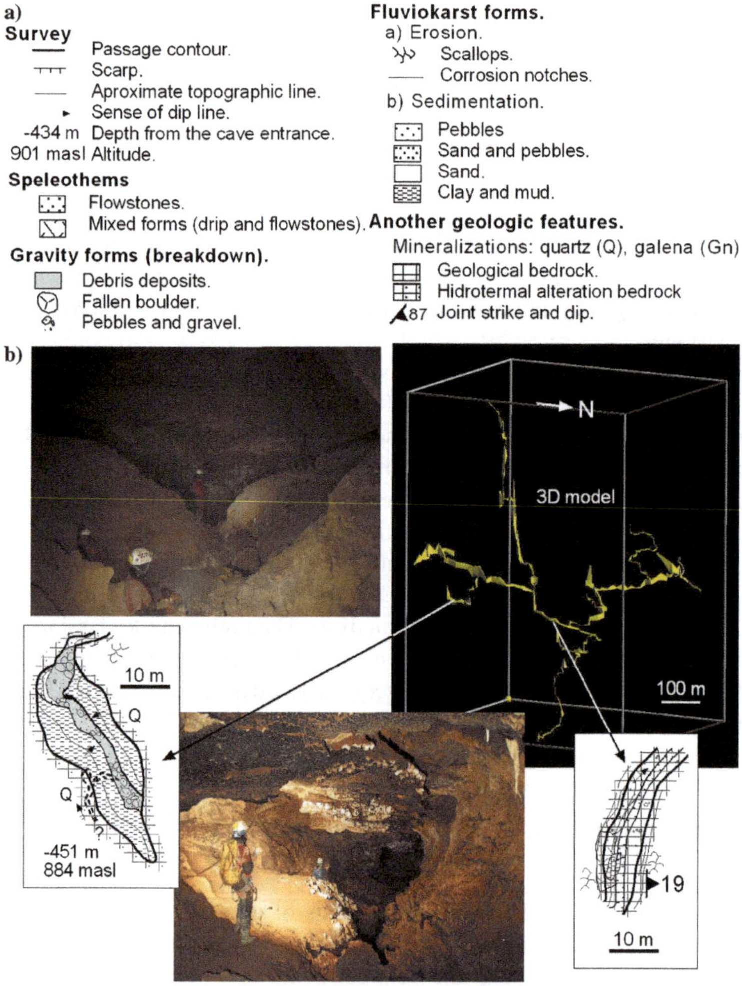


**Fig. 2** **a** Part of the legend of the geomorphological map of Torca Teyera. **b** 3D Model, zooms and pictures of selected zones of the geomorphologic map

tive and canyon shaped passages, as well as solution runnels. Gravity features have been recognized in the inactive levels, mainly corresponding to debris and fallen boulders and pebbles. Finally, several groups of speleothems have been recognized: dripstones, flowstones and mixed forms are the more relevant. Locally, pool deposits and coraloids forms have also been found. The geomorphol ogical map also includes other geological aspects such as evidences of mineralization, altered bedrock, structural data or outcrops of volcanic rocks.

The geological mapping and cross-section of Fig. 1a and b show qualitatively the relationships between the cave and the tectonics. The preferred orientations of the passages correspond to the fault systems (SE-NW, SW-NE and N-S direction) and the bedding (NW-SE). Seven families of joints have been recognized (Fig. 1c), also showing the most frequent directions and inclinations of the passages: (1) passages that bend 20° to the N, (2) passages that bend 20° to the NE and (3) subvertical passages. The comparison of the families of joints with the main orientations and inclinations of the cave highlights that the galleries whose route is inclined 20° to the North (1) are controlled mainly by the intersection between families 5 and 6. The passages that dip to 20° to the NE (2) are ruled by the intersection between families 1, 2, 5 and 7. The subvertical passages (3) are conditioned by the families 1, 3, 4 and 6, as well as their intersections. The families 2, 3 and 5 were recognized by Borreguero (1986). The seven joint families described can be established as the main factor controlling speleogenesis.

## 4 Conclusions

The speleological surveys of caves are good geological documents as these surveys contain information about cave morphometry, hydrology and geomorphology. However, their development is complex due to the adversity of the environment.

The 3D model is an approximation to the geometry of the karst system as a whole. In detail, the model represents a good approximation only in passages and shafts whose section is subcircular because the model used an octagonal section to approximate the shape passage. Where the passages display a canyon morphology modeling is worse because each station only uses a constant value of widths.

The cave geomorphologic mapping defines the spatial distribution of endokarstic features, and therefore, of the processes conditioning the karst evolution. Some limitations of the method are conditioned by the problems derived from the survey: the elements are better represented in the cave floor. The loss of information on walls can be solved by shoot down on the cave limits. The geological mapping of the cave environment and the cross-section represent a good approximation about the qualitative structural control of the cave, showing the relationship between the structure and the endokarst. The influence of structural control in cave morphology is evaluated by using stereographic projection of dip and direction of joints and passages. As a result, it is shown that the factors controlling speleogenesis are the position of the joints and the intersection between them. Therefore, the use of survey data together with structural data can give a good approximation to evaluate the structural factors that control the speleogenesis in vertical caves.

**Acknowledgements** This research has been funded through "CN-06-177 contract of Asturias Government-Oviedo University" CALIBRE project (CAVECAL) (CGL2006-13327-C04/CLI) provided by Ministerio de Educación y Cultura, and GRACCIE project (CONSOLIDER PROGRAM) (CSD2007-00067) provided by Centro de Investigación Científica y Tecnológica. The authors acknowledge Juan Bahamonde, Óscar Merino, Asociación Deportiva Gema, G.E. Polifemo and GES Montañieros Celtas for their help.

## References

Borreguero M (1986) Special Picos: Puertos de Ondón. Neuchâtel, Suiza (in press)

David E (2009) Visual Topo. http://vtopo.Free.fr Accessed 8 November 2009

Ford D, Williams P (2007) Karst hydrogeology and geomorphology. Wiley. England.

Jiménez-Sánchez M, Aranburu A, Martos E, Domínguez-Cuesta M (2006) Cuevas prehistóricas como Patrimonio Geológico en Asturias: métodos de trabajo en la cueva de Tito Bustillo. Trabajos de Geología 26:163–174

Kambesis P (2007) The importance of cave exploration to scientific research. J of Cave and Karst Studies 69:46–58

Merino-Tomé O A, Bahamonde J R, Colmenero J R et al (2009) Emplacement of the Cuera and Picos de Europa imbricate system at the core of Iberian-Armorican arc (Cantabrian zona, north Spain): New precisions concerning the timing of the arc closure. Geological Society of Am 121(5–6):729–751

# Tracking the Space-Time Evolution of a Karstic System with Fast Dissolution Kinetics Developed in Gypsum Formations

A. Charmoille, F. Bech, X. Daupley, and P. Gombert

**Abstract** The wood of la Tussion, near Paris is affected by a certain number of surface collapses of the sinkhole type. The lack of knowledge on the phenomenon for the formation and evolution of these sinkholes of natural origin is a factor, which prevents the application of suitable protective solutions. To develop this knowledge, INERIS in July 2009 engaged into a hydrogeological and geotechnical study of this area. This study has the goal of evaluating the "ground movement" risk. As these are evolutive phenomena, it is necessary to have extensive knowledge of the hydrogeological operation of the site, the subterranean water flows being at the origin of the dissolution. The results of this study give the possibility of better targeting the areas to be treated and the nature of the work to be applied to allow safe access to the visitors of this park.

## 1 Introduction

The surface impact of the development of in-depth karstic networks, in gypsum layers may prove to be very significant for human infrastructures (Klimchouk et al. 1996; Klimchouk 1996; Gutiérrez and Cooper 2007). One of the most striking examples relates to the new town of Puilatos, near Saragossa (Spain) which had to be abandoned in 1985 subsequent to surface collapses which affected the buildings and infrastructures of this town. The underlying strongly karstified Miocene gypsum layer is responsible for these collapses. The towns of Ripon and Darlington (Cooper and Waltham 1999; Cooper 2006), in England, are also "textbook cases" of karstic phenomena with fast evolution, generating at the surface a large number of collapses (deep collapsed cones, subsidence areas), which have a strong impact on the buildings and engineering structures (bridges and roads). The subsidences

A.Charmoille, F.Bech, X.Daupley, P.Gombert
INERIS – Institut National de l'Environnement industriel et des Risques, Parc Technologique Alata, BP2, 60550 Verneuil-en-Halatte, France, e-mail: arnaud.charmoille@ineris.fr

B. Andreo et al. (Eds.), *Advances in Research in Karst Media*
DOI 10.1007/978-3-642-12486-0, © Springer 2010

and collapses have been active since 1834. In 1995 and subsequently in 1997, a collapsing chimney with a diameter of 10 m and a depth of 5.5 m reached the surface and caused the destruction of a dwelling (Cooper, 2006). In the Paris Basin, the main surface collapses having an impact on strongly urbanized zones, are related to the dissolution of Bartonian and Lutecian gypsum. The hydrogeological behavior is very complex there, characterized by several superposed aquifers, occurring in the dissolution process. Studies by Toulemont (1981 and 1987) and of Thierry et al. (2008) have in particular shown the influence of pumping underground waters over several years on the acceleration of the dissolution and of the creation of subterranean voids, responsible for collapses and subsidences at the surface. This is a detailed study on the mechanisms driving the dissolution process within a typical dissolution context of the Paris Basin.

## 2 Presentation of the Site

Located on the district of Villepinte at about fifteen kilometres to the North-East of Paris, the wood of la Tussion extends over a morphologically not very hilly plain. This 19-hectare wood is limited in its northern part by the town of Villepinte and in the southern part by a railway track and then by the Ourcq canal which flows from East to West for returning to Paris. This wooden area, located in a strongly urbanized sector of the Paris region, is used by the general public as an environmental recreation park. The geological substratum is formed by Bartonian (infra-gypseous marls, limestone of St-Ouen and sands of Beauchamp) and Lutetian (marls and loose stones, coarse limestone) formations. These layers are hidden at the outcrop by a quaternary cover of silts or modern alluvia. The stratigraphic column of Fig. 1 shows the whole of the recognized soils to a depth of 60 m. Several of these horizons contain gypsum levels illustrated in red in Fig. 1 and recognized by borings. It is important to note that the thickness of the gypsum layers from the lower Bartonian is not sufficient for explaining the voids observed at the surface. From a hydrogeological point of view, the alternation of permeable levels and of impermeable levels makes it difficult to delimit the different aquifers. It is generally recognized that to a depth of 60 m, three aquifers are present in the sector. From top to bottom, these are:

- the groundwater level of infra-gypseous marls, the most superficial;
- the aquifer of the limestones of St-Ouen;
- the aquifer of the marls and loose stones.

## 3 Experimental Protocol and Foreseen Procedure

To better understand the hydrogeological mechanisms which have an influence on the creation of dissolution cavities under the wood of la Tussion, several complementary investigations were provided. First of all, a full search in the archives and at relevant organisms is being accomplished to rebuild the history of the occupa-

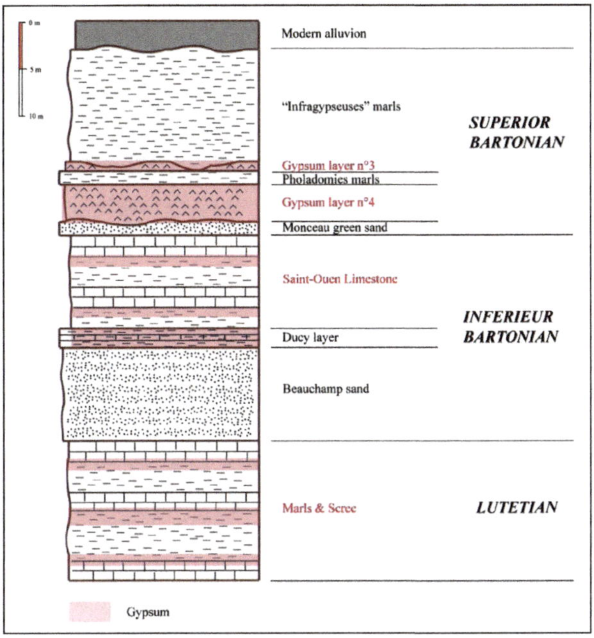


**Fig. 1** Stratigraphic column of geological layers present at the vertical of the wood of la Tussion

tion of the ground and of the use of the subsoil in this sector, notably as regards underground water resources. This first step will be followed by:

*Hydrogeological investigation*: Continuous piezometric tracking of the main aquifers present to a depth of 60 m, continuous tracking of the physico-chemical parameters in several points of the two most superficial aquifers, space time monitoring (24 points sampled every 3 months) of the chemical composition of the three main aquifers, monitoring the chemical composition and the physico-chemical parameters of the Ourcq canal, in situ dissolution tests, tracing tests, and modelling of chemical equilibria.

*Geotechnical investigations*: An inventory of all the sinkholes present in the wood and a characterization of their geometry will be performed as a set up for topographic monitoring over several months.

## 4 Results

Piezometric measurements conducted from March to October 2009 show a drop in the piezometric level by 0.5 m during this period. In the central part of the wood, where the majority of the recent sinkholes are found, a piezometric depression of 6 m is observed with respect to the eastern half of the wood and of 2 m relatively to the western half (Fig. 2). This depression observed during the whole duration of the monitoring, extends towards the South beyond the investigated areas.

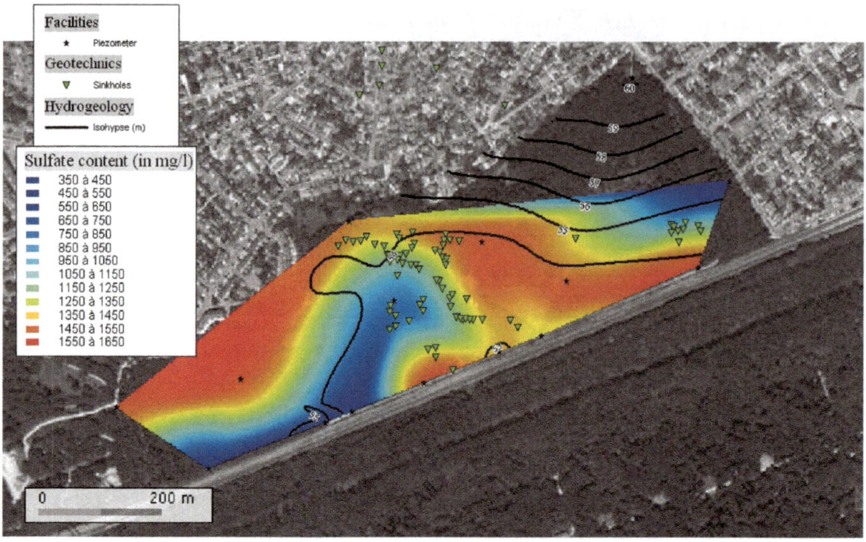


**Fig. 2** Aerial photograph of the wood of la Tussion (www.geoportail.fr) with superposition of piezometry, sulphate ion concentrations and water table of infra-gypseous marls, in November 2009. The main collapses inventoried at the wood are also illustrated

With the sampling campaigns intended for analyzing the water from the aquifers in presence, four main hydrochemical facies may be differentiated:

- the first represents hyper-calcium sulphated waters for which the concentrations reach on the average 1300 mg/L for $SO_4^{2-}$ ions and 590 mg/L of $Ca^{2+}$ ions. The analyzed samples originate from the surface stratum of infra-gypseous marls.
- the second displays average concentrations of 500 mg/L and 270 mg/L respectively. These calcium sulphated waters were also sampled in the aquifer which is the closest to the so-called surface of the infra-gypseous marls.
- the third belongs to the aquifer of the St-Ouen limestones and shows respective concentrations of 200 mg/L and 130 mg/L; these are calcium bicarbonate waters.
- the fourth with a calcium bicarbonate facies is represented by not very mineralized waters of the canal which have an average concentration of 55 mg/L of sulphate ions and 95 mg/L of calcium ions.

## 5 First Hydrogeological Interpretations

The subsoil of the wood of la Tussion consists of an inter-stratified complex of marls, gypsums and limestones. It is therefore often difficult to identify the aquifer reached by the different pre-existing piezometers on the site. The first investigations from a hydrochemical point of view allowed clear differentiation of the waters originating from the aquifer of the St-Ouen limestones and those originating from the water table of infra-gypseous marls. The waters from the St-Ouen limestones are, because of the nature of their reservoir rock, less loaded with sulphate and calcium ions than those originating from the most superficial groundwater table. If the results of the samplings carried out in the infra-gypseous marl water table are analyzed, there appear two clearly distinct areas, which are also differentiated from the point of view of their electrical conductivity. The areas to the East and the West of the wood have sulphate concentrations of the order of 590 mg/L with electrical conductivities fluctuating around 1.5 $mS\,cm^{-1}$. They are separated by a band where the sulphate concentrations reach 1300 mg/L on the average with conductivity of the order of 2.8 $mS\,cm^{-1}$.

If the main collapses presently inventoried in the wood of la Tussion, are superposed to the acquired hydrogeological data, it appears that the maximum sulphate concentration area (1400 mg/L) corresponds to the most active area from the point of view of collapses: this is also the sector where draining of the groundwater is the most significant (piezometric depression). Increase in the sulphate concentrations therefore expresses the existence of an active dissolution area (Poirot et al. 2007). This statement is consistent with the conceptual model established by Toulemont (1981), which shows that the active dissolution area is located at the limit of the sulphate saturation fringe: once water is saturated with sulphates, dissolution is then stopped. The different tracking operations set into place and the complementary investigations, which will be conducted, should allow the specification of the hydraulic engine of this dissolution mechanism. However, with the first conducted investigations, a first conceptual diagram of the hydro-geotechnical operation of the wood of la Tussion may be drawn up.

## 6 Conclusions, Perspectives

This study began in 2009 and will continue until 2011. The already conducted investigations have allowed a better understanding of the hydrogeological operation of the site. Thus, in the first 15 m under the topographic surface, two aquifers were clearly differentiated by their chemical signature. One of the important points, which will have to be solved, concerns the delimitation and the evolution of the active dissolution area detected in the groundwater of the infra-gypseous marls and marked by an increase in sulphate concentrations. What is the origin of the water, which is undersaturated with sulphates, which feeds this aquifer and which contributes to permanent activation of this active dissolution area? Are the filtering waters only responsible for this? On the other hand, does the Ourcq canal play a role, being a large fresh water reservoir? But considering the hydrogeological context focus of attention will be on the possible hydraulic connection with deeper aquifer which could allow flow rising of low-sulfate waters.

**Acknowledgements** The Department of Seine-Saint-Denis who participate in financing this study, facilitate the access to the investigation area and to the available information. IGC (Inspection Générale des Carrières) (General Quarry Inspection Body) who accompany this study as experts on this theme and have carried out a certain number of borings providing access to the deepest substrata.

## References

Cooper AH (2006) Gypsum dissolution geohazards at Ripon, North Yorkshire, UK, IAEG2006 Field Trip Guide Ripon. British Geological Survey:1–13

Guerrero J, Gutiérrez F, Lucha P (2003) Paleosubsidence and active subsidence due to evaporite dissolution in the Zaragoza area (Huerva River valley, NE Spain): processes, spatial distribution and protection measures for transport routes. Engineering Geology 72:309–329

Gutiérrez F, Cooper AH (2001) Evaporite Dissolution Subsidence in the Historical City of Calatayud, Spain: Damage Appraisal and Prevention. Natural Hazards, Kluwer Academic Publlishers, 25:259–288

Klimchouk A et al (1996) Gypsum karst of the World: a brief overview. Int. Journal of Speleology. 25 (3-4), Chapter II.1:159–181

Klimchouk A (1996) Hydrogeology of gypsum formations. Int. Journal of Speleology. 25 (3–4), Chapter I.6:83–89

Poirot N, Laouafa F, Daupley X (2007) Etat de l'art et éléments d'appréciation des processus physico-chimiques. INERIS, Rapport d'étude DRS-07-86102-17332A:1–63

Thierry P., et al (2008) 3D geological modelling a turban scale and mapping of ground movement susceptibility from gypsum dissolution: The Paris example (France). Engineering Geology réf. doi :10.1016/j.enggeo.2008.12.010:1–11

Toulemont M (1981) Evolution actuelle des massifs gypseux par lessivage. Cas des gypses lutétiens de la région parisienne, France. Bulletin de Liaison des Ponts & Chaussées, Spécial X, janvier 1981:35–47

Toulemont M (1987) Les risques d'instabilité liés au karst gypseux lutétien de la région parisienne – Prévision en cartographie. Bull. de liaison P et Ch, n0150/151, réf. 3192:109–116

# Hypogene Speleogenetic Evidences in the Development of Cova des Pas de Vallgornera (Mallorca Island, Western Mediterranean)

J.J. Fornós, A. Ginés, J. Ginés, F. Gràcia, A. Merino, J. Cifre, and F. Hierro

**Abstract** In the southern part of the island of Mallorca, and developed in Upper Miocene reefal limestones, the Cova des Pas de Vallgornera is an exceptional coastal cave. Littoral mixing dissolution processes represent the most important speleogenetic mechanism to be considered in the eogenetic karst platform where it develops. Nevertheless, part of the cave consists of an extensive network of galleries that show morpho-sedimentary features pointing to a possible participation of hypogene speleogenesis. Its morphological assemblage illustrates the typical coastal karstification, a noticeable meteoric water recharge and a possible deep recharge of hypogenic origin. Solutional features related to rising flow are abundant. The presence of vents and some related speleothems (crusts and rims) must be highlighted, together with Mn and Fe rich deposits hosting several minerals not observed until present in other caves of the region. Given the monotonous surface geology around the cave, it is suspected that ascending hypogene solutions could have reacted with the host rock to form a unique mineral assemblage. These deep-seated phenomena could be associated with the feeble geothermal anomalies existing in the area, related to SW-NE faults.

## 1 Introduction

Cova des Pas de Vallgornera, located on the coast of Mallorca Island (W-Mediterranean), is an extensive maze cave partially drowned by brackish phreatic waters. It lies in the natural area of Migjorn, a littoral karst region built up by an Upper

J.J. Fornós, A. Ginés, J. Ginés, J. Cifre, F. Hierro
Universitat de les Illes Balears, Palma (Mallorca), Spain, e-mail: joan.fornos@uib.es

F. Gràcia, A. Merino
Federació Balear d'Espeleologia, Palma (Mallorca), Spain

B. Andreo et al. (Eds.), *Advances in Research in Karst Media*
DOI 10.1007/978-3-642-12486-0, © Springer 2010

Miocene reefal limestone sequence. Until recently, the speleogenetic mechanisms associated with the littoral mixing zone were considered responsible for the excavation of most caves in Migjorn area (Ginés and Ginés 1992; Gràcia et al. 2007) being remarkable the extensive participation of breakdown processes in its present-day volumetric configuration (Ginés and Ginés 2007).

Recent discoveries of new passages and breakdown chambers are evidences of a complex speleogenesis. Apart from the dissolution in the freshwater-seawater mixing zone, large quantities of allochthonous infillings have been found related to an important surface recharge. Moreover, solutional features observed in the new sections of the cave include rising channels, ceiling channels, feeders and cave rims (Merino 2007); these morphologies together with the mineralogy of some deposits, may be tangible evidence of hypogenic processes (Ginés et al. 2009).

## 2 Geological Setting and Cave Description

Upper Miocene tabular deposits, constituted by a complex reef sequence (Fig. 1), outcrop all along the southern and eastern coast of Mallorca, shaping the flat karst region called Migjorn where the cave is located (Fornós and Gelabert 2004). This

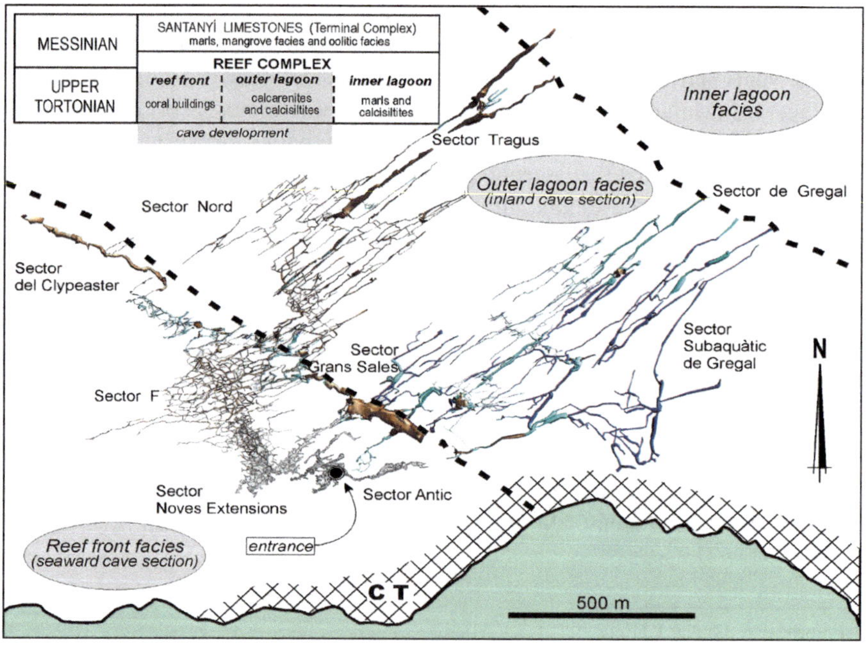


**Fig. 1** The plan pattern of Cova des Pas de Vallgornera shows clear relationships with the facies disposition within the Upper Miocene Reef Complex. The Messinian Terminal Complex (**CT**) outcrops only on the sea cliffs of the area

post-orogenic littoral carbonate platform that fringes and onlaps the folded mesozoic basement, has been affected by Neogene normal faults, disposed in a SW-NE trend, having produced a relative uplift of the area.

The cave has a current development longer than 63,000 m (Fig. 1). The only existing entrance to the cave connects with a series of breakdown chambers quickly attaining the phreatic level that shows an assemblage of aquatic galleries decorated with speleothems. A narrow passage at one of the furthest ends allows the connection with important extensions, which begin with several great chambers disposed in a NW-SE trend; the dimensions attained by the biggest of them are 200 per 80 m. The inner part of the cave shows an irregular maze pattern where several main galleries can be differentiated, running parallel more than 1 km in a SW-NE direction (Merino et al. 2008).

The general morphology of the endokarstic system can be divided into two well-individualised parts: a seaward or outer part of the cave where breakdown and spongework passages are dominant and, on the other hand, an inland or inner part characterised by joint-guided galleries (Fig. 1) shaped with dissolution forms. Breakdown processes are ubiquitous all along the cave, especially in the seaward part of the cave.

As a function of the lithological and hydrogeological characteristics of these young carbonate rocks, a clear dichotomy in the morphogenetic assemblages arises between those parts of the cave excavated in the reef front facies and the inner galleries developed in the back reef facies (Ginés et al. 2009). Cave zones showing generalised breakdown processes occur in the highly porous reef front facies, where the corals are omnipresent being affected by intense differential dissolution phenomena.

## 3 Morphological and Mineralogical Evidences for Hypogene Speleogenesis

The presence of rising channels is frequent in the galleries, being related to solutional ascending processes (Klimchouk 2007). These forms range from millimetres to some decimetres in width (Figs. 2a and b) and develop vertically along the overhanging walls of the galleries. Some of these morphologies are similar to the bubble trails (Palmer 2007; Audra et al. 2009), consisting in rising channels originated by the release of dissolved gases (presumably $CO_2$, in this case) that contribute to the aggressiveness of shallow phreatic waters. As far as morphologies of larger dimensions, small chambers and galleries in the cul-de-sac occur in the inner passages of the cave, showing in their floors impenetrable vertical conduits that could have acted as feeders (Klimchouk 2007).

Particular Mn and Fe rich blackish deposits occurring as detritic fine grained sediments occur on the floors and on the subhorizontal walls of the passages, alternating sometimes with reddish clay and silty materials. These deposits have also been also observed as very thin black coatings, more or less crystallized and/or indurated, covering the walls of galleries or filling the rock fissures or, even forming bubbles like hard crusts that develop over its rocky hollowed floors.

Cave rims and vents have been identified (Merino 2007) in the inner galleries of the cave (Fig. 2c). Cave rim morphology resembles a shell; both sides of these projections are strikingly different. While the outside is rough and coralline, the inside is smooth, like the inside of the tube below the rim. Small and whitish crystallizations on the rim's edge have been spotted. Rims occur in the upper borders of feeder-like solutional vertical hollows or vents which are narrow inaccessible pas-

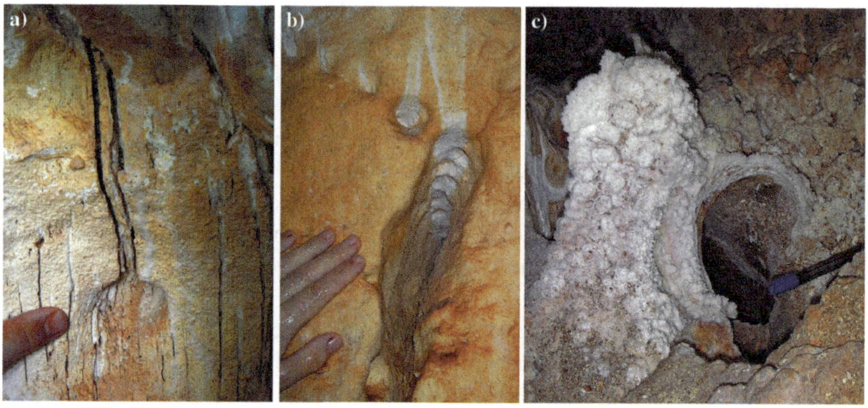


**Fig. 2** Morphological features related to hypogene speleogenesis consist of sharp solutional ascending grooves whose widths are millimetric (**a**), as well as wider rising channels (**b**). Cave rims (**c**) grow on the upper part of vertical vents occurring on the floors of some galleries

**Fig. 3** BSE images of some minerals present at the cave rims and vent walls: **a** nordstrandite, **b** strontianite, **c** ankerite, calcite and barite and **d** celestine

sages that link two different levels of galleries. Its walls are smooth and covered by weathered limestone and mineral deposits.

The nearby areas of the vents normally show mineral crusts (Fig. 2c) and scattered mineralizations, different from the common speleothem minerals found in the mallorcan cave environment (Onac et al. 2005). They form minute crystalline precipitates as well as degradative weathering products on the rock walls with non-frequent chemical elements, where a series of rare mineral species have been identified (Fig. 3). Huntite, monohydrocalcite, strontianite, siderite, todorokite, celestine, and various clay minerals, apart from gypsum are very abundant (Merino et al. 2009). There are also completely unexpected minerals such as barite, nordstrandite, maghmite or paralstonite. Besides, on the outermost part of the vent, tiny, well-formed transparent celestine crystals have been observed mainly on the wall bedrock where gypsum has also been found.

## 4 Discussion and Conclusions

Cova des Pas de Vallgornera remains clearly individualised with respect to the normal trends of the endokarst in Migjorn region. The plan development is really complex, in general terms, and fits in the maze caves category as defined by Palmer (1991, 2007).

Some morphological features as well as specific deposits point to the involvement of other speleogenetic mechanisms working together with the coastal mixing zone processes (Ginés et al. 2009). Hypogene speleogenesis seems to have played an important role in the excavation of the cave.

Whereas the spatial disposition and morphology in the seaward section of the system is like those in other coastal mixing caves of Migjorn, in the inland passages the pattern corresponds to a shallow phreatic 2D maze, in which the presence of solutional features presumably corresponding to a basal recharge is remarkable. The pattern of the inner galleries of Cova des Pas de Vallgornera seems to correspond to a hydrogeological situation leading to the generation of a shallow phreatic maze, related to a local boost of the water aggressiveness. This situation is compatible with a hypogenic recharge, associated with local geothermal anomalies, and the mixing with the coastal aquifer installed in this eogenetic karst platform.

Geothermal anomalies in the unconfined aquifer of the southern part of Migjorn region are linked to the neogene extensional faults that would have come to sustain the participation of hypogene speleogenesis in shaping the Cova des Pas de Vallgornera. Thermal waters with temperatures up to 51.6 °C have been documented in the unconfined sea level controlled aquifer, in the vicinity of Llucmajor village (López and Mateos 2006). The basal recharge documented in the area must be related to regional groundwater circulation coming from the mesozoic folded basement outcropping in the central mountains of the island.

One of the most conspicuous facts is the presence of deposits rich in Mn and Fe, elements that are moderately soluble in anoxic deep waters and easily precipitate as oxides and hydroxides in oxigenated shallow aquifers (Palmer 2007). Also, the chemical composition of minerals forming the rims and crusts with the presence in quite high amounts of non-frequent chemical elements such as $S^{3+}$, $Si^{4+}$, $K^{+}$, $Sr^{2+}$, or even $Ti^{4+}$, $Ni^{2+}$, $Zr^{4+}$, $Cu^{2+}$, $Cr^{3+}$, $Ba^{2+}$, and La in different proportions points towards an hypogenic influence into the crusts and cave rims. Their genesis must be related to the participation of episodic deep- seated processes in the geomorphological evolution of the cave.

**Acknowledgements** This work has been supported by MICINN-FEDER research projects CGL2006-11242-C03-01/BTE and CGL2009-07392.

## References

Audra P, Mocochain L, Bigot JY, Nobécourt JC (2009) The association between bubble trails and folia: a morphological and sedimentary indicator of hypogenic speleogenesis by degassing, example from Adaouste Cave (Provence, France). Int J Speleol 38(2):93–102

Fornós JJ, Gelabert B (2004) Baleares. In: Vera, JA (ed) Geología de España. Sociedad Geológica de España – IGME. Madrid

Fornós JJ, Pomar L, Ramos-Guerrero E (2002) Balearic Islands. In: Gibbons, W and Moreno, T (eds) The Geology of Spain. The Geological Society. London

Ginés A, Ginés J (1992) Las Coves del Drac (Manacor, Mallorca). Apuntes históricos y espeleogenéticos. Endins 17–18:5–20

Ginés A, Ginés J (2007) Eogenetic karst, glacioeustatic cave pools and anchialine environments on Mallorca Island: a discussion of coastal speleogenesis. Int J Speleol 36(2):57–67

Ginés J, Ginés A, Fornós JJ, Merino A, Gràcia F (2009) About the genesis of an exceptional coastal cave from Mallorca Island (Western Mediterranean). The lithological control over the pattern and morphology of Cova des Pas de Vallgornera. In: White, WB (ed) Proc. 15th Int. Congress Speleol., 1: 481–487. Kerrville, USA

Gràcia F, Fornós JJ, Clamor B, Febrer M, Gamundí P (2007) La Cova de sa Gleda I. Sector Clàssic, Sector de Ponent i Sector Cinc-Cents (Manacor, Mallorca): geomorfologia, espeleogènesi, sedimentologia i hidrologia. Endins 31:43–96

Klimchouk, AB (2007) Hypogene speleogenesis: hydrogeological and morphogenetic perspective. National Cave and Karst Research Institute, Special Paper 1, Carlsbad, New Mexico

Klimchouk AB, Ford DC, Palmer AN, Dreybrodt W (eds) (2000) Speleogenesis. Evolution of karst aquifers. National Speleological Society. Huntsville, USA

López JM, Mateos RM (2006) Control estructural de las anomalías geotérmicas y la intrusión marina en la plataforma de Llucmajor y la cubeta de Campos (Mallorca). Instituto Geológico y Minero de España. Madrid. Serie Hidrogeología y Aguas Subterráneas 17:607–613

Merino A (2007) Solutional sculpturings and uncommon speleothems found in the Cova des Pas de Vallgornera, Majorca, Spain. NSS News 65(9):14–20

Merino A, Fornós JJ, Onac BP (2009) Preliminary data on mineralogical aspects of cave rims and vents in Cova des Pas de Vallgornera, Mallorca. In: White, WB (ed) Proc. 15th Int. Congress Speleol., 1:307–311. Kerrville, USA

Merino A, Mulet A, Mulet G, Croix A, Gràcia F (2008) La Cova des Pas de Vallgornera (Llucmajor, Mallorca) alcanza los 55 kilómetros de desarrollo topográfico. Endins 32:33–34

Onac BP, Fornós JJ, Ginés A, Ginés J (2005) Mineralogical reconaissance of caves from Mallorca Island. Endins 27:131–140

Palmer AN (1991) Origin and morphology of limestone caves. Geol Soc Am Bull 103:1–21

Palmer AN (2007) Cave geology. Cave Books, Dayton, Ohio

# A Disappearing Wetland of Karst Origin: the Laguna de la Alberca (Ronda, Málaga)

J.J. Durán, A. García de Domingo, and L. Linares

**Abstract** A number of 19th-century topographical documents show a lake called Laguna de la Alberca near the village of Cuevas del Becerro in the Serranía de Ronda mountains (province of Málaga). Subsequently, evidence of this wetland disappears and there are no cartographic documents or written references to it. Early 20th-century topographical maps at the scale of 1:50,000 do not include it, nor does the catalogue of Spanish lakes prepared by L. Pardo (published in 1948). It does not appear in the inventory of wetlands carried out in Spain in the final decades of the 20th century. Nor does it appear in the lists of wetlands considered "disappeared", i.e. dried up and no longer functioning as such. In other words, the Alberca lake is, for all practical purposes, a wetland that was known and registered nearly two centuries ago, and that at some time in recent history "disappeared" from the official records and has been passed over by naturalists and scientists. This paper includes the first data on the hydrogeological characterisation of the Alberca lake and offers an explanatory model of its origin and karstic functioning, as well as pointing to the possible causes of its historical disappearance.

## 1 Introduction

The Laguna de la Alberca appears in a number of 19th-century maps, such as the map of the provinces of Malaga and Cádiz (Alaber and Mabón 1853) and the map of the province of Málaga (Martín Ferrero 1855), near the Cuevas del Becerro locality (Fig. 1) in the Serranía de Ronda mountains (Málaga). It subsequently disappeared

J.J. Durán, A. García de Domingo
Instituto Geológico y Minero de España (IGME), C/ Ríos Rosas, 23, 28003 Madrid, Spain,
e-mail: jj.duran@igme.es; a.garcía@igme.es

L. Linares
Academia Malagueña de Ciencias, c/ Moratín, 4, 1°, 29015 Málaga, Spain,
e-mail: luislinares@telefonica.net

B. Andreo et al. (Eds.), *Advances in Research in Karst Media*
DOI 10.1007/978-3-642-12486-0, © Springer 2010

from the official Spanish maps. There is no reference to the lake on the map called Teba (number 1.037) of the 1916 topographic map of Spain at the 1:50,000 scale, though there is a point with the caption "*sima*" (sinkhole) at the location of the sinkhole into which the Arroyo del Tejedor stream flows and that now functions as a drain for the lake area (Fig. 2a). The 1960 topographic map (1:50,000 scale) does not represent the lake as such, but it does define a depression delimited by a curve at a level of 700 m a.s.l. (Fig. 2b). A small pond, still existing today, is shown to the extreme northwest of an extension to this depression (which did not appear in the 1916 topographic map). This map also shows the loss of the Arroyo del Tejedor stream in the extreme northeast of the depression. The last topographic map available at the scale of 1:50,000, published in 1996, does not show the depression mentioned above, although the small pond does appear in the northwest, together with the symbols representing a floodable zone.

It does not appear with the name Laguna de la Alberca in the catalogue of Spanish lakes published in the mid-20th century (Pardo, 1948). It is also not mentioned

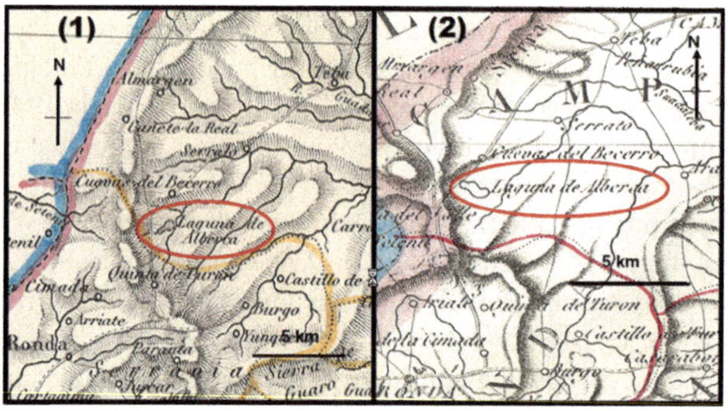


**Fig. 1** (1) Map of the province of Malaga (Martín Ferrero 1855), (2) Map of the provinces of Cádiz and Málaga (Alaber and Mabóm 1853), showing the Laguna de la Alberca

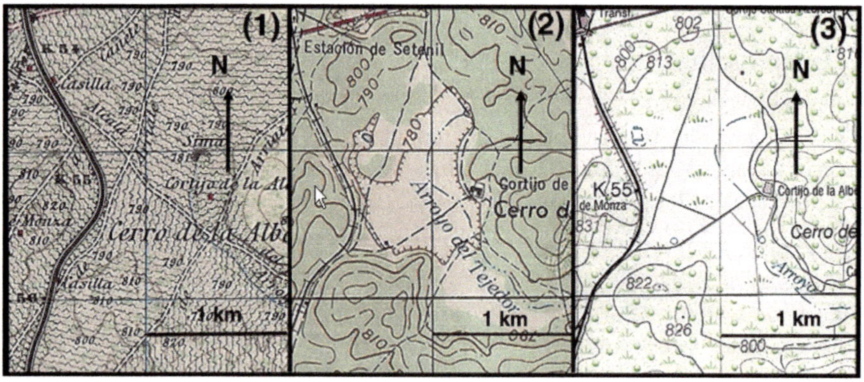


**Fig. 2** Topographical maps at a scale of 1:50,000 of the area around the Laguna de la Alberca: (1) 1916; (2) 1960; and (3) 1996

in any database of current wetlands, nor in that prepared by the Ministry of the Environment (*Base Documental de los Humedales Españoles*) [Documentary Base of Spanish Wetlands] updated in February 2007, or in that by the Government of Andalusia (*Inventario de Humedales de Andalucía*) [Inventory of Wetlands in Andalusia], 2009. It also does not appear in the lists of wetlands that have dried up or disappeared in recent decades. The result of recent investigations carried out on the basis of analysis of the historical maps mentioned above, with the help of currently available technological tools (satellite images) and field work, allowed the location of the lake, and characterisation in hydrogeological terms, which are the basic aims of this work.

## 2 Geological and Hydrogeological Characterization

The depression of the Alberca lake is filled by Quaternary sediments of limited thickness (between 0 and 3 metres), deposited over biocalcarenites, conglomerates and marls dated to the Tortonian/Messinian periods. The unit fossilises an important paleorelief formed on oolitic and pisolitic limestones from the Lower-Middle Jurassic of the Internal Subbetic. The impermeable sediments that limit the aquifers are constituted for clays and evaporites (Triassic) and flyschs and clays (Cenozoic). The calcareous levels of the Jurassic are extremely karstified, with extensive field karren at the surface. The Upper Miocene biocalcarenite levels are also extremely karstified. Hydrogeologically, this sector is characterised by the presence of two aquifers: an upper one, formed on levels of biocalcarenite from the Upper Miocene, is porous, basically due to intergranular porosity, and to karstification and fracturation; and a lower aquifer, formed by Jurassic limestone, intensely karstified in this zone. Overall, it is a unique hydrogeological system that is free, complex, heterogeneous, anisotropic and occasionally multilayered (*Atlas hidrogeológico de la provincial de Málaga* [Hydrogeological Atlas of Málaga], 2007). Over the last years, the groundwater table level of this aquifer has fallen as a consequence of exploitation.

## 3 Characterization of the Laguna

The Laguna de la Alberca is at the coordinates 36°51′21.32″N/5°06′34.10″W (UTM X: 311975.58/Y: 4081054.03) at an approximate altitude of 775 m a.s.l., with the water layer extending to around 55 hectares at the times of greatest flooding. The lake basin has a long N-S axis of more than 1000 m and a shorter E-W axis of 700 m. The maximum limits of the basin have been established in accordance with the topographical map of 1965 and geomorphological field criteria (Fig. 2-2). The maximum capacity of the lake basin with the defined perimeter is less than 1 $hm^3$.

The relief of this sector is extremely hilly. This lake is etched by the final stretches of the Arroyo del Tejedor stream, whose meander ends in a sinkhole at the northern border of this closed depression (coordinates UTM X: 0312291/Y: 4081256), represented in all the modern topographical maps. The final stretch of this stream has been channelled and artificially dug within the Quaternary deposits that make up the sedimentary infill of the lake, with the aim of draining this former damp area and drying the adjoining fields (Fig. 3). There are indications of other human intervention to desiccate the lake, such as ditches dug transversally to the stream and a small construction over the sinkhole, which prevents it from being obstructed and enables it to be maintained and cleaned.

The groundwater table level of the detrital aquifer (in October 2009), measured in a well (X: 0311085/Y: 4079949) at the western edge of the lake, is situated at 76.5 m a.s.l., i.e. 13.5 m below the surface of the lake basin at this point. An old well (around 12 m in depth) has been found in the southern sector of the lake above this detrital aquifer. It was built at the start of the 19th century, and used with the aid of a chain pump. It is now dry. The groundwater table level of the Jurassic aquifer (also in October 2009) in the well supplying Setenil, located in the Cortijo del Tejedor some two kilometres SE from the one described above, is at 776 m a.s.l. when the water level into the well is static. These data indicate that the aquifers are connected in this zone and form a single aquifer system and that the polje in which the intermittent lake is located is now not connected directly with the phreatic zone aquifer, and is far above the groundwater table level.

The Alberca lake occupies an area on the recharge zone of these aquifers, while the system discharges towards the Alcalá del Valle-Setenil zone, some 8 km NW, most probably towards the springs in the Arroyo de los Molinos sector.

The sinkhole of the former lake is located over detrital levels from the Upper Miocene very near the Jurassic limestone. It has not yet been explored, although it

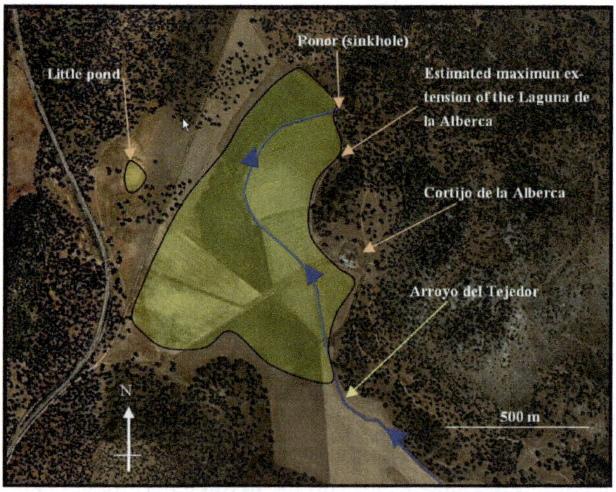


**Fig. 3** Satellite image of the Laguna de la Alberca area, showing the perimeter of the maximum level of the Laguna de la Alberca. Note the agricultural works inside the lake basin

appears that there is a penetrable sub-horizontal cavity in its initial stretch of about one metre in diameter, which follows the angle of dip of the formation. According to the local inhabitants, currently the lake is flooded every eight to ten years at times of exceptionally intense precipitation, for a period of about 15–20 days.

The origin of the lake is related to the process of karstic alimentation of the aquifer, located above a large exokarstic landform (polje) developed over the detrital levels of the Upper Miocene. The water that feeds the lake is superficial, and proceeds from the flow from the Arroyo del Tejedor stream that recharges the aquifers via the sinkhole. The estimated annual volume of this stream for the whole water basin is around 3.5 h $m^3$/year, according to an annual average precipitation of 750 mm. Currently the lake is ephemeral, and is flooded for only short periods when the precipitation reaches significant levels and results in a flow from the Arroyo del Tejedor stream that the sinkhole cannot absorb. Occasionally, there can also be extensive flooding when the Quaternary clay sediments that make up the deposits of the lake are saturated by persistent precipitation and the natural or forced drainage is insufficient to evacuate the amount of water that has fallen into the basin. Satellite images and aerial photos of the zone show the presence of lines that for the most part converge towards the banks, most probably corresponding to old human intervention designed to drain the depression and limit the periods during which the lake was flooded (Fig. 3). These lines do not reach the western border of the depression, where the small lake floods more often and for longer times in periods of heavy rain, possibly because the drainage work does not affect this sector.

## 4 Discussion and Conclusions

The old lake of La Alberca, whose natural hydric regime is currently affected by man-made drainage in its basin, is located on a karst depression (polje) developed over thick detrital materials of the Upper Miocene. This former lake is a wetland that can be categorised as “in a recharge zone”. It is an epigenetically formed lake that is seasonal in nature, with short periods of flooding at times of greatest precipitation. The recharge is exclusively from precipitation into the lake basin itself and from the surface runoff provided by the Arroyo del Tejedor stream, while the discharge is through evapotranspiration from the lake area, and above all through the sinkhole at the northern end.

From the historical point of view there is evidence that the Alberca lake existed, possibly as a permanent lake, given that it was reflected in the historical maps of the $19^{th}$ century before 1880. The rainfall records of the best historical series available in the south of Spain indicate a period of much more abundant precipitation than the annual average between 1850 and 1905. These facts (the high levels of annual precipitation and the cartographical representation of the lake in contemporary maps) support the idea that the lake was permanent, or with a certain regular annual stability at the end of the 19th century. So why did it disappear from the topographical records at the start of the 20th century? The most probable hypothesis is that it disappeared for climatic reasons. In areas that are relatively nearby in the southeast of Spain (such as, for example, Doñana in the province of Huelva), there is evidence of the disappearance of lakes at the start of the 20th century due to the sustained reduction of annual precipitation over time (Sousa Martín et al. 2004). This fact of nature could have encouraged agricultural occupation of the lake basin, which was flooded only in favourable hydrological situations, in years with high rainfall levels. Drainage work and artificial extension and maintenance of the sinkhole that drains the polje in which the lake is located, to prevent its obstruction and water retention, have increased the hyperannual periods during which the lake is dry. This has led to the current situation in which it is only filled in extraordinary periods of extremely high rainfall, about every 8 to 10 years. It is possible that this change to a highly sporadic filling can explain the disappearance of the lake from the 20th century maps. Climatic change, linked to human activity in the lake basin to increase arable land, have meant that this wetland has until now disappeared from view for the scientific community.

## References

Atlas Hidrogeológico de la provincia de Málaga [Hydrogeological Atlas of Málaga] (2007) Instituto Geológico y Minero de España-Diputación de Málaga. Madrid, 3 vols. + 1 map, 220 pp

Base Documental de los Humedales Españoles, [Documentary Base of Spanish Wetlands] (2006) Ministerio de Medio Ambiente y Medio Rural y Marino. Version 4 (April, 2006). Madrid. Updated at February 2007.

Consejo Superior Geográfico (1965) Edición Militar del Mapa Topográfico Nacional 1:50,000. Hoja n° 1037 (Teba). [Military Edition of Nacional Topographical Map, 1:50,000 Scale]

Dirección General del Instituto Geográfico y Estadístico (1916) Edición Militar del Mapa Topográfico Nacional 1:50,000. Hoja n° 1037 (Teba). [Military Edition of Nacional Topographical Map, 1:50,000 Scale]

Inventario de humedales de Andalucía (2009) Consejería de Medio Ambiente. Junta de Andalucía. Sevilla. [Inventory of Wetlands in Andalusia] Updated at 13/05/2009.

Mapa de las provincias de Malága y Cádiz (1853) Engraving by R. Alabern y E. Mabóm. Printed in Barcelona (Calle de los Baños n° 15) by J. Vázquez. Published in: Madrid, Librería Española; Barcelona, Librería Española. [Map of the provinces of Málaga and Cádiz]

Mapa de provincia de Málaga (1855), by Martín Ferreiro. Engraving by Alabern. Edited by Gaspar y Roig, Madrid [Map of the province of Málaga]

Pardo L (1948) Catálogo de los lagos de España. Instituto Forestal de Investigaciones y Experiencias. Ministerio de Agricultura. Madrid, 522 pp

Servicio Geográfico del Ejercito (1996) Cartografía Militar de España 1:50,000. Hoja n° 1037 (Teba) [Military Cartography of Spain, 1:50,000 Scale]

Sousa Martín A, García Murillo P, González Pérez M, García Barrón L (2004) La desaparición de las lagunas del entorno de Doñana (Huelva, España) y su relación con cambios climáticos recientes. In: García Codrón JC, Diego Liaño C, Fernández de Arróyabe Hernáez P, Garmendia Pedraja C, Rasilla Álvarez D (eds.), El Clima entre el Mar y la Montaña. Asociación Española de Climatología y Universidad de Cantabria, Serie A, n° 4, Santander, 715–734 pp

# Geomorphological Characteristics of the Karst-Related Lakes in Gypsum in the Arcas and River Moscas Lake Complexes (Cuenca Province)

M. Martínez-Parra, A. De la Hera, E. López-Pamo, M.J. Moreno, E. Montero, and E. Santofimia

**Abstract** The Arcas and River Moscas lake complexes, protected natural areas in Cuenca province, are located in interstratified karst developed in Cretaceous-Tertiary boundary gypsum. These sites are in the central part of the province, near the town of Cuenca, each covering an area of some 3 $km^2$ and distributed between the municipalities of Arcas del Villar, Cuenca, Fuentes, Valdetórtola and Villar de Olalla. This paper presents certain geomorphological aspects established in the framework of a broader study on the hydrogeology of the two complexes, fed chiefly by groundwater discharge from this gypsiferous aquifer, whose hydrodynamic behaviour is determined by its fractures and faults.

## 1 Geographical and Geological Introduction

The study area is in the eastern-central region of the Iberian Peninsula, in Castilla-La Mancha, 8 km from the town of Cuenca, with annual average rainfall of 600 mm and an average temperature of 12 °C. It corresponds to the two lake complexes of Arcas-Ballesteros and the River Moscas, of which the former is protected as a natural reserve and the latter is in the process of being declared a micro-reserve.

The Cuenca mountains (Cordillera Ibérica) consist of Jurassic and Cretaceous carbonate materials, strongly folded and fractured, forming a large anticlinorium whose core consists of Palaeozoic and Triassic materials (gypsiferous clay, conglomerate and dolomite) and its flanks of Jurassic and Cretaceous carbonate mate-

M. Martínez-Parra, A. De la Hera, E. López-Pamo, E. Santofimia
Instituto Geológico y Minero de España (IGME), C/ Ríos Rosas, 23, 28003 Madrid, Spain,
e-mail: m.martinez@igme.es

M.J. Moreno, E. Montero
Autonomous Agency for Natural Areas in Cuenca Province, Regional Government of Castilla-La Mancha, Lorenzo Goñi, 5, 16004 Cuenca, Spain

B. Andreo et al. (Eds.), *Advances in Research in Karst Media*
DOI 10.1007/978-3-642-12486-0, 

rials, with folds at its end defining NW-SE oriented depressions, filled with Tertiary detrital deposits.

The lake complexes studied have developed in marl, gypsum and carbonate Cretaceous-Tertiary boundary materials, defined in the area as Fm. Villalba de la Sierra marls and clays. Stratigraphic observation from base to top shows 120 m of gypsum and marl, 90 m of alternating limestone, massive gypsum and marl, and 50 m of gypsum. These materials are part of a NW-SE oriented syncline, on whose western flank, coinciding with the anticlinal axis, the Arcas-Ballesteros lake complex occurs, while the River Moscas lake complex occurs 8 km away on the eastern flank.

## 2 Hydrogeological and Geomorphological Characteristics

The Cretaceous-Tertiary gypsiferous formations form different aquifers: one, at the surface, associated with gypsum and marls and the other associated with massive and deepest gypsum horizons. The massive gypsum does not emerge in the lake complexes, and the karst is developed in interstratified form, which may be classed in Klimchouk's classification (Klimchouk 1996) as type ISK underlying karst, evolving without offering an exterior aspect except when it affects the covering layer. These lake complexes have been studied from a microbiological perspective (Armengol 1997; Camacho 1997 and Rodrigo 1997) but less so in hydrogeological terms (Sanz 2002; Martínez-Parra 2009; De la Hera et al. 2009).

The Arcas-Ballesteros lake complex occupies an area of some 2.52 $km^2$, consisting of 38 dissolution and collapse dolines, funnel-shaped and mostly with water. Their size is variable, with areas ranging from little more than 100 $m^2$ to nearly 4000 $m^2$. Some of these dolines are very close to each other and have merged, forming uvalas. From the permanent ponds, drainage channels run to the River San Martín, with a flow rate gauged for the Barraganes lake in July 2009 at 14 l/s. This flow shows a groundwater feed from the gypsiferous aquifer, and its confinement.

The depth of the lakes is greater in the central part of the complex (10–14 m), indicating a preferential dissolution and/or circulation area, while the lakes near the river San Martín are less deep, at 1.5–4.4 m. From the bathymetric measurements taken by Camacho (1997) it was determined that 60% of the lakes are flat-bottomed and the remaining 40% funnel-shaped. From the bathymetric readings taken by IGME in July 2009 in the Arcas-4 and Barraganes-1 lakes, it was determined that the former has a shallow water body (0–1 m) in more than half of its area, and at the centre there is a collapse chimney, with a tapered shape and steeply sloping sides. The Barraganes-1 lake, at a lower elevation, has a funnel shape (Figs. 1, 2).

In the River Moscas lake complex there are less abundant morphologies than in the previous complex, and less density. The size of these lakes ranges from a little more than 600 $m^2$ to 13,000 $m^2$, a larger size than for any of the lakes in the Arcas-Ballesteros complex. Three centres may be distinguished in this complex: Ojo de Corba-Atalaya (1.4 $km^2$), and the ponds of Las Zomas and Mohorte.

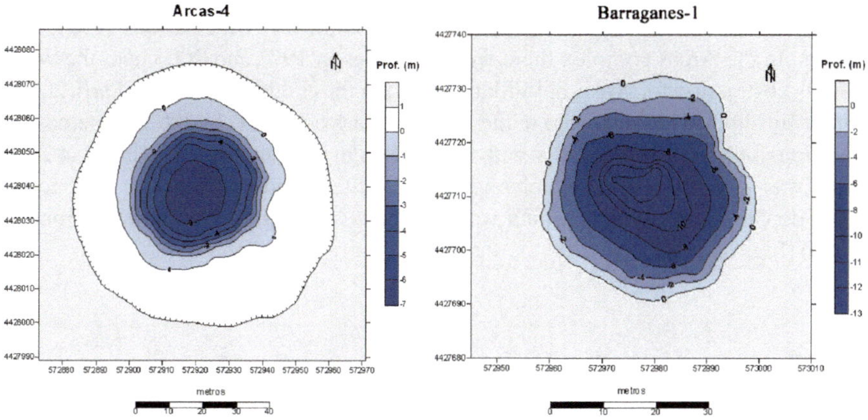


**Fig. 1** Bathymetry of Arcas-4 and Barraganes-1

**Fig. 2** Barraganes-1 (*left*) and doline created in April 2009 near Ojo de la Corba (*right*)

The geomorphologies present are determined by the location on the flank of the syncline, a subsidence area threaded by the River Moscas, where few dissolution dolines are without water: only the old Tamariz lake, now dry, and the doline that formed nearby in 2002. As in the other lake complex, there is drainage from the permanent lakes to the River Moscas, and also between them; thus between Ojos de Corba 2 and Ojos de Corba 1, situated at a slightly lower elevation, there is a flow of 27 l/s (July 2009).

The Las Zomas Lake is the deepest and most irregular of all. Its morphology, according to Armengol (1997), is characteristic of a collapse doline, with a depth of 6.7 m and a small area in the middle that sinks sharply down to 15.1 m, forming a chimney. The lake's location, separate from the rest of the Fuentes area, at the foot of a carbonate slope with an inclination of more than 30°, indicates a possible mechanical deformation of these materials, and so the lake may have been able to take advantage of that fractured area to form and to provide a preferential circulation and emergence area from the confined gypsiferous and carbonate levels.

There is much karst activity in both lake complexes, with various recent collapses. In the Arcas complex there were collapses in 1978 and 2003 near the water supply abstraction for Arcas del Villar, damaging the conduit (Martínez-Parra 2003). Other subsidence has occurred in the River Moscas complex. In 2002 a tapered dissolution doline formed, initially with water, though it is currently dry. Early in 2006 there was a small collapse, which on 3 April 2009, caused a doline very close to the Ojos de Corba lakes, from which water emerged naturally (Moreno and Montero 2009) (Fig. 2).

## 3 Influences on Doline Formation

The formation of dolines and especially collapse holes associated with intrastratal karst is linked to the presence of fractures and faults, among other factors. The valley of the River Moscas seems to follow a set of fractures, oriented NW-SE and then changing direction to WNW-ESE. In turn, the lakes and dolines have developed together with the elongated WNW-ESE relief, which may indicate the existence of faults and fractures. These discontinuities reach the aquifer and form a preferential circulation route, allowing the water associated with the karst and gypsiferous aquifers to emerge.

The Arcas-Ballesteros area is in the anticlinal axis of the fold, a distended area in which amalgamated distensive NE-SW and NW-SE fractures occur, forming a mesh of fractures determining the lakes' development (Fig. 3).

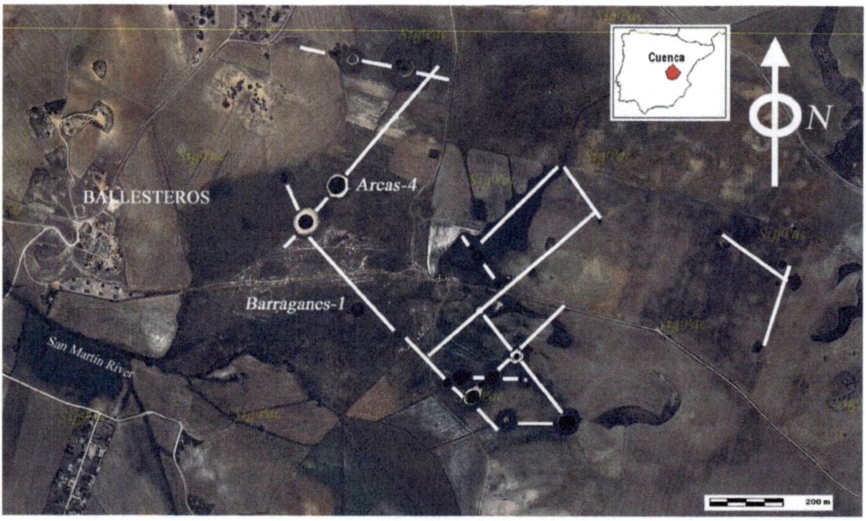


**Fig. 3** Dolines and their relationship with fractures and faults (*white line*) in the Arcas-Ballesteros complex

# 4 Conclusions

The development of karst in interstratified gypsum and its associated morphologies giving rise to the Arcas and River Moscas lake complexes is closely linked to the distensive fracturing phenomena affecting Mesozoic and Cenozoic materials, leading to the formation of collapse dolines and the emergence of groundwater from confined aquifer formations.

**Acknowledgements** This study is part of the activities covered by the Agreement signed by IGME and the Cuenca Provincial Council for 2008–2011.

# References

Armengol J (1997) Caracterización estructural del zooplancton de las lagunas cársticas de Cuenca, con especial atención a su distribución vertical. Doctoral Thesis. Valencia University.

Camacho A (1997) Ecología de los microorganismos fotosintéticos en las aguas microaerobias y anóxicas de la laguna de Arcas. Doctoral Thesis. Valencia University.

De la Hera A, Martínez M, Lopez-Pamo E, Santofimia E (2009) Primeros resultados sobre el estudio hidrogeológico del sinclinal de Fuentes y su relación con los complejos lagunares de Arcas y del río Moscas (Cuenca, España). Meeting on The Role of Groundwater in Wetland Functioning. Zaragoza 22–23 October 2009. International Association of Hydrogeologists-Spanish Group

Klimchouk A (1996) The typology of Gypsum karst according to its geological and geomorphological evolution. International Journal of Speleology 25(3/4), Gypsum Karst of the World (Eds. Alexander Klimchouk, David Lowe, Anthony Cooper & Ugo Sauro). pp. 49–60

Martínez-Parra M (2003) Informe hidrogeológico para la mejora del abastecimiento público de agua potable a la localidad de Arcas del Villar (Cuenca). Spanish Geological Survey (IGME). Internal report

Martínez-Parra M (2009) Hidrogeología de la Serrania de Cuenca. Thesis in preparation. Madrid Autonomous University

Moreno MJ, Montero E (2009) Informe sobre la evolución geomorfológica del complejo lagunar del río Moscas. Autonomous Agency for Natural Parks (Cuenca). Internal report

Rodrigo MA (1997) Limnología comparada de las lagunas de dos sistemas cársticos de Cuenca. Bacterias fotosintéticas de la laguna de la Cruz y la Laguna Arcas-2. Doctoral thesis. Valencia University.

Sanz E (2002) El karst en yesos en Fuentes (Cuenca). Karst and Environment. Nerja Cave Foundation, Nerja-Malaga. 15–17 September 2002. Book of Proceedings, pp. 351–358

# Konya Karapınar Obruks (Sinkholes) of Turkey

G. Günay, İ. Çörekçioğlu, S.O. Eroskay, and G. Övül

**Abstract** Konya plain in Central Anatolia features various unique karstic formations from different ages. The most peculiar of these formations are the obruks, which are mainly found within the lacustrine limestone of Neogene age. The unconfined groundwater in the region outcrops in the form of obruk lakes. The Neogene formations at the upper part and the basement limestones are hydraulically linked. The Neogene formation is more porous because of cavities, and the pervious nature of the basement is due to karst conduits. The obruks have been formed by the combined effects of the carbonates, structures, karstification, and the tectonic history of the area.

## 1 Introduction

"Obruk" as a term is the name that is used to refer to the cylindrical sinkholes in Konya plain in Central Anatolia. Some of these are larger formations and are therefore referred to as "obruk lakes," such as those in Gökhöyük, Meyil and Kızören.

In this paper, the origins of obruks are studied as well as the possibilities of benefiting from them in this semi-arid region. To explore the irrigation potential

G. Günay
Retired Prof. from Hacettepe University, Consultant on Hydrogeology and Engineering Geology, e-mail: ggunay@ttmail.com

İ. Çörekçioğlu
Retired Hydrogeologist DSI Regional Directorate, Konya Turkey

S.O. Eroskay
Retired Prof. from Kültür University

G. Övül
Retired Geophysicist

B. Andreo et al. (Eds.), *Advances in Research in Karst Media*
DOI 10.1007/978-3-642-12486-0, © Springer 2010

of the obruks, which is highly in demand in the region, the State Hydraulic Works Groundwater Department initiated an intensive drilling program. Konya DSİ (State Hydraulic Works) Regional Directorate Groundwater Division launched a drilling campaign for 100 wells for the 1976–77 irrigation season in the vicinity of Çumra, each one of the wells aiming to exploit a yield of 100 l/sec.

## 2 Geology and Hydrogeology

The local basement of the mountains, which surround the Konya plain, consists of schists, phyllites, quartzites, crystallized limestones, as well as of penecontemporaneous volcanics with fresh physiographic features and magmatic rocks. On top of these layers, basal conglomerate and the partly recrystallized and noticeably fractured Bozdağ limestone with colors ranging from light to dark grey overlie the plain, respectively. The limestone formation was dated to the Permian age with the help of some fossils. The Bozdağ limestone overlaps a thick sequence of light-colored, compact, solid and Jurassic-Cretaceous-aged limestone, which can also be seen along the Taurus mountain range in the south (Fig. 1).

The plain also features an extensively-exposed Cretaceous ophiolitic system and thick Tertiary formations. In the east, there are distinguishable sedimentary basins of the Paleocene, Eocene and Oligocene ages, as well as lacustrine deposits of the Miocene and Pliocene ages, which cover large areas in the plain. The overall thick-

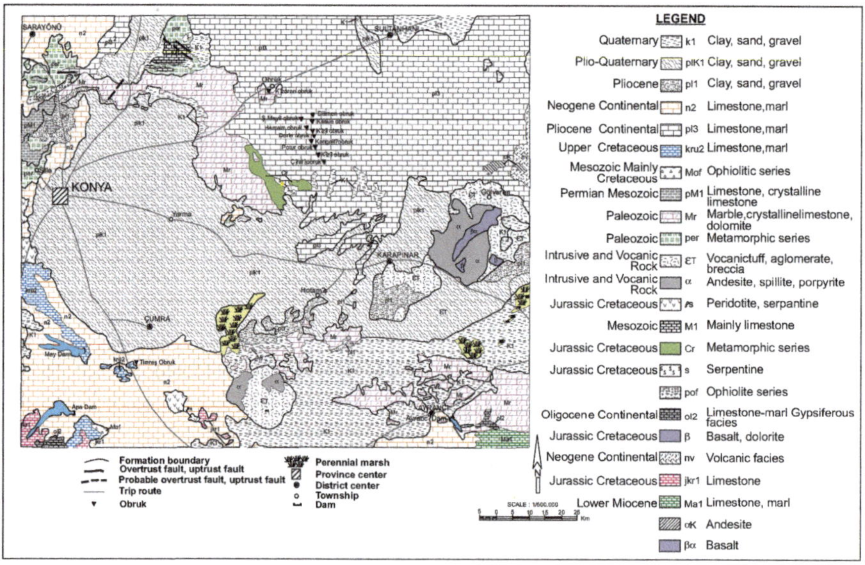


**Fig. 1** Simplified Geological Map of the Konya Closed Basin (After Çörekçioğlu 1994)

ness of the plain with the abovementioned layers is about 350 m, with local variations (70 m at Çıralı and 150 m at Kızören). The thickness of the plain is even less near Tuzgölü lake.

The water levels of some of the obruks, such as Timraş, Kızören, Meyil, are measured twice a day. In this study, the relationship between precipitation, replenishment, etc. was studied at the former obruk. The precipitation data are obtained from the records of the local meteorological station. To have a better understanding of the regional precipitation trends, data from Konya, Aksaray, Karaman and Karapınar stations are also taken into consideration.

## 2.1 The Origins of the Obruks

Obruks are located in the Neogene limestone and seem to align with the definite trends, and are concentrated over the fault lines. The latter situation can be observed at the Kızören, Meyil and Çıralı obruks. Even though the obruks located in the Obruk Plato between Karapınar and Sultanhanı (Gökhöyük, Apasaraycık, Belkuyu, etc) have a NW-SE direction, these do not overlap with observable faults. They also stretch along the strike of the Neogene limestone, where the origins of their formation are only speculative. However, this alternative explanation calls for an assumption about the higher permeability for the lower limestone as well as the structural lowering of the upper limestone (Neogene).

It is understood from the analysis of the water samples from various depths that there was a gradual enrichment in terms of $H_2S$. The reason for this enrichment may be either a slower movement or being recharged for a long stretch of time. The direction into which the waters flow is defined by the local variations that took place during the rising movement.

## 2.2 Morphometry of Obruks

Almost all obruks developed within the Neogene lacustrine limestone (Fig. 1), except for the Kızılca, Bellikuyu, and Tahtalı Obruks that developed within the limestone of Mesozoic age. These obruks are dry. Figure 2 shows illumination image of digital elevation model of Obruk Plateau, Konya Closed Basin. Neogene limestone is generally horizontal and sometimes intercalated with clay and marl. The obruk development is intensified in the area called Obruk Plateau, which is located between Sultanhanı Village (Kızören) and Karapınar County at the south and extends over an area of about $120\,km^2$. However, there also are some significant obruks out of the Obruk Plateau area such as the Apasaraycık Obruk, which developed near Apa Dam in the Neogene conglomerates southeast of the mapping area and the Timraş Obruk near Çumra south of the Konya, which developed in connection with the Mesozoic limestone which is exposed near the obruk lake.

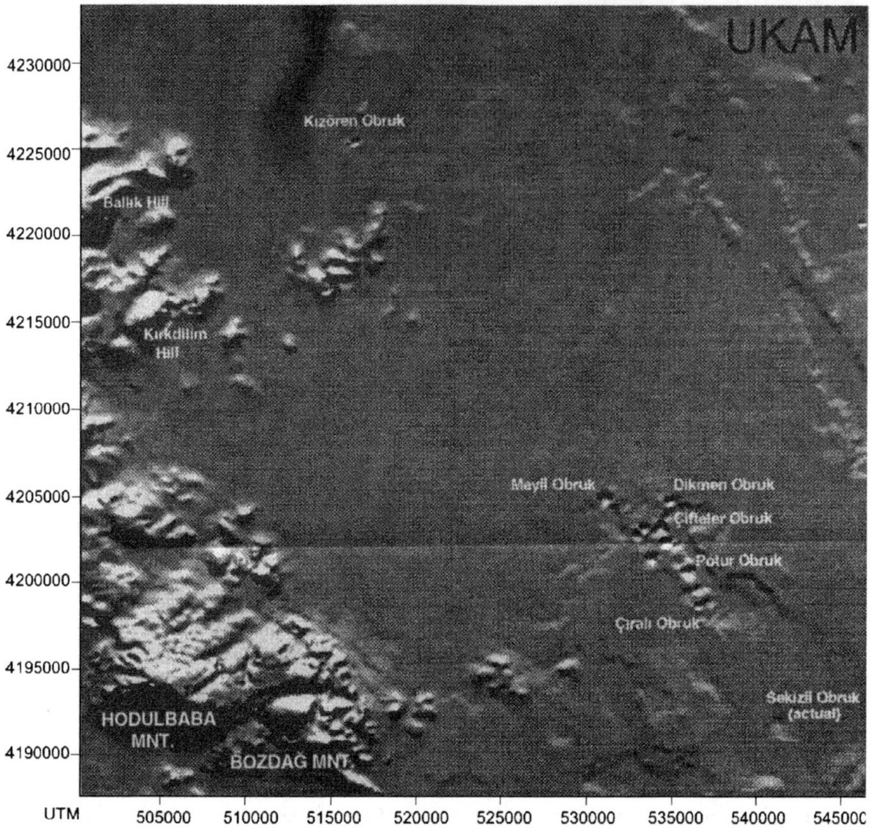


**Fig. 2** Illumination image of digital elevation model of Obruk Plateau, Konya Closed Basin. (UKAM 1995)

## *2.3 Recent and Old Obruks*

Akviran Obruk has been formed with a sudden collapse in 1977. It is 60 m deep with water at its bottom. It has a conical shape and the diameter at the top is only 15 m, while the diameter at the bottom is 30 to 35 m. The wall is made of silt, clay and marl. Nebili Obruk developed in 1983 in silts and marls with a diameter of 10 m. This obruk again gets wider towards the bottom. It is 70 m deep and hits the water table. However, fragments due to the sudden collapse and the following landslide cover the water table.

Sekizli Obruk area includes more than one obruk, which developed recently. One of the obruks was formed in 1983, starting with a hole of 10–15 cm. After a few months it suddenly collapsed with the effect of heavy rainfall. The lithology in this obruk is made up of silt, clay and marl. Its diameter is 5 m and it is 8.5 m deep. Another obruk in Sekizli area occurred in May 1995. It is a huge obruk with a diameter of about 45 m and a depth of about 60 m. The lithology is again clay, silt and marl. Sudden collapses still continue in this area.

Most typical and good examples of the older obruks are the Meyil, Kızören, Çıralı, Timraş, Karain, Dikmen, Yunus, Hamam, Kızıl, Yarım, Fincan, Potu, Kangallı, Yılanlı, Zincanca and Kayalıbaşı obruks, the first three of which hit the water tables. These obruks are arranged in NW-SE direction. Their diameters vary between 8 m to 700 m. The deepest one has a depth of 80 m. Meyil, Dikmen, Karain, Kızıl, Kangallı and Çıralı obruks have been formed within the limestones. Çifteler and Ak obruks have developed within marly, silty and clayey deposits (Fig. 1). Head of water in Meyil Obruk has decreased from 34 m to 30 m in the last 5 years. Long-term water level and precipitation records in the Meyil Obruk have revealed that the water level is strongly influenced by precipitation.

Kızören obruk is one of the most striking structures encountered in large numbers in the Closed Basin of Konya. The Kızören obruk, which is a collapse type of doline, has an approximately elliptical shape with a long axis of 180 m and short

**Table 1** Isotope values from Konya Obruk waters

| Name of Sample | Date of Sampling (for $O^{18}$ and Deuterium) | $O^{18}$ (‰) | Deuterium (‰) | Date of Sampling (for Tritium) | Tritium (TU) |
|---|---|---|---|---|---|
| Timraş | 25.10.1999 | −7.30 | −44.59 | 24.12.1999 | $11.20 \pm 1.20$ |
| Meyil Obruk | 25.10.1999 | −2.03 | −22.56 | 24.12.1999 | $10.15 \pm 1.20$ |
| Kızören | 25.10.1999 | −9.84 | −66.18 | 24.12.1999 | $6.30 \pm 1.20$ |
| Apasaraycık | 25.10.1999 | −5.78 | −32.47 | – | – |
| Çıralı Obruk | – | – | – | 24.12.1999 | $4.25 \pm 1.15$ |

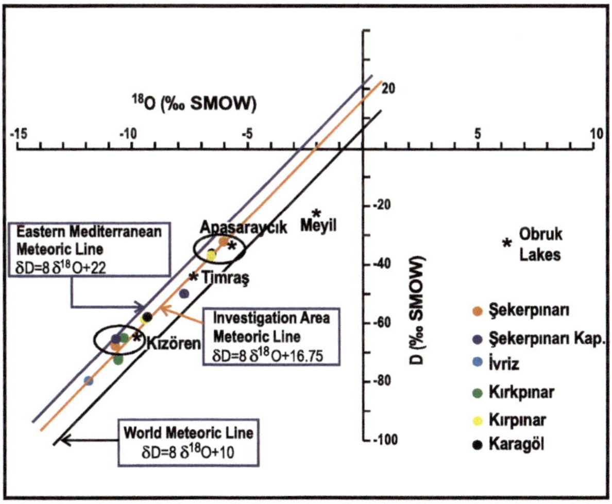


**Fig. 3** Oxygen 18 and Deuterium relationship

axis of 150 m. The maximum depth of the sinkhole is 145 m from the surface of the water. Although the water level in the sinkhole fluctuates during winter and summer depending on the ground water recharge, the change in water level generally does not exceed 1–2 m.

Obruk Lakes waters were collected in the year of 1999 and analysed by the State Hydraulic Works establishment (DSİ) Technical Research Department Laboratories in the year 2000 (Table 1, Fig. 3).

In Fig. 3, $O^{-18}$ and Deuterium relationship can be seen in the graph. Generally, between the Eastern Mediterranean Line and World Meteoric Line there is the "Investigation Area Line" which is fit to the $\delta D = 8\,\delta O^{18} + 16.75$ line. All the springs around the investigation area such as Şekerpınarı, İvriz, Kırkpınar, Kırpınar and Karagöl are on the same line. Also Kızören, Timraş, Apasaraycık obruks are over the same line, except Meyil obruk. Meyi obruk values correspond to evaporated water.

## 3 Conclusions

The distribution of the obruks on the surface do not seem to be accidental, as they tend to align with the surface traces of the faults in the region. These have been formed by the combined effects of the carbonates, structures, karstification, and the tectonic history of the area. It can be stated that the cumulative effects of precipitation, recharge, and some tributaries recharge the groundwater tables as is seen today. Due to the fact that these water table variations are still only on the surface of the obruk lakes, there is reason to believe that intervention from beyond the borders of the plain is also a factor. Yet another factor is the crystallized Permo-Trias aged karstic limestone that is found at the basement.

## References

Çörekçioğlu İ (1994) Karapınar kuzeyi obruklar sahası karst hidrojeolojisi incelemesi, DSİ Bölge Md. Konya

Günay G (1976) Konya Sarıcalar dolayının jeolojisi ve yeraltısuyu olanaklarının izotop yöntemlerinden de yararlanılarak incelenmesi. Doktora tezi. İÜFF Mecmuası, İstanbul

Eroskay SO (1976) The factors influencing the Konya obruks and their groundwater potentials evaluation, IÜFF Mecmuası seri B (1976) 41(1–4):5–14, İstanbul

UKAM (1995) Guide Book, Int. Symp. and field Seminar on Karst Waters and Environmental Impacts. 10–20 Sept. 1995, Beldibi, Antalya, Turkey

# Hydrogeology and Geomorphology of the Calar del Espino (Betic Cordillera, Southern Spain), a Highly Interesting Scientific and Experimental Karstic System

F. Moral

**Abstract** Sierra de Segura is an extensive mountain range, about 3000 $km^2$ in area, with 800–1200 mm/year of precipitation and groundwater resources assessed as 236 $hm^3$/year. Calar del Espino forms a small and representative hydrogeological unit that discharges almost all groundwater flow through the El Tejo spring. Carbonate rocks nearly 500 m thick are exposed over 6 $km^2$ in the core of a synclinal structure and constitute a hanging aquifer. The summit of Calar del Espino (1600–1700 m asl) is a karstic plateau corresponding to a remnant of the Segura post-Miocene paleosurface. Hydrograph analysis and physical and chemical characteristics of the water lead to the conclusion that the El Tejo karstic system shows a heterogeneous degree of karstification, little variability in the physical and chemical characteristics of the water, high dynamic reserves, and a long water residence time.

## 1 Introduction

Groundwater drained from Sierra de Segura aquifers feeds the sources of the main rivers of southern Spain, the Guadalquivir and Segura Rivers. Appropriate management of these valuable water resources, including ecological function protection, requires a better understanding of the groundwater flow and hydrochemical processes in carbonate aquifers. Under natural flow regime conditions, one of the more common methodologies in karstic hydrogeology is to study the functioning of the spring to make inferences regarding the infiltration and groundwater processes, the hydraulic parameter fields and the degree of karstification of the aquifer (Kiraly 2002).

The aims of this research are to characterise the hydrodynamic and hydrochemical behaviour of the El Tejo spring, the main discharge point of the Calar del Espino

F. Moral
Departamento de Sistemas Físicos, Químicos y Naturales, Universidad Pablo de Olavide, Carretera de Utrera, Km 1, 41013 Seville, Spain, e-mail: fmormar@upo.es

B. Andreo et al. (Eds.), *Advances in Research in Karst Media* 
DOI 10.1007/978-3-642-12486-0, 

aquifer, using the Mangin's method, and to describe the geological and geomorphological features that determine the hydrogeological function of this system.

## 2 Description of the Study Area

The Calar del Espino (Fig. 1) is in the northern Sierra de Segura at the headwaters of the Tus and Madera Rivers, both tributaries of the Segura River. The climate is typical of Mediterranean mountains, with hot dry summers and high levels of precipitation from October to May, sometimes as snow (mean annual precipitation, about 1200 mm). The mean annual temperature ranges between 8.5 and 11.5 °C depending on altitude.

From a geological point of view, the Calar del Espino belongs to the Inner Prebetic Domain, composed of sedimentary rocks deposited on the southern Iberian margin from the Triassic to the Miocene. In the study area, outcrops consist of alternating Lower Cretaceous sandstones, marls and carbonates, a 400-m-thick sequence of Upper Cretaceous dolostones and limestones, and finally Miocene limestones and carbonate sandstones less than 100 m thick. The geological structure of this area consists largely of west-verging gentle folds that strike approximately north-south. The Calar del Espino is a synclinal structure with its axis 250 m to the east of the summit line (Fig. 2). Because the axis plunges slightly to the south in the northern part of the Calar del Espino and to the north in the southern Calar del Espino, the Upper Cretaceous-Miocene rocks have a basin-like geometry.

One of the more conspicuous geomorphic characteristics of northern Sierra de Segura is the marked contrast between the steepness of the slopes on the summit

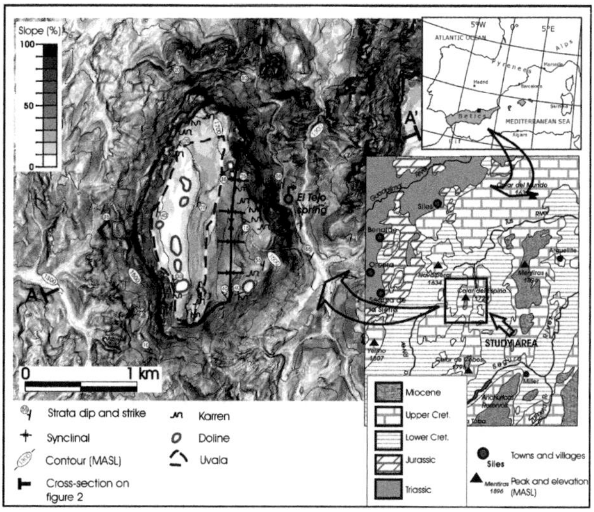


Fig. 1 Location and geomorphologic map of Calar del Espino carbonate aquifer

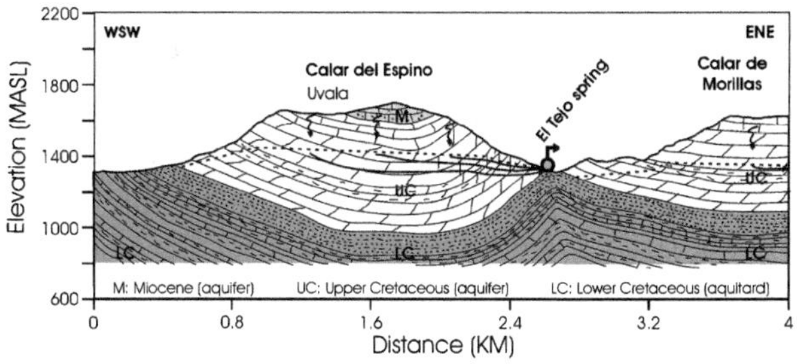


**Fig. 2** Hydrogeological cross-section (location on Fig. 1) of Calar del Espino aquifer

and the hillsides (Fig. 1). The Calar del Espino summit area, located between 1550 and 1720 m asl, has gentle slopes with grades generally lower than 40%. In fact, the summit is a plateau formed on limestones that favour the development of various exokarstic forms such as karren fields, dolines, and a large uvala. In contrast, the hillsides have frequent scarps and slopes, which have grades generally greater than 50%. The summit karstic plateau matches in altitude and morphology with the summits of other mountains in the Sierra de Segura. Therefore, these summits are considered the remnants of a paleosurface that is being dissected by the surface drainage network. The Sierra de Segura paleosurface likely originated close to the Upper Miocene base level, shortly after the region emerged during the Alpine orogeny, followed by a period of uplift to 1500–1900 m asl and fluvial erosion (Moral 2005). North of the Segura River, extensive erosion has produced an excellent example of inverted topography. Here deep fluvial valleys have been excavated along the anticlinal axes and the synclinal cores now form the mountainous regions, on top of which remnants of the Segura paleosurface are preserved.

Lithologic, topographic and climatic conditions of this region have allowed the development of numerous exokarstic forms, especially dolines, on most of the Segura plateau. There are several thousands of dolines in Sierra de Segura, all on the paleosurface, with the largest number in the western region of the plateau, which experiences the highest precipitation.

## 3 Hydrogeology and Monitoring of the Calar del Espino

Carbonate outcrops spread over about 1220 km$^2$ of Sierra de Segura, which is approximately 40% of its total surface area. Approximately one third of these outcrops are on the karstic plateau that represents the main recharge area of aquifers in this region. The degree of erosion of the sedimentary cover and the lithological and structural features allow two hydrogeological sectors separated by the upper Segura River valley to be determined: the "Calares" sector, to the north and west, and the Plateau sector, to the south and east (Moral 2005). In the "Calares" sector, the main aquifer formations (Upper Cretaceous-Miocene carbonates) are deeply dissected by fluvial networks, often to depths greater than the base of these units. In short, these permeable formations have been divided into numerous hydrogeological units that generally hang above the local base level.

The Calar del Espino unit has the appropriate size, geometry and hydric behaviour to be an excellent experimental hydrogeological system. More than 90% of the discharge occurs through the El Tejo spring and recharge occurs by partial infiltration of rainfall and snow melt, therefore it is possible to take accurate measurements of the different components of the water budget by installing a gauging station in El Tejo spring and a weather station on the Calar del Espino summit.

El Tejo spring, located at 1320 m asl, discharges approximately 28 L/s. Its waters have a steady temperature of 10 °C and a low saline content. Water hydrochemical facies vary from calcium-bicarbonate to calcium-magnesium-bicarbonate in type. The groundwater is in equilibrium with an atmosphere that contains about 4470 ppm of $CO_2$ and is close to being saturated with calcite and dolomite (Table 1).

**Table 1** Physical and chemical characteristics of the El Tejo spring water. The $HCO_3^-$, $Ca^{2+}$ and $Mg^{2+}$ contents are given in mg/L

| Date | $Q$ (L/s) | $T$ (°C) | EC (µS/cm) | pH | $HCO_3^-$ | $Ca^{2+}$ | $Mg^{2+}$ | log $P_{CO_2}$ | $SI_{calcite}$ | $SI_{dolom.}$ |
|---|---|---|---|---|---|---|---|---|---|---|
| $n$ | 30 | 32 | 32 | 25 | 30 | 30 | 30 | 24 | 24 | 24 |
| Aver. | 27.9 | 10.0 | 328 | 7.66 | 243.6 | 53.4 | 15.1 | −2.36 | 0.10 | −0.21 |
| Max. | 76.1 | 10.7 | 386 | 8.03 | 282.0 | 68.0 | 18.0 | −1.85 | 0.40 | 0.43 |
| Min. | 15.0 | 9.0 | 266 | 7.15 | 152.0 | 39.0 | 11.0 | −2.72 | −0.39 | −1.17 |
| STD | 15.4 | 0.4 | 25 | 0.24 | 25.4 | 6.4 | 1.7 | 0.24 | 0.24 | 0.46 |
| CV | 55.3 | 3.9 | 7.8 | 3.1 | 10.4 | 12.0 | 11.6 | | | |

Noteworthy changes in the spring's physical-chemical parameters are related to the hydrodynamic state of the spring. It is observed that the following variations take place during periods of high flow (Fig. 3): a) a decrease in mineralisation and in partial pressure of $CO_2$, b) a minor drop in water temperature, and c) a fall in magnesium content.

According to Mangin's method for recession hydrograph analysis (Mangin 1975), the El Tejo spring has a baseflow coefficient that ranges between 0.0009 and 0.0016 per day. The dynamic volume, that is the volume of water in storage in the aquifer above the level of the outflow spring, is about 1.2 $hm^3$.

Mangin (1975) proposed two parameters to classify karst aquifers into five groups, which are the regulation power parameter ($k$) defined as the ratio between the dynamic volume and the spring discharge volume over an annual hydrological cycle, and the parameter $i$ that represents the time lag between infiltration and recharge. In the case of the El Tejo spring, these parameters have a mean value of 1.16 years and 0.43, respectively. Figure 4 shows the hydrodynamic parameters of the El Tejo spring in comparison with other karstic systems of the Pyrenees, Basque Country and Spanish Mediterranean regions (Moral 2005).

In relation to the main springs of Sierra de Segura, the water from the El Tejo spring is a little more saline and more saturated in dolomite. Because the kinetics

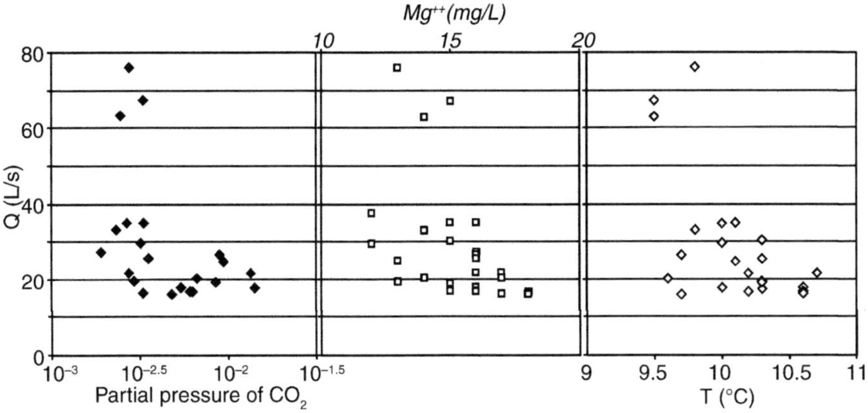


**Fig. 3** The physical-chemical characteristics of water versus discharge flow at the El Tejo spring

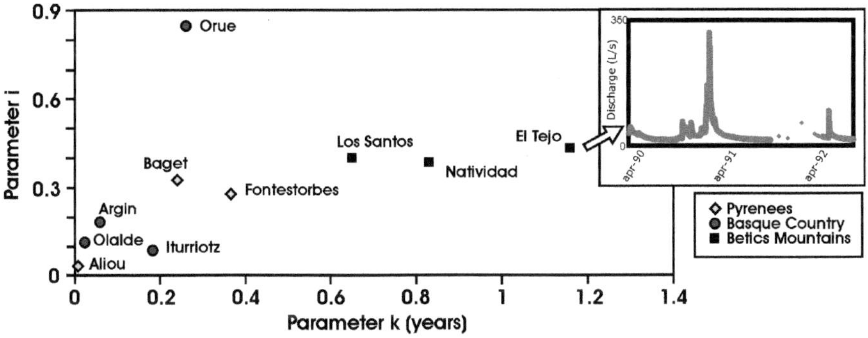


**Fig. 4** El Tejo hydrograph and spring classification according to Mangin's (1975) method for recession hydrograph analysis

of dolomite dissolution are much slower than those of calcite dissolution, one can deduce that water from El Tejo spring is relatively evolved from an hydrochemical point of view, and therefore has a relatively long residence time in the aquifer. In addition, the small fluctuations in the physical-chemical characteristics of water suggest diffuse groundwater flow is more dominant than conduit flow (Shuster and White 1971). Nevertheless, the form of the hydrograph and the physical-chemical changes during periods of high flow, which have to be connected with the arrival at the spring of recently infiltrated water, all suggest the El Tejo system has a karstic conduit network.

The low baseflow recession coefficient and the high regulation power parameter ($k$) of the aquifer also demonstrate the existence of mostly diffuse flow, substantial groundwater reserves, and thus a long residence time for water in this aquifer. A value of parameter $k$ greater than 0.5 would characterise a non-karstic aquifer according to Mangin (1975); nevertheless, other authors have pointed out that there is no theoretical reason for accepting this limit (El-Hakim and Bakalowicz 2007;

Liu et al. 2009). Consequently, Moral (2005) stated that certain factors affecting parameter $k$, such as recharge and length of baseflow recession, are determined by climatic conditions, which would explain higher parameter $k$ values in less rainy Mediterranean regions.

Finally, it is important to emphasise the outstanding development of exokarstic forms on limestone outcrops in the upper part of the Upper Cretaceous-Miocene carbonate beds, whereas these forms are not present on the underlying dolomites, which outcrop at the Calar del Espino hillsides. This apparent dual behaviour of the El Tejo system could be caused by the heterogeneous degree of karstification in the aquifer, greater in the upper limestones, because of the recent fluvial downcutting, followed by the fall of the regional base level and the aquifer water tables. This process subsequently displaced groundwater flow to the lower less karstified zones of the carbonate aquifer.

## 4 Conclusions

The Calar del Espino karstic aquifer has sufficiently well known hydrologic functioning and the appropriate geometric, geological and morphological characteristics to undertake detailed research of its hydrodynamic and hydrochemical behaviours and to accurately measure the major components of its water cycle. In spite of the numerous exokarstic forms on the upper limestones and the existence of a karstic conduit network demonstrated by the hydrodynamic and hydrochemical behaviours of the El Tejo spring, this hydrogeological system has relatively high fissure and intergranular porosity and a long residence time for water.

Making use of more widespread classification methods for karst aquifers based on spring responses, the results seem contradictory at first sight. The implementation of Mangin's hydrograph analysis reveal the non-karstic character of the Calar del Espino aquifer. However, the annual spring discharge volume, used to calculate the parameter k, is dependent on the recharge rate, and therefore on the climatic conditions. Evidently, karstic systems from Mediterranean regions, which are generally less rainy, tend to have higher values of parameter $k$ than the Oceanic climate regions where the classification scheme was originally defined. Improvements in the knowledge and classification methods with respect to karstic aquifers requires increased study of karst spring behaviour (hydrograph, hydrochemical characteristics, water temperature, etc.) in a wider variety of climatic, geologic and geographic regions. More and more new data suggest that karstic aquifers do not necessarily have low storage capacities, short residence times, and highly variable physical-chemical parameters, as has traditionally been believed.

## References

El-Hakim M, Bakalowicz M (2007) Significance and origin of very large regulating power of some aquifers in the Middle East. Implication on karst aquifer classification. J Hydrol 333:329–339

Kiraly L (2002) Karstification and groundwater flow. In: Gabrosek F (ed) Evolution of karst: from prekarst to cessation. Zalozba ZRC, Postojna-Ljubijana

Liu L, Shu L, Chen X, Oromo T (2009) The hydrologic function and behavior of the Houzhai underground river basin, Guizhou Province, southwestern China. Hydrogeol J. doi:10.1007/s10040-009-0518-z

Mangin A (1975) Contribution à l'étude hydrodynamique des aquifères karstiques. Thèse Univ Dijon. 1st part: Ann Spéléol 29-3:283-332. 2nd part: Ann Spéléol 29(4):495–601. 3rd part: Ann Spéléol. 30(1):21–124

Moral F (2005) Contribución al Conocimiento de los Acuíferos Carbonáticos de la Sierra de Segura (Alto Guadalquivir y Alto Segura). Tesis Univ Pablo de Olavide, Sevilla

Shuster DI, White WB (1971) Seasonal fluctuations in the chemistry of limestone springs: a possible means from characterizing carbonate aquifers. J Hydrol 14:93–128

# Hydrochemical Variations of the Huanglong Spring-Fed Travertine-Depositing Stream in the Huanglong Ravine, Sichuan, SW China, a World Natural Heritage Site

H. Wang and Z. Liu

**Abstract** Huanglong Ravine belongs to the transition zone that separate the Qinghai Highland from the Sichuan Basin, Sichuan province, SW China. Travertine deposits over a width of $\sim 250$ m for a length of 3.5 km from some springs uprive along the ravine. Methods of automatic hydrochemical logging and in-situ titrating combined with indoor analysis were used to understand the dynamic hydrochemical variations of the Huanglong Spring-fed travertine-depositing stream. The studies about the hydrochemistry of the spring-fed travertine-depositing stream are mainly focused on the spatial variation, lucubrating the travertine and discussing the characters of spatio-temporal variations and its controlling mechanism on the basis of departed analysis about spatial variations and the influence of human activities on the stream and the travertine-deposition have important meanings not only to the developing mechanism of karst, but also to the protection and exploitation of the world natural heritage site's travertine landscape. It was found that the deposition of travertine was due to huge $CO_2$ degassing from water, which led to the decrease in $pCO_2$ and conductivity, and increase in pH and $SI_C$ from the Huanglong Spring to downstream. However, the downstream hydrochemical evolution was interrupted by the snow-melting water in the ravine through dilution effect, and the downstream spring waters through concentrating effect. On the other hand, the chemistry of the Huanglong Spring was stable on the diurnal scale. However, in the spring-fed pools downstream, $pCO_2$ and EC were lower, and pH and $SI_C$ were higher in daytime than in nighttime, which shows that the deposition of travertine was quicker in daytime than in nighttime. It was found that this was caused by the combined action of higher water temperature and higher aquatic algae photosynthesis at daytime. In addition, it was found that the phosphate concentration of stream water increased remarkably downstream in the tour midseason in the Huanglong Ravine, showing the existence of water pollution by the tourism activities.

H. Wang, Z. Liu
State Key Laboratory of Environmental Geochemistry, Institute of Geochemistry, Chinese Academy of Sciences, 46 Guanshui road, Guiyang 550002, China,
e-mail: whjing2006@hotmail.com

B. Andreo et al. (Eds.), *Advances in Research in Karst Media*
DOI 10.1007/978-3-642-12486-0, 

# 1 Introduction

Travertine is a freshwater carbonate that is deposited in open-air conditions in limestone areas (Ford and Pedley 1996) and the chemical sedimentary deposit forms from spring-fed stream waters rich in dissolved carbon dioxide and calcium bicarbonate. So far, very little is known about the temporal variations of the hydrochemistry and the travertine deposition. Moreover, the controlling mechanisms behind are also inexplicit. More practically, is there any contamination caused by tourism activity and how to distinguish the influence of the climate and the human activities on the travertine deposition?

This study is significant for two major reasons. Firstly, it is the first major attempt in the world heritage site to reveal diurnal variations of hydrochemistry in a travertine-depositing stream under different biological conditions. Secondly, it extends the understanding of contemporary hydrochemical processes in travertine-depositing streams, and thus provides more information for sampling design and interpretation of travertine paleoenvironment.

# 2 General Settings of the Study Area

Huanglong Ravine is about 360 km NW of the province capital, Chengdu, Sichuan Province, on the southern slopes of the Minshan Mountains that separate the Qinghai Highlands from the Sichuan Basin. Its elevation ranges 3100 to 3600 m above sea level. The geology consists of Paleozoic carbonate rocks exceeding 4000 m in thickness, overlain by about 1000 m of Mesozoic clastic rocks plus Cenozoic alluvial gravels, glacial moraines and travertine (Liu et al. 1995). In Huanglong Ravine, the travertine deposits are built up over a width of $\sim 250$ m for a length of 3.5 km (Fig. 1), in the form of flowstones with many rimstone dams over 1 m in height

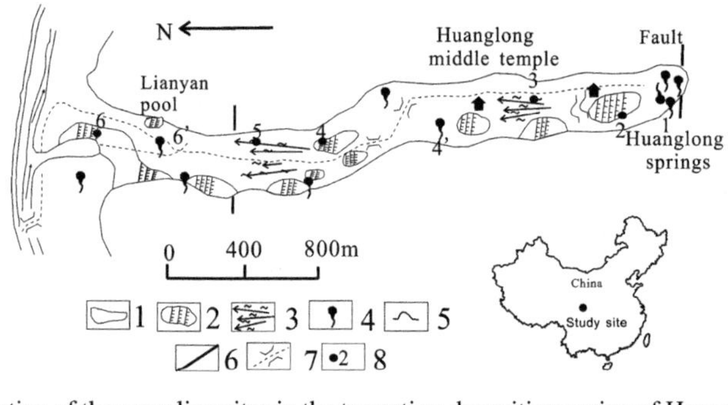


**Fig. 1** Distribution of the sampling sites in the travertine-depositing ravine of Huanglong, Sichuan, SW China. Note: 1. boundary of the travertine scenery; 2. rimstone dam and pool; 3. stream-bed; 4. spring; 5. travertine cave; 6. fault; 7. trail; 8. sampling site

impounding the surface stream. The surface of travertine deposits is covered with bryophytes and cyanobacteria of various colors.

The average annual precipitation is 759 mm and the mean temperature 1.1°C. Groundwater issues along a fault zone at an altitude of about 3580 m asl, with a water temperature about 5°C higher than that of annual mean air temperature. The surface stream is a mixture of this groundwater with glacier and/or snowmelt water from higher mountains (the highest peak Xuebaoding: 5588 m asl).

## 3 Methods

To understand the general chemistry of water, the water samples were collected by syringes with 0.45 μm Minisart filter and analyzed in the State Key Laboratory of Environmental Geochemistry, Institute of Geochemistry of Chinese Academy of Sciences. Concentrations of $K^+$, $Na^+$ and $Mg^{2+}$ were determined by atomic absorption spectrometry, those of $Cl^-$, $PO_4^{3-}$ and $SO_4^{2-}$ by ion chromatography.

Water temperature, pH and specific conductivity of water at each sampling site were measured in-situ with hand-held water quality data logger (WTW 350i), with resolutions of 0.01 pH, 0.1°C and 1 μs/cm, respectively. A Greenspan CTDP 300 multi-channel data logger (Liu et al. 2007) was used in Huanglong Spring. Rainfall, water stage, water temperature, pH and specific conductivity were monitored every 15 min, with resolutions of 0.5 mm, 0.001 m, 0.1°C, 0.01 pH, and 0.01 μs/cm, respectively. In situ titrating was used to measure the $[Ca^{2+}]$ and $[HCO_3^-]$ of water semimonthly with Aquamerck Alkalinity Test and Hardness Test.

## 4 Results and Explanations

### *4.1 General Hydrochemical Compositions of the Huanglong Spring*

The general hydrochemical compositions of the Huanglong Spring (sampling site No.1, Fig. 1) shows that $Ca^{2+}$ and $Mg^{2+}$ are the major cation in the spring water, with molarity percentage of $>95\%$, and $HCO_3^-$ is the major anion in the spring water, with molarity percentage of $>90\%$. Thus, the hydrochemical type of spring water is $HCO_3$–Ca, Mg, which reflects the control of the Carboniferous to Permian limestone bedrock at the site.

The spring water has very high concentration of calcium and bicarbonate. They are more than 240 and 750 mg/L, respectively, and the corresponding $CO_2$ partial pressure is more than 12,500 Pa. Meanwhile, the spring is near equilibrium with respect to calcite, and even slightly undersaturated with respect to calcite, so there is scarcely any calcite deposition nearby.

## 4.2 Temporal Hydrochemical Variations of the Huanglong Spring

There is neither clear seasonal nor diurnal variation in hydrochemistry of the Huanglong Spring. This indicates high regulation capacity of the system on the hydrochemistry. On the other hand, it may also reflect the features of endogenic karst system where all the hydrochemical indexes are relative stable and less influenced by the climate (Liu et al. 2003). This situation contrasts with the epikarst springs whose hydrochemistry shows both clear seasonal and diurnal variations, which are determined by the climate change.

## 4.3 Spatial and Diurnal Variations of the Hydrochemistry of the Travertine-Depositing Stream

After about 3.5 km downstream, the $CO_2$ partial pressure of water decreased from 14,000 Pa to 100 Pa, pH increased from 6.3 to 8.4, and the calcite saturation index increased from near zero in the spring to more than 1.0. Correspondingly, calcium carbonate, the travertine's major component, was deposited. With the deposition of travertine along the streambed, the concentration of $Ca^{2+}$ and $HCO_3^-$ of water decreased from 240 and 750 mg/L to 100 and 300 mg/l, respectively.

However, the general patterns of the spatial hydrochemical variations were interrupted by the occurrence of the downstream springs Nos. 4′ and 6′ along the stream. For example, the increase of the concentration of the $Ca^{2+}$ and $HCO_3^-$ at the downstream sample sites 4 and 6 was due to the input of the spring water into the stream. However, the abnormal decrease of the concentration of $HCO_3^-$ and $Ca^{2+}$ at site No. 3 was not the result of the travertine deposition but due to the dilution effect of a surface stream fed by the snow-melting water (Liu et al. 1995), which was evidenced by the decrease of $SI_C$ at site No. 3.

After the spring water emerged, it was changed from a closed system to an open system. Thus, its hydrochemistry experienced not only spatial change, but also remarkable temporal change. All parameters such as temperature, $pCO_2$, pH, $SI_C$ and conductivity at Lianyan pool show distinct diurnal variations, i.e., $pCO_2$ and EC (specific conductivity) were lower and pH and $SI_C$ were higher in daytime than at nighttime. The pH increase in pools may be caused mainly by photosynthesis of algae, phytoplankton and aquatic plants in the pools.

To know the quantitative influence of water temperature and aquatic plants on the diurnal variations of hydrochemistry, the influence of water temperature on Henry constant by calculation (Lu et al. 2006) in Lianyan pool were estimated: the percentage of the influence of water temperature in total influence of the water temperature and aquatic plants on $pCO_2$ diurnal variations was only 19%; the influence of photosynthesis of aquatic plants was 81%.

### 4.4 Spatial Variations of Concentrations of Phosphate in Tourism Off-Season and Midseason: Influence of Human Activity

The concentration of phosphate had little variations along the ravine in the tourism off-season, with values close to the backgrounds (spring No. 1 and surface water) and even decreased at site No. 2 possibly due to the utilization of phosphate by aquatic plants. However, the concentration of phosphate increased abruptly from sampling site No. 4 downstream in the tourism midseason (e.g., September 2007), indicating the existence of the pollution. It was thus inferred that the entrance of the phosphate pollution to the stream was between sites No. 3 and No. 4. Based on the field survey, it was found that it is very likely the tourism restaurant (with toilets behind) between sites No. 3 and No. 4 contributed to the phosphate pollution.

According to Drysdale et al. (2002), the deposition of calcite would be restrained if the concentration of phosphate is more than 1 μmol/L (about 0.095 mg/L). So the decrease in travertine deposition in the middle-to-down stream that happened recently (Guo et al. 2002) may be related to the increase of the phosphate concentration caused by increased tourism activities, especially since the 1990s when Huanglong was listed by UNESCO as a world natural heritage site in 1992.

## 5 Conclusions

By examining the spatial and temporal variations of the hydrochemistry in a travertine-depositing stream at Huanglong Ravine, Sichuan, SW China, it was found that the deposition of travertine was due to huge $CO_2$ degassing from water, which led to the decrease in $pCO_2$ and electrical conductivity, and increase in pH and $SI_C$ from the Huanglong Spring to downstream. However, the dilution and concentration effects caused by the snow-melting water and the springs, respectively, in the Huanglong Ravine interrupted the regular spatial variations of hydrochemistry of the stream. On the other hand, the chemistry of the Huanglong Spring was stable on the diurnal scale. However, in the spring-fed pools downstream, $pCO_2$ and EC (electrical conductivity) were lower, and pH and $SI_C$ were higher in daytime than at nighttime, which shows that the deposition of travertine was quicker in daytime than in nighttime. It was found that this was caused by the combined action of higher water temperature and higher aquatic algae photosynthesis during daytime.

In addition, it was found that the phosphate concentration of stream water increased remarkably downstream in the tourism midseason in the Huanglong Ravine, indicating the existence of water pollution by the tourism activities. The increase of phosphate concentration may be one of the reasons for the decrease in travertine deposition rates during the past decades, which needs to be taken care of and given more comprehensive study in the future for the protection of the world-famous travertine landscape.

## References


Drysdale RN, Taylor MP, Ihlenfeld C (2002) Factor controlling the chemical evolution of travertine-depositing rivers of the Barkely karst, north Australia. Hydrol Proc 16:2941–2962

Ford TD, Pedley H (1996) A review of tufa and travertine deposits of the world. Earth-Science Reviews 41:117–175

Guo J, Peng D, Yang J (2002) Research of water circulation and the travertine formation causes of Huanglong in Songpan county. Sichuan geological paper 22:21–26

Liu Z, Dreybrodt W (2007) The Kinetics of Karst Processes and Environment. Beijing, China

Liu Z, Svensson U, Dreybrodt W et al (1995) Hydrodynamic control of inorganic calcite precipitation in Huanglong Ravine, China: Field measurements and theoretical prediction of deposition rates. Geochimica et Cosmochimica Acta 59:3087–3097

Liu Z, Yuan D, He S et al (2003) Origin and forming mechanisms of travertine at Huanglong Ravine of Sichuan. Geochimica 32:1–10

Lu B, Liu Z, Liao C et al (2006) The influence of aquatic plants on diurnal variations of hydrochemistry in karst system – A case in the Guilin Karst Experimental Site. Carsol Sinica 25:335–340

# The Role of Sculpted Forms Along Endokarstic Active Conduits in the Development of Fluviokarstic Canyons. The Rio Puron Cave Conduit (Spain)

J.A. Ortega Becerril, G. Garzón Heydt, and J.J. Durán

**Abstract** The Purón River valley shows a special configuration due to convergence of surficial and groundwater processes in a phreatic conduit. Typical karst processes of dissolution–precipitation have been replaced by fluvial activity, where mechanical water erosion morphologies are dominant. All this seems to relate to a stage of abandonment of the underground environment in favour of the surficial domain. River incision in rock canyons is accomplished in many cases by phreatic conduits evolution, its collapse and opening to the surface. During the initial stages of this process, river piracy and superficial changes in the drainage network occur by capturing in first place the groundwater system. In this context of fluvial activity, numerous sculpted forms, such as potholes, are found in the conduit which is the result of changes and evolution in the cave into a more evolved morphology, pointing to river stream power concentration along certain conduit sections.

## 1 Introduction

Bedrock rivers are actually one of the most outstanding items in fluvial geomorphology and specially canyon incision and genesis is being discussed from a scientific standpoint in the framework of recent quantitative data (Wohl and Tinkler 1998). In this sense, previous theories, as proposed by several authors for development by entrenchment of pre-existing river networks like superimposition, antecedence, or stream piracy, need to be analysed in the context of river incision capability according to stream power, climatic, lithological and tectonic constraints. One remarkable issue is gorge incision from karstic conduits exposed at the surface by opening, collapse and exhumation (fluviokarstic model). These theories are based on the inves-

J.A. Ortega Becerril, G. Garzón Heydt
Facultad de Geología, Depto. Geodinámica, Universidad Complutense de Madrid, Avda. Jose Antonio Novais, 2, 28040 Madrid, Spain, e-mail: jaortega@geo.ucm.es

J.J. Durán
Instituto Geológico y Minero de España (IGME), C/ Ríos Rosas, 23, 28003 Madrid, Spain

B. Andreo et al. (Eds.), *Advances in Research in Karst Media* 
DOI 10.1007/978-3-642-12486-0, © Springer 2010

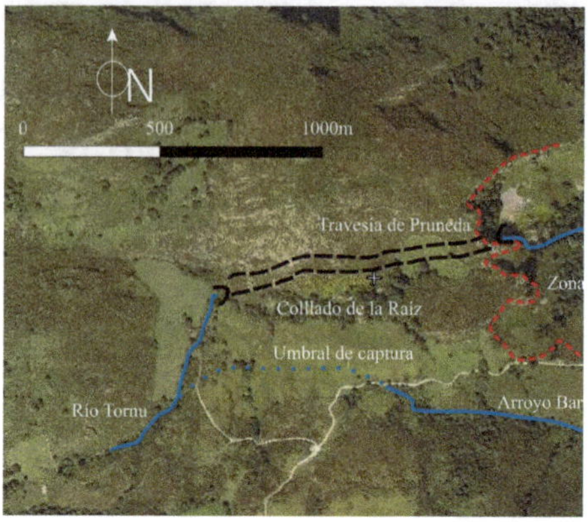


**Fig. 1** Orthophotography showing the piracy through the active endokarstic conduit of the Pruneda cave (*Red dashed line* indicates backwards incision, *black dashed line* represents the endokarstic conduit)

tigation of limestone canyons where karst processes are evident, for example by the presence of water sources or its disappearance in sinks. The occurrence of several morphological features present in some river canyons might lead to the interpretation of their underground origin. Endokarstic conduits play in this sense an interesting role; as they offer a tangible example of how common cave processes, as the opening of a conduit by solution or the formation of speleothems, suffer the transformation from being the dominant elements to play a secondary role. The supergenic river transforms the conduit environment, re-excavating the channel, creating inner and secondary channels, eroding the stream bed and precipitation features, and introducing new elements more appropriate to surficial environments such as sculpted bedforms, potholes and grooves until they define a step and pool morphology.

The Puron River (Cantabrian Mountains, Northern Iberia) shows an endokarstic conduit with well-preserved morphologies in which the dominance of fluvial erosion over the precipitation or dissolution features is apparent. The short conduit, about one kilometer long, crosses the obstacle of a limestone ridge in the way of the river flow into sea direction (Fig. 1). Inside the karstic passage, there is a good demonstration of fluviatile morphologies such as different types of potholes, scallops, longitudinal and transverse grooves and inner channels. However, there are still remains of the former karstic environments, with varied speleothems and more than one conduit levels. Morphologically, it seems to represent the capture of an underground drainage network that results in the surficial piracy of the Purón River with transformation of the fluvial network. This represents a good example of the complex evolution of a mixed-type canyon (underground and surficial system) in a karst terrain.

## 2 Fluviokarstic Morphology and River Evolution

The development of gorges in soluble rocks, mainly carbonates, is easily explained due to preferential incision by corrasion and fluvial mechanical action. Valley entrenchment proceeds to capturing cave systems that lose their activity as groundwater conduits evolve into epiphreatic conduits or even loose their connection with the groundwater system. These cavities sink by collapse generating incised, but open, stream channels with large blocks in the riverbed. The transition between river areas with karstic or with fluvial domain has been referred by Phillips et al. (2004) showing that, at regional scale, the presence of either one or the other type do not necessarily depend on factors such as lithology, structure or topography. Instead, changes in the slope are responsible for triggering the predominance of one process over the others.

In the Purón River case, head tributaries (Tornu and Bardales Streams) follow hercinic quartzite outcrops, running over them as they prevent water infiltration. Once a soluble carbonatic level, streams infiltrate into the ground developing the Pruneda cave. More active backwards erosion involves the genesis of a rough relief with larger gradient, steeper walls, more gravitational slope activity and sometimes, the developing of canyons or gorges. In this case, the Tornu Stream, tributary of the Purón River, with a better incised, steeper, narrower and youthful channel, from the Pruneda spring on, has captured the Arroyo Bardales basin, that shows a wider and deeper valley with a clearly more mature and evolved morphology (Fig. 2). Capture is evident from the map (Fig. 1), featuring capture elbows and linear valley morphology. In this respect, the natural prolongation of the Bardales Valley at its head area is the Tornu River Valley, which follows the same direction. The Tornu River infiltrates at a blind valley that instead of flowing into its natural outlet, Arroyo Bardales, is captured along an underground reach and leaves through the Pruneda spring to follow the Purón River Valley. One of the genetic assumptions about river piracy in permeable zones is that stream captures do not occur on the superficial drainage network, but essentially by the groundwater system. The active Pruneda conduit in the Purón River might be an extreme example of a kárstic favoured river capture at its initial phase, in which the ancestral Bardales Stream pours its flow into the Tornu River, more active in its incision. The valley entrenchment and opening is observed in the spring environs where slope processes are very active (debris talus and fallen blocks in the riverbed) and the channel mor-

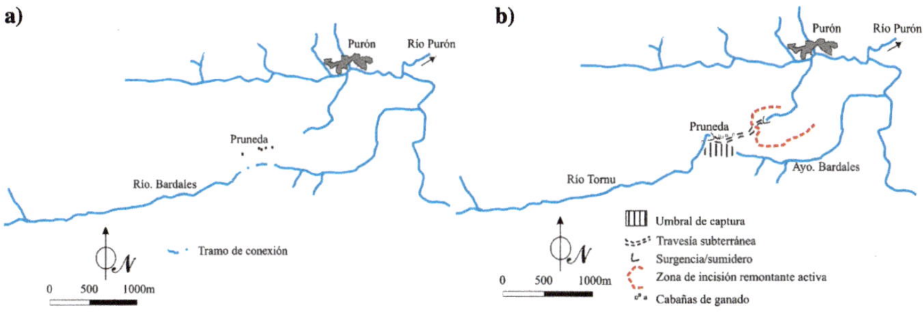


**Fig. 2** Evolution of stream piracy through the endokarstic conduit. **a** Headward erosion that captures the upper stream and produces a shift in the river network to catch the groundwater of the conduit. **b** Situation prior to piracy, with the stream flowing in W–E direction according to preferential alignments defined by Ordovician materials

phology is disorganized and incipient (several knickpoints in the riverbed, no channel incision, sudden changes in river course direction and control by stratification). Downstream, the stream shows a better evolved morphology, developing an incised bedrock channel reshaped from previous potholes that collapsed forming a defined internal channel.

## 3 Sculpted Forms in Endokarstic Conduit

Ford (1965) showed the occurrence of fluvial origin erosive forms linked to the development of an endokarstic conduit. The potholes implied streambed degradation on the cavity, showing different erosion levels. Ford (1965) suggested that a possible cause for these different levels could be stream piracy, but he considered this cause as unlikely, taking in account several changes in underground flow, and finally assigns a climatic origin to explain the situation. More recent papers (Springer et al. 2006) justify streambed paleolevels emergence as result of degradation of the previous ones and the development of an inner channel. In fact, further research work by Handcock et al. (1998) concludes that although sculpted forms may be important components of channel erosion, that is a stochastic phenomena whereby sculpted forms appear and disappear throughout the channel in such a way that the net effect results in bed lowering.

Inside the Purón conduit, there are reaches with different morphologies (Fig. 3). The conduit follows preferential weakness paths, such as bedding planes, giving sharp turns in its course due to jointing interference. In straight sections the bed shows a step-pool morphology with the presence of numerous potholes of varying size and morphology. Large open potholes with cylindrical morphology are predominant, like the ones described by Lorenc et al. (1994); Springer et al. (2006). More evolved typologies are also found, such as breached and lateral open potholes (Fig. 4), which imply an active bedrock entrenchment with an inner channel formation. In this sense, Ford (1965) states that the breaching processes in potholes are in opposition to the formation of an inner channel, but recent investigations relate the origin of inner channels with pothole development followed by breaching (Wohl 1998). The chained potholes reach represents a change of local slope although the slope of each of them is low. This high slope reach inside the conduit involves the concentration of stream power and energy required to erode the streambed. The articulation on this section is carried out by a couple of knickpoints, leaving a more active segment (energy concentration) between two waterfalls. These forms erode the conduit bedrock and have excavated an internal channel that collects ordinary flow. On top of it, appears a channel expansion able to contain floodwaters and showing erosion forms in speleothems, flooding marks and deposits as well as rock erosional features such as scallops. Depending on the reaches, it might work like a closed conduit with siphon activity. The interference level between floods and pre-existing karstic features is visible in some lateral flowstones, which show scallops marks, retouching the feature and leaving deep marks and polished surfaces

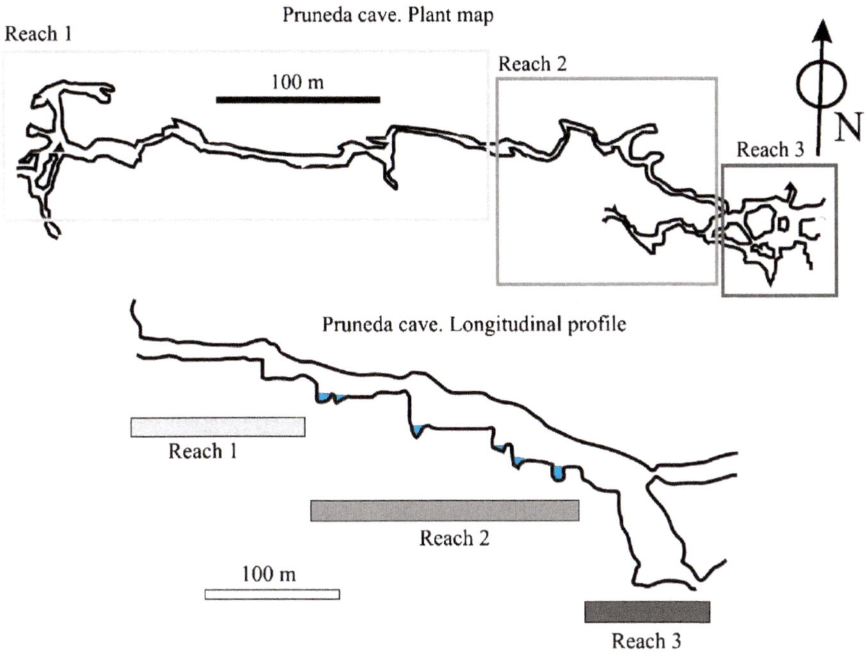


**Fig. 3** Ground map of Pruneda cave endokarstic conduit with the three reaches showing their morphological differences (*upper sketch*), and longitudinal profile of the channel and the three analyzed reaches (*lower sketch*)

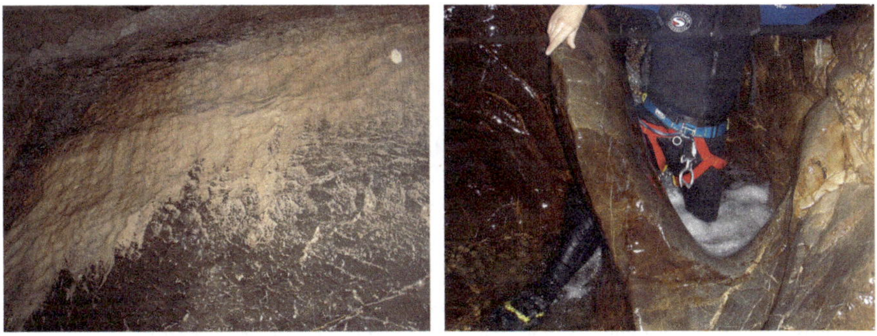

**Fig. 4** Flood eroded speleothems showing scallops over flowstone inside the endokarstic conduit (*left*). Pothole breaching forming a pillar in reach 2 (*right*)

on the limestone (Fig. 4). The underground drainage network shows some erosional and evolutive features such as the anastomosing conduits described by Phillips et al. (2004), or vertical incision forms as the ones described by Jaillet et al. (2004). Both of them are considered as features that show the dominance of fluviatile over karstic processes.

# 4 Conclusions

Relationship between karstic underground or surficial processes, mainly fluvial and gravitational ones, is not always easy to be established. The profuse pothole presence indicates a continuing work of fluvial action, and their evolved typologies and features imply the predominance of fluvial mechanical abrasion over karstic denudation. The underground passage shows areas of more active incision like potholes and sections separated by abrupt turning elbows similar to open gorge features occurring in bedrock rivers. The presence of anastomosing riverbed patterns, large precipitation features eroded and reshaped by water action imply the pure karstic features abandonment in favour of the fluvial processes. The gorge opening in this area is subject to the occurrence of previous phenomena such as the diversion of the underground drainage network and changes on the superficial, network due to river piracy. The results are anomalous surficial landforms such as broad misfit valleys with a small watershed, abrupt drainage pattern orientation changes, wind gaps or piracy thresholds and very active backwards erosive streams. Ongoing research and intended future work would require the dating of speleothems which interact in some pothole paleolevels to estimate the time of abandonment, and to relate it with other morphological works developed on the study area, especially with coastal changes.

**Acknowledgements** This work was funded by project N° BSCH-UCM. Gr58/08

# References

Ford DC (1965) Stream potholes as indicators of erosion phases in limestone caves. Bull. of the Nat. Spe. Soc., vol 27, N° 1, 27–32.

IGME 1981. Mapa Geológico de España, E 1: 50.000 (hoja N° 32). Llanes.

Jaillet S, Pons-Branchu E, Brulhet J, Hamelin B (2004) Karstification as geomorphological evidence of river incision: the karst of Cousance and the Marne valley (eastern Paris Basin). Terra Nova, 16:167–172.

Lorenc MW, Muñoz Barco P, Saavedra J (1994) The evolution of potholes in granite bedrock, W Spain. Catena, 22:265–274.

Phillips JD, Martin LL, Nordberg VG, Andrews WA (2004) Divergent evolution in fluviokarst landscapes of central Kentucky. Earth Surf. Proc. and Land., vol.29(7):799–819.

Springer GS, Tooth S, Wohl EE (2006) Theoretical modelling of stream potholes based upon empirical observations from the Orange River, Republic of South Africa. Geomorphology, 82:160–176.

Tinkler KJ, Wohl EE (eds) (1998) Rivers over rock: fluvial processes in bedrock channels, Am. Geophys. Union, Geophysical Monograph 107: Washington; 1–18.

Wohl EE (1998) Bedrock channel morphology in relation to erosional processes. In: Tinkler KJ, Wohl EE (eds). Rivers over rock: fluvial processes in bedrock channels, Am. Geophys. Union Geophysical Monograph 107: Washington; 133–151.

# Karst Geosites in NE Italy

F. Cucchi, F. Finocchiaro, and L. Zini

**Abstract** Karsts areas are "landscapes of special geological and geomorphologic interest which call for conservation" and can therefore be defined as potential geosites according to the definition given by Cleal et al. [2]. This communication briefly describes the geological, geomorphologic and hydrogeologic features of the most important areas in North-Eastern Italy where highly karstified – both on the surface and underground – carbonate rocks outcrop. Some of them perfectly exemplify geological and geomorphologic events and their own geological history, thus serving as cultural models not only at a local but also at a global level [1]. Typical examples of karst geosites are the Alpine Karst environment of Mt Canin and the Classical Karst near Trieste, as well as the Cansiglio – Cavallo Plateau, the limestone turbidites of Julian Prealps Flysch, the glaciokarst valleys of the Carnic Alps and the karren of the Devonian limestone.

## 1 Introduction

Friuli Venezia Giulia is extraordinarily rich in intensely karstified areas, some of which – Trieste Karst and Mt. Canin – enjoy worldwide recognition for their epigean and hypogean forms. The region has a total area of 7850 $km^2$. Carbonate rocks crop out over approximately 1900 $km^2$, affecting around 5000 $km^2$ of mountains and hills in the region (Fig. 1). More than 7000 caves have been discovered and listed in the regional cave registry so far. This means that their average density amounts to more than 3 caves per karstifiable $km^2$. In some areas this density reaches peak values of 70 caves/$km^2$, for instance close to Fernetti, in the Karst and of 264 caves/$km^2$ on the plateau of Col delle Erbe, in Mt. Canin Massif. Equally widespread are epigean

F. Cucchi, F. Finocchiaro, L. Zini
Dipartimento di Scienze Geologiche, Ambientali e Marine, University of Trieste, Via E. Weiss 2, Trieste, Italy, e-mail: cucchi@units.it;finofu@units.it;zini@units.it

B. Andreo et al. (Eds.), *Advances in Research in Karst Media* 393
DOI 10.1007/978-3-642-12486-0, © Springer 2010

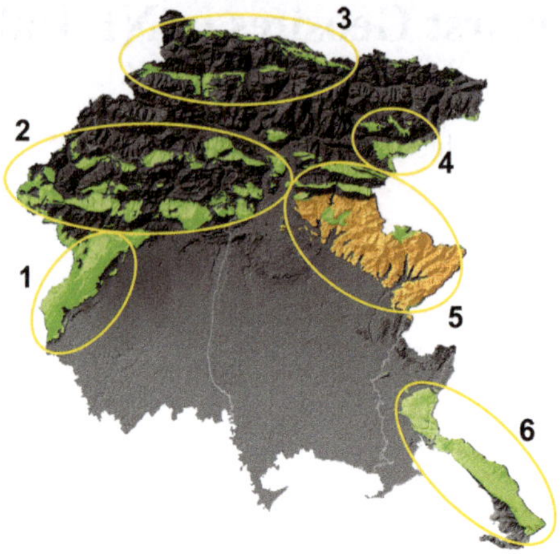


**Fig. 1** Main karst areas in Friuli Venezia Giulia Region. *Green*: limestone; *orange*: calcarenite and siltite; 1: Cansiglio-Cavallo; 2: Carnic Prealps; 3: Carnic Alps; 4: Julian Alps; 5: Julian Prealps; 6: Classical Karst

karstic forms such as dolines, karrens and grize, which are often of remarkable size and typology (Cucchi 2002).

## 2 Geological and Geomorphological Characteristics

The rocks that crop out in the region are almost totally sedimentary and belong to a stratigraphic succession, which ranges in age from the Silurian period up to the present time. This sequence of formations – ideally placed one on top of the other – forms a stratigraphic column with a thickness of over 15,000 m. Terrigenous rocks as sandstone, argillites, siltites, conglomerates and breccias and carbonate rocks as limestone and dolostone, prevail. Subordinate, even though quite widespread in limited areas, are evaporitic rocks such as gypsum, vacuos dolomite breccias. The distribution pattern of these rocks follows a West-East elongated strip; the age of the rocks changes from the most ancient to the most recent while progressing from North to South. This distribution is due to the geodynamic structure of the area, characterised by overthrusts and imbricated plates.

The Palaeozoic Carnic Range is the hallmark of the northernmost area of the region and extends along a belt between 10 and 20 km wide. The stratigraphic scheme includes pre-Ercinian sequences, which comprise alternations of continental deposits, delta plain deposits and shallow seawater, karst has not developed. The sequences include stratified limestone with Orthoceras which date back to the Silurian period and coralliferous cliff limestone from the Devonian period. This limestone, which can reach thicknesses of hundreds of meters, is often intensely karstified.

The succession of limestone, dolostone, siltites, sandstone from the Lower and Middle Triassic does not present significant karstic phenomena. It is in the Upper Triassic period that important epigean and hypogean karstic phenomena start to develop, with the Dolomia Principale Fm. (Noric), composed of massive dolostone and stratified dolomitic limestone, and with Megalodon gray limestone of Dachstein's limestone Fm. (Retic), fairly stratified and reaching thicknesses of up to 800 m. The platform sequence continues with the well-stratified, and sometimes oolitic, grey limestone of Calcari grigi del Friuli (Lias), with oolitic and crinoidal flint-rich limestone (Vajont Limestone and Cellina Limestone Fm.), and the strongly stratified limestone of the Soccher Limestone Fm. (Dogger–Malm), where karstic phenomena are marked and widespread. Overall thickness exceeds 500 m. The most karstifiable and karstified formations in Friuli Venezia Giulia – with intense and diffuse karst – date back to the Cretaceous period. Examples include Aurisina Limestone Fm., with rudist limestones and massive biogenic, typical slope and reef deposits with thicknesses ranging in the Classical Karst sector up to approximately one thousand meters.

Intensely karstified are also Cenozoic limestones, such as the Liburnic Group limestone and the nummulitic and alveolinid limestone, with fascinating caves opening in the whole karst area. Equally interesting for karst are the limestone, sandstone and conglomerate interbeddings that can be found in complex successions of Paleocene and Eocene Flysch, mainly siliciclastic.

## 3 Main Karst Areas

The Cansiglio – Cavallo Massif rises with steep slopes from the western Friuli plain to reach the height of the plateau over 1000 m (Fig. 1). Karst phenomena are present throughout the massif, though they create different forms depending on the lithology, the geology and the structure of the area: the characteristics of the karstic morphologies that can be found on the plateaus are halfway between low-height and Alpine karst phenomena. Forms are often mixed and also show traces of glacial activity. The area features two large poljes and numerous limestone outcrops with different karstification levels and many underground and surface karst forms. As for the latter, dolines prevail: these are located at the borders of the eastern side of the plateau and are symmetrical, close to each other and more then 20 m deep, with slopes sometimes characterised by wonderful karren. A total of 250 caves can be found in this area, the most important of which are the Bus de la Genziana cave (587 m deep) and the Col de la Rizza abyss, (800 m deep and 3.5 km of overall extension). Also the springs of River Livenza, with water discharges exceeding 16 $m^3 s^{-1}$, are of karstic origin (Gorgazzo, Santissima, Molinetto and other minor springs).

The dominant tectonic style of *Carnic Prealps* is characterized by imbricated thrusts: karst phenomena have given origin to both surface forms, which often give the landscape its typical undulated trend with frequent dolines, and to large cave systems, which drain underground water towards water-rich springs. The number of known caves is 650. The central and southern area (approximately 100 $km^2$) is one of the regions with the highest density of karst phenomena. From a geological point of view, this area mainly comprises carbonate rocks among which Cretaceous rocks can be found. These rocks are extremely pure limestone and cover approximately half of the area under examination. These lithologies are characterised by intense karstification, also caused by a favourable tectonic situation, and create stunning karst forms (Fig. 1).

Highly karstified is also the area near Gerchia's plateau, located at elevation between 500 and 700 m a.s.l. The area is rich in frequently elongated dolines with coalescing rims and karren. Quite interesting from a morphological point of view is the deep karstic gully that crosses the plateau, approximately 1 km long, which runs within the walls of Torrent Cosa, 250–300 m high. Caves (many of which are springs or sinkholes) amount to more than one hundred. Among them, the La Val – Noglar system ought to be mentioned, with a development of almost 7 km.

Carnic Alps, which include 332 known caves, feature two intensely karstified areas where Devonic limestone crops out with karst landforms, which proves the interaction among lithology, tectonics and glacial activity. The top of Mt. Pal Piccolo is a plateau, which features the greatest number of karst landforms of the whole area. Here, it is possible to see rather frequent small closed valleys, dolines, swallow holes and karren. This 1-km$^2$ area features 51 caves (Fig. 1).

The Julian Alps, where thick highly karstifiable Mesozoic limestone successions crop out, host wide outstanding karst areas, as Mt. Canin. Here, can be found all the karst landforms typical of alpine mountain environments, which are often matched by impressive hypogean systems: there are 25 caves exceeding 500 m of development and 300 m of difference in altitude. The deepest abysses have recently been discovered in the Slovene area: five of them exceed 1000 m and one has the longest vertical development of the world, featuring a single shaft of 643 m. The whole Canin range is one of the most stunning examples of Alpine Karst in Italy and in Europe, both for the variety of and the richness in epigean forms, and for the morphology and dimensions of the caves. The 180 km$^2$ of the karst area between Italy and Slovenia are theatre to thousands of caves (2031 in Italy only): these numbers are quite significant also if compared to other karst areas in the world (Fig. 1). The Dachstein limestone characterises almost the whole massif and features the most impressive deep karst phenomena. This limestone is approximately 700–800 m thick and can be further divided into a more decidedly calcareous association (the upper part of the formation) and a dolomite-limestone one (lower part), which causes a diversification in the hypogean forms. The most frequent landscape is characterised by mountain rocks exposed to intense weathering: furrows prevail along the steepest side of bedding planes, where banks are thicker and less jointed; crevasses abound where rock jointing has enabled karst to penetrate deeper. Kamenitzas and karren give the exposed surfaces a suggestive character. Among epigean macroforms, the most characterising morphologies are snow shafts and dolines, often with very steep or even vertical walls, where collapse or typical dissolution dolines.

Deep karst is characterised by caves with a prevailing vertical development. Noteworthy among them are the Col delle Erbe complex (7 entrances, a development of over 36.8 km, tens of underground shafts that reach up to a $-935$ m depth) and the Foran del Mus Complex (24 entrances, a development of 15 km and $-1110$ m). At present, six are the caves that are known to exceed a depth of 1000 m. The caves of Mt. Canin show the classical morphologies of high-mountain abysses:

vadose forms (waterfall-shafts and structural gullies) that intercept sub-horizontal active or no longer active systems (phreatic tunnels with circular or ellipsoidal section). Here, rarer – but often massive – morphologies are collapse caves. Frequent are also the "hints" of recent tectonic movements (the area is highly seismic). As far as hydrology around Mt. Canin is concerned, between Italy and Slovenia there are nine perennial springs and a series of associated temporary springs with an overflow function. Some of these are exploited for drinking and for hydroelectric purposes and are valuable water resources for the whole area.

Julian Prealps are composed of Mesozoic platform, slope and basin carbonate rocks and a large and thick Maastrichtian-Eocene turbiditic succession (4000 m thick). Besides the typical karst landscape – characterised by the lack of a surface hydrographic network, by doline-rich plateaus, large limestone pavements and deep abysses – hills rise crossed by numerous torrents and streams, where the vegetation is luxuriant and hypogean complexes with a great morphological variety can be found. The largest cave in this area is Grotta Nuova di Villanova, with a total plan of 8020 m (Fig. 1).

The Classical Karst is the karstic area par excellence, where highly karstifiable limestone crops out, giving rise to the greatest possible variety of epigean and hypogean karst forms. This characteristic has turned the area into the worldwide symbol of karst phenomena. Timavo springs, already sung by the Latins, stand for violent discharge of underground water, which are in turn fed by the underground River Timavo (Reka). This river is swallowed into Skocjanske Jame (Slovenia) swallow hole and after more than 40 km of unknown underground path, reappears in S. Giovanni di Duino springs (Fig. 1).

The Classical Karst is a large morphokarstic unit, which extends to the South-East of the River Isonzo as far as Postojna (Slovenia). It is a quasi-rectangular plateau with a karstified thickness of at least 500/600 m. The plateau has an area of approximately 600 $km^2$, it is 15 km wide and has the longest side trending SE-NW for some 40 km. From a geological point of view, it belongs to the "Karst-Friuli carbonate platform", a northern offshoot of the "Adria Plate". The platform consists in a thick sequence of prevailingly carbonate rocks, dating back to the Triassic at the bottom and to the Eocene period at the top of the succession, covered by a turbiditic siliciclastic sequence also known as Flysch Fm.

Classical Karst is the product of relatively mature karst phenomena, which have now been evolving for almost 10 million years: initial surface forms are now almost unrecognizable. Caves have preserved rare primary morphologies, altered by filling deposits, collapses, extremely varied speleothems, which, in turn, hide vadose entrenchments due to variations of base levels and adaptation to tectonic movements. The spring system is essentially composed of: the Timavo springs in San Giovanni di Duino, Doberdò and Pietrarossa lakes, the minor springs that feed the Lisert and Moschenizze canals (approximately 20 $km^2$) and the marine-coastal springs scattered along the shoreline of the Gulf of Trieste, going from Aurisina to Duino (approximately 7 km). These waters come from different but hydraulically interconnected aquifers, with an estimated average water discharge of 40 $m^3/s$ and peak values of approximately 175 $m^3/s$. Speleological and subaqueous explorations have revealed a complex system of flooded caves, which extends over more than 1900 m up to a depth of – 83 m a.s.l.

In the Italian Classical Karst sector (less than 300 $km^2$), more than 3000 caves are known (over 150 of which develop for more than a hundred m and 6 of which for thousands of m), plus some 80 dolines extending for more than 100 m and limestone pavements and karren developing for some tens of $km^2$. The most representative epigean forms are undoubtedly the polje with the lake of Doberdò, the Riselce doline near Sgonico and the karren of Borgo Grotta Gigante. Choosing among the large number of caves is rather difficult: the Grotta Gigante cave has been open to the public since 1908 and is listed in the Guinness World Record for being the largest tourist cavern in the world; the Trebiciano Abyss is the most famous vertical cave, with a branch of the Timavo flowing at its bottom; the Claudio Skilan Cave is the largest one, and features large caverns and stunning speleothems.

Val Rosandra, located in the Province of Trieste at the southern border of the Classical Karst, is a very deep valley excavated in Tertiary limestone, whose morphology is influenced by lithology and tectonics. The Rosandra River excavates gullies, embedded meanders, swirl holes and creates waterfalls and rapids in white limestone. The representative hypogean system is that of Mt. Stena: this is a vast and articulated complex exceeding 7 km of development, a fascinating example of karst influenced by geological and environmental conditions. The interaction among geology, vegetation and geographic location makes Val Rosandra a unique geosite at a world level.

## References

1. Carton A, Cavallin A, Francavilla F, Mantovani F, Panizza M, Pellegrini GB, Tellini C (1994) Ricerche ambientali per l'individuazione e la valutazione dei beni ambientali, Il Quaternario – Italian Journal of Quaternary Sciences, 7(1):365–372
2. Cleal CJ, Thomas BA, Bevins RE, Wimbledon WAP (1999) GEOSITES an international geoconservation initiative, Geol. Today, 15:64–68
3. Cucchi F (2002) Il carsismo. In: Società Geologica Italiana (ed) Alpi e Prealpi Carniche e Giulie, BE-MA, Milano, 107–108

# Engineering Geology in Karst

# Uncovered Caves and Their Conservation During the Construction of Motorways Over Classical Karst and Other Types of Slovenian Karst

M. Knez and T. Slabe

**Abstract** More than 350 caves, an important part of Slovenian natural heritage, opened on 70 km of recently constructed motorways over the Kras. The karstologists take part in the planning and building of motorways. Newly discovered caves are studied and as much as possible there is a great effort to preserve them. They are an important trace of the Classical karst aquifer development. Smaller entrances to shafts and old caves are closed by concrete slabs and caves cut by road-cuts by stone walls. More important caves near the border of the motorway are accessible by concrete pipes and the largest cave, developed around the tunnel is approachable by doors in the wall of the tunnel. New knowledge gathered by planners and builders of motorways is used in planning and implementation of other interventions on karst and for protection of the karst heritage.

## 1 Introduction

One of the major ongoing projects in Slovenia is to link the country with modern expressways. Almost half of Slovenia is karst and more than half of its supply of water comes from karst aquifers. Slovenia is the home of the Classical Karst (Kras) region, which gave its name to numerous world languages for the type of landscape that develops on carbonate rock and where the science of karstology began to develop. The ancient word for stone gave the origin to the ancient name for the region. Comprising an important part of Slovenian natural and cultural heritage, the sensitive karst landscape demands good knowledge and serious effort for its preservation.

For a number of years, karstologists have cooperated in the planning and construction of expressways in the Kras region (Kogovšek 1993; Slabe 1996; Knez and Slabe 2004, 2007). In the selection of expressway and railway routes, the main consideration is the integrity of the karst landscape and therefore the routes chosen avoid the more important surface karst features (dolines, poljes, collapse dolines,

M. Knez, T. Slabe
Karst Research Institute ZRC SAZU, Postojna, Slovenia, e-mail: slabe@zrc-sazu.si

B. Andreo et al. (Eds.), *Advances in Research in Karst Media* 
DOI 10.1007/978-3-642-12486-0, 

karst walls) and already known caves. Special attention is devoted to the impact of the construction and use of expressways on karst waters. Expressways should therefore be impermeable so that runoff water from the road is first gathered in oil collectors and then released clean onto the karst surface.

The authors studied the impact of traffic routes on karst waters. Kogovšek (1993) determined the contents of polluted water flowing daily from the expressways. Small quantities of stagnant water found in caves along the expressways contained traces of mineral oils.

During the construction of expressways karstological monitoring was also performed. Newly revealed karst phenomena as an important part of the natural heritage was studied and advice was given on how to preserve them if the construction work allows it. At the same time the new findings are of great help to the construction companies. A number of new findings on the formation and development of the karst surface, epikarst, and the perforation of the aquifer have been acquired. More than 350 caves were opened on the 70 km section of expressway built in Classical Karst in the last few years.

## 2 Planning

From the karstological viewpoint, the planning of roads requires an evaluation of the karstic relief, karstic underground and hydrological singularities, and an evaluation of the road lay-out alternatives presented. Numerous karstic phenomena are exposed in any construction location in the Karst: dolinas, filled or empty caves, and segments of old or recent drainage paths through the Karst. Many deep karst phenomena (caves) have already been exposed by denudation and can be recognised from the surface. At the present time, unroofed caves discovered during road construction are the focus of special attention. The authors are aware that a good karstological study of the area across which the road is planned allows a good choice of road lay-out and is one of the basic starting points for planning construction in this unique and vulnerable landscape.

The first step is to collect data on surface karstic phenomena, especially in dolinas, collapse dolinas, blind valleys and other morphological landforms, using published literature, archives and various collections. Later, fieldwork enables to the definition criteria for the mapping of the area for the selected road lay-out. In the field the important rock segments from a karstological aspect are evaluated. The known entrances into underground chambers are marked on a map and, if necessary, they are supplemented with new ones. On the basis of surface mapping and the genetic interpretation of morphologically clear and denuded caves that are visible in the relief, a preliminary projection of the underground caves is made. If necessary, plans are made for surplus material deposition on the basis of surface mapping.

Field hydrogeological mapping is performed. If the need arises tracing experiments are carried out. Hydrological maps are produced and upgrades are made to the existing ones with the results of fieldwork and tracing experiments. The following is a short summary of the basic guidelines for road planning:

- the selection of the road lay-out is based on a comprehensive assessment of the karst, with an emphasis on local characteristics.
- the selected road lay-out must bypass even isolated unique karstic phenomena.
- one of the priority objectives of planning is the preservation of the karstic aquifer.

## 3 Karstological Monitoring During Construction

The removal of soil and plant cover from the karstic relief and extensive groundwork during the digging of the roadbed and tunnels exposed surface, epikarstic and underground karstic phenomena. The task was to study these phenomena as part of natural heritage, to propose methods for their preservation and, of course, to inform the constructors of any new findings to assist them in overcoming obstacles to construction (Fig. 1).

Karstic relief is sectioned by dolinas and unroofed caves. Dolinas are filled with soil to a higher or lesser degree. Vertical shafts and fissures are located at the bottom of dolinas, allowing the water to run off. The soil must be removed from the dolinas and the bottom reinforced by heaping rocks to form a vault. This is necessary because openings of shafts are usually smaller than the caves below them. The next step is to fill the sinkhole with layers of gravel. Unroofed caves are old caves that emerged on the surface during the lowering of the karstic relief and are without the top section of their ceilings. The epikarst is interlaced with fissures, particularly evident in cretaceous limestone and somewhat less evident in paleogenetic limestone. A number of fissures opened up on dolina bottoms and slopes.

During the construction of 70 km of motorway in recent years in the Karst, over 350 caves have been exposed. Caves (Fig. 1) can be classified according to the development of the aquifer: old caves through which water flowed when the karstic

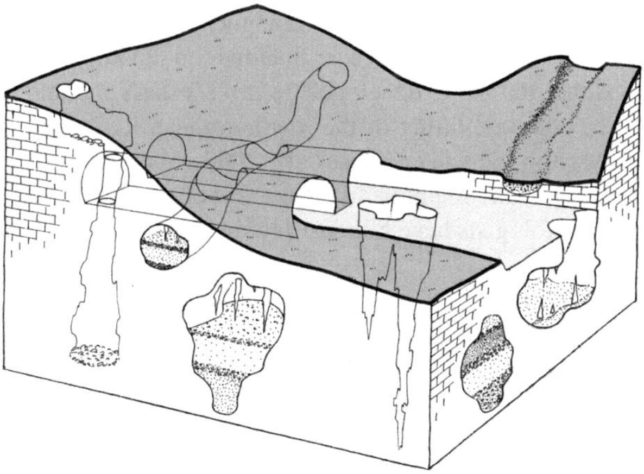

**Fig. 1** Caves, typical for all stages of Classical Karst development, found in today's epikarst

aquifer was surrounded and covered to a high level with flysch; and shafts through which water flowed vertically from the permeable karstic surface into underground water. The deepest shaft measured 110 m. Old caves are either empty or filled with sediment; the latter represents almost two thirds of caves, while one third are already without ceilings.

Kartologists attempted to preserve as many caves as possible. This was simplest with shafts. The small entrances were closed with concrete lids. Those old caves with stable ceilings were preserved as well. Caves that were opened due to blasting and were located in hollowed rock had to be blasted again and filled in. Caves that were cut through during the digging of road banks, with entrances located in the banks, were closed with rock walls. The most interesting and well-preserved caves were protected in their entirety; although they are located beneath the motorway, they are accessible. Access to these caves is through concrete pipes that finish at the edge of the road with a closed shaft. The largest cave system in the tunnel LC-S647 is almost entirely preserved. Below traffic belt passages are connected with large concrete pipes. In the side of the tunnel there is a special door to enter the cave. The impact of different types of detonation in caves was studied; this information will be useful in further construction and in the preservation of karstic phenomena.

## 4 New Findings on Karst Development Obtained During Motorway Construction

Unroofed caves (Knez and Slabe 2007) are a special and frequent karstic landform. This significant surface karstic landform is a known phenomenon that has not yet been fully researched. The share of this landform in the Karst is larger than previously thought, with insufficient attention being paid to it.

A large proportion of caves were filled with sediment. The sediment was most frequently fine-grainy flysch flood sediment interspersed with layers of rubble. Sediment samples were collected for the paleomagnetic research work. It was found that the sediments in some caves are older than the top of Olduvai chron; therefore, it was concluded that the caves developed before Messina age and that they were fossilised after the repeated filling of the Mediterranean Basin with water, i.e. approximately 5.2 Ma ago (Šebela and Sasowsky 2000). Sediment dating points to the oldest periods of Karst karstification and it was concluded that the oldest caves are much older than karstologists have hitherto assumed.

## 5 Motorway Construction on Low and Covered Karst of Dolenjska

Expressway construction on the low and covered karst of Dolenjska has helped in uncovering Slovenia's natural heritage. The karst formed in a characteristic fashion below a surface covered by layers of sediment of various thicknesses. The subsoil shaping of carbonate rock placed its stamp on the entire epikarst and vadose zones. The surface is dissected into subsoil stone forests (Fig. 2), surface areas of karren are smaller, and large caves and shafts filled with fine-grained sediment have been discovered. Large areas of subsoil stone forests are difficult to discern before starting excavations or geophysical studies. A special subsoil karren shaped by the oscillating level of underground water was uncovered by excavations.

**Fig. 2** Subsoil stone forest excavation (*left*) and caves in breccia above flysch (*right*)

## 6 Construction of Motorway on Breccia Slopes

The geological, geomorphological, speleological, and hydrological diversity of Slovenia's karst has been demonstrated by the studies of the karstification of breccia that formed beneath the western slopes of the Mount Nanos high karst plateau. In most cases water percolating diffusely through the permeable surface of scree material or breccia to the more or less impermeable flysch bedrock creates young karst phenomena.

Characteristic types of caves (Fig. 2) developed in the young and very porous breccia, which is consolidated only in places, lying on the more or less slanting flysch, an impermeable bedrock. The true karst caves are small and their development was enabled by the sediment that was deposited on their bottoms and which as a rule fills them. They formed in a locally and periodically flooded zone and are paragenetically enlarged. The largest caves formed above the contact with the impermeable flysch bedrock where the largest streams join. Their shape reflects the varying degrees of consolidation of the breccia. Along fissures that are the consequence of the sliding of breccia and scree material down the slanting bedrock of frequently saturated flysch, fissure caves formed across the slope.

Motorway constriction was due to geological characteristics extremely demanding task.

## 7 Conclusions

The collaboration of karstologists in motorway construction in the Karst has been advantageous. However, it is important that there is cooperation in the planning and construction process, as well as in the latter monitoring of the impact of motorways on the environment. This means that cooperation is necessary during the whole process of intervening in this vulnerable karstic landscape; only in this way will the goals of preserving natural heritage and deepening basic knowledge of the development of the formation of the karst and of motorway construction in this unique environment be achieved. There are different known karst types, each one demanding a different approach; this is why cooperation with constructors must be constant and ongoing. These findings across most of Slovenia in the past ten years have been implemente. Cooperation between road planners, constructors and karstologists is an example to be followed in the planning and implementation of other interventions into the different types of karst.

## References

Knez M, Slabe T (2004) Highways on karst. In: J Gunn (ed) Encyclopedia of caves and karst science, Fitzroy Dearborn, New York, London

Knez M, Slabe T (2007) Kraški pojavi razkriti med gradnjo slovenskih avtocest. Založba ZRC Publishing, Ljubljana

Kogovšek J (1993) Water composition flowing off our roads. Ujma 7:67–69

Slabe T (1996) Karst features in the motorway section between Čebulovica and Dane. Acta Carsol 13:221–240

Šebela S, Sasowsky ID (2000) Paleomagnetic dating of sediments in caves opened during highway construction near Kozina, Slovenia. Acta Carsol 29:303–312

# Understanding the Leaks in Chabrouh Dam Through Detailed Hydrogeological Analysis of the Qana Plateau (Lebanon)

I. Bou Jaoude, R. Karanouh, N. Momjian, A. Chehadeh, and S. Cheikh Hussein

**Abstract** The Qana Plateau is an 8 $km^2$ semi-isolated plateau in Mount Lebanon. The Chabrouh dam is located on the eastern flank of the Qana Plateau. Leaks of up to 200 L/s were observed around the right embankments of the dam basin on the Qana Plateau side. Karstification is now documented to a depth of 1565 m a.s.l. and hence leaks are focused in those lower beds, flowing along structurally controlled conduits. Hydrographs of the three major springs (Hadid, Qana and Terrache) show clear karstic characteristics and a major increase in their discharge for the year 2009 revealing a possible contribution from the Chabrouh dam basin. An equivalent porous medium mathematical model of the plateau further highlighted the possible contribution from the basin to the surrounding springs.

## 1 Introduction

Since Phoenician times the people that occupied the Eastern coast of the Mediterranean Sea adapted to karstic environments prevalent along the coastal zone. In Lebanon the need to manage the large losses of water to the sea, resulting from fast overland flow, necessitated looking into one solution building dams. Constructing such big structures in a karstic environment and retaining the water in their basins has long been proven difficult, with numerous examples around the world (Ertunç 1999, Milanovic 2004 and Mohammadi et al. 2007). Carbonate rocks along with their associated karstic environment form approximately 65% of Lebanon's surface area (Fig. 1), so it is not surprising that dams may be partially if not fully constructed on such rocks in Lebanon. The Qaraoun dam completed in 1954 was the

I. Bou Jaoude, R. Karanouh, N. Momjian, A. Chehadeh, S. Cheikh Hussein
Department of Geology, American University of Beirut (AUB), P.O. Box 11-0236/2010, Beirut, Lebanon, e-mail: iboujaoude@gmail.com

I. Bou Jaoude, R. Karanouh
Spéléo club du Liban (SCL), P.O. Box 70-923, Antleias, Lebanon

B. Andreo et al. (Eds.), *Advances in Research in Karst Media* 
DOI 10.1007/978-3-642-12486-0, © Springer 2010

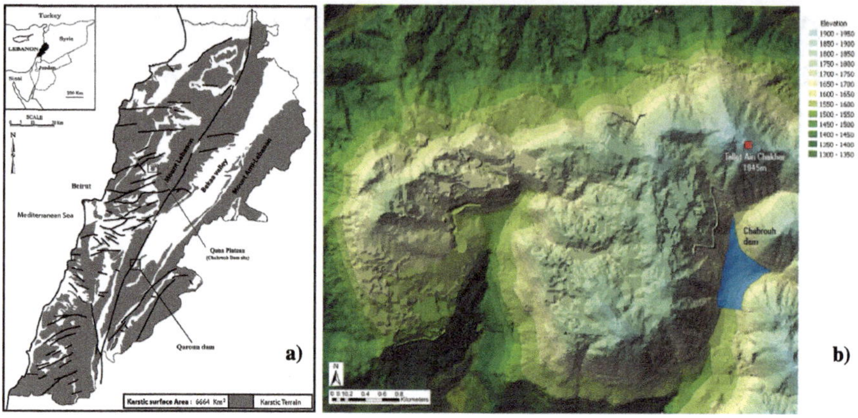


**Fig. 1** **a** Simplified structural map of Lebanon showing the location of the Chabrouh and Qaraoun Dams. **b** Colored topographic map of the Qana plateau showing the Chabrouh dam and basin

first of the big hydroelectric structures to be built in Lebanon with documented seepages (Dubertret 1984). After the war construction resumed and the Chabrouh dam was completed in 2006 with a 12.2 $km^2$ basin and water level in the basin reaches 1613 m a.s.l.

The Chabrouh dam and its basin lie to the east of the Qana plateau (Fig. 1). The plateau is semi-isolated and covers an area of approximately 8 $km^2$ in central Mount Lebanon. The plateau's elevation rises from approximately 1600 m a.s.l. to 1945 m a.s.l. (Fig. 1).

During investigation stages of the potential Chabrouh dam site in the 1970s, Majdalani (1977) argued that karst rocks in the Chabrouh valley make building such a big structure and retaining the water in the basin difficult. Bou Jaoude (2006) revealed that channeling of water through structurally controlled karstic conduits from the Chabrouh basin into the surrounding springs might occur. During construction between 2002 and 2006 large karstic cavities were observed at the right abutment at different levels beginning at 1565 m a.s.l. (Kallas, personal communication, 2009). This paper investigates the nature of the leaks in the Chabrouh dam basin.

## 2 Geological/Hydrogeological Background

The Lebanese segment of the N-S trending Dead Sea Transform Fault is a major restraining bend. Secondary structures such as the E-W and the ENE-WSW faults are strike slip faults with small dip slip component. In the Qana plateau the effects of these secondary structures are tertiary scale dip slip faults trending NW-SE (Fig. 2). These structures and associated smaller scale fractures act as passageways for water and enhance karst development (Bou Jaoude, 2006). The Qana plateau is composed of geologic rocks ranging in age between Cenomanian, at its highest elevation, and

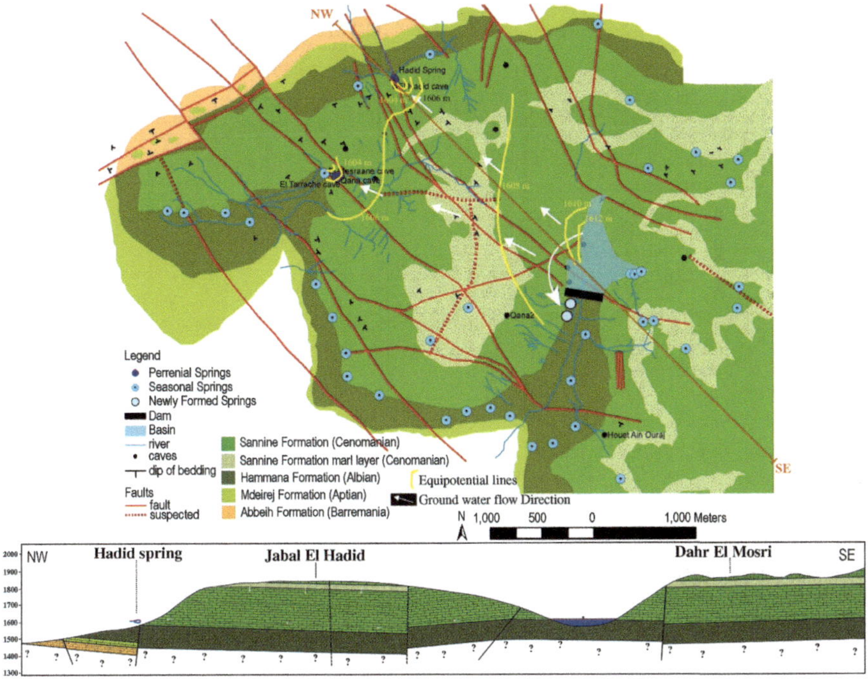


**Fig. 2** Simplified geological map and cross section of the Qana plateau showing the location of the major springs, newly formed springs on the *right embankment* of the Chabrouh dam, the output of the mathematical model and predicted groundwater flow direction

Albian, at its lowest. The Cenomanian rocks are composed of interbeds of marl, marly limestone and limestone beds and form the major karstic aquifer in the study area. The Albian rocks consist of blue-greenish marl, limestone and volcanic tuffs. These act as the underlying aquiclude. There is a transition zone between the Cenomanian and the Albian rocks made up of interbeds of marl, marly limestone and limestone. This acts as a low conductivity semi-aquifer layer and is the discharge zone of most major springs in the area.

## 3 Monitoring Spring Discharge

The three major springs in the Qana plateau are the Hadid, Qana and Terrache springs. All of these springs issue from the transition zone in the lower levels of the Cenomanian aquifer at an elevation of approximately 1600 m a.s.l. The discharge of the springs was monitored for a full year between 2008 and 2009 and hydrographs showed a typical karstic nature (Fig. 3). The spring's hydrographs are typically karstic in nature with steep recession and possibly three recession coefficients.

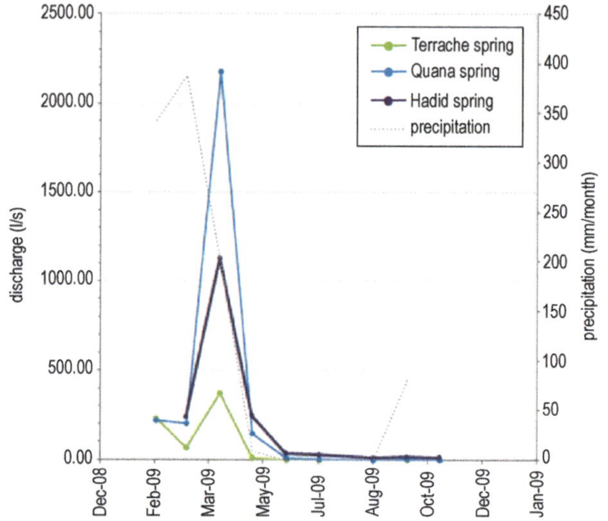


**Fig. 3** Hydrographs of Hadid, Terrache and Qana springs with monthly precipitation rates for the year 2009. Precipitation data obtained from DGCA (2009)

The discharge measurement obtained by Majdalani (1977) and by Bou Jawdeh et al. (2001) for the Qana spring for the month of December was approximately 57 L/s and 10 L/s, respectively. In comparison to the value, 91.87 L/s measured in 2009 for the same month and approximately the same amount of precipitation, showed a 38 and 89% increase.

The discharge measurement obtained by Majdalani (1977) and by Bou Jawdeh et al. (2001) for the Hadid spring for the month of July was approximately 17 L/s and 15 L/s, respectively. In comparison with 28.27 L/s obtained in 2009 for the same month and approximately the same amount of precipitation it showed a 39 and 47% increase, respectively.

**Table 1** Basic sensitivity analyses for the calculation of the catchment area

| Name of Spring | Avg. Discharge (20–33%*)($m^3/s$) | Average Precipitation** | Infiltration Rate*** | Catchment Area ($km^2$) | Bou Jaoude (2006) Catchment Area ($km^2$) |
|---|---|---|---|---|---|
| Qana | 0.4352–0.71808 | (1379 to 1600 mm/y) | 70–80% | 11–27 | 3 to 5 |
| Terrache | 0.074672–0.1232088 | | | 2–5 | |
| Hadid | 0.226448–0.3736392 | | | 6–14 | 3 to 5 |

* Percent from the maximum discharge according to exponential decay of springs hydrographs
** Average precipitation obtained from DGCA (2009) and added to it estimated snow fall values
*** Values based on UNDP (1970)

A basic sensitivity analyses is used to evaluate and compare possible extent of the catchment area for the three main springs in the Qana plateau for the year 2009 with earlier estimates. Considering an exponential decay, the average discharge of the springs will be in the order of 20–33%. Infiltration rates in the Cenomanian aquifer in Lebanon are based on research conducted by the UNDP (1970). All the values obtained (Table 1) were at least two fold higher than the catchment obtained from previous studies conducted before filling the basin in the dam. They were even larger than the total size of the plateau.

# 4 Observed Leakages

Before filling the Chabrouh basin several small karstic cavities were observed on the right embankment of the dam. During construction several karstic zones were observed on the right embankment of the dam (Kallas, personal communication) to a depth of 1565 m a.s.l. After filling the basin in 2006 several springs were observed on the right embankment discharging approximately 200 L/s. One spring is issuing from a 1-m limestone bed in the lower Albian Formation in the lower levels of the transition zone.

# 5 Conceptual Model/Mathematical Model

When looking at the possible interaction between the Chabrouh basin with the Qana plateau and its surrounding springs for the year 2009 several conditions might occur, but one model was devised from available data and used for the mathematical model (Fig. 4). Considering the head in the Cenomanian aquifer is at its minimum level that is a bit higher than the 1600 m a.s.l., the elevation of the major springs in the plateau. It is below the level of the water in the Chabrouh basin, which is 1613 m a.s.l., hence, a head difference of 13 m. This probably results in water recharging the aquifer from the basin and eventually feeding the surrounding major springs. This

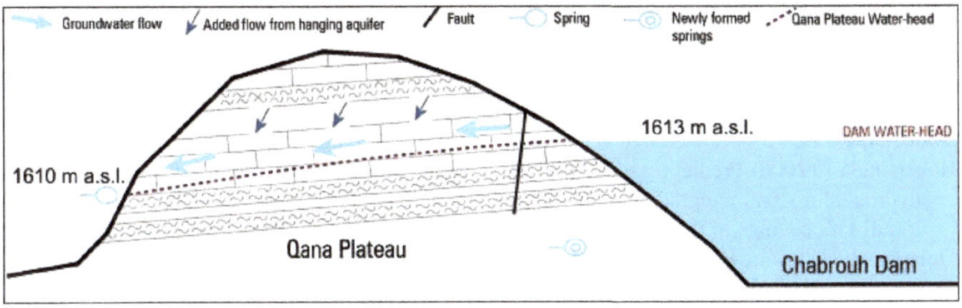


**Fig. 4** Conceptual model and its associated mathematical model

is evident from the unprecedented increase in the discharge of the springs for the year 2009.

To simulate groundwater flow in the Qana plateau and to analyze effect of the Chabrouh basin on the surrounding springs, an equivalent porous medium model using MODFLOW was devised. The Cenomanian aquifer is given a hydraulic conductivity of 0.0038 m/s, porosity of 0.15 and total porosity of 0.2. Knowing the nature of the plateau, being almost closed, with no recharge from surrounding aquifers, recharge into the aquifer from precipitation was assigned according to the values obtained from DGCA (2009) shown in Fig. 3. The lower limit of the aquifer was the underlying Albian aquiclude. A constant head boundary condition was assigned for the Chabrouh basin with head at 1613 m a.s.l. The springs are considered the major outlets with the assigned discharge rates of the year 2009. Because no wells are present in the plateau, the initial head conditions of 1620 m a.s.l. were assumed to be present in the plateau, considering the flooding condition during the wet season. The equipotential lines for the dry season obtained from the model (Fig. 2) showed the Chabrouh basin is leaking into the aquifer and feeding the Hadid and Terrache springs (Fig. 3). With the limited data available on the condition of the groundwater in the plateau except for the out flowing springs it will be difficult to calibrate and validate the model. However, tracer testing is the next step in this ongoing project.

## 6 Conclusion

Geological investigation showed karstification in the Cenomanian and Albian rocks to reach an elevation of approximately 1565 m a.s.l. After filling the dam in 2006 not only small springs were observed on the right embankment but also a noticeable increase in discharge in the Qana, Hadid and Terrache springs. Indicating a possible leakage from the Chabrouh dam basin into the Qana plateau. This is further highlighted by the conceptual and mathematical models.

**Acknowledgements** The authors would like to thank Mr. G. Haddad and Dr. W. Labaky.

## References

Bou Jawdeh I, Metni M, Nader F (2001) Identifying preferential karstic routes within the Qana Plateau, central Lebanon: A multidisciplinary approach. Bou Jaoude I, Badaoui H, Karanouh R (ed) Middle East Speleological Symposium. Beirut. Lebanon

Bou Jaoude I (2006) Predicting the effect of Chabrouh dam reservoir on the surrounding karstic hydrogeology: An integrated scientific approach. Goldscheider N, Mudry J, Savoy L, Zwahlen F (ed) Proceedings of the 8th conference on limestone hydrogeology. Neuchâtel, Switzerland

Dubertret L (1984) La Retenue De Qaraoun, Au liban: un example de retenue sur calcaires fissure's karstique. Hydrogeology of karstic terrains case 1:167–173

DGCA (Directorate General of the Civil Aviation) (2009) Meteorological Center, Rafic Hariri International Airport, Ministry of Public works and Transportation, Precipitation Data

Ertunç A (1999) The geological problems of the large dams constructed on the Euphrates River (Turkey). Engineering Geology 51:167–182

Kallas L (2009) Personal Communication

Majdalani M (1977) Geology and Hydrogeology of the Faraya-Afqa area, Central LEBANON. (Master thesis) American University of Beirut, Lebanon

Milanovic P T (2004) Water Resources Engineering in Karst. CRC Press LLC, USA

Mohammadi Z, Raeisi E, Bakalowicz M (2007) Method of leakage study at the karst dam site. A case study: Khersan 3 Dam, Iran. Environmental Geology 52:1053–1065

UNDP (1970) Etude des Eaux Souterrains, Liban. Nation Unies, New York, USA

# Decision Support Procedure for Constructing Karst Underground Reservoirs – a Case Study on Perućac Karst Spring (Western Serbia)

I. Jemcov, S. Milanović, and P.T. Milanović

**Abstract** One of the most complex systems of tapping structure in karst is construction of an underground dam and reservoir to provide artificial water storage and a karst controlled groundwater discharge regime. Examples of underground dams are not so numerous; and, therefore, the risk component in their building is much emphasized. In this paper, a proper procedure for building a support system for constructing underground karst reservoirs was proposed, focusing on the example of the Perućac spring, located in Dinaric karst, in Western Serbia, and analyzing possible effects of building an underground reservoir and selecting tapping using this procedure.

## 1 Introduction

Some experiments and case studies of underground dams in karstified rocks prove that an artificial underground storage system may be a realistic way for technically and economically practical groundwater management (Yoshikawa and Shokohifard 1993). Examples of the comprehensive studies on underground dam designs and their construction, as well as their positive outcomes are recorded in China (Daoxian 1990; Lu 1986) and Japan (Yoshikawa and Shokohifard 1993). One of the largest and most complex research projects on underground damming is the Ombla Power Plant (Milanović (1988).

Building underground dams requires a detailed research and this is its main drawback. Furthermore, even if very complex investigations are involved, the risk com-

I. Jemcov, S. Milanović
University of Belgrade, Faculty of Mining and Geology, Department of Hydrogeology, Djusina 7, 11000 Belgrade, Serbia, e-mail: igor@jemcov.com; milanovicsasa@sezampro.yu

P.T. Milanović
Strumicka 19, 11000 Belgrade, Serbia, e-mail: petar.mi@eunet.rs

B. Andreo et al. (Eds.), *Advances in Research in Karst Media*
DOI 10.1007/978-3-642-12486-0, © Springer 2010

ponents are unavoidable. Therefore, one needs to analyze possible effects of building an underground reservoir in very early stages of the exploration. An initial step in these studies involves calculating storage space in the karst aquifer under natural conditions, based on groundwater budget. Further, simulation groundwater exploitation conditions were applied and, finally, a procedure for selecting proper taping structure as result of optimization, was proposed.

## 2 General Characteristics of the Study Area

The Perućac spring represents a typical karst aquifer of Dinaric karst in Western Serbia. The spring is located at the right bank of the Drina River at the foothill of the Tara Mountain. Its catchment is developed mainly in Triassic limestone and dolomite that is surrounded by the impervious Palaeozoic rocks (Fig. 1). The spring outlet is based at the contact between sandstone and overlaying karstified limestone. The catchment area of the Perućac spring was estimated to be approximately 75 km$^2$. The average annual precipitation in the period between 1991 and 1998 was 984 mm. Furthermore, the mean discharge in this period of time was $Q_{\text{av}} = 1.53\,\text{m}^3/\text{s}$, taking values in the range between $Q_{\min} = 0.45\,\text{m}^3/\text{s}$ and

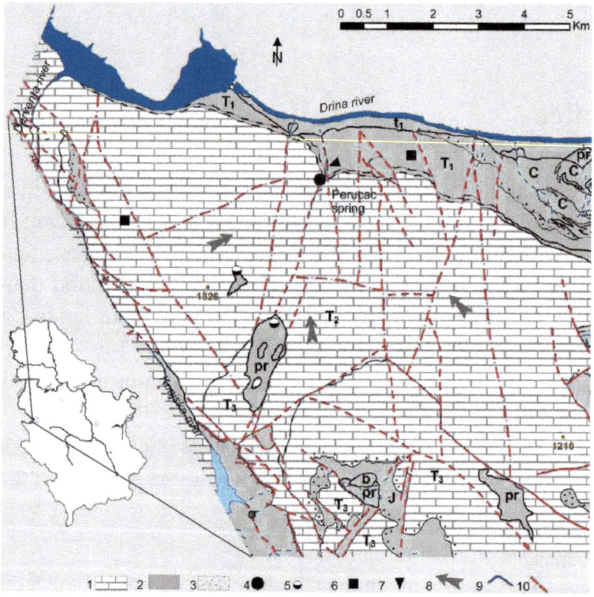


**Fig. 1** Simplified hydrogeological map of the Perućac spring. 1. karst aquifer; 2. non-karstic rocks; 3. alluvial aquifer; 4. karst spring; 5. ponor; 6. precipitation station; 7. gauging station; 8. main directions of the groundwater flow; 9. river; 10. fault

$Q_{\text{max}} = 9.82\,\text{m}^3/\text{s}$. The spring is partially used as a water supply for the nearby settlement and the local fish farm and its minimum consumption is 250 L/s.

For studying properties and hydrodynamic behavior of the Perućac karst aquifer several methods were used, e.g. hydrograph analysis, dye test, geophysical methods, borehole logging, water pressure tests (Lugeon tests), etc.

By applying the time series analysis, the obtained result shows that the Perućac karst spring has a large storage capacity and a well structured system which has a significant influence on the spring discharge (e.g. high shares of base-flow component, delay of impulse response and high attenuation effect). The effective infiltration as a percent of the total precipitation was estimated to be 60% (Jemcov and Petrič 2009). The analysis of several boreholes located along the proposed dam route shows that the limestone is highly permeable (Fig. 2). The most common permeability values are larger than 10 Lu (Milanović and Milanović 2000).

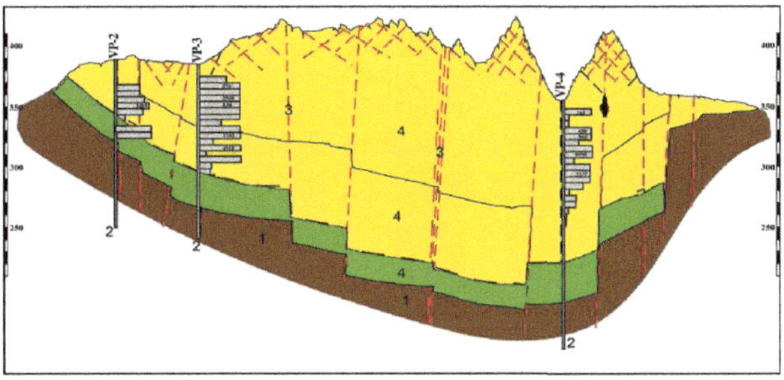

**Fig. 2** Cross-section along boreholes. 1. impervious rock; 2. borehole; 3. fault; 4. karstified limestone

# 3 Method for Estimating the Extraction Capacity

General characteristics of the Perućac karst source indicate a significant potential for groundwater management. This can be achieved by an increase of its storage capacity and an appropriate control of its groundwater discharge regime (Milanović and Milanović 2000).

Prior to proposing the possibility for the tapping structure, the utility regime needs to be estimated. This step includes estimation of the karst groundwater budget and the storage in the karst aquifer under natural conditions. Based on the inflow-outflow relation, the values of changes in storage ($\Delta Vi$) were obtained by adding and varying the continuous cumulative values to the initial storage level ($Vi = \Delta Vi + Vi - 1$). The groundwater storage ranges from $7.5\times106$ to $11.45\times106\,\text{m}^3$ between the beginning and the end of the recession period (Fig. 3).

The next step is to simulate the groundwater releasing conditions using the concept of discharge regime control. Quantifying the optimal water extraction involves many uncertainties, which need to be analyzed. These include randomness origi-

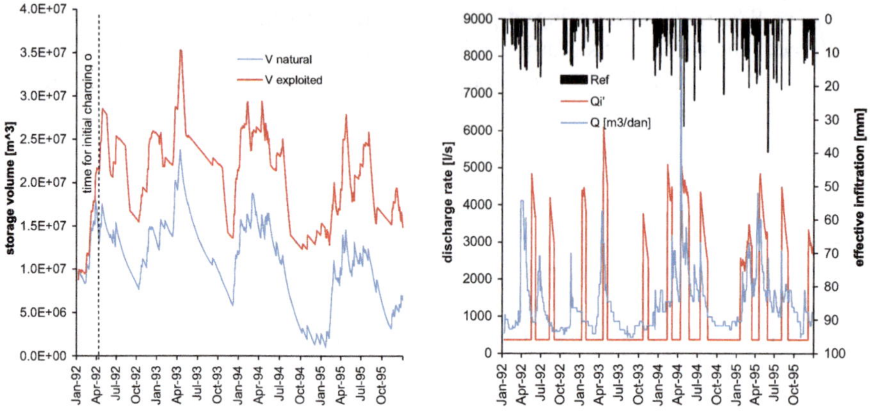


**Fig. 3** Estimated state of storage and simulated exploitation conditions in the karst water reservoir of the Perućac source (*right*) (selected period 1992–1995)

nating from the estimation of the maximum operation level (reservoir's volume), as well as the recession coefficient of the future, artificial reservoir. The maximum operational groundwater level under artificial conditions is estimated to be 87 m higher than the minimum level of the spring outflow (Milanović and Milanović 2000). This provides an opportunity for constructing an underground dam and providing a multipurpose ($Qi$) underground water reservoir for both permanent (water supply, fish farm, ecol. criteria etc.) and temporary users (power plant, irrigation etc.) – $Qe = \Sigma Qi$. The power plant would be partially operational in decrements of 5 m from the maximum water level, while the rest of the storage, $Qe = 370$ l/s, would be used for water supply, fish farm and ecological criteria. The state of the storage in KHS, given the conditions of the exploration, can be described by the following relation:

$$Ve_i' = Vp_i - \left[ \underbrace{\left( \sum_{1}^{i} Qe_i' - Qp_i' \right)}_{\text{current water deficit}} + \underbrace{(Vp_{i-1} - Ve_{i-1}')}_{\text{previously formed waer deficit}} \right] \tag{1}$$

$\Sigma Qe_i' = Ve_i\ i - 1 \times \alpha'$; $Qp$ – discharge of the karst spring under natural conditions; $Vp_i$ – state of storage in natural conditions of the discharge regime; $Qe_i$ – simulated discharge regime under condition of exploitation; $Qp_i'$ – fictive, changed, discharge in the karst spring as a consequence of changes in the storage in case the exploitation is instantly canceled; $Qe_i'$ – fictive discharge in the karst spring $Ve_i'$ – state of storage in the karst system in the exploitation regime, which corresponds to fictive discharge – $Qp_i'$.

According to the previous relation, the water level and state of the storage in the karst reservoir cumulatively reduces due to different users. The key issue in the previous calculation might be calculating the fictive discharge in the karst spring – $Qp_i'$;

The increase of the reservoir's storage will lead to the change in the discharge. This can be resolved by introducing a (fictive) coefficient of the recession ($\alpha'$), which is inversely proportional to the state of the storage – $\alpha' = Qp_i/Ve_i$ (Milanović 2004). A representative value of the fictive recession coefficient of the future underground reservoir is the maximum value under natural conditions.

To assess the water volume that reduces the reservoir's level by 5 m from the top height, the water level data in piezometers are used. During the flood period, a level decrease of 5 m induces a discharge of $6.15 \times 10^6\,m^3$, which could be estimated as the minimum operational value used for the hydropower plant. Further, to estimate the maximum water level in the underground reservoir, zone fluctuations at the outlet were used. The largest change in the water level is approximately 37 m. The calculated water volume that is discharged in the period between the high and low water season is approximately $8.83 \times 10^6\,m^3$. The additional storage that is artificially provided is estimated to be $1.3 \times 10^7\,m^3$.

Based on the applied analysis, continuous operation of power plant varies from 18–56 days, while time of filing up of karst reservoir until exceeding maximal water level is significantly higher and varies from 35 to 164 days (Fig. 3).

## 4 Selection of the Proper Tapping Structure

Based on the previous analysis, possible technical solutions for the tapping structure were the solutions that are mainly related to constructing of arch-like non-permeable barrier (diaphragm wall, grout curtain and plugging of karst channels) connected to the impermeable sandstones;

**A-1** Alternative: consider some improvements of the existing, old, tapping structure without any influence on the regime control; this example does not consider multipurpose solutions. Thus, the costs of the consumed electric power will not be reduced.

**A-2** Alternative: consider diaphragm wall in front of the spring (Fig. 4). This solution presents a combination of the reservoirs, surface-underground and provides the largest storage space;

**A-3** Alterative: consider positive cut-off (diaphragm wall) that is directly connected to the rock massif (Fig. 4). This is quite opposite from the previous example as the solution is a strictly underground karst reservoir with a smaller storage space. However, it is more economical and less vulnerabile;

**A-4** Alternative: consider grout curtain and plugged karst channels that are located at least 30 m behind the spring (Fig. 4).

The last step in this procedure is the selection of proper tapping structure in process of optimization (Jemcov et al. 2002). The goal of the optimization is to determine the best solution among all feasible and suitable solutions, given the adopted criteria. The optimal solution represents a compromise between intentions (criteria) and possibilities (limitations). A multi-criterion optimization provides optimal system parameters subject to several criteria.

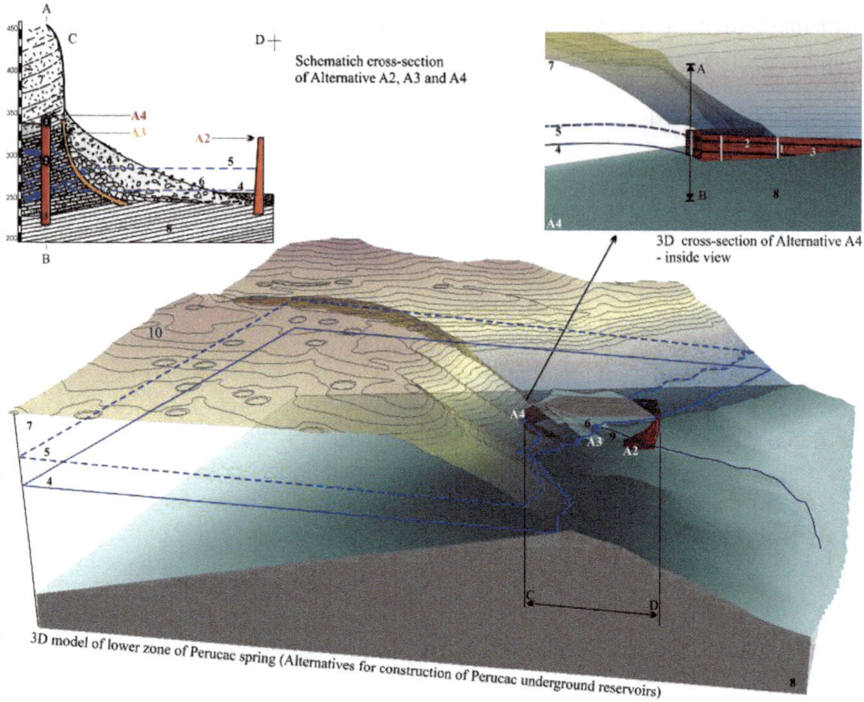


**Fig. 4** Presentation of alternatives in 3D hydrogeological model: 1. Boreholes; 2. Grouting galleries; 3. Grout curtain (underground dam); 4. GWL under natural conditions; 5. Calculated max. GWL; 6. Perućac spring; 7. Karstified limestone; 8. Impervious rock; 9. Debris – spring zone; 10. Tara mountain; A2 – surface dam in front of the main outlet zone; A3 – positive cut-off (diaphragm) wall connected to impervious rock; A4 – Grout curtain (underground dam)

The optimization procedure can be defined by the following transcription: $O : (S, Q, G) \rightarrow x^*$ ($O$ – transcriptional function; $S$ – system description; $Q$ – optimization criteria; $G$ – limitations; $x^*$ – best solution). The $Q$ criterion, in a multicriterion optimization, involves several criteria functions $(F(x))$ and preferential structure $(P)$: $Q = (F(x), P)$. To evaluate and optimize a given tapping struc-

**Table 1** Values of criterion functions and combination of criterion weights

| Alternative | f-1 ($\times 10^6$ €) | f-2 | f-3 | f-4 | f-5 |
|---|---|---|---|---|---|
| A-1 | 0.20 | 1.02 | 1 | 3 | 2 |
| A-2 | 48 | 3.08 | 3 | 7 | 5 |
| A-3 | 35 | 0.63 | 4 | 6 | 3 |
| A-4 | 19 | 0.52 | 5 | 6 | 1 |
| max −fi+ | 25 | 0.52 | 5 | 3 | 1 |
| min −fi− | 48 | 3.08 | 1 | 7 | 5 |

| Combination | f-1 | f-2 | f-3 | f-4 | f-5 |
|---|---|---|---|---|---|
| 1 | 1 | 1 | 1 | 1 | 1 |
| 2 | 1 | 2 | 1 | 1 | 2 |
| 3 | 1 | 3 | 2 | 2 | 3 |
| 4 | 2 | 3 | 1 | 2 | 2 |
| 5 | 1 | 2 | 2 | 1 | 2 |

ture, the following criteria were used: Investments costs – ***f-1***; Economical price of water – ***f-2***; Quality and protection conditions – ***f-3***; Time needed for system construction – ***f-4***; Ecological criteria – ***f-5***.

Multi-criteria optimization is subject to five criteria, against which four different solutions are compared. Based on the calculation, the results of this analysis, it was concluded that A-4 is the best solution for the water supply system, which proved to better in all scenarios (Table 1).

## 5 Conclusion

In this paper the obtained result shows that the Perućac karst spring is a good candidate for constructing the artificial underground reservoir. The proposed procedure for selecting the proper tapping structure should be a strong support to decision makers whose interests are the optimal usage of karst groundwater resources and minimization potential risks for implementation in future projects.

## References

Daoxian Y (1990) The Construction of Underground Dams on Subterranean Streams in South China Karst. Institute of Karst Geology. Guilin. China

Jemcov I, Dokmnanović P, Stevanović Z, Milanović S (2002) An example of management of groundwater resources under complex hydrogeological conditions. Groundwater and Human Develop. XXXII IAH Mar del Plata. Argentina

Jemcov I, Petrič M (2009) Measured precipitation vs. effective infiltration and their influence on the assessment of karst systems based on results of the time series analysis. J. Hydrology 379:304–314

Lu Y (1986) Some Problems of Subsurface Reservoirs Constructed in Karst Regions of China. Inst. of Hydrogeology and Eng. Geol. Beijing, China

Milanović PT (1988) Artificial underground reservoirs in karst: Experimental and project examples. Proceedings of the IAH 21st Congress. Guilin, China

Milanović PT, Milanović S (2000) Possibilities and conditions for Perućac Underground storage Marmaris. IHP-V Tech. Doc. in Hydrology. No 49, Vol. II., 289–294. UNESCO Paris, 2001

Milanović PT (2004) Water resources Engeneering in karst. CRC press

Yoshikawa M, Shokohifard G (1993) Underground dams: A new technology for groundwater resources development. Proceedings of Articles of International Karst Symposium. Shiraz, Iran

# Aeration Zone in Karst – Properties and Investigations

P.T. Milanović

**Abstract** Aeration, vadose or unsaturated zone in karst aquifer is the zone above the water table (the zone between land surface and piezometric level measured in stand pipe piezometers). This zone can be divided into two sections: an upper section permanently water free and a deeper section temporarily saturated by water. During dry periods, the deeper section is temporarily an aeration zone; however, during the rainy season it is the most dynamic part of the karst aquifer. Within this section, there exist karst channels, caverns, underground flows with free surface (underground rivers), and underground siphonal lakes. An aeration zone plays an important role during the construction of dams, reservoirs and tunnels in karst and can cause various kinds of problems. A particular problem appears if the dam sites and reservoirs are located above the aeration zone or when tunnels cross huge empty caverns or temporarily saturated aeration zones. Due to an abrupt rising of the water table the pressure of air trapped in the karst channels and siphons significantly increase. As consequence, the water tightness of the reservoir bottom can be endangered. The induced seismicity, also, has been documented at the vicinity of some large temporary springs. Some common investigation methods in karst are not successful in the case of aeration zone. The most effective methods in karst as groundwater level monitoring and tracer tests are not useful. Speleology is the most common method but only if karst channels are accessible. TV logging was successfully applied at different locations. Experiments with smoke and ear as tracers also provided good results. Some of those methods are presented in this article.

## 1 Introduction

A deep aeration zone below dam sites and reservoirs means a deep water level. Deep water table requires deep grout curtains or large surface remedial works. Due

P.T. Milanović
Strumicka 19, 11000 Belgrade, Serbia, e-mail: petar.mi@eunet.rs

B. Andreo et al. (Eds.), *Advances in Research in Karst Media*
DOI 10.1007/978-3-642-12486-0, 

to these circumstances in some cases the dam sites are relocated or construction of dams and reservoirs was abandoned.

Distinctive examples are: Abolabas dam (Iran) – water table 52–55 m below dam site bottom; Ourkis dam (Algeria) – empty caverns 47–52 m below dam site foundation; Reservoir Vrtac (Montenegro) – water table about 50 m below reservoir bottom; Reservoir Havasan (Iran) – water table 12–15 m below the river level; Akkopru Reservoir (Turkey) – water table $\sim$ 80 m below reservoir bottom and Reservoir Hutovo (Herzegovina) – water level more than 100 m below reservoir bottom.

The caves and karst channels with strong natural air movement are common in all karst regions. Intensity of natural air current varies with season, time of day and outside weather conditions. Air current from deep boreholes also is common phenomenon in karst. This phenomenon indicates good connection of karst channels in the aeration zone including connection with land surface.

Air current in the aeration zone of a karst aquifer can be the consequence of different events: temperature difference of surface air and air in karst channels and caverns; very fast water table fluctuation; and squeezing of air bubbles from water. Air circulation can be created artificially, as well.

Due to saturation by water a large volume of air is pressed out from caverns and channels. Aeration of karst porosity is registered in some piezometric boreholes in the form of fast air current. In the case of some aeration pipes the measured airstream velocity was 15 m/s (Bagarić 1980).

The water itself contains some quantity of air in the form of bubbles. This part of the air released from the water and penetrated toward the aeration zone provoking micro pulsation of pressure in piezometers.

In solving the problem of water tightness of the reservoirs, the bottom of which is being covered with alluvial sediments, the great problem appears because of air captured in the karstified bedrock. The appropriate solutions are aeration pipes. To check operational capabilities of those pipes one possibility is smoke tracer artificially injected into the karst porosity using strong fans.

## 2 Air Current as a Consequence of Water Table Fluctuation

During the dry season when the water table in a karst aquifer is close to minimum all above rock mass has characteristics of an aeration zone. Karst channels and caverns in aeration zone were created during earlier stage(s) of the karst aquifer evolution process. Some of those features are filled with cave deposits (clay, sand, pebbles) and blocks of limestone but many of them are free for air circulation.

However, in periods of a wet season, porosity in the zone of ground water fluctuation is temporarily filled with water. This zone is characterized by fast fluctuation of water table. The karst aquifer reacts very fast after high precipitation, sometimes as fast as in 10–15 hours, and in some cases even faster. With high rain showers the water table reacts in less than four hours. Sometimes the water table increases 90 m in only 10 hours. The difference between maximal and minimal levels of the

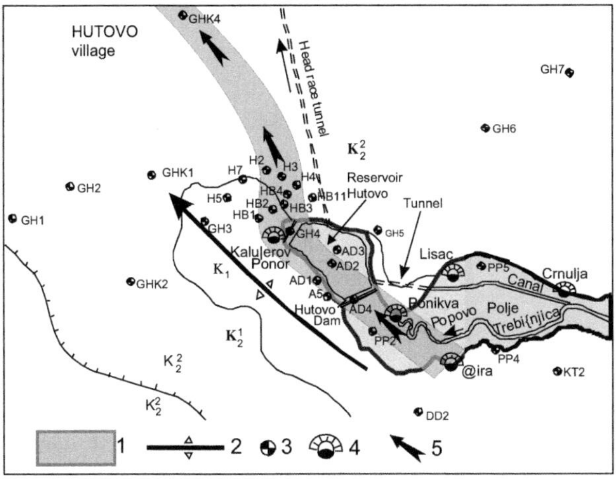


**Fig. 1** Reservoir Hutovo. 1. Karstified zones with underground flows, 2. Axis of anticline, 3. Borehole, 4. Large ponor, 5. Direction of underground flows

water table can be great. The largest fluctuation amplitudes of the water table were measured in piezometers located in the eastern Herzegovina region, from 281 to 312 m.

The phenomenon of extreme groundwater fluctuations was observed at some piezometric boreholes in the area of Hutovo reservoir (Fig. 1).

The reservoir bottom, at the very end of Popovo polje, is covered with alluvial deposits, the thickness of which increases from flanks toward the middle part of the polje, where it reaches about 30 m. In natural conditions the flood level in this part of the polje was about 40 m (Milanović 2007). Paleorelief has a typical karst topography – a lot of sinkholes and ponors. The area of the reservoir was losing water, under natural conditions, through 75 registered ponors in the alluvium. Beneath this part of the polje the water table fluctuates about 100 m. Experiments with the radioactive isotope Bromine 82 showed the underground flow to have the average velocity in the range of 45–55 cm/s. It was found that the velocity is greatest immediately after the water sinks in because the water tends to reach the base of erosion as rapidly as possible since the energy gradients are highest in the initial section of the route.

The main direction of underground flows was defined by very detailed mapping, geophysical investigations and number of tracer tests using dye and radioactive tracers. The karst channels and caverns along this zone were detected, by drilling and radioactive borehole logging, to the depth of 230 m. Two different levels of underground flows were defined.

The measurements of air current and water table fluctuations were organized in the period of water table rising. the continuous uprising of the water table does not mean the air current is constantly in one direction. In some boreholes the direction of air current alternately changes. Periodically, the air blows out of the hole and sucks into the hole (Fig. 2).

**Fig. 2** Diagram of cyclic changing of air current direction from piezometric borehole A-5 during rapid water table increasing (Kovačina and Skopljak 1978)

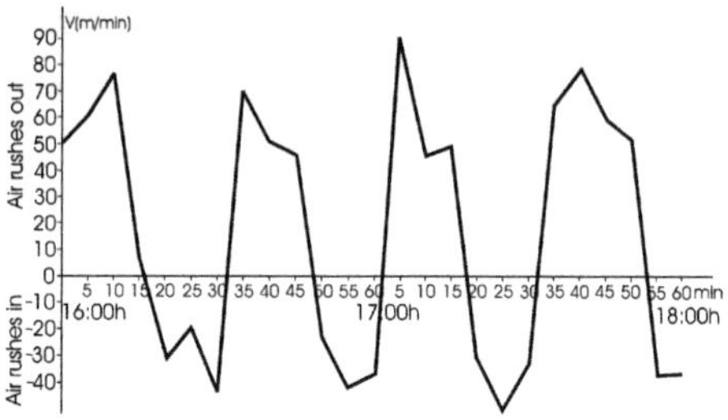


Due to huge swallowing capacity of ponors in the wide area of Hutovo Reservoir the water table rises very fast. the problem appears to be below the reservoir. All natural connections with surface were plugged during treatment of the reservoir bottom. The air was trapped in blocked channels. At a few places the air under pressure demolished protective measures (compacted clay blanket and PVC foil). Part of the air escapes through the aeration pipes and boreholes.

From two boreholes in this area a strong air current was registered. From both holes the velocity of air current was almost periodically. Nevertheless, as the boreholes are relatively close ($\approx$ 1600 m) the velocity of water table rising was considerably different. In 24 hours the water table in A-4 rose 21 m, however in A-5 only 1.5 m.

Periods of changing of air current in borehole A-5, for GWL (Ground Water Level) depth of 44.73 m varies between 13 to 22 min (Fig. 2). The maximal measured velocity of air current was about 110 m/min, and maximal velocity in the period of the air sucked into borehole was 95 m/min.

In the case of A-4 the periods of air current direction changing are: for GWL depth of 75.61 m the changing period is $\approx$ 25 min; and for GWL depth of 54.12 m the changing period is $\approx$ 27.5 min.

There is strong functional relation between volume of air, which is squeezed out from karst voids and velocity of ground water increase.

## 3 Investigations in Aeration Zone

To analyze connections of inaccessible karst channels in an aeration zone the gaseous tracers and air can be used only. The basic principle of this method consists of injection of smoke or some other gases into the system of karst voids located at the aeration zone.

Tracer diffusion in the air is 1000 times higher than the water. Consequently, the tracer dispersion through the cavities is higher comparedi with water. Water and air circulate through the karstified rock toward the points of minimum potential. In the case of water circulation, the potential minimum is at the points of lowest elevations, while in the case of air circulation it is at the points on the ground surface.

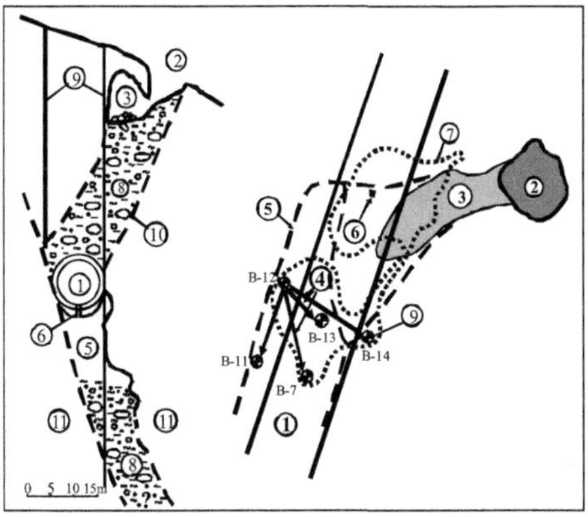


**Fig. 3** Investigations by smoke tracer of tunnel tube damage. 1. Tunnel tube, 2. Collapse at surface, 3. Cavern, 4. Connections established by smoke tracer, 5. Cavern contours below tunnel, 6. Opening in tunnel invert, 7. Cavern contour above tunnel 8. Unconsolidated cave deposits, 9. Boreholes, 10. Cavern wall, 11. Solid limestone

The used gas tracer must be lighter than air, easily detectable at the ground surface in high dilution (radioactive or smelling gases), adsorbed by water or water vapor in negligible quantity, and if possible, it should be visible (like smoke, in different colors).

The smoke tracer was used as a successful investigation method in case of heavily damaged headrace tunnel tube of the Reversible Power Plant Capljina (BiH), and for investigations of aeration zone below the Hutovo Reservoir (Fig. 3).

During the first year of tunnel (8 km long, 8 m diameter) operation under high pressure, indications of possible damage at the tunnel lining have been noticed. Among different investigation methods (drilling, water table monitoring, measuring of leakage, geological analysis, speleological investigations, TV logging and tracing by dye) the smoke tracer was applied (Milanović et al. 1987). To identify the cave shape and its interconnections around the endangered tunnel tube, the army smoke cans were lowered into the cave through a borehole. The smoke appearance from other boreholes showed that the washing out process formed mutually connected empty space below, at the side and above the tunnel tube.

The smoke tracer has been used during investigations of very karstified bottom of Hutovo reservoir. Two chemical compounds, which produce large quantities of smoke, were used as tracers. The investigation included the aeration zone, which was connected with the surface, in natural conditions, by a number of ponors at reservoir bottom. After water tightness treatment, by compacting of clayey layer at the reservoir bottom, blanketing of reservoir banks by shotcrete and plugging of ponors by inverted filter and PVC foil, only open natural connection was the main ponor Ponikva out of the reservoir (Fig. 4). Artificial connection with the aeration zone was established with some piezometric boreholes (∅ 5.1 cm) and aeration pipes (∅ 20–60 cm) (Milanović 2004).

Three strong fans, with a total capacity of 21 $m^3$/s (35 KW) were installed at the opening of the vertical manmade shaft, which is connected with Ponikva ponor. Total of about 350,000 $m^3$ of air were injected into the underground (Bagarić et al. 1980). Totally five smoke cans were thrown into the karst channel. The emerging

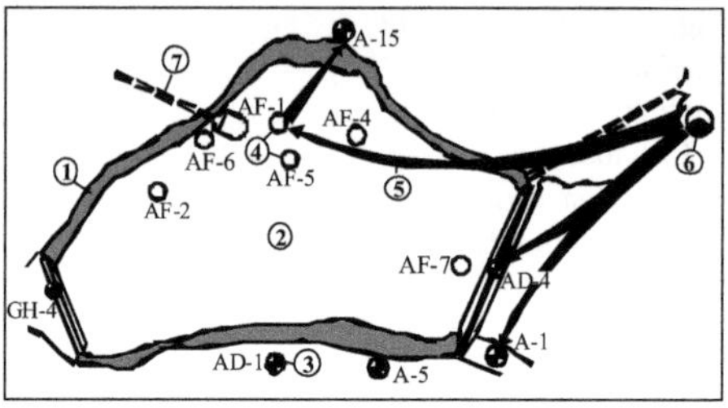


**Fig. 4** Hutovo Reservoir. 1. Reservoir bank, 2. Reservoir bottom, 3. Borehole, 4. Aeration pipe, 5. Underground connection, 6. Ponor Ponikva, 7. Tunnel

smoke was observed at the aeration pipe, at a distance of 1100 m away, six hours after the fan was started. The velocity of emerging smoke current was 5.5 m/s. The smoke also appeared from piezometers A-4 and A-5 in relatively small quantities. The smoke current from piezometers was 0.5 m/s.

In the next experiment, the smoke was injected into the aeration zone through the aeration pipe. The fan capacity for aeration tube was 0.24 $m^3/s$. Three smoke cans were activated at the bottom of the aeration pipe. During 10 hours of continuous fan work about 8200 $m^3$ of air was injected into the underground, but there was no indication of smoke appearance in the surrounding piezometers and aeration tubes. However, the next day the strong smell of smoke appeared from the borehole A-15, about 120 m fare from the aeration pipe.

## References


Bagarić I, Kovačina N, Milanović P (1980) The use of gas tracers for defining the position of conduits with important transport characteristics in the aeration zone of karst aquifer. Naš Krš. Bulletin of Speleological society, (in Serbian language) BiH. Sarajevo. Kovačina N, Skopljak E (1978). Micropulzation of pressure in piezometrs. Yugoslav-US research project "Karst Hydrology and Water Resources", not published. Sarajevo

Milanović P, Vučić M, Jokanović V (1987) A cavern around powerplant headrace tunnel tube. Groundwater effects in geotechnical engineering, Balkema, Roterdam

Milanović P (2006) Karst of Eastern Herzegovina and Dubrovnik litoral. ASOS, Belgrade

Milanović P (2004) Water Resources Engineering in Karst. CRC Press LLC, Boca Raton, Florida

# Mapping Buried Karst Features with Capacitive-Coupled Resistivity System (CCR) and Ground Penetrating Radar (GPR)

C. Neukum, C. Grützner, R. Azzam, and K. Reicherter

**Abstract** Surface near karst features in catchment areas of drinking water abstraction points are possible hotspots for groundwater contamination. In karst areas with absent or patchy cover sediments, karst features can be mapped with traditional techniques or by image processing of satellite images or aerial photographs. Karst features which are buried below thin soil or sediment cover may not necessarily be identified with these techniques. In a karst area in western Germany, geophysical surveys have been tested to identify karst features below a relatively thin soil cover, which have been refilled for agricultural purposes. Resistivity measurements (OhmMapper) and ground penetrating radar (GPR) have been applied on a site, where the presence of karst features (dolines) is historically proven. The results of the OhmMapper survey show distinct distribution of resistivity for those transects where high resistivity is interpreted as limestone and low resistivity is interpreted as soil, sediments and filled dolines. The results of the GPR survey are weak, and the structures of the filled dolines are not clearly represented by the GPR logs. This is probably due to the wet hydrological conditions during the measurement campaign. Capacitive-coupled resistivity measurement (CCR) is a suitable technique to identify surface near buried karst features. The CCR measurement can therefore be a useful supplement during the delineation of groundwater protection areas in karst areas where buried karst features are supposed.

## 1 Introduction

Capacitive-coupled resistivity measurements (CCR) have been used in geosciences for around 40 years (Timofeev et al. 1974; Ziekur and Grinat 2007). The method is

C. Neukum, R. Azzam
RWTH Aachen University, Engineering Geology and Hydrogeology,
Lochnerstraße 4–20, 52064 Aachen, Germany, e-mail: neukum@lih.rwth-aachen.de

C. Grützner, K. Reicherter
RWTH Aachen University, Neotectonics and Natural Hazards, Lochnerstraße 4–20,
52064 Aachen, Germany

B. Andreo et al. (Eds.), *Advances in Research in Karst Media* 
DOI 10.1007/978-3-642-12486-0, 

applied to a wide range of investigations in geosciences like groundwater prospection, permafrost research, archaeology, etc. Ground penetration radar (GPR) has been employed for detection of buried faults, underground tectonics, karst features, sedimentary structures, etc. for about 25 years (Goodman 1994; Neal 2004; Mochales et al. 2008). In this work, both investigation techniques have been applied to identify buried karst features in a catchment area, where groundwater is used for public water supply.

## 2 Test Site

The test site is in western Germany around 70 km west of Cologne and a few kilometres south of Aachen in the close vicinity of the border to Belgium (Fig. 1). The test site is in a karst area at the northern limb of the "Venn Sattel", which forms a large-scale NE-SW striking anticline in the northern part of the Eifel. It consists of rocks from Cambrian to Carboniferous age. The surveyed cross sections are within a karst catchment area of Lower Carboniferous age where groundwater is used for public water supply. The limestones show several phases of karstification, which results in a very high permeability of the aquifer (Kasig 1980; Dienst 1996). The limestone in the test site is covered mainly by Haplic Luvisols, Stagnic Albeluvisol and Regosols, which, in turn, cover loess sediments of several decimetres thickness.

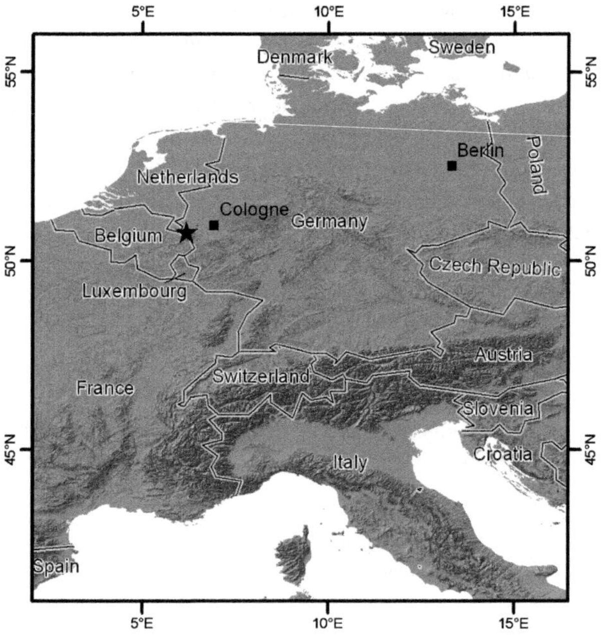


**Fig. 1** Location of the test site in Germany

Collapsed dolines have been refilled by farmers with local soil material to keep the land useable for agricultural purposes.

# 3 Methods

## *3.1 Capacitive-Coupled Resistivity System (CCR)*

The OhmMapper is a capacitive-coupled geoelectric system. The electrical field is transmitted by a capacitive coupling, and the specific resistivity of the underground is measureable to a depth of around 20 m. The method operates at frequencies between 8 to 32 kHz using a dipole transmitter and several receiver units. The investigation depth depends on the distance between transmitter and receivers. The benefits are a higher data collection rate compared to traditional galvanic-coupled geoelectric measurements across long profiles and the possibility to gauge sealed or frozen surfaces with high resistivities. Furthermore, sticking electrodes into the ground is not necessary anymore. Five receivers with two 2.5-m dipole cables for each unit have been used for this study. A penetration depth of approximately 5 m could be reached. Data inversion has been done with RES2DINV as there is a lack of special processing software for line electrodes. From these studies it is known that better inversion results may be achieved with a customised virtual dipole length that differs from the one used in the field. However, this approach is much more time

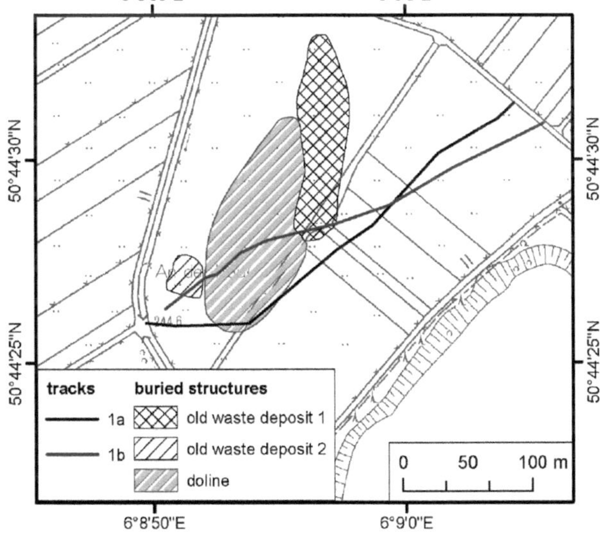


**Fig. 2** Tracks of CCR and GPR measurements and location of a buried karst feature with adjacent anthropogenic soil structures

consuming. As a high spatial resolution is not needed for this geological investigation, the standard technique was used. Two CCR profiles have been recorded. One along track 1a and one along track 1b (Fig. 2). Due to agricultural infrastructure it was not possible to cross the buried karst feature in other directions.

### 3.2 Ground Penetrating Radar (GPR)

GPR is an active and non-destructive geophysical method for near surface investigations. At the frequency range of 10 MHz to 2 GHz, spatial resolution is up to some centimetres while the penetration depth is up to some 10 m, depending on the antenna. The propagation of electromagnetic waves in solid media is mostly influenced by dielectric permittivity and conductivity; hence, GPR is most suitable when those parameters change (because of water content, salinity, porosity, cavity, grain size or saturation). Radar waves are reflected at interfaces between materials with different dielectric permittivities. The depth of the imaged structures can be computed from the two-way-traveltime of the waves when the wave velocity is known.

For identifying the buried karst structure a GSSI 100 MHz antenna was used with the SIR-3000 and a hand-held GPS. All data processing was done with ReflexW by Sandmeier Software, Karlsruhe, Germany. The processing steps applied were move start time, background removal, gain adjustment and averaging. Two GPR profiles along track 1b (Fig. 2), one from west to east and another into opposite direction were recorded each over a distance of around 330 m.

## 4 Results

The CCR logs of both tracks show distinct vertical and horizontal distributions of the underground resistivity (Fig. 3). Track 1a shows a sudden vertical decrease of high resistivities after 80 m that lasts for around 20 m. From 100 to 200 m the resistivity is then increasing to higher values and from 200 to 220 m a decrease of high resistivities is measured. From 220 m to the end of the profile the resistivity is on a more or less equal level. The profile of track 1b shows a sudden decrease in resistivity at around 50 m, which keeps on an equal level for some more 50 m. Between 110 and 120 m the resistivity is decreasing over a short spatial distance and increases sharply over the next 5 m, which might be corresponding to the old waste deposit 1. After this sharp peak it decreases again to lower resistivities keeping the level for the next 20 m. Between 160 and 190 m, a three-forked peak of higher resistivities is appearing after which the resistivities are decreasing again to be continued on a more or less equal level to the end of the profile. The distinct spatial distribution of the inverse model resistivity is interpreted as the uneven top of the limestone. The overlying soils consist of a high percentage of clay and silt, therefore having a lower resistivity.

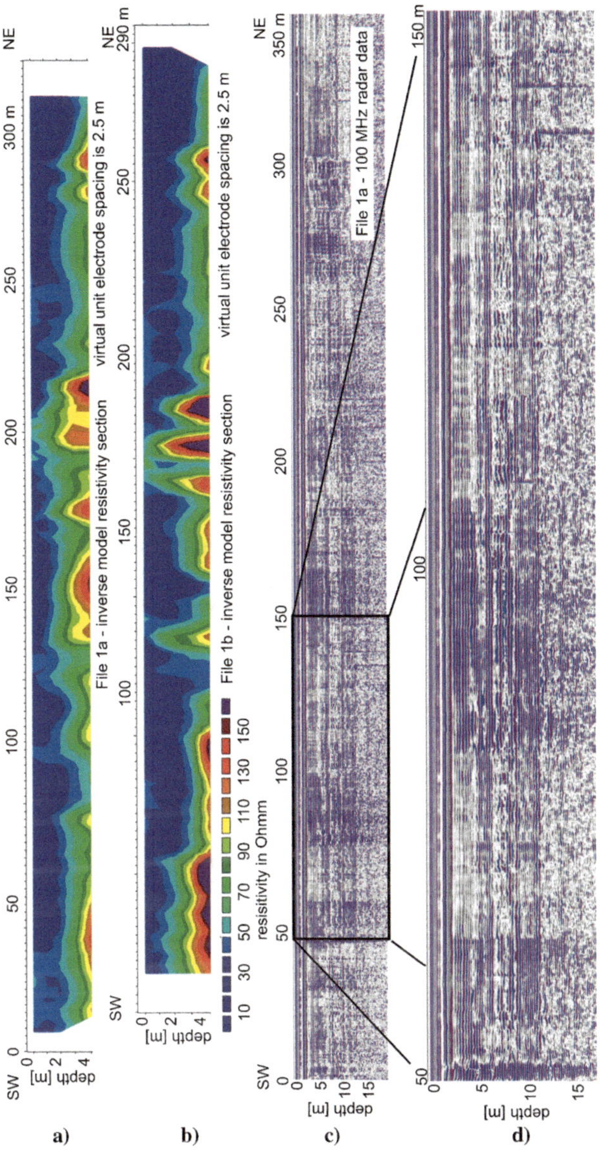


**Fig. 3** OhmMapper records: **a** track 1a, **b** track 1b. GPR records: **c** track 1a, and **d** detailed view of profilemeter 50 to 150

The GPR logs show a poor data quality and no distinct reflector over the whole tracked profile. The uneven surface, which is indicated by the CCR logs, cannot be affirmed by the GPR measurements. This might be due to the wet surface conditions during recording (high attenuation of electromagnetic waves). Amplitude variations along the profile (darker and lighter areas in the radargramm) might point to karst structures but do not coincide with the CCR profiles. Those features might be caused by surface effects as well.

## 5 Conclusion

The GPR technique was not successfully applied under the humid climate conditions. A future application under dry conditions will give more information on the usefulness of GPR for the identification of buried karst structures. Studies in more arid areas showed satisfying results in the past (Mochales et al. 2008).

The measurements show that CCR is a suitable technique to identify surface near buried karst features. Using different experimental setups allow a deeper insight into underground structures, which might be necessary in more developed karst systems. The results of the CCR measurements on buried karst structures are comparable to galvanic-coupled geoelectric measurements (Neukum 2006). The main advantage of CCR is a more simple experimental setup, which results in more time and cost effective geoelectric logging. The CCR measurement can therefore be a useful supplement during the delineation of groundwater protection areas in karst areas where buried karst features are suspected.

## References

Dienst MA (1996) Hydrogeologische und wasserwirtschaftliche Untersuchungen in paläozoischen Kalksteinen südlich Aachen [hydrogeological and water managmenet related investigations of the paleozoic limestone south of Aachen]. Mitteilungen zur Ingenieurgeologie und Hydrogeologie, 60, Department of Engineering Geology and Hydrogeology, RWTH Aachen University, 135 p.

Goodman D (1994) Ground-penetrating radar simulation in engineering and archaeology. Geophysics 59(2):224–232

Kasig W (1980) Zur Geologie des Aachener Unterkarbons [Geology of the Aachen Lower Carboniferous]. Habilitation RWTH Aachen University

Mochales T, Casas AM, Pueyo EL, Pueyo O, Román MT, Pocoví A, Soriano MA, Ansón D (2008) Detection of underground cavities by combining gravity, magnetic and ground penetrating radar surveys: a case study from the Zaragoza area, NE Spain. Environ Geol 53:1067–1077

Neal A (2004) Ground-penetrating radar and its use in sedimentology: principles, problems and progress. Earth Sci Rev 66:261–330

Neukum C (2006) Ermittlung eines Validierungsparameters zum Vergleich von Vulnerabilitätskonzepten in Karstgebieten [Determination of a validation parameter to compare vulnerability concepts in Karst areas]. PhD Thesis Univ. Karlsruhe Germany

Timofeev VM, Rogozinski AW, Hunter JA, Douma M (1994) A new ground resistivity method for engineering and environmental geophysics. Proceedings of the Symposium for the Application of Geophysics for Environmental and Engineering Applications, Environmental and Engineering Geophysical Society, 701–715

Ziekur R, Grinat M (2007) Ein Vergleich von Widerstandsmessungen mit einem Multielektrodensystem und dem OhmMapperTM. 22. Kolloquium Elektromagnetische Tiefenforschung, Hotel Maxičky Děčín Czech Republic, October 1–5

# Karst-ALEA: a Scientific Based Karst Risk Assessment for Underground Engineering

M. Filipponi and P.-Y. Jeannin

**Abstract** This paper presents the Karst-ALEA ground investigation method. The method consists of linking various existing evidence derived from geological and geomorphological data to a probability of karst occurrence. This allows assigning a risk to some specific 3D volumes of a karstic rock mass.

## 1 Introduction

Many recent tunnel constructions around the world showed that uncertainties in the geology of the rock volume being tunnelled through, including those related to karst processes, are a major issue, as they may lead to economic, social, security-related and environmental problems. Problems are increased due to an insufficient understanding of speleogenesis (e.g. Marinos 2001).

In previous years, different techniques were developed for the local detection of voids during the excavation and operating phases and control of encountered karst structures (Marinos 2001). However, today no method for the prediction of karst occurrences at a more regional-scale is available (i.e. for the planning phase).

The lack of such a methodology of regional karst risk assessment for underground construction is due to gaps in understanding speleogenesis but also to a missing transfer of knowledge from karst research to the applied domains.

Most ground investigation methods and classifications that are presently available do not include speleogenetic considerations and consider the karst occurrence

M. Filipponi
Laboratoire de géologie de l'ingénieur et de l'environnement (GEOLEP), Swiss Federal Institute of Technology Lausanne (EPFL), Switzerland, e-mail: marco.filipponi@epfl.ch

P.-Y. Jeannin
Swiss Institute of Speleology and Karstology (SISKA), Switzerland

B. Andreo et al. (Eds.), *Advances in Research in Karst Media*
DOI 10.1007/978-3-642-12486-0, © Springer 2010

as randomly distributed in space (Zhang et al. 1993; Veni 2005). This makes any inter/extrapolation of observed/detected karst features almost impossible.

The main question is to know whether there is a developed network of conduits and if there is, where it is and in some cases what their characteristics are (e.g. size of the karst conduits; their filling: water/air/sediment; hydrogeological condition: phreatic/vadose).

This paper presents a scientific based method for karst risk assessment for underground engineering. The method is still in validation and development. However, the application of the method on first case studies was encouraging, showing that the proposed method was able to correctly delimitate zones of increased karst risk.

## 2 Speleogenetical Background

Understanding of the spatial distribution of karstic voids significantly progressed during the last two decades. This chapter presents a short state of the art of the principles of karst development, which is relevant for karst risk assessment. For a complete overview of cave development and concepts presented here, be referred to the literature.

Authors investigating cave networks generally agree that conduits do not develop randomly but are related to discontinuities within the rock mass. Karstic rock masses are pervaded by a network of discontinuities (joints, faults, bedding planes and beds). These are the primary flow paths for groundwater and guide cave development. Furthermore, karstification along these discontinuities is again not random but selective. Recent research results, based on the 3D analysis of cave systems around the world, showed that at regional scale only a few discrete lithostratigraphical horizons (i.e. between 3 and 5 horizons) guide more than 70% of the phreatic conduits (e.g. Filipponi et al. 2009) (Fig. 1). Detailed investigation of these particular horizons (also named "inception horizons," Lowe 1992) showed that most of the time they have a thickness of some centimetres to decimetres and are identifiable (Lowe 1992; Filipponi 2009).

The identification of the position of inception features in a rock mass will provide a substantial increase of information for delimiting zones of increased probability of karst occurrences.

From a hydrogeological point of view, a karstic rock mass can be subdivided into four speleogenetic zones (inception, gestation, phreatic development and vadose development zone) with distinct speleogenetic processes as well as a specific dissolution void distributions (Filipponi 2009).

In the inception zone ("deep phreatic conditions") it is assumed that the flow and dissolution rate are very slight and therefore also the size of the dissolution voids.

The conduit size distribution changes dramatically within the gestation and phreatic development zone (near the water table). Depending on the geometry of the inception features relative to the re- and discharge areas and the duration of the hydrological phase (time between two valley incision events, i.e., phases where the

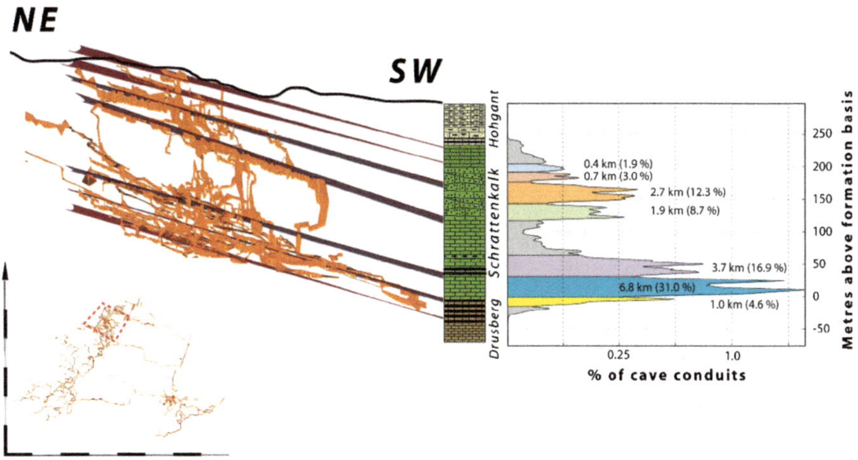


**Fig. 1** Horizontal projection of one part of the Siebenhengste cave system (Switzerland, length 154 km). Up to seven potential "inception horizons" can be identified within the Schrattenkalk limestone formation. Total thickness of the limestone sequence is nearly 180 m

hydrologic boundary conditions are more or less constant), different geometries and characteristics (size distribution) can be assumed (Filipponi 2009).

Flow in the vadose cave development zone is ultimately controlled by gravity i.e. is mainly vertical. The vertical cave conduit distribution within this zone (below the epikarst) can be considered as almost homogeneous in altitude; on the contrary, the development of the meandering passages between shafts is often related to inception horizons.

Because, the hydrogeological boundary conditions of most cave systems changed during time (mainly the position re- and/or discharge areas by valley incision events), there was observed in many cave systems a superposition of different phases (i.e., coexistence of voids at different karstification state are close together). Each fossil spring is thus correlated to a speleogenetic phase and corresponding passages (Palmer 1987). A study of evolution of the springs in geologic time helps to improve considerably the prediction.

These two evidences, the existence of inception features and the enhancements on the understanding of the development of underground karst structures in time and space due to changes in hydrogeological boundary conditions, will significantly improve the design and interpretation of ground investigation (Fig. 2). Essentially, it is now possible to quantify the probability of karst occurrences inside a karst massif by identifying the few inception features that guide the karstification in regional scale, reconstructing the hydrogeological history and winnowing different speleogenetical zones.

Although the reconstruction of the "true" geometry (i.e., the exact position of the karst conduits and their characteristics) is not possible yet, the method provides a substantial increase of information on the potential for encountering karst occurrences in underground engineering projects, at least in a probabilistic way.

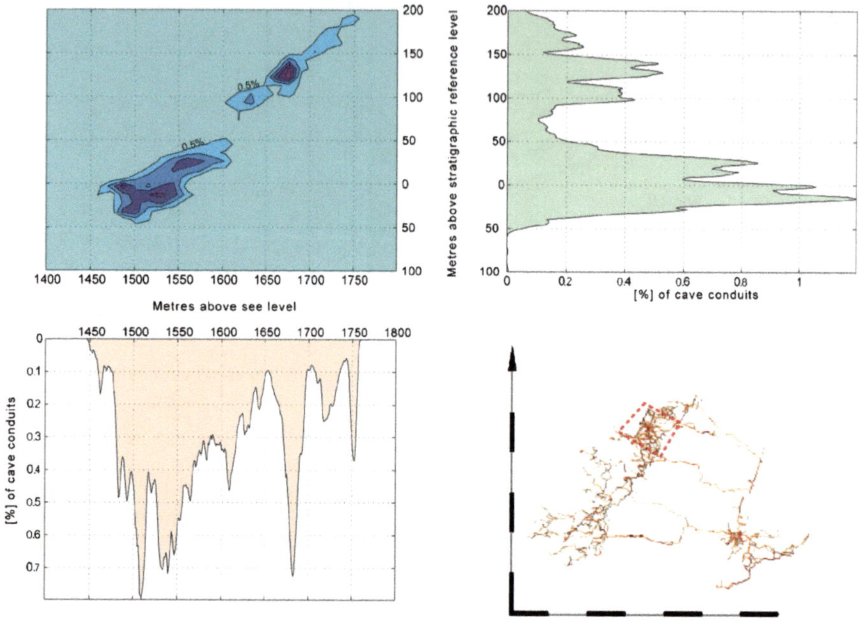


**Fig. 2** The frequency map shows that an increased number of conduits developed at the intersection of the inception horizons with the paleo-water tables. *Top left*: Frequency map of the cave segments (%) relative to the altitude ($x$-axis) and the distance to the lithological reference surface ($y$-axis), contour lines interval 0.5% of conduits. *Top right*: Histogram of the conduit distribution relative to the stratigraphic reference horizon. *Bottom left*: Histogram of the conduit distribution relative to their elevation. *Bottom right*: Cave map of the Siebenhengste Cave System with the analysed zone (*rectangle*)

## 3 The Description of the Method

The hazard assessment is based on the assumption that a significant amount of dissolution features occur along the inception features and that the delimitation of the speleogenetic zones will characterise the probability of conduit size distribution along them. In other words, the concept of the speleogenetic zones (e.g. Filipponi 2009) can be reformulated into different probability of karst occurrence and characteristics zones and assign them hazard levels (Fig. 3). Thereby, it is possible to subdivide a karstic rock mass into areas of different risk levels.

Basically the method consists of five steps:

1. Identification of potential inception features: Inception features can be identified by 3D analysis of known cave systems by field observations or the interpretation of borehole logs and tests (Filipponi et al. 2009). *A 3D probability-model of the inception features will be produced.*
2. Reconstruction of the landscape history and the related speleogenetic zones: The landscape history can be reconstructed by interpretation of speleological

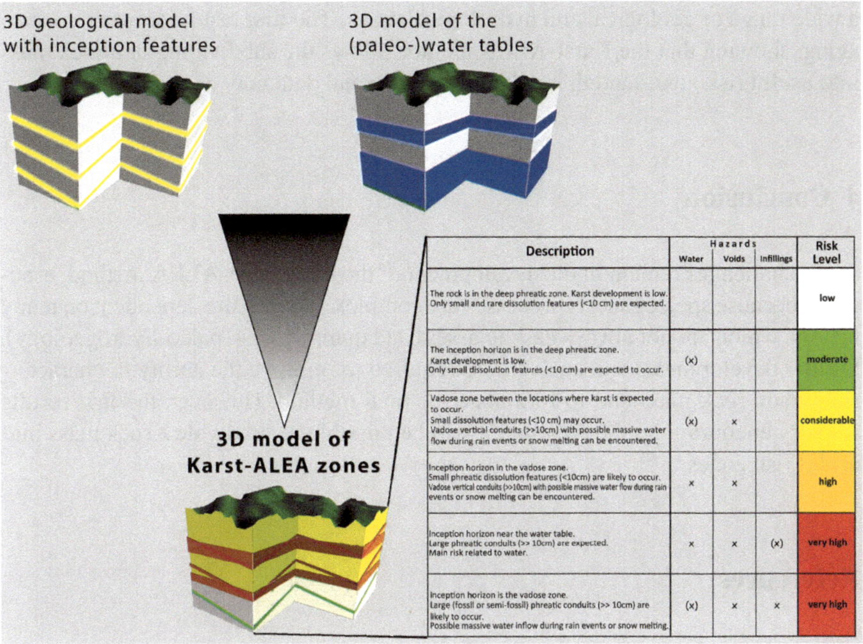


| Description | Hazards: Water | Hazards: Voids | Hazards: Infillings | Risk Level |
|---|---|---|---|---|
| The rock is in the deep phreatic zone. Karst development is low. Only small and rare dissolution features (<10 cm) are expected. | | | | low |
| The inception horizon is in the deep phreatic zone. Karst development is low. Only small dissolution features (<10 cm) are expected to occur. | (x) | | | moderate |
| Vadose zone between the locations where karst is expected to occur. Small dissolution features (<10 cm) may occur. Vadose vertical conduits (>>10cm) with possible massive water flow during rain events or snow melting can be encountered. | (x) | x | | considerable |
| Inception horizon in the vadose zone. Small phreatic dissolution features (<10cm) are likely to occur. Vadose vertical conduits (>>10cm) with possible massive water flow during rain events or snow melting can be encountered. | x | x | | high |
| Inception horizon near the water table. Large phreatic conduits (>> 10cm) are expected. Main risk related to water. | x | x | (x) | very high |
| Inception horizon is the vadose zone. Large (fossil or semi-fossil) phreatic conduits (>> 10cm) are likely to occur. Possible massive water inflow during rain events or snow melting. | (x) | x | x | very high |

**Fig. 3** Schematic description of the Karst-ALEA method, based on the combination of a 3D model of inception features and a model for the hydrogeological (paleo-)conditions

data or the interpretation of geomorphological data (e.g. identification of erosion/accumulation terraces). *A 3D probability-model of the speleogenetic zones will be produced.*

3. Delimitation of areas of different karst occurrence probabilities: based on the identified inception horizons and the reconstruction of the history of the speleogenetic zones, areas of different probabilities of karst occurrences can be deliminated. Therefore, the probability-models of step 1 and 2 will be settled together. *A 3D probability-model of karst occurrences will be the result.*
4. Characterization of hazards: For the different probability areas, delimited in step 3, the possible hazards are listed and evaluated, considering the current prevailing hydrological boundary conditions. The list includes not only hazards for the construction, but also for the environment.
5. Risk assessment: The combination of the deliminated probabilities for karst occurrence (step 3) with the possible hazards (step 4) leads to a risk assessment (Fig. 3).

It should be noted that this method is not applied on a plan or a longitudinal section analysis but based on 3D models (i.e. 3D geological models as well as 3D models of the landscape reconstruction).

A validation of the method in terms of utility and practicability is currently done as back analysis of already constructed tunnels. The tunnels were selected to cover a wide range of geological and hydrological cases. The first results are very encouraging, showing that the Karst-ALEA method allows the subdividing of a rock mass into useful risk zones and that only a few additional data may be required.

## 4 Conclusion

Some challenges remain in the development of the final Karst-ALEA method, especially because speleogenesis processes are complex and they are dependent on many factors, which are not always easy to assess and quantify (e.g. paleo-hydrogeology). Further developments and analysis are required to improve the ability to characterize karstic rock mass and to formulate the final method. However, the first results are very encouraging, showing that it will be possible to subdivide a rock mass into useful risk zones.

## References


Filipponi M (2009) Spatial analysis of karst conduit networks and determination of parameters controlling the speleogenesis along preferential lithostratigraphic horizons. Thesis, Ecole polytechnique fédérale, Lausanne

Filipponi M, Jeannin P-Y, Tacher L (2009) Evidence of inception horizons in karst conduit networks. Geomorphology 106:86–99

Lowe DJ (1992) The origin of limestone caverns: in inception horizon hypothesis. PhD Thesis, Manchester Polytechnic, United Kingdom

Marinos PG (2001) Tunnelling and mining in karstic terrain: An engineering challenge. In: Beck and Herring (eds): Geotechnical and Environmental Applications of Karst Geology and Hydrology, 3–16

Palmer AN (1987) Cave levels and their interpretation. Bulletin of the National Speleological Society of America 49, 50–66

Veni G (2005) Lithology as a predictive tool of conduit morphology and hydrology in environmental impact assessments. Proceeding of the Sinkholes and the Engineering and Environmental Impacts of Karst Conference, 46–56

Zhang Q, Tian S, Mo Y, Dong X, Hao S (1993) An expert system for prediction of karst disaster in excavation of tunnels or underground structures through a carbonate rock area. Tunnelling and Underground space Technology 8(3):373–378

# The Impacts of Quarrying in the Apulian Karst (Italy)

M. Parise

**Abstract** The karst region of Apulia (S Italy) has been interested for many centuries in quarrying activities, with regard to extraction of the main carbonate rocks that were used as ornamental and building stones, strongly contributing to creation of Apulian architectural styles, from "romanico" in the Bari area to "barocco leccese" farther south. Rocks were quarried not exclusively at the surface, but also through extensive systems of underground quarries. Apart from the negative effects directly derived from quarrying, other effects have to be addressed after extraction, when the sites are abandoned without any reclamation work provided, often becoming illegal landfills. This paper, through selected examples, intends to highlight environmental degradation related to presence of quarries in Apulia, and the frequent mismanagement of karst. In particular, some cases will illustrate destruction of caves (even quite remarkable in size and depth) and loss of the karst landscape.

## 1 Introduction

High fragility and vulnerability of the karst environment is due to a number of intrinsic characteristics of karst (White 1988; Ford and Williams 2007). Apulia, an almost entirely karst region of southern Italy, is not an exception to the rule. On more than one occasion the effects of human activities resulted producing severe changes to the karst ecosystems with serious negative effects to the region (Parise and Pascali 2003; Calò and Parise 2006). Among human activities, quarrying is certainly one of those producing the worst impacts on the natural landscape. Development of quarries causes immediate, sometimes irreversible, changes to the original environment: loss of surface karst features, destruction of the epikarst, diversion of the surface hy-

M. Parise
CNR-IRPI, Bari, Italy, e-mail: m.parise@ba.irpi.cnr.it

B. Andreo et al. (Eds.), *Advances in Research in Karst Media*
DOI 10.1007/978-3-642-12486-0, © Springer 2010

drography are only some of the possible negative effects that have to be considered (Gunn 1993, 2004).

Apulia (Fig. 1) has been involved for many centuries in quarrying activity, regarding the extraction of Cretaceous limestone, Miocene marly calcarenites, and Plio-Pleistocene calcarenites, that are the most widespread lithotypes in the region. These rocks were used as both ornamental and building stones (Andriani and Walsh 2003), strongly contributing to create the architectural styles typical of Apulia, from "*romanico*" in the Bari district to "*barocco leccese*" farther south (Fig. 2). They were quarried not only at the surface, but also through development of extensive

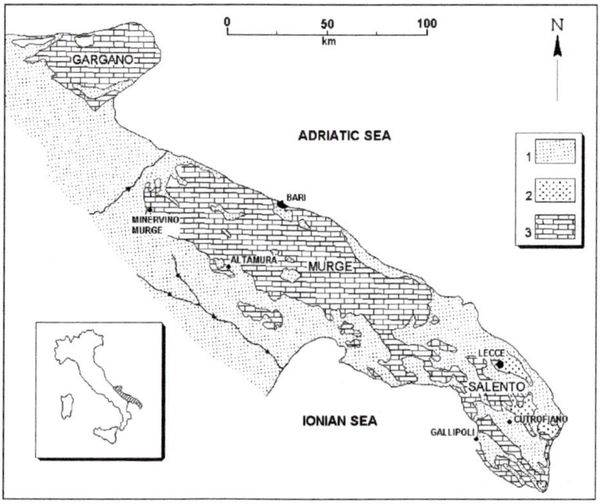


**Fig. 1** Geological sketch of Apulia, and localities mentioned in the text. Key: 1) alluvial deposits, clays and calcarenites (Pliocene-Pleistocene); 2) bioclastic carbonate rocks (Paleogene) and calcarenites (Miocene); 3) platform carbonate rocks (Cretaceous)

**Fig. 2** Examples of the use of carbonate rocks in Apulia: the Cretaceous limestones are used for many churches of "*romanico*" style in the Bari district, and for "*trulli*", the typical Apulian countryside house (*left*). Miocene marly calcarenites have been, on the other hand, widely utilized for the "*barocco*"-style churches in Lecce (*right*), farther south

systems of underground quarries. In many sectors of the region, in addition to use of the rocks as ornamental stones for the main religious buildings and monuments, they were also utilized for common houses in the towns, and for the rural architecture that includes "*trulli*", a typical cone-shaped country house which characterizes wide sectors of Murge, in central Apulia (Fig. 1), and the widespread dry-stone walls used to delimit the fields (Laureano 2001).

Underground quarrying historically developed throughout all the Apulia region, due to a number of reasons, the main ones being: i) the presence of the rocks with the best characteristics at a certain depth, not exposed at the surface; ii) the need to preserve surface lands to be used for agriculture. Thus, subterranean galleries and passages were excavated to reach the rock levels to be cultivated for extraction: this mostly occurred by digging vertical shafts from the surface. Even though generally the levels to be excavated were present at shallow depth, in some cases it was necessary to excavate deep shafts, down to 35–40 m. Such works, starting from the XIX century, were undoubtedly a nice piece of engineering ability for that time, and represent today, for those sites where the tunnels are still safe and sound, interesting examples of industrial archaeology. In many cases, however, after a first phase of rational and careful exploitation of the underground rocks, that even considered carefully the need to leave pillars of certain size to provide support to the underground galleries, a more confused and less controlled phase followed, without control in the development of the subterranean voids. This brought about several events of instability (Fig. 3), not only involving the underground environment but also that which transferred their effects to the surface, due to upward propagation of falls from the ceiling, and to direct opening of sinkholes above the tunnels; As a consequence, several quarries were abandoned, and sometimes partially filled by solid wastes. With time, when quarries were located at the outskirts of towns, the urban expansion that occurred during the XX century, combined with the loss of memory of the presence of the subterranean features, resulted in construction of buildings just above the underground galleries. Many towns of Apulia are today in such a situation, and in the last years have been affected by instability events that brought to the attention of public opinion the risk posed by these "hidden" and "forgotten" structures.

**Fig. 3** Instability in calcarenite underground quarries: *left*: extensive failures at Altamura; *center*: falls from ceiling and walls in a quarry at Cutrofiano; *right*: wall failures in a quarry at Gallipoli

## 2 Case Studies

Minervino Murge is in inland Apulia, near the border with Basilicata. Cretaceous limestones characterize the landscape, in sub-horizontal bedding. These rocks have been affected by at least two phases of karst that led to filling of the palaeo-caves by terra rossa and bauxite deposits. Many quarries in the Cretaceous limestones have been opened starting from the 1950s, and the areas of extraction have progressively increased since then (Fig. 4).

Even though in 1986 a specific law was promulgated to protect caves in Apulia, opening and progressive expansion of quarries brought rapid destruction of many karst sites and caves. Hostility of quarry workers towards cavers was very strong, and even when a cave was found in an advancing quarry, all the requests to visit and explore the cave were denied. The main problem was that the aforementioned law was never really enforced, which made impossible the safeguarding of the caves discovered during quarrying (Fig. 5). In this way, many caves, even of large size, have been lost.

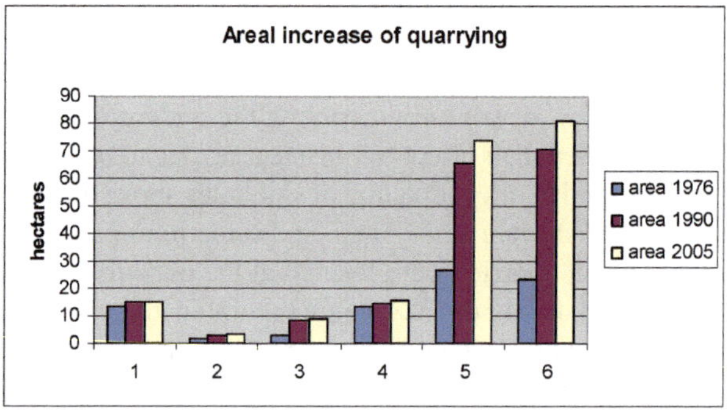


**Fig. 4** Areal evolution of quarries in the Minervino Murge territory (time span 1976–2005). Numbers on the $x$-axis refer to six 1:5000 scale maps: 1) Minervino Murge; 2) La Contessa; 3) Minervino Murge Faro; 4) Minervino Murge sud; 5) Mass. Di Palma; 6) Monte Scorzone

**Fig. 5** Surface and underground quarrying: large caves destroyed by quarrying at Minervino Murge (*left*), and an access shaft to underground calcarenite quarry at Cutrofiano (*right*)

One of the best examples of underground quarrying is found near Cutrofiano, a small town in the Lecce province where Pleistocene calcarenites have been largely quarried and used as building stones (Parise et al. 2007). Calcarenites are overlain by sandy-silts and clays, ranging in thickness from 5 to 35 m; this was the main reason to start underground quarrying, by digging deep shafts (Fig. 5) to pass the surface deposits and reach the calcarenites, without losing any value for the agricultural land. Quarrying was the main activity in the area, and practically all the families of the local community had at least one person involved in working underground. At the time the first quarries were opened (first decades of the XX century), they represented with no doubt excellent pieces of engineering work. Advancement of the quarries proceeded manually, but with astonishing precision: the galleries were stopped before reaching the sectors where, at the surface, the main roads ran, or at property limits.

There was a great regularity in width and height of the galleries: the overall plan of the quarries was in chessboard pattern or with staggered pillars, but size of the pillars was always carefully planned and maintained, for security reasons. Later on, in the peak phase of request for building stones, further quarries developed, and in some cases the work was not carefully planned as before: greater size galleries were locally excavated, property limits overcome, and the first signs of instability were registered, consisting in detachment from the ceiling or the walls (Fig. 3), and in some cases failures propagating upward until reaching the ground surface, where sinkholes developed. Many quarries consequently had to be abandoned, and, later on, once the work in the underground environment became no more convenient economically, the galleries were used to dump solid and liquid wastes, a common practice, unfortunately, both in surface and underground quarries of Apulia (Delle Rose et al. 2007).

Recently, within reclamation programs, part of the extensive network of subterranean galleries at Cutrofiano was explored, with particular focus on instability, both for direct failures within the quarries and the possibility of propagation at the

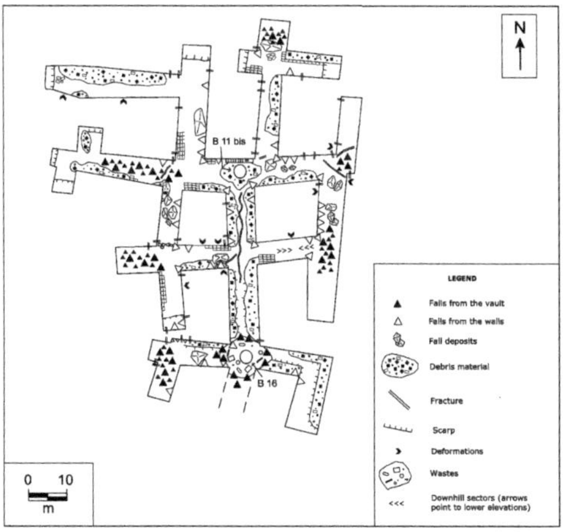


**Fig. 6** Survey of an underground quarry at Cutrofiano, showing the main features of instabilities

surface. Maps of surveyed galleries were produced (an example being shown as Fig. 6) to illustrate the many observed evidences of instability. These include, in addition to fall deposits deriving from both the vault and the walls, also evidence of deformations (rock wedges protruding from the walls, open fractures, etc.) that are significant premonitory signs of incipient failures.

It has to be noted that, notwithstanding the recent age of the quarries, no map showing their spatial development was available at the technical office of local authorities. It was as if the memory of the underground work was gone, even though it represented the main activity for most of the men of the previous generations.

## 3 Conclusions

Quarrying activity is without a doubt one of the most dangerous anthropogenic activities in karst, for direct and indirect effects. The quarrying industry has the ability to alter the land surface more comprehensively than any other peacetime human activity, determining the formation of new anthropogenic karst landforms (Gagen and Gunn 1987). The latter, consisting of newly produced rock slopes, are young landforms which are out of equilibrium with the surrounding environment, and therefore likely to evolve rapidly from their form, through rock falls, toppling failures, and detachment of rock blocks. Due to high fragility of karst, and the extreme vulnerability of its precious water resources, quarrying in karst areas should be carefully planned and controlled. This was not the case, unfortunately, for Apulia, where a number of environmental disasters have been produced because of uncontrolled development of quarries, with severe and irreversible damage to karst. Tens of caves destroyed, loss of peculiar, sometimes unique, ecosystems, loss of cultural and natural landscape are only some of the negative effects registered in the last decades. Many problems arise, in particular, at the end of the extraction activity, as no reclamation work is generally carried out, which results in further pollution. It should be fundamental, on the other hand, to recover the site, in the attempt to put it back within the local natural landscape.

## References

Andriani GF, Walsh N (2003) Fabric, porosity and water permeability of calcarenites from Apulia (SE Italy) used as building and ornamental stone. Bull Eng Geol Environ 62:77–84

Calò F, Parise M (2006) Evaluating the human disturbance to karst environments in southern Italy. Acta Carsologica 35(2):47–56

Delle Rose M, Parise M, Andriani GF (2007) Evaluating the impact of quarrying on karst aquifers of Salento (S Italy). In: Parise M, Gunn J (eds) Natural and anthropogenic hazards in karst areas: recognition, analysis and mitigation, Geol Soc London, sp publ 279:153–171

Ekmekci M (1993) Impact of quarries on karst groundwater systems. In: Gunay G, Johnson AI, Back W (eds) Hydrogeological processes in karst terranes, IAHS publ 207:3–6

Ford D, Williams P (2007) Karst hydrogeology and geomorphology. Wiley, New York

Gagen P, Gunn J (1987) A geomorphological approach to restoration blasting in limestone quarries. In: Beck BF, Wilson WL (eds) Karst hydrogeology: engineering and environmental applications. Proc. 2nd Multidisc Conf Sinkholes Environ Impacts Karst, 457–461
Gunn J (1993) The geomorphological impacts of limestone quarrying. Catena 25:187–198
Gunn J (2004) Quarrying of limestones. In: Gunn J (ed) Encyclopedia of cave and karst science. Routledge, London, 608–611
Hobbs SL, Gunn J (1998) The hydrogeological effect of quarrying karstified limestone: options for protection and mitigation. Quart J Eng Geol 31:147–157
LaMoreaux PE, Newton JG (19867) Catastrophic subsidence: an environmental hazard, Shelby County, Alabama. Environ Geol Water Sci 8(1/2):25–40
LaMoreaux PE, Powell WJ, LeGrand HE (1997) Environmental and legal aspects of karst areas. Environ Geol 29:23–36
Laureano P (2001) Water atlas. Bollati Boringhieri
Parise M, Pascali V (2003) Surface and subsurface environmental degradation in the karst of Apulia (southern Italy). Environ Geol 44:247–256
Parise M, Donno G, De Pascalis A, De Pascalis F, Inguscio S (2007) Subsidence and sinkholes related to quarrying in karst. Geophys Res Abs 9:01460
White WB (1988) Geomorphology and hydrology of karst terrains. Oxford Univ. Press, Oxford

# Influence of Karst Phenomena on Water Inflow to Zn-Pb Mines in the Olkusz District (S Poland)

J. Motyka and M. Czop

**Abstract** Lead and zinc ore deposits of the Mississippi Valley type (MVT), located in the Olkusz ore district (south Poland) are connected with strongly karstified Triassic carbonates, mainly dolomites. In this paper changes in the groundwater inflow quantities to the Zn-Pb mines in the Olkusz region are discussed: Boleslaw, Olkusz and Pomorzany. The mining activity in the Olkusz region is connected with probably highest groundwater inflows in the World. Occurrence of the karstic systems within Triassic rocks is a reason for extremely high water inflow to the underground mining workings, obtaining the value of 40 $m^3$/min in Boleslaw mine, about 95 $m^3$/min in Olkusz mine and about 300 $m^3$/min in Pomorzany mine. The main problem in the dewatering of the Triassic ground is the presence of the water-filled karstic systems. After the cutting of the karstic channels or caverns by the mining workings an abrupt increase of the water inflow to the mine dewatering systems is observed. Karstic forms, as an intrinsic drainage systems for fissures and matrix porosity, are responsible for enlargement of the recharge zone of mines.

## 1 Introduction

Lead and zinc ore deposits in the Olkusz region are occurring in the northeast part of the Upper Silesian Coal Basin periphery. These deposits belong to Mississippi Valley type (MVT) group of the Zn-Pb ores. According to Leach and Sangster (1993), MVT ores are epigenetic type, precipitated from dense basinal brines at temperatures ranging between 75 and 200 °C, typically in platform carbonate sequences and lacking genetic affinities to igneous activity. Hosted rocks for the Olkusz Zn-Pb ores are Middle Triassic strongly karstified carbonates, mainly ore-bearing dolomites.

J. Motyka, M. Czop
AGH – University of Science and Technology, Faculty of Geology, Geophysics and Environmental Protection, Department of Hydrogeology and Engineering Geology, Krakow, Poland, e-mail: motyka@agh.edu.pl; mariucz@agh.edu.pl

B. Andreo et al. (Eds.), *Advances in Research in Karst Media* 
DOI 10.1007/978-3-642-12486-0, 

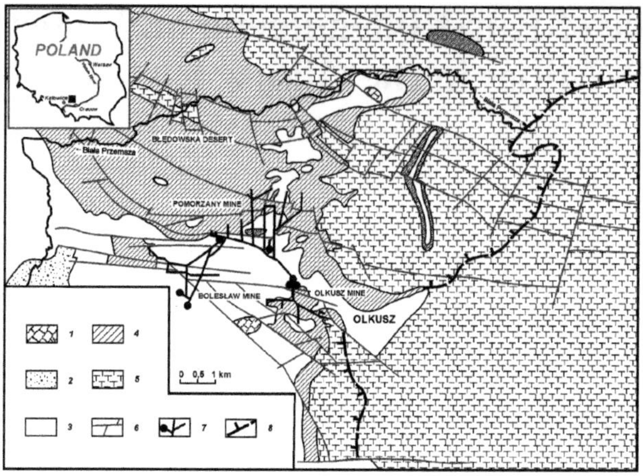


**Fig. 1** Geological map of the Olkusz region Legend: 1) D-C1 carbonates; 2) P-conglomerates; 3) T1,2 – carbonates; 4) T3 – clays, claystones; 5) J3 – limestones; 6) main faults; 7) mine galleries; 8) range of T3 under J3

Mining of the ores in the Olkusz region has a very long tradition and according to historical sources dates back to the 12th century. Nevertheless, archaeological data indicate that exploitation of the lead ores with silver admixture was operating in late Neolith. After running the exploitation in the saturation zone, water was the serious factor, which make difficulties for execution of the underground mining workings. Drainage of the ore-bearing carbonate rocks was started in the 16th century, by using drain adits. The parts of these old workings are existing and still operating. Intensive drainage of the Triassic carbonates began at the end of the 19th century with application of the steam-driven dewatering pump in the few small mines. After 2nd World War the Boleslaw mine, which was flooded in the 1930s of the 20th century, was activate. Moreover, the two new mines: Olkusz and Pomorzany were built in the Olkusz region (Fig. 1).

## 2 Groundwater Flow Systems Within Triassic Aquifer

Four main groundwater aquifers are present in the Olkusz Zn-Pb region: 1) Quaternary, 2) Jurassic, 3) Triassic and 4) Palaeozoic (Devonian and Carboniferous). The Quaternary aquifer is of porous type, built up of fluvioglacial sands with gravels, debris or rarely dust, clays or silt clays insertions. Jurassic, Triassic and Palaeozoic aquifers are of the karst-fissures type, consisting of carbonate rocks: dolomites and limestones. Groundwater flow conditions within Triassic main groundwater aquifer require the consideration in the regional and local scales and also for the karstic system, including the voids, which form the hydraulic network. According to Motyka (1998), within the Triassic carbonates it is possible to separate four systems of the particular spaces with differences in geometry, mainly dimensions: 1) porous space (matrix), 2) fissures, 3) caverns and 4) filled forms.

Matrix porosity of Triassic limestones is 0.002–0.08, with arithmetic mean of 0.017. The hydraulic conductivity of the matrix range from $3.42 \times 10^{-12}$ to $1.21 \times 10^{-8}$ m/s, with a geometric mean of $3.8 \times 10^{-10}$ m/s. The porous space in limestones has a very low specific yield, ranged from 0 to 0.0091, with the arithmetic mean equal to 0.00064. The fissures porosity range from 0.00023 to 0.044 with arithmetic mean equal to 0.0052. The mean hydraulic conductivity of the fissures network and interbedding planes of the limestones massif is $8.7 \times 10^{-4}$ m/s for the saturated zone. Mean cavern porosity in these rocks is estimated at about 0.002 and the hydraulic conductivity of the cavern space is about $1.6 \times 10^{-1}$ m/s (Motyka 1998). The same author estimates mean porosity of filled forms in Triassic carbonates which is about 0.01 for limestones and about 0.015 for dolomites. Mean value of hydraulic conductivity filled forms in these carbonates in the natural conditions is estimated as $2.8 \times 10^{-9}$ m/s.

Matrix porosity of the Triassic dolomites in the same area ranges from 0.0054 to 0.34, with an arithmetical mean value of 0.109. Hydraulic conductivity of the matrix ranges from $2.1 \times 10^{-11}$ to $4.9 \times 10^{-6}$ m/s, and specific yield ranges from 0 to 0.15 with an arithmetic mean equal to 0.028. The fissures porosity in the dolomites range between 0.004–0.016 with the mean value of 0.0037. The mean hydraulic conductivity of the fissures network and bedding planes is $1.03 \times 10^{-3}$ m/s for the shallow weathered zone and $5.5 \times 10^{-4}$ m/s for the saturated zone. Mean cavern porosity of dolomites is estimated at about 0.006, and the mean hydraulic conductivity is about $1.5 \times 10^{0}$ m/s. The hydraulic properties of filled forms are not known; however, Motyka (1998) estimates the porosity of filled forms for dolomites is about 0.015.

## 3 Water Inflow to Zn-Pb Mines

After 2nd World War in the Olkusz area three Zn-Pb mines were active: "Bolesław", Olkusz and Pomorzany.

Boleslaw mine was built at the beginning of the 19th century. Archival data indicate that water inflow to underground workings of the Boleslaw mine range from 22 $m^3$/min to 39 $m^3$/min. The mine was closed and flooded in 1931 due to economic reasons connected with the Great Depression. After 2nd World War, at the beginning of the 1950s of the 20th century, Boleslaw mine was dewatered. Systematic measurements of the water inflow to the Boleslaw mine were started in the second half of 1958. Temporal changes in the water inflow to the Boleslaw mine were strictly connected with opening out of the adjacent Zn-Pb ores. Dewatering of the Olkusz mine, which was started in August 1958, has only the insignificant influence on the water inflows to the Boleslaw mine. From November 1973 the rapid and significant decrease in the amount of the water inflows to the Boleslaw mine was observed. The reason for these changes was the beginning of the Pomorzany mine dewatering.

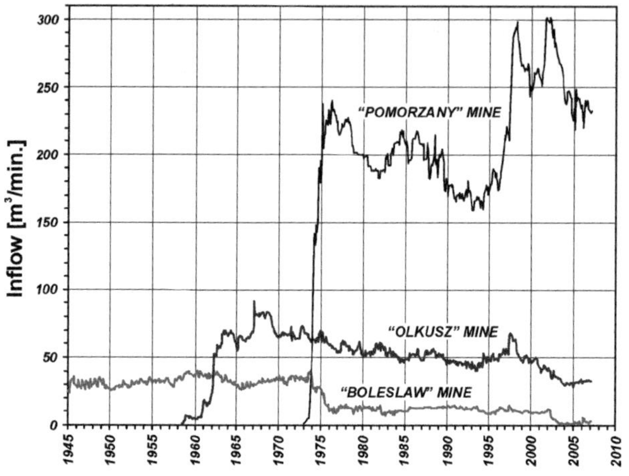


**Fig. 2** Groundwater inflows to the Zn-Pb mines in the Olkusz region

Inflow to the Boleslaw mine from about 30–34 m$^3$/min observed at the beginning of the 1970s in the 20th century was dropped to the value range from 8 m$^3$/min to 15 m$^3$/min in 1990 and only about 6–8 m$^3$/min in 1995. In 1995, the liquidation process of the Boleslaw mine was started and actually the mine underground workings are partly flooded and the main shaft is used as an intake of water for technological purposes. Depending on the requirements, the intake discharge ranges from 0.7 m$^3$/min to 5.6 m$^3$/min (Fig. 2).

The construction of the Olkusz mine was started in 1957 by shafts sinking. The water inflows at this beginning stage of mine dewatering were about 8 m$^3$/min. In 1960, the horizontal mine galleries were drifted for opening out and exploitation of the Zn-Pb ore. After cutting of the Lower Triassic dolomites, underlying ore-bearing carbonates, water inflows to the Olkusz mine increased to the value of about 10 m$^3$/min. As a consequence of the Middle Triassic limestone cutting the further increase of water inflow to about 30 m$^3$/min was observed (Fig. 2). In April 1962, the main exploratory drift cut the system of the caverns in the Middle Triassic limestones. The inflow of water from these karstic forms reached the value of about 20 m$^3$/min and the total inflow to the Olkusz mine increased to about 52 m$^3$/min. As a consequence of the progress in mining exploration in the next 2 years, further increase of the water inflow to the Olkusz mine was observed. Between May 1963 and August 1964, the amount of the water inflowing to the dewatering system of the described mine reached the value of about 70 m$^3$/min. In the following period, the inflow was changed with both short duration decrease and also increase. In February 1967, the extremely large water inflow to the mine workings occurred from the karstic cavern system. The measured discharge of this inflow was about 37 m$^3$/min and total inflow to the Olkusz mine increased to the value of about 90–95 m$^3$/min. After draining of the water gathered in the cavern system, total inflow decreased rapidly to about 82 m$^3$/min and after the next year was stabilized in the interval between 68 $m^3$/min and 72 $m^3$/min. With effect from 1972, in spite of significant fluctuation of about 10 $m^3$/min, the distinct trend of the total inflow decrease was observed. The reasons for these changes were the starting of the drainage of the Pomorzany mine and long-term hydrological drought in Poland for the period 1984–1996. The minimal total inflow for this time interval of 41 $m^3$/min was measured in August 1993. In July 1997, the extreme rainfall was the reason for the serious flood in South Poland. This phenomenon has an impact on the increase in total inflow to the Olkusz mine, which in September 1997 reached the value of about 68 $m^3$/min (Adamczyk and Motyka 2000). The peak-inflow decreased rapidly to the value of about 50 $m^3$/min. In the last time sequence, the decreasing trend of the total inflow to Olkusz mine was observed, from the second half of 2003 to the present inflow which dropped to the value of about 28–33 $m^3$/min.

Construction of the Pomorzany mine was started in 1969 by the drifting of the opening-out heading within Permian conglomerates, underlying Triassic carbonates. In the beginning phase of the mine construction, the inflows from Permian rocks were insignificant, with discharge of about a few litres per minute. That is why the measurements were not started before July 1972. Maximal inflow from Permian conglomerates of about 5 $m^3$/min was observed in August 1973. In October 1973, the underground workings were drifted within karstified Triassic carbonates. During the next seven months the total inflow to the Pomorzany mine increased from about 5 $m^3$/min to above 100 $m^3$/min (Fig. 2). The increasing trend was continued and in October 1975 total inflow reached the maximal value of about 240 $m^3$/min. Rapid increase of the inflows was connected with the opening of the following water-bearing karstic channels and caverns. In the next years, after attained maximum, the total inflow to Pomorzany mine decreased and the minimal value of about 185 $m^3$/min was measured in July 1983. Opening the new region of the Zn-Pb ore in 1983–1984 was connected with a decrease of the inflow to the value of about 220 $m^3$/min. Long-term hydrological drought in Poland in the period 1984–1996 was the reason for the systematic decrease of the water inflow to Pomorzany mine. In April 1994, inflow reached the minimal measured value, only 160 $m^3$/min. Heavy rainfalls in the second half of 1996 and extreme rainfall amounts in July 1997 had a strong influence on the amount of water inflowing to the dewatering system of the Pomorzany mine. This situation was similar to the previously described example of the Olkusz mine. In Pomorzany mine the maximal measured inflow, about 300 $m^3$/min, was observed in May 1998. After this peak point in inflow, the stabilization of the inflow was observed on the level of about 250 $m^3$/min. The situation with rainfall-induced increase of the inflows to the mine was observed also at the turn of years 2001 and 2002. As a consequence of heavy rainfalls in the summer of 2001 after 1 year, in August 2002, the total inflow to the Pomorzany mine attained the value of about 300 $m^3$/min the situation observed after this peak point was analogical to the stated in the period 1997–2000. In February 2004, the inflow decreased to the value of about 230–250 $m^3$/min and is holding in this level to the present.

# 4 Conclusions

Occurrence of the water-bearing karstic channels and caverns within Triassic carbonates is a reason for the extremely high inflows to the Zn-Pb mines in the Olkusz region. the necessity of the pumping of large amounts of water has a significant impact on the Zn-Pb deposit exploitation costs. Discharge and characteristics of the inflows to the Zn-Pb mines in Olkusz region are strongly influenced by occurrence of the karstic forms. In a stage of the deposit opening any situation of the karstic system incision was connected with abrupt increase of the inflows to the mine. Discharge of the individual inflows from caverns were reached about 40 $m^3$/min for Olkusz mine and about 100 $m^3$/min for Pomorzany mine.

Temporal changes in the inflows from karstic systems are dependent on the type of groundwater aquifer, which is a recharge source for particular inflow. A fissured-karstic aquifer is a source for the extremely large inflows with rapid decrease of the discharge and following stabilization on the value result from recharge of the cavern system by fissures. A porous-karst-fissured aquifer has an additional source of water, which is gathered in pore spaces. Inflows from this type of aquifer have a relatively longer period of high discharges. A typical feature of the porous-karst-fissured aquifer with occurrence of the filled forms is an initial low discharge, which increases gradually according to the intensity of the tunnelling process. After draining of the water accumulated within karstic systems and porous space, the inflow discharge decreased slowly to the level resulting from the conductivity of the fissure system.

The prognosis of the discharge of the particular inflow (up to 1.6 $m^3$/s), especially from the mine face, is extremely difficult. The best methodology, and also very expensive, for safe mining exploitation is the drilling of the exploratory, protection hole. But this method results only in determination of the existence of karstic channels without any information about volume of the connected system and also volume of water gathered within system. In the past, the geophysical and isotopic methods were used, but it was found that they are completely useless for these purposes.

**Acknowledgements** This work was supported by research program no. 11.11.140.139 of the Faculty of Geology, Geophysics and Environmental Protection, AGH University of Science and Technology, Krakow.

# References

Adamczyk Z, Motyka J (2000) Water inflow to lead and zinc mines in the Olkusz area (SW Poland). Przegl. Geolog. 48(2):171–175 (in Polish)

Leach DL, Sangster DF (1993) Mississippi Valley-type lead-zinc deposits. In: Kirkham RV, Sinclair WD, Thorpe RI, Duke JM (eds.) Mineral deposit modeling. (pp. 289–314). Geol. Assoc. Can. Spec. Pap., 40

Motyka J (1998) A conceptual model of hydraulic networks in carbonate rocks, illustrated by examples from Poland. Hydrogeol. Jour. 6:469–482

# Research on Caves

# A Model of Karstification in Extended Limestone Plains by Mixing Corrosion

W. Dreybrodt and F. Gabrovšek

**Abstract** When water from the surface of a limestone plain seeps through the vadose zone in fractured rock down to the water table of an unconfined aquifer containing water saturated with respect to calcite, mixing of these waters causes renewed dissolution of limestone. A model coupling flow in the fractures with dissolution rates describes the evolution of karstification by dissolutional widening of the fractures. At the beginning of karstification a domain of surface water saturated with respect to calcite at a $p_{CO_2}$ of 0.05 atm floats on the phreatic water saturated with respect to calcite at a $p_{CO_2}$ of 0.001 atm. At the border of these domains their waters mix and create dissolutional widening of the fractures by mixing corrosion. A channel evolves along the border migrating downstream by about 100 m in 100,000 years. Below this channel a zone of fractures with aperture widths up to one cm has originated. The change of the hydraulic conductivity in the mixing zone shifts the border of the domains, allowing the channel to migrate downstream. Below it the zone of widened fractures is invaded by saturated phreatic water and dissolution stops. This process continues at the downstream part of the conduit. In summary, this modeling reveals an effective mechanism of intense karstification on large karst plains.

## 1 Introduction

There has been much progress in understanding the processes that form caves and karst systems by modeling speleogenetic evolution (Dreybodt 1988; Gabrovšek and Dreybrodt 2001; Dreybrodt and Gabrovšek 2002; Dreybrodt et al. 2005). However, the variety of parameters, which determine the evolution of caves, is so large that

W. Dreybrod, F. Gabrovšek
Karst Research Institute SASA, Titov Trg 2, 6230 Postojna, Slovenia, e-mail: gabrov@zrc-sazu.si

W. Dreybrodt
Karst Processes Research Group, Institute of Experimental Physics, University of Bremen, 28359 Bremen, Germany, e-mail: dreybrodt@t-online.de

B. Andreo et al. (Eds.), *Advances in Research in Karst Media* 
DOI 10.1007/978-3-642-12486-0, © Springer 2010

models can only give views through keyholes. This is also true for the variety of geological settings. Waters that enter limestone can be aggressive with respect to calcite and readily form caves. It is also possible that they are saturated with respect to calcite. When saturated waters with different chemical compositions mix, new aggressivity arises and cave evolution can proceed distant from the input of the water, which feeds the aquifer (Bögli 1964; Dreybrodt 1988; Dreybrodt et al. 2005).

In this work, a model of cave evolution in a large karst plain with an unconfined aquifer in fractured limestone rocks below is presented, where water saturated with respect to calcite flows with a flow rate of several meters per year. Under such conditions the water is saturated with respect to calcite and dissolution of limestone is excluded. If, however, some river or lake on the plain feeds water with high $p_{CO_2}$ into this aquifer, although this water after seeping through the vadose zone has become saturated with respect to calcite, mixing with the phreatic water of the aquifer will initiate karstification by mixing corrosion in the zone where the waters mix. In this study, which structures of karst porosity and of caves originate under such conditions were investigated.

## 2 Concept of Geological Setting

Figure 1 shows a flat limestone plain, which receives water from a distant mountain area. After some distance an almost horizontal water table is established with a flow rate $Q_{in}$ evenly distributed to the fractures of the aquifer. This will be used as an idealised boundary condition in the modeling domain.

Sufficiently far away from the recharge area a river or a lake infiltrates water seeping to the water table with flow rate $q_{in}$. Somewhere remote from the input of surface water the water table drops to some base-level river. To limit the size of the modeling domain to a reasonable extent, it is assumed that a constant head $h_{out}$ at the right hand side boundary where evenly distributed flow as in the initial state of evolution is expected. Basically, this fixed boundary condition is an approxima-

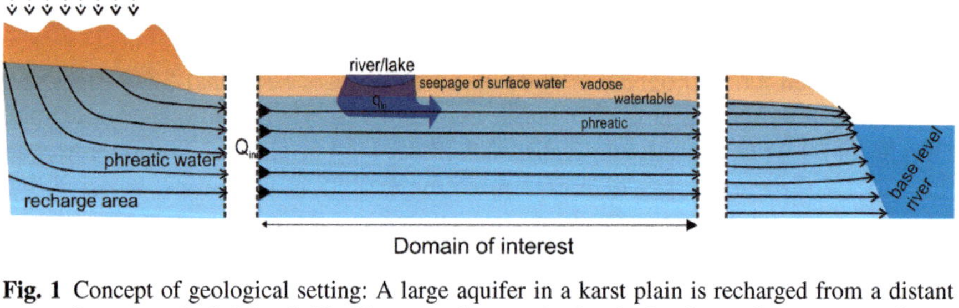


**Fig. 1** Concept of geological setting: A large aquifer in a karst plain is recharged from a distant mountain area on the *left hand side*. The water flows under low hydraulic gradient to a very distant base-level river on the *right hand side*. The *middle part* is a small part of the entire aquifer, where surface water from a river or lake seeps through the vadose zone down to the phreatic water

tion, which is valid only as long as the evolving distribution of widening fractures, redistributing the flow and focussing it, stays sufficiently far from this boundary. Therefore, all runs must be terminated when the distance from the structure arising by dissolution to the right hand side boundary is equal to its length.

## 3 The Model

Figure 2 depicts the modeling domain consisting of a grid of rectangular cells of 4 m by 0.5 m representing fractures of 1 m width. The lower boundary is impermeable. The left hand boundary has a constant input $Q_{in} = 1.6 \cdot 10^{-2}\,\text{cm}^3\,\text{s}^{-1}$ into each fracture in its lower phreatic part. This amounts to a flow rate of about 1 m/year. At the right hand side boundary, a fixed head $h_{out} = 0$ is imposed. At the upper boundary, which is open to flow, a region between 280 m and 310 m with input $q_{in} = 5 \cdot 10^{-2}\,\text{cm}^3\,\text{s}^{-1}$ per fracture represents the source of surface water corresponding to an infiltration rate of about 400 mm/year.

This water is saturated with respect to calcite at a partial pressure of $CO_2$, $p_{CO_2} = 0.05$ atm, when it encounters the water table of the aquifer. The water in the aquifer entering from the left hand side boundary is saturated with respect to calcite at a $p_{CO_2}$ of 0.001 atm. The aperture widths of the fractures are created by a lognormal distribution with average aperture width of 0.02 cm and $\sigma = 0.02$ cm. Basically, the domain represents a stripe of the aquifer 1 m wide, 75 m deep, and 1200 m long. The position of the water table is determined by the inputs $Q_{in}$ and $q_{in}$.

To calculate the widening of the fractures in each time step linear and non-linear rate equations were used based on experimental data and theoretical considerations (Buhmann and Dreybrodt 1985; Eisenlohr et al. 1999; Kaufmann and Dreybrodt 2007) and the well documented modeling techniques were applied as reported by Dreybrodt et al.; 2005 and Gabrovšek and Dreybrodt 2001.

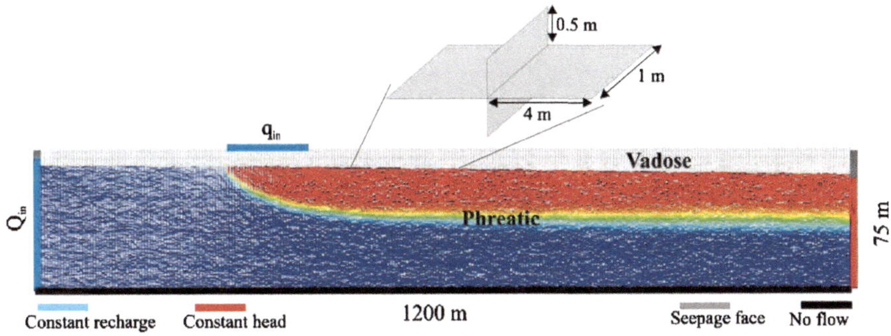


**Fig. 2** Modeling domain at the onset of karstification. At the top of the domain recharge by a river or lake feeds the flow rate $q_{in}$ into each fracture within the *blue region*. The *light blue color* represents the vadose zone at the onset of karstification. The *dark blue region* contains phreatic water. The *red region* represents the surface water from the river/lake. The *colored rim* between these two regions represents the mixing zone of phreatic and surface water (see text)

## 4 Results and Discussion

Figure 3a,c,e,g,i depict the dissolution rates by a colour code and the aperture widths of the fractures by a bar code at various times of karstification. Each step in thickness of the bar code designates an increase in aperture-width by a factor of two. Red is a widening of $4.7 \cdot 10^{-3}$ cm/year and blue is $1 \cdot 10^{-9}$ cm/year. As can be seen from the black areas of fractures, dissolution is absent where the phreatic and surface waters do not mix. The mixing zone of these waters is seen in the blue-green area, where mixing corrosion creates moderate rates of widening. Figure 3b,d,f,h,j show the equilibrium concentrations $c_{eq}$ of the water. Blue is the equilibrium concentration $c_{eq,phreatic}$ of the phreatic water and $c_{eq,surface}$ (red) that of the surface water. In the regions of these colours dissolution is absent, corresponding to the black areas in Fig. 3a,c,e,g,i. Where the waters mix the equilibrium concentrations change between $c_{eq,phreatic}$ (blue) and $c_{eq,surface}$ (red). Dissolution rates are maximal in the green region. Figure 3b,d,e,h,j also depict the flow rates by a linear bar code in units of $q_i/q_{max}$. $q_i$ is the flow rate through fracture $i$ and $q_{max}$ is the maximal flow rate occurring in the net.

The evolution after 15,000 years is shown by Fig. 3c,d. A small channel with high dissolution rates has developed at the upstream side of the input region of surface water.

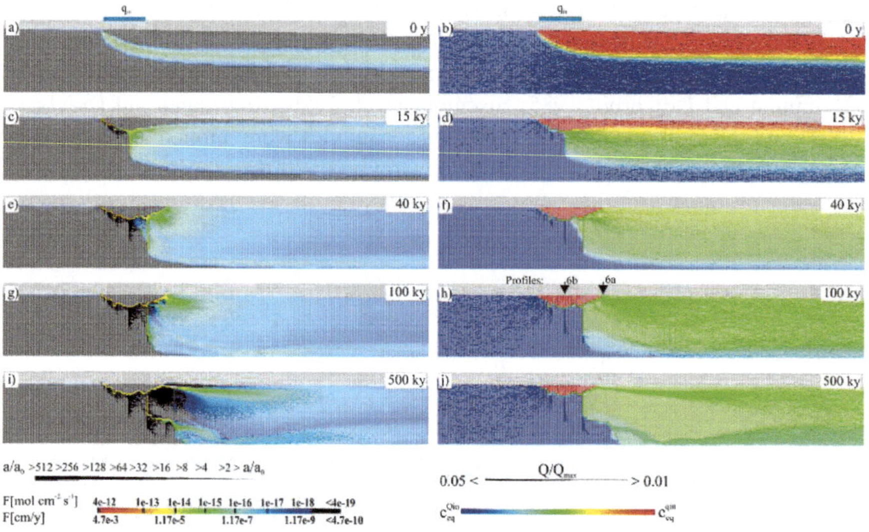


**Fig. 3** Evolution for the standard scenario: The *left hand side* (a,c,e,g,i) represents fracture aperture widths (*bar code below*) and widening of the fractures in cm/year (*color code below*). The *light grey areas* show the vadose zone. *Black regions* do not experience solutional widening. The *right hand side* (b,d,f,h,j) illustrates flow rates (*bar code*) and the values of $c_{eq}$ of the water. *Blue regions* contain phreatic water from $Q_{in}$ with concentration $c_{eq}^{Q_{in}}$, *red regions* are filled with surface water with $c_{eq}^{q_{in}}$. The colors *light blue* to *orange* show the regions, where the waters mix (see text)

This channel of low hydraulic resistance attracts both phreatic and surface water, and due to mixing the dissolution rates are high. The mixed water flows out of the channel dispersing into the downstream fractures of the aquifer. Dissolution rates are still high close to the exit of the mixed water. As the zone of mixing widens they decrease. At the border between the surface water (black-green-blue) a narrow zone of moderate dissolutional widening (green) at the beginning in Fig. 3a and yellow in Fig. 3b extends downstream.

Another interesting feature is the evolution of a vertical shaft close to the exit of the channel. This shaft attracts phreatic water, which mixes with the water from the channel to create significant widening again.

After 40,000 years this shaft has become so permeable that solely phreatic water enters the shaft and flow is directed upwards to the mixing zone in the channel. Consequently, Fig. 3e shows black zones of already widened fractures and deeper shafts, where dissolution is absent. This is also seen in Fig. 3f, where these regions are invaded by saturated phreatic water (blue). The channel has migrated downstream and at its exit a new shaft develops.

After 100,000 years (Fig. 3g,h) the downward evolution of this shaft terminates because phreatic water enters into its bottom. The bar code for the flow shows that surface water and phreatic water are directed towards the channel leaving a zone of low flow rates below. Most of the water flows through the channel and from there is dispersed back to the aquifer, where the porosity has increased evenly. After 500,000 years (Fig. 3i,j) a parallel channel starts to evolve at the end of the shaft and water from the ends of both channels is dispersed into the aquifer. Two fans with high flow focus water to the right hand side of the modeling domain, leaving large regions of the aquifer with low flow rates. At this time the boundary conditions at the exit are still valid. After 750,000 years, however, regions of low flow reach this boundary and the conditions of constant head break down. Therefore, the run was terminated after 500,000 years. In summary, cave conduits evolving close to the water table, leaving significant cavernous structures below them were found.

The question must be asked, whether the formation of shafts just arises accidentally by the realisation of the net. Therefore a new statistical realisation of the net with identical statistical properties was created. Although details are different, the general structures obtained are similar to the evolution of the scenario discussed above. Therefore, these can be regarded as characteristic for karst evolution in such speleogenetic settings.

## 5 Conclusion

The aim of this work has been to elucidate karstification in an extended limestone aquifer with low hydraulic gradient and filled with water saturated with respect to calcite, which receives water also saturated with respect to calcite from a local input on the surface above.

If both waters have different chemical compositions with respect to $CO_2$ and $Ca^{2+}$-concentrations, they gain new solutional capacity, when they mix. Downstream from the local input a domain is established, which contains solely surface water. At its border with the domain of the phreatic water both types of water mix in the intersections of the fractures and consequently these fractures experience widening of their aperture widths. These changes of hydraulic conductivity shift the border dividing the two domains and consequently also the zone of mixing. A channel forms at the upstream border of the local input region. This channel continues to receive surface water and phreatic water and grows continuously in width and length. Below it mixing due to the complex flow conditions is also present and horizontal and vertical karstificaction is active. Due to the increasing hydraulic conductivity in these regions saturated phreatic waters invade them and solutional widening of fractures comes to an end. As the active upper channel migrates downstream, new zones of karstification below it become active and stop growth when they are filled by phreatic water later on. After 500,000 years the runs were terminated. During this time the leading channel has propagated several hundred meters downstream and has reached aperture widths of several 10 cm. Below this channel a zone of fractures with aperture widths of about 1 cm extends down to a depth of several 10 m. At the end of the channel a zone of mixed water fills the entire downstream part of the aquifer. Close to the exit a fan protrudes with dissolutional widening of about $10^{-6}$ cm/year.

In summary this study reveals an effective mechanism of intense karstification on large karst plains, which is caused solely by mixing corrosion and proceeds in time scales of 100,000 years.

## References

Bögli A (1964) Mischungskorrosion: Ein Beitrag zum Verkarstungsproblem. Erdkunde 18:83–92

Buhmann D, Dreybrodt W (1985) The kinetics of calcite dissolution and precipitation in geologically relevant situations of karst areas.1. Open system. Chemical Geology 48:189–211

Dreybrodt W (1988) Processes in karst systems: physics, chemistry, and geology. Springer Berlin

Dreybrodt W, Gabrovšek F (2002) Basic processes and mechanisms governing the evolution of karst. In: Gabrovšek F (ed) Evolution of karst: from prekarst to cessation, Carsologica Zalosba ZRC Ljubljana

Dreybrodt W, Gabrovšek F, Romanov D ( 2005) Processes of speleogenesis: A modeling approach. Carsologica, 4. Založba ZRC, Ljubljana

Eisenlohr L, Meteva K, Gabrovšek F, Dreybrodt W (1999) The inhibiting action of intrinsic impurities in natural calcium carbonate minerals to their dissolution kinetics in aqueous $H_2O$–$CO_2$ solutions. Geochim et Cosmochim Acta 63:989–1001

Gabrovšek F, Dreybrodt W (2001) A model of the early evolution of karst aquifers in limestone in the dimensions of length and depth. Journal of Hydrology 240:206–224

Kaufmann G, Dreybrodt W (2007) Calcite dissolution kinetics in the system $CaCO_3$–$H_2O$–$CaCO_3$ at high undersaturation. Geochim et Cosmochim Acta 71:1398–1410

# Isotopic ($^{13}C$) Signature of $CO_2$ Sources in the Vadose Zone of a Mediterranean Karst (Nerja Cave Site, Southern Spain)

I. Vadillo, J. Benavente, F. Carrasco, A. Soler, and C. Liñán

**Abstract** This study is based on in situ measurements of the soil and the vadose zone ($< 60$ m) in a Mediterranean karst experimental site near Nerja Cave (a show cave in dolomite marbles in South Spain). $CO_2$ concentrations in depth were registered in boreholes drilled in experimental site. The $CO_2$ content generally increases with depth. Measurements indicate average vadose air $CO_2$ concentrations of nearly 40,000 ppm, with a maximum of nearly 60,000 ppm. In this context, the cave itself appears to be a vadose subsystem above the groundwater level, with significantly lower $CO_2$ concentrations (a few thousands of ppm maximum) due to its ventilation. The $\delta^{13}C$–$CO_2$ data of the vadose air point to an origin of the gas mainly related to microbiological processes associated to the consumption of dissolved organic matter in the groundwater surface. This gas can diffuse or flow laterally, upward or downward through karst conduits. Interactions between air masses of surface origin (relatively dry, with variable temperature and low $CO_2$ content) and typical vadose attributes (relatively high $CO_2$ content, near-saturated humidity and 21 °C temperature) produce clear ascendant or descendant air fluxes inside the boreholes, especially those that cross significant karst voids.

## 1 Introduction

$CO_2$ partial pressure in both soil air and the vadose zone of carbonate aquifers is the master variable that controls the development of karstogenetic processes (Atkinson

I. Vadillo, F. Carrasco, C. Liñán
Centre of Hydrogeology of University of Malaga, Department of Geology, Faculty of Science, University of Malaga, Malaga 29071, Spain,
e-mail: vadillo@uma.es

J. Benavente
Water Research Institute, University of Granada, 18071 Granada, Spain

A. Soler
Department of Crystallography, Mineralogy and Mineral Deposits, Faculty of Geology, University of Barcelona, 08028 Barcelona, Spain

C. Liñán
Nerja Cave Foundation. Carretera de Maro s/n. 29787, Nerja (Málaga), Spain


DOI 10.1007/978-3-642-12486-0, 

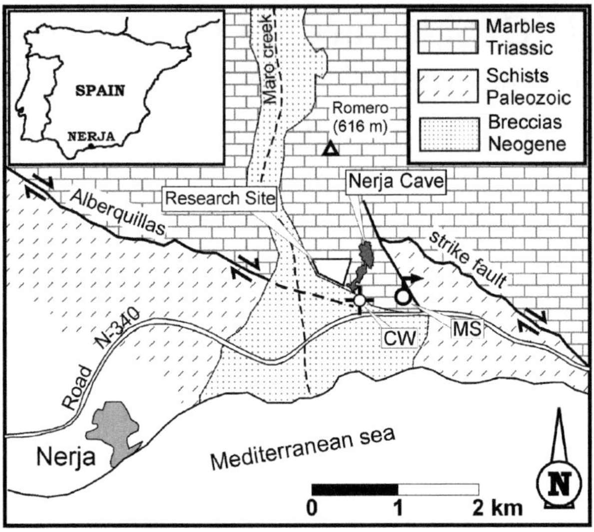


**Fig. 1** Location and general geological sketch of the study area. Nerja Cave, research site and some points of hydrogeological interest (Maro spring-MS and Cave well-CV) are also shown

1977; Bourges et al. 2001) and can be up to two orders of magnitude greater than open atmosphere (Keller and Bacon 1998; Walvoord et al. 2005). This increase is related to soil biological activity; however, the concentration of $CO_2$ is not always explained solely by a biogenic origin. Therefore, it is necessary to make direct measurements of the $CO_2$ content, considering its time evolution and variations with depth and isotopic signature in the vadose zone of carbonate aquifers (Wood and Petraitis 1984; Cerling et al. 1991; Keller and Bacon 1998; Hamada and Tanaka 2001; Walvoord et al. 2005).

This study is based on the results of experimental measurements conducted near Nerja Cave (Malaga province, South Spain), less than 1 km from the Mediterranean shore (Fig. 1). Nerja Cave is an important show cave, visited by some 500,000 people each year mostly during summer months.

This work presents the results of the monitoring from 2006 to 2007 of the subsurface air $CO_2$ concentrations in the research site near Nerja Cave (Fig. 1). Systematic measurements of the $CO_2$ content of the air in the scarce karst soil, and especially in the vadose environment, are based on a number of boreholes, most at 30 m depth. The results of these measurements can be compared with those obtained inside the cave, where the main environmental variables are routinely monitored.

## 2 Methodology

The area is placed in the Alpujarride Complex of the Betic Cordilleras (Fig. 1). Nerja Cave is developed in Triassic dolomitic marbles. It is underlain by a thick sequence of schists, mostly Palaeozoic. To the South of the area there is a regional fault, the

Alberquillas fault (Fig. 1), which has proven to be active in recent geological times, acting both as a normal and a strike-slip fault.

The grade of fissuring and karstification of the Triassic carbonate formation conform an aquifer of regional importance (the Alberquillas aquifer).

A significant proportion of the carbonate outcrops in the study area are bare rock. This soil cover is scarce and patchy, rarely exceeding 15 cm thick. The natural vegetation consists mainly of shrubs and pine trees.

The research site is northwest of the cave entrance, and extends between an approximate altitude range of 160–280 m a.s.l. (Fig. 1). Its main installations consist of nine boreholes (points 1 to 11). All boreholes have depths between 15 m (points 1 and 4) and 30 m (points 3, 5, 6, 8, 9 and 11). Most of the boreholes crossed karst openings and cavities during drilling. These are especially important in boreholes 1, 4 and 11, representing around 50% of the total length drilled.

From 2006 to 2007, a VAISALA GM70 device, adapted for use in downhole borehole logging, provided data of temperature, relative humidity and $CO_2$.

Four sampling campaigns were carried out between March 2005 and September 2006 for collecting air $\delta^{13}C$–$CO_2$ samples. $\delta^{13}C_{CO_2}$ samples were obtained after manual gas extraction with syringe and introduced in evacuated Exetainer® Vials. Notation is expressed in terms of δ per mil relative to the V-CDT.

## 3 Results and Interpretation

Five campaigns for measurements of $CO_2$ content in depth were carried out in July, October and November 2006 and February, May and June 2007 (Fig. 2). All of the measurements were made between 10:00 h and 19:00 h. The objective was to identify seasonal variations induced by the different external air temperatures, which approximately ranged from 15 °C (November and February) to 30 °C (July).

In general, the gas concentration increases with depth, but the shapes of the curves vary from one point to another. The maximal $CO_2$ contents are found in the bottom parts of the boreholes and are variable depending on the points, ranging from 55,000–60,000ppm, at points 9 and 11, to 15,000–25,000 ppm at point 1. The minimal values are typical of the open atmosphere, and appear at all points although reaching different depths. For most of the points displaying regular graphs, the highest gas contents were found in the campaign of July 2006 and the lowest in November 2006 or February 2007. Three of the boreholes (1, 4 and 11) display more stepwise changes. There were also noticeable air flows, ascendant or descendent, in some of these installations, when opened for measuring. The sudden increase of the gas content at certain depths at points 1, 4 and 11 is roughly coincident with the depth of the roof of significant cavities crossed. At point 4, for example, the ascendant flow is characterised by relatively high $CO_2$ content (25,000 ppm) through the column.

The lowest $CO_2$ contents are generally associated with the lowest relative humidity range (40–80%). This may suggest an incoming flow from the external at-

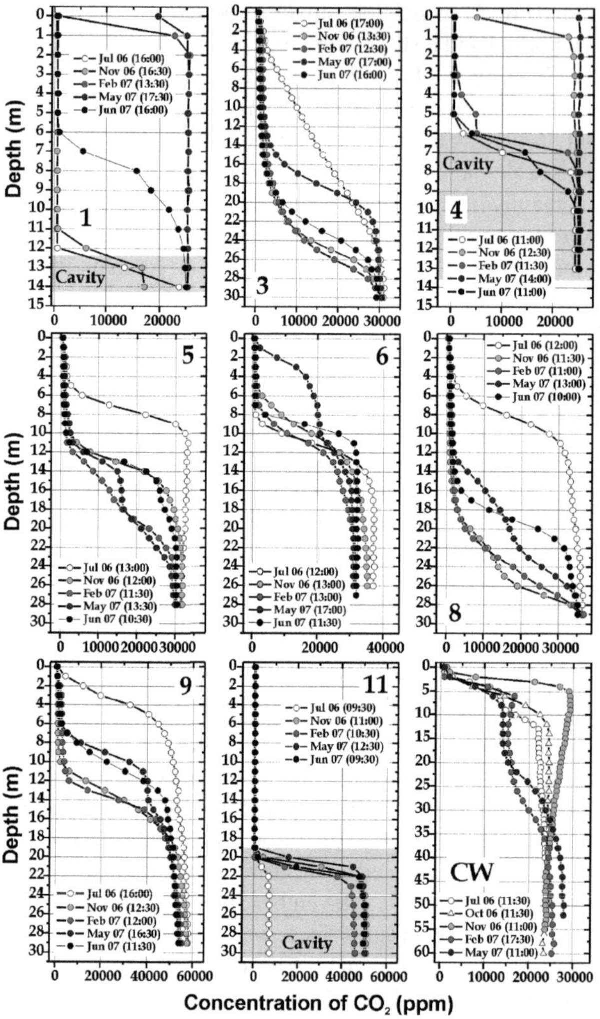


**Fig. 2** Graphs of air $CO_2$ content vs. depth in the monitoring boreholes in different measurements campaigns. Main cavities have been indicated in *grey*

mosphere air, which could also explain the $CO_2$ values. This is found mostly in the early summer (external temperatures of 25–30 °C) and suggests the influence of soil gas, probably originating from the downward migration of the peak of gas production in the early spring. In November and February, the fronts of $CO_2$ contents of more than 20,000 ppm are found at greater depths (18 m on average, against 10 m in July) associated with high, near-saturation humidity values and with a very stable temperature near 21 °C.

Combined data of $CO_2$ concentration and $\delta^{13}C$–$CO_2$ (Fig. 3) show a wide range of both $CO_2$ concentration (460 ppm to 53,640 ppm) and $\delta^{13}C$–$CO_2$ (−10.5‰ to

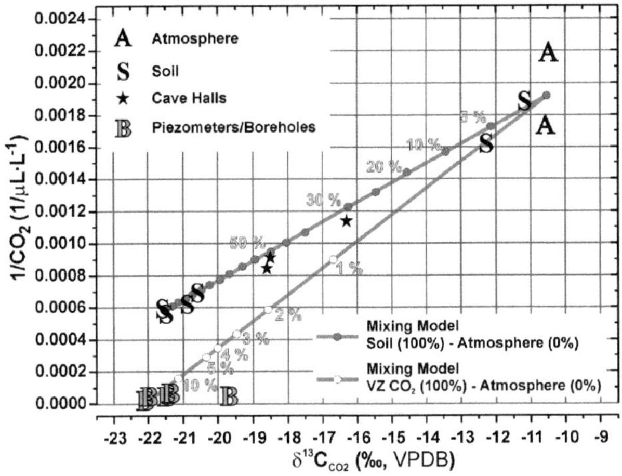


**Fig. 3** Plot of $\delta^{13}C$–$CO_2$ vs. $1/CO_2$ of samples collected in atmosphere, soil, cave hall and piezometers/boreholes in the vadose zone of the Nerja cave area. Mixing models between Atmosphere and Soil and Atmosphere and Vadose Zone (VZ) $CO_2$ have been calculated

−22.1‰). Two groups of samples are distinguished: (i) samples from the atmosphere (A), soil (S) and cave halls, and (ii) samples from piezometers 4, 9, a well near the cave and another a few km to the East (B). The first group of samples shows $CO_2$ concentrations below 2,000 ppm and presents the wider range in $\delta^{13}C$, from near typical atmosphere values (∼10.5‰) to soil values of −22‰, typical of region where C3 plants predominates (Cerling et al. 1991). These two extreme values are the end-points of a mixing line and between them are located the air samples of the Nerja Cave. The second group of samples (piezometers/boreholes) categorizes the air with higher $CO_2$ concentrations (>15,000 ppm) and $\delta^{13}C$ values near −22‰. Within this group of points is found the maximum $CO_2$ concentration (∼54,000 ppm).

Three $CO_2$ sources are suggested in the unsaturated zone of the aquifer: atmosphere, soil and a third one. Soil $CO_2$ results from biological processes involving plants and/or degradation of organic matter. This surface-origin is the main contributor of $CO_2$ to the soil air, but the scarce soil of this Mediterranean region does not seem to be able to produce such high concentration $CO_2$ values of the vadose zone.

The most plausible process seems to be the microbial consumption of dissolved carbon in groundwater. Many field studies support this additional deep $CO_2$ source originating at or immediately above the water table (Hendry et al. 1993; Keller and Bacon 1998; Wood and Petraitis 1984; Wood et al. 1993; Walvoord et al. 2005). This hypothesis is supported by the very high dissolved organic carbon sampled immediately below the water level (∼10–30 mg/L) and the $\delta^{13}C$–$CO_2$ values in depth (−22‰) that are similar to the microbial respiration in the soil (Cerling et al. 1991), indicating a microbial degradation as the main source for the very high concentrations of $CO_2$.

## 4 Conclusions

The research done at the Nerja experimental site has allowed the identification of $CO_2$ concentrations that are frequently in the range of 20,000 to 40,000 ppm in the vadose zone (<60 m), with maximum values near 60,000 ppm. The study of the vertical variation of the air through the boreholes shows a general trend of a gradual increase in $CO_2$ content with depth. The highest $CO_2$ contents in the vadose zone are shallower in the summer and deeper in the winter.

$\delta^{13}C$–$CO_2$ values indicate the presence of three $CO_2$ sources in the unsaturated zone of the aquifer: atmosphere, soil and a third one (with the highest $CO_2$ contents and $\delta^{13}C$–$CO_2 = -22‰$) related to microbial degradation of the dissolved organic matter in groundwater. Carbon isotopic results indicate that the $CO_2$ sampled inside the cave comes mainly from atmosphere and soil, and that contribution of high $CO_2$ contents coming from vadose zone (>15,000 ppm) is very low or negligible.

**Acknowledgements** This research was funded by the Cueva de Nerja Foundation. The authors would like to thank the Serveis Cientificotècnics of the University of Barcelona (Spain) for their isotopic analytical support.

## References

Atkinson TC (1977) Carbon dioxide in atmosphere of unsaturated zone- Important control of groundwater hardness in limestones. J. Hydrol. 35(1–2):111–123.

Benavente J, Vadillo I, Carrasco F, Soler A, Liñán C, Moral F (2010) Air $CO_2$ contents in the vadose zone of a Mediterranean karst: field measurements and hydrochemical implications (Nerja Cave experimental site, Southern Spain). Vadose Zone Journal. 29:647–659.

Bourges F, Mangin A, D'Hulst D (2001) Le gaz carbonique dans la dynamique de l'atmosphère des cavités karstiques: l'exemple de l'Aven d'Orgnac (Ardèche). C.R. Acad. Sci., Ser. IIa: Sci. Terre Planets 333:685–692.

Cerling TE, Solomon DK, Quade J, Bowman JR (1991) On the isotopic composition of carbon in soil carbon dioxide. Geochim. Cosmochim. Acta, 55:3403–3405.

Hamada Y, Tanaka T (2001) Dynamics of carbon dioxide in soil profiles based on long-term field observation. Hydrol. Processes 15:1829–1845.

Hendry MJ, Lawrence JR. Zanyk BN, Kirkland R (1993) Microbial production of $CO_2$ in unsaturated geologic media in a mesoscale model. Water Resour. Res. 29:973–984.

Keller CK, Bacon DH (1998) Soil respiration and georespiration distinguished by transport of vadose $CO_2$, $^{13}CO_2$ and $^{14}CO_2$. Global Biogeochem. Cycles 12:361–372.

Walvoord MA, Striegl RG, Prudic DE, Stonestrom DA (2005) $CO_2$ dynamics in the Amargosa desert: fluxes and isotopic speciation in a deep unsaturated zone. Water Resour. Res. 41, doi:10.1029/2004WR003599.

Wood BD, Keller CK, Johnstone DL (1993) In situ measurement of microbial activity and controls on microbial $CO_2$ production in the unsaturated zone. Water Resour. Res. 29:647–659.

Wood WW, Petraitis MJ (1984) Origin and distribution of carbon dioxide in the unsaturated zone of the southern High Plains. Water Resour. Res. 20:1193–1208.

# Effect of Ventilation on Karst System Equilibrium (Altamira Cave, N Spain): an Appraisal of Karst Contribution to the Global Carbon Cycle Balance

S. Sánchez-Moral, S. Cuezva, A. Fernández-Cortés, D. Benavente, and J.C. Cañaveras

**Abstract** Altamira cave air $CO_2$ concentration, and both cave and air $\delta^{13}C$ values seasonally vary indicating the cave behaves as a $CO_2$ reservoir or source in winter and summer, respectively. During the '$CO_2$-reservoir phase' the $\delta^{13}C$ of the cave air is lower, because it is influenced by the infiltration of soil-derived organic carbon-rich waters. In the '$CO_2$-source phase' the $\delta^{13}C$ of the cave air is heavier due to the ventilation of the cave and the mix with the air from the external atmosphere and soil. The $^{13}C/^{12}C$ analyses confirmed the importance of the external soil as the $CO_2$ source for the underground system as well as the effect of ventilation on system equilibrium.

## 1 Introduction

For the carbon present in the three principal phases of underground systems (rock, water, and air), every phase change that comes about includes isotopic fractionation processes that modify the proportions of each stable isotope. Thus, each chemical reaction implies a particular discrimination factor that bestows a characteristic isotopic signal on the newly generated phase. The characterization of this isotopic signal for each of the new phases involved in underground $CO_2$ transfer processes between the interior and exterior of the underground environment is useful in the evaluation of the degree to which different processes are important. Likewise, the study of the behaviour of $CO_2$ in karstic environments is critical because its contribution to the global carbon cycle balance (Yuan 1997; Kowalski et al. 2008).

S. Sánchez-Moral, S. Cuezva, A. Fernández-Cortés
Dpto. Geología, Museo Nacional de Ciencias Naturales, CSIC, 28006 Madrid, Spain

D. Benavente, J.C. Cañaveras
Dpto. Ciencias de la Tierra y del Medio Ambiente, Univ. Alicante, Campus San Vicente del Raspeig, 03080 Alicante, Spain, e-mail: jc.canaveras@ua.es

B. Andreo et al. (Eds.), *Advances in Research in Karst Media*
DOI 10.1007/978-3-642-12486-0, © Springer 2010

The aim of the study is to characterize the interchange processes between the underground (endokarst) and the external environment, with special emphasis on the study of exchange cycles of atmospheric $CO_2$.

Altamira Cave is world-famous for possessing a remarkable collection of Palaeolithic rock paintings and engravings. It is a 270 m-long downward-trending cave located in the unsaturated water zone of a tabular polygenic karst system. The depth of the cave ranges from 4 to 21 m. The climate is Atlantic and humid and the pluvial regime is moderate to high.

## 2 Methods

Annual microclimatic series for different sectors of the cave and exterior have been obtained. Studied climate parameters in cave atmosphere include: air temperature, rock temperature, $CO_2$ on air, and relative humidity. Above the cave the measured parameters also include $CO_2$ fluxes of the ecosystem, soil temperature and soil relative humidity. Karstic waters were monthly sampled for both geochemical analysis and infiltration rate studies. The $^{13}C/^{12}C$ isotopic analysis of air from the cave interior, soil pore space and exterior were done by the GEOTOP Isotope Laboratory (Québec, Canada). The analyses of solid (rock and speleothems) and water $^{13}C/^{12}C$ isotopic signatures were done by the Stable Isotope Analytical Service of the University of Salamanca and Zaidin Station (CSIC, Granada), respectively.

## 3 Results

Annual average temperature in the cave is 14°C, and relative humidity is approximately 100% the whole year. Cave air shows lower $CO_2$ concentration during summer when the external temperature is higher than cave air temperature (Fig. 1). In winter, $CO_2$ concentration in the cave is approximately eight-ten times higher than in summer, then the cave acts as a $CO_2$ reservoir. The underground air renewal prevails during spring-summer seasons, when the exterior air temperature is constantly above the cave air temperature. Under these dry environmental conditions the air exchange between the cave and the outer atmosphere is favoured, with an intense degasification process taking place such that the $CO_2$ levels drop to minimum concentrations of 1000 ppm. Degassing processes involve a net emission of $CO_2$ to the atmosphere. From November to May the outside temperature is frequently below the cave air temperature, and intense rainfalls and high values of atmospheric relative humidity are registered. Under these environmental conditions, the membranes covering the cave (host rock and soil) tend to a full state of water saturation; so, a limited air exchange prevails between the cave and exterior atmosphere through the network of fissures and pores. When the confined conditions are reached (especially during the rainy periods), the microfracture networks, carry diphasic infiltration (water plus air) delaying the air gaseous transfer between the cave atmosphere and the exterior. During this stage the cave behaves as a $CO_2$ reservoir, reaching the maximum mean levels: 5000–6000 ppm of $CO_2$. Therefore, recharge of $CO_2$ in the cave is closely related to infiltration water input after rainfall. This gravitational water percolates through the soil zone and host-rock porosity and is recharged of $CO_2$ due to organic activity in soil and water-rock interaction. During the summer time, the cave degasification is produced except when rainfall events occur. The main $CO_2$ cave recharge coincides with the beginning of the rainy season, at the end of September (Fig. 1).

The data (Table 1) invokes for a seasonal variation of the $\delta^{13}C$ in the $CO_2$ of the interior of the cave. During the '$CO_2$-reservoir phase' the $\delta^{13}C$ of the cave air is light, because this is influenced by the soil contribution. In the '$CO_2$-source phase' the $\delta^{13}C$ of the cave air is heavier due to the ventilation of the cave and the mix with the air from the external atmosphere and soil. The $\delta^{13}C$ value of the infiltration waters presents a seasonal variation, similar to that observed in the cave air (Fig. 2). The $\delta^{13}C$ of the water (averaging −12.6‰), varies according to the degree of water/rock interaction with less negative values corresponding to greater time

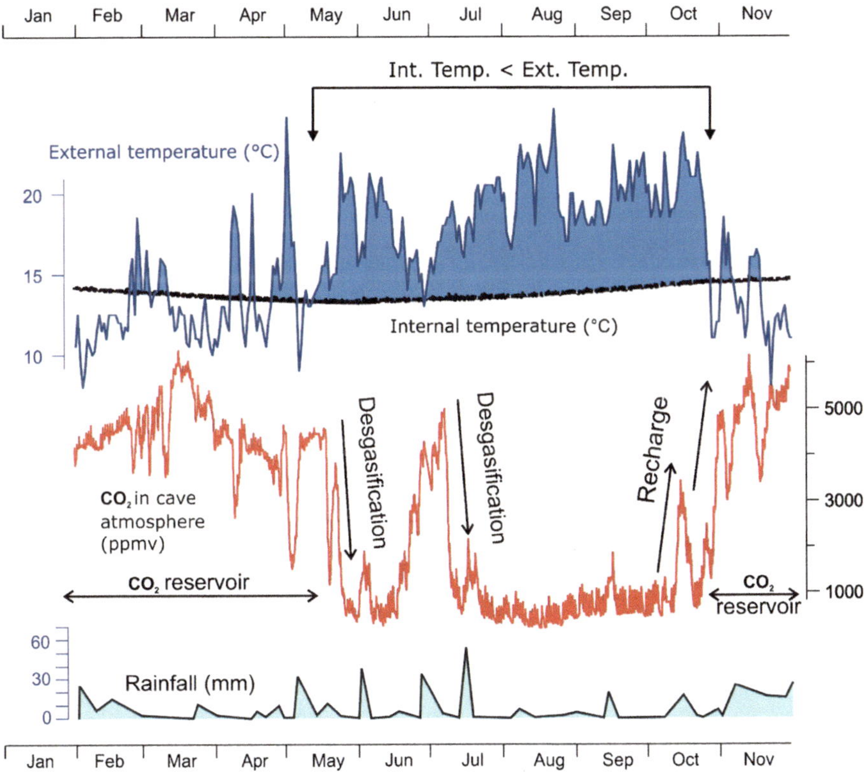


**Fig. 1** Time series (*annual cycle*) of the main microclimatic parameters of Altamira cave in relation to the external atmosphere

**Table 1** Isotopic signal of air (average values, 2005)

| Cave environmental dynamics | $CO_2$ reservoir | $CO_2$ source |
|---|---|---|
| | T ext < T cave | T ext > T cave |
| AIR $\delta^{13}C$ | MARCH | SEPTEMBER |
| Cave | −21.9 | −18.0 |
| Soil | −21.4 | −18.7 |
| External atmosphere | −9.7 | −10.4 |

spent reaching the cavity due to enhanced influence of the hostrock ($\delta^{13}C = +2‰$ limestone, $+3‰$ dolomite). The mean $\delta^{13}C$ of rapidly infiltrating water ($-12.9‰$) corresponds to the fractionation signature (coefficient of $+9.1‰$ at 14°C; Emrich et al. 1970) of $CO_2$ dissolution and formation of $HCO_3$ (Fig. 3). Slow steady fluxes result from dolomite reactions and show heavier mean values ($\delta^{13}C = -11.4‰$). The speleothems show $\delta^{13}C$ values varying from −4 to −13.5‰. In detail, the currently growing speleothems show an averaged value −8.4‰.

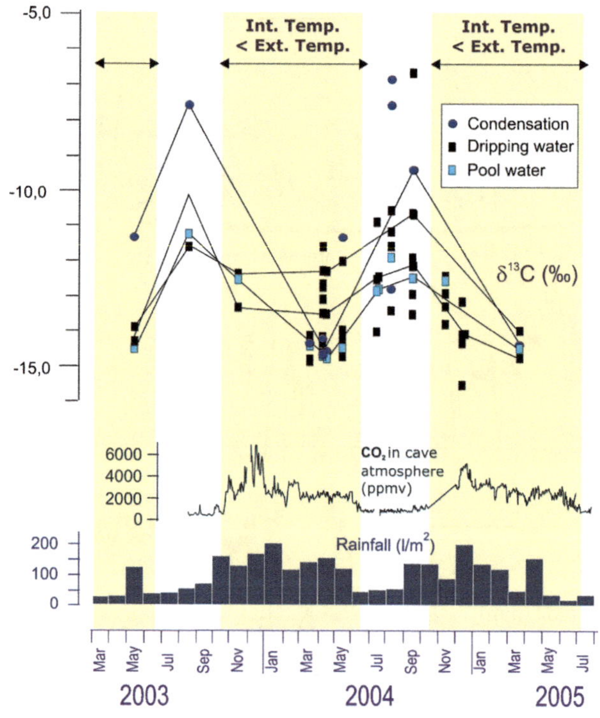


**Fig. 2** Time series (*annual cycles*) of the aqueous phase $\delta^{13}C$ isotopic signatures in relation to cave air $CO_2$ concentration and rainfall

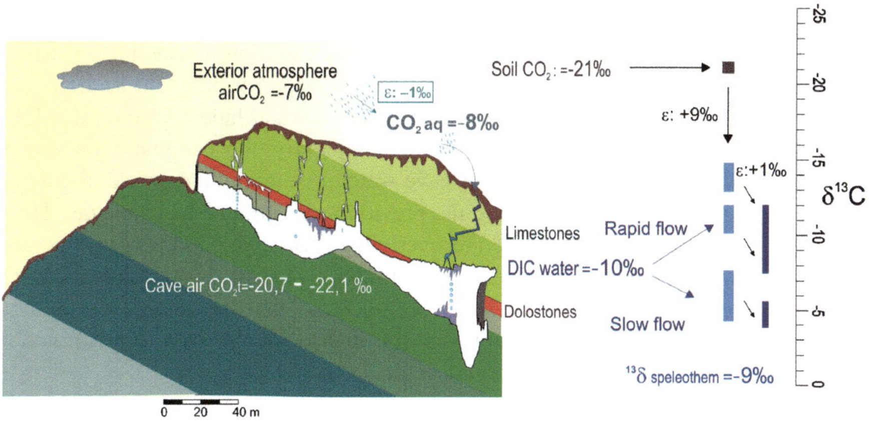


**Fig. 3** $^{13}C/^{12}C$ isotopic composition of $CO_2$ from air (external atmosphere, soil pore space and cave atmosphere), water and solid (speleothems and host-rock)

# 4 Discussion and Conclusions

The $^{13}C/^{12}C$ analyses for air, water and rock/speleothem in Altamira Cave (Spain) confirm the importance of the external soil as the $CO_2$ source for the underground system, as well as the effect of ventilation on system equilibrium. The similarity between the typical C3 vegetation signal of the air in the soil pore space ($\delta^{13}C$ $-21‰$) (Deines et al. 1980) and the air in the different points sampled in spring inside the cave ($\delta^{13}C = -18.7$ to $-21.4‰$) reflect a clear connection within the gas phase between the exterior and the interior (Table 1). These data evidence that karst underground atmospheres accumulate carbon dioxide, which has their origin in the overlying soils, and that carbon dioxide should be considered in the annual balances of atmosphere/soil flows in different ecosystems. The cyclic variation of cave air $CO_2$ concentration indicating the cave behaves as a $CO_2$ reservoir or source in different season is clearly outstanding in Altamira cave, as, the same as in other shallow karst systems (Fernández-Cortés et al. 2009).

The annual cycle of the aqueous phase isotopic signatures (Fig. 2), in addition to the rate of water/rock interactions, presents a seasonal variation, similar to that observed in the cave air, showing the influence of the gaseous exchange between cave and exterior atmosphere. During summer, periods of enhanced ventilation and reduced $CO_2$ concentration are observed to coincide with heavier isotopes (enriched in $\delta^{13}C$). This suggests an increased external influence, but also that the cave water is more prone to degasification; the kinetic isotopic discrimination makes the water heavier than equilibrium isotopes. This kinetic effect is more pronounced in waters with slow infiltration rates, where water from condensation mixes with the slow drip and the time scale for interaction between cave water and air – prior to drip – is enhanced.

The isotopic signature of gaseous $CO_2$ inside the cave also is influenced by ventilation and degasification of the cavity, due to the invasion of isotopically heavy air from the exterior atmosphere and the fractionation (enrichment of light isotopes in the gaseous part) caused by the entrance of water laden with organic $CO_2$ and its subsequent gasification (Fig. 3). Consequently, the isotopic signature of the speleothems precipitated from this cave water, essentially composed of calcite, should conform to expectations from the previously mentioned mechanisms. Thus, it ought to reflect the influence of the overlying vegetative canopy – in type and dynamic – and also the importance of the ventilation process. Indeed, it is seen that the speleothems have values ($\delta^{13}C = -4$ ato $-13.5‰$) that are congruent with their precipitation in isotopic equilibrium with the interior of the cave (discrimination coefficient of $+0.4‰$ at 14°C, Labonne et al. 2002).

**Acknowledgements** This research was supported by CGL2006-11561/BTE project. All Altamira Cave Research Centre and Museum staff is acknowledged for their collaboration throughout the whole research period.

## References

Deines P (1980) The isotopic composition of reduced organic carbon. In: Fritz P and Fontes JCh (eds) Handbook of Environmental Isotope Geochemistry 1:329–406.

Emrich K, Ehhalt D and Vogel JC (1970) Carbon isotope fractionation during the precipitation of calcium carbonate. Earth Planet Sci Lett 8:363–371.

Fernández-Cortés A, Sánchez-Moral S, Cuezva S, Benavente D, Abella R (2009) Characterization of trace gases fluctuations on a "low energy" cave (Castañar de Íbor, Spain) using techniques of entropy of curves. Int J Climatol. In press. doi:10.1002/joc.2057.

Kowalski AS, Serrano-Ortiz P, Janssens IA, Sánchez-Moral S, Cuezva S, Domingo F, Were A, Alados-Arboledas L (2008) Can flux tower research neglect geochemical $CO_2$ exchange? Agric For Meteorol 148(6–7):1045–1054.

Labonne M, Hillaire-Marcel C, Ghaleb B, Goy JL (2002) Multi-isotopic age assessment of dirty speleothem calcite: an example from Altamira Cave, Spain. Quat Sci Rev 21:1099–1110.

Yuan D (1997) The Carbon Cycle in Karst. Z Geomorph N F 108:91–102.

# Influence of Daily Visiting Regime in Tourist Cave at Different Seasons

J. Cuevas-González, A. Fernández-Cortés, M.C. Muñoz-Cervera, J.M. Andreu, and J.C. Cañaveras

**Abstract** Tourist caves with a daily regime of visits are exposed to severe alteration of their microclimatic balance. The record of microclimatic parameters in Canelobre Cave (Alicante, SE Spain) has allowed the recognition of two successive seasonal periods along each annual cycle with different microclimatic behaviour: (i) a warm period, dominated by stratification of air masses with progressive increase of microclimate parameters; and (ii) a cold period, dominated by ventilation of cave and approach to atmospheric values. These dynamics determine the influence of microclimate changes in the evolution of dripping water chemistry. The present-day visiting regime perturbs dripping water chemistry and the natural dynamics of growth of speleothems. For this reason, adequate management of visits considering microclimatic behaviour of the cave, could significantly minimize alterations to its tourist use.

## 1 Introduction

Environments with a very stable microclimate, like in a cave, dissolution/precipitation processes are essentially controlled by small imbalances among the $pCO_2$ values of air and infiltration water. Therefore, microclimatic variations (natural or man-made) play a key role in $CO_2$ degasification of waters and, consequently, in

J. Cuevas-González, M.C. Muñoz-Cervera, J.M. Andreu, J.C. Cañaveras
Dpto. Ciencias de la Tierra y del Medio Ambiente, Univ. Alicante, Campus San Vicente del Raspeig, 03080 Alicante, Spain, e-mail: jaime.cuevas@ua.es

J. Cuevas-González, M.C. Muñoz-Cervera, J.C. Cañaveras
Laboratorio de Petrología Aplicada, Unidad asociada UA-CSIC, 03080 Alicante. Spain

A. Fernández-Cortés
Dpto. Geología, Museo Nacional de Ciencias Naturales, CSIC, 28006 Madrid, Spain

B. Andreo et al. (Eds.), *Advances in Research in Karst Media* 
DOI 10.1007/978-3-642-12486-0, 

their saturation state and mineral precipitation rates (speleothem deposition). Microclimatic data obtained in Canelobre cave by means of an in-situ microenvironmental parameter automatic recorder and geochemical analysis of dripping waters, have allowed the observation of human-induced alterations in the saturation state of mineral phases in the cave.

## 2 Study Site

Canelobre Cave, located in the Sierra del Cabeçó d'Or (Alicante, SE Spain), constitutes a clear example of massive tourist use of a natural karstic georesource. This cave is the more daily visited in the province of Alicante and one of the most visited in Spain, with an annual affluence of near 60,000 visitors (165 people/day). Summer months (mainly July and August) accumulate about 38% of annual visits, reaching daily highs of over 700 people/day (Cuevas-González et al. 2009). Visits begin with the opening of an artificial tunnel, which has an access door that often remains open, and the switch-on of an electrical illumination system. This daily pattern of visits results in the effect of increased $CO_2$ concentration (visitors breathing) and temperature (illumination system) that produces daily cycles of variable magnitude. Also, the cave has a parallel use as a venue for events (e.g.: musical concerts). These episodes of massive assistance, up to 300 visitors, are easily recognizable in the microclimate record because of the notable highlight on daily cyclic oscillations (Cuevas-González et al. 2009).

In the exterior, the annual average temperature is 17.6°C, and the average environmental humidity is 58%. Mean rainfall within Canelobre cave area is generally lower than 400 mm/year (semiarid climate). The absence of precipitations is very lingering during summer, registering scarce rainfall events during August and September.

## 3 Methods

Microclimatic data were obtained by a real-time in-situ microenvironmental parameters recorder (Andreu et al. 2007) that records $CO_2$, temperature, relative humidity and atmospheric pressure data. A weather station installed outside simultaneously records temperature, relative humidity, atmospheric pressure and rainfall data. Parallel, monthly dripping water sampling was carried out. Hydrochemical analyses were performed in MNCN/CSIC laboratories (Madrid) using Ionic Capillarity Electrophoresis equipment. Calcite saturation index (S.I.) and $pCO_2$ in equilibrium with water have been calculated using PHREEQC v. 2.12 software (Parkhurst and Appelo 1999).

# 4 Results

## *4.1 Cave Air*

Mean temperature inside the cave is 17.8°C, with highest thermal oscillations in winter and minimum at the end of summer (September and October) (Fig. 1). Relative humidity (R.H.) values vary from 60 to 100% with higher oscillation between December and March, coinciding with exterior lower temperatures. From May to October, R.H. shows high values close to saturation (Fig. 1).

Cave air $CO_2$ concentration (annual average: 493 ppm) is usually near average concentration outside (360–380 ppm). Higher monthly average values are during summer, registering local increments by effect of massive visits, which may even double the annual average concentration (Fig. 1). General chaotic behaviour of cave air $CO_2$ concentration results from the visiting regime, which does not follow a previously defined visits schedule (Fernández-Cortés et al. 2009).

The cave shows a high energy exchange with exterior, mainly controlled by difference of temperatures between exterior and interior of the cave and for the visit regime. Two periods are discriminated along annual cycle (Fig. 1):

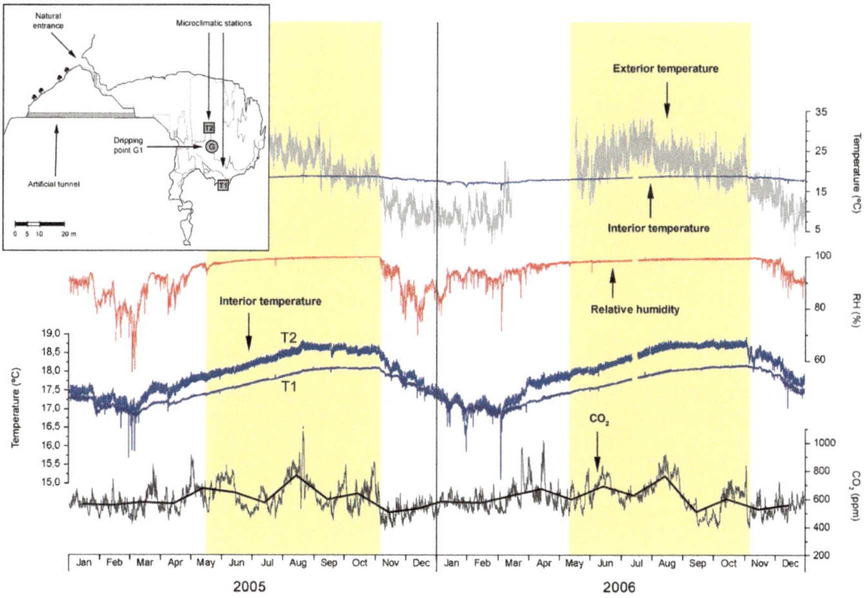


**Fig. 1** Microclimate pattern in Canelobre cave in 2005–2006. *Shady area* corresponds to period of stratification of cave atmosphere, where relative humidity and temperature tend to be stabilized in their annual maximum values. *Black line* over $CO_2$ record shows monthly mean values. *Left top*: sketch of the cave showing position of microclimatic stations and G1 dripping point

- Period of preferential stratification of air (summer; from May to October). Outside temperature is higher, so thermal fluctuation into the cave is lower. Stable microclimate prevails with restricted gaseous exchange with outside atmosphere and a gradual thermal increase caused by daily tourist use. Under these conditions, the cave air is stratified by density according to a thermal gradient and RH values close to 100% are reached.
- Period of preferential renovation (ventilation) by convective air circulation (winter-spring; from November to April). External air temperature is lower than in the cave and the exchange of air masses is favoured, causing the entrance of denser air that leads to a higher variability in cave air parameters.

## 4.2 Water

Dripping waters are of calcium-bicarbonate type, and the calculated oversaturation values in calcite and aragonite indicates that speleothem formation is active (Fernández-Cortés et al. 2008). Hydrochemical behaviour of a representative dripping point G1 (Table 1, Fig. 2) shows annual variations in $Ca^{2+}$ concentrations, with summer values around 58 mg/L and winter values around 45 mg/L.

This decrease is well correlated with electric conductivity (E.C.) and saturation index in calcite (S.I.) values: i) high values of both parameters in summer are a consequence of a longer time of residence of the infiltration waters in the epikarst zone, where high structural heterogeneity have been recognized (Fernández-Cortés et al. 2007), and, ii) low values in winter-spring are caused by a decrease in drip water saturation in calcite due to preferential precipitation of this mineral and/or increased infiltration by meteoric water (Fernández-Cortés et al. 2008).

Canelobre cave fits the standard pattern concerning a well-mixed ‘diffuse flow’ cave drips, where high summer cave air $pCO_2$ depresses calcite deposition, while low winter $pCO_2$ promotes degassing and enhances deposition rates (Fernández-Cortés et al. 2009 *in press*). The coincidence of high values of $CO_2$ produced by visits during winter-spring can cause an alteration in carbonate precipitation processes.

**Table 1** Monthly values of hydrochemical parameters for the sampling point G1 during 2006

| Date | E.C. (μS/cm) | S.I. | $Ca^{2+}$ (mg/L) | $HCO_3$ (mg/L) |
|---|---|---|---|---|
| 24-Feb-06 | 268 | 0,45 | 44,3 | 91,46 |
| 30-Mar-06 | 265 | 0,26 | 42,6 | 103,66 |
| 26-Apr-06 | 309 | 0,65 | 55,7 | 137,20 |
| 6-Jun-06 | 328 | 0,62 | 58,9 | 155,49 |
| 12-Jul-06 | 341 | 0,64 | 58,0 | 158,54 |
| 25-Sep-06 | – | 0,59 | 58,2 | 170,73 |
| 25-Oct-06 | 311 | 0,48 | 52,0 | 140,24 |
| 30-Nov-06 | 295 | 0,07 | 46,5 | 109,76 |
| 20-Dec-06 | 280 | 0,02 | 44,5 | 102,44 |

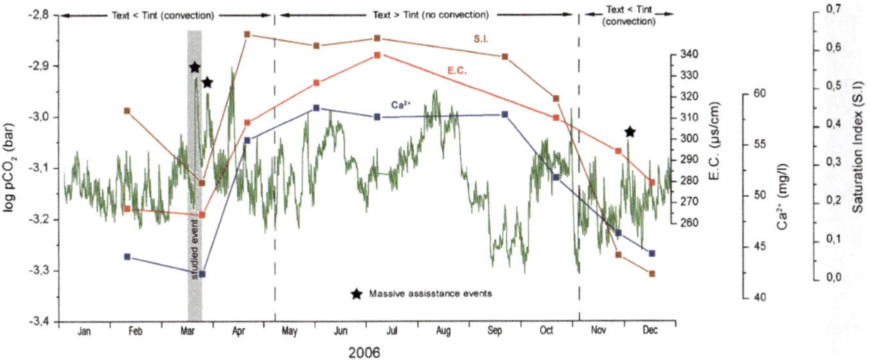


**Fig. 2** Record of cave air $pCO_2$ and cave water $Ca^{2+}$, S.I. and C.E. from dripping point G1, during 2006. The division of convection and non-convection periods is extracted from Fig. 1

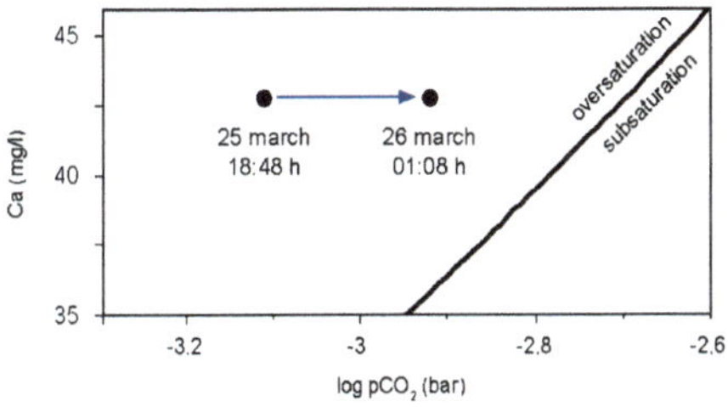


**Fig. 3** Idealized diagram of $pCO_2$ increment effect (expressed as log $pCO_2$ of cave air) on the saturation state with respect to calcite of dripping water at sampling point G1. Equilibrium boundary is calculated for 17°C (based on Langmuir 1997)

An episode of massive visits (March 2006) has been chosen to perform a model of balance of phases in karst systems (Fig. 2). This episode caused an increment (during 5–6 hours) of 337 ppm in cave air $CO_2$ concentration (Fig. 3).

## 5 Discussion

Calcite and aragonite oversaturation, induced by cave ventilation and the consequent precipitation of carbonate minerals, indicates the existence of a limit for mineral precipitation marked by water $pCO_2$ value. When air $pCO_2$ decreases below that limit the mineral precipitation rate increases causing a stabilization of mineral oversaturation (Fernández-Cortés et al. 2008). Degasification from water to air is a slow process conditioned by the slow kinetic of $HCO_3^- \rightarrow CO_2 + H_2O$ reaction. However, incorporation of air $CO_2$ to cave water is a process that reaches equilibrium in a few minutes (Cuezva 2008).

An abrupt increase in cave air $CO_2$ concentration in winter-spring, when $CO_2$ values are lower due to preferential renovation by convective air circulation, will produce a fast increase in cave water $pCO_2$ that will force a disequilibrium toward subsaturation conditions (Fig. 3). This fact provokes a slowing down of mineral precipitation and, in extreme cases, could lead to dissolution of carbonate crystals.

## 6 Conclusions

During an annual cycle in Canelobre cave, the rate of gaseous exchange between cave and outside atmosphere determines the values of cave air $pCO_2$ and, in consequence, the degree of mineral saturation of infiltration waters to reach the balance with cave air $pCO_2$. Analysis of microclimatic and hydrochemical data shows how during the winter-spring, where generalized ventilation of cavity takes place, corresponds with the time of preferential mineral precipitation.

Tourist caves with a daily regime of visits are exposed to severe alteration of their microclimatic balance. Two of the main factors that cause this perturbation are: (i) contribution of carbon dioxide produced by visitors through breathing and, (ii) in case of cavities with an artificial access for visitors, the frequent connection that occurs with the outer atmosphere through this access. While the second factor can be minimized with new infrastructures, the first factor can only be dropped by restricting tourist access to the cave. This measure commonly conflicts with the economic value (resource) that a show cave represents.

Cave management plans should include conservation of natural variation patterns of the cave air $CO_2$ concentration, which has a fundamental role in both speleogenetic (morphological) and speleothemic (mineral-forming) processes.

**Acknowledgements** Financial support has been provided by MCI (CGL2008-05929/BTE). Authors thank Ayto. de Busot and cave staff for their interest in the project and their cooperation.

## References

Andreu JM, Cañaveras JC, Cuevas J, García del Cura MA, Hernández Bravo JA, Muñoz-Cervera MC, Soler V (2007) Caracterización microclimática de la Cueva de Canalobre (Alicante). In: Durán JJ, Robledo PA, Vázquez J (eds). Cuevas turísticas: aportación al desarrollo sostenible. Publicaciones del IGME, Serie Hidrogeología y Aguas Subterráneas, n° 24, pp. 105–114.

Cuevas-González J, Fernández-Cortés A, Andreu JM, Cañaveras JC (2009) Condiciones de ventilación e influencia antrópica en la Cueva del Canelobre (Alicante): alteración microclimática en una cavidad turística. In: Durán JJ, Lopez-Martinez J (eds) Cuevas turísticas, cuevas vivas. IGME, Madrid.

Cuezva S (2008) Dinámica microambiental de un medio kárstico somero (Cueva de Altamira, Cantabria): microclima, geomicrobiología y mecanismos de interacción cavidad-exterior. Ph. D. thesis, Universidad Complutense de Madrid.

Fernández-Cortés A, Sánchez-Moral S, Cañaveras JC, Cuevas-González J, Cuezva S, Andreu JM (2009). Variations on seepage water geochemistry induced by natural and anthropic microclimatic changes: Implications for the speleothems growth conditions. Geophysical Research Abstracts, V.11: EGU2009-5868.

Fernández-Cortés A, Cuevas-González J, Andreu JM, Cañaveras JC, Sánchez-Moral S (2008) Influencia de los cambios microclimáticos en la hidroquímica de los goteos en la Cueva del Canelobre (Alicante). Geotemas 10:1561–1564.

Fernández-Cortés A, Cuevas-González J, Cañaveras JC, Andreu JM, Sánchez-Moral S, García del Cura MA y Hernández-Bravo JA (2007) Variación espacio-temporal de Ca-Mg-Sr en el agua de goteo de la Cueva del Canelobre (Alicante): ejemplo de procesos de infiltración en ambientes kársticos semiáridos. Geogaceta 43:83–86.

Langmuir D (1997) Aqueous environmental geochemistry. New Jersey, Prentice-Hall, Inc.

Parkhurst DL, Appelo CAJ (1999) User's guide to PHREEQC (v.2) – a computer program for speciation, batch-reaction, one-dimensional transport, and inverse geochemical calculations. Water-Resources Investigations Report 99-4259, USGS.

# Atmospheric Model in the Llamp Shaft. Garraf – Spain (2008–2009)

R. Cano, X. Font, P. Cociña, and A. Sanmartí

**Abstract** An atmospheric model is demonstrated using the shafts of Garraf and its function as an example of the karstic dynamic in the vadose zone. The breakdown of the atmospheric composition in the Llamp shaft is shown and the hydro-atmospherics relationship ($CO_2$ – vadose water) is indicated.

## 1 Introduction

The work has been carried out in the Garraf Massif. The temporal evolution of the hypogeal atmospheres has provided information on the relation between changes in the atmosphere and the availability of $CO_2$, and also the levels of corrosion.

For the realisation of this study an agreement has been reached between the Geology Department of the UB, the Parc Natural de Garraf and the Unio Muntanyenca Eramprunyà de Gavà.

The area of the study is in the southeast of Garraf, karstic massif forming part of the Serralada Costero Catalana, situated between the Llobregat River, the Mediterranean and the Penedes Plain, 20 km south of the city of Barcelona. The dump for solid, urban residues for Barcelona, closed in 2007, is located within this Massif. The investigations have been carried out in the shaft of Llamp. The Llamp shaft (70 m deep) is situated where the greatest density of shafts is found. It is made up of

R. Cano
Geologist and speleologist, Unió Muntanyenca Eramprunyà (Gavà), Spain,
e-mail: raulcanorelucio@yahoo.es

X. Font
Dpt. Geoquimica, Petrologia i Prospecció Geològica, Facultat de Geologia,
Universitat de Barcelona, Spain

P. Cociña, A. Sanmartí
Speleologist, Unió Muntanyenca Eramprunyà (Gavà), Spain

B. Andreo et al. (Eds.), *Advances in Research in Karst Media*
DOI 10.1007/978-3-642-12486-0, © Springer 2010

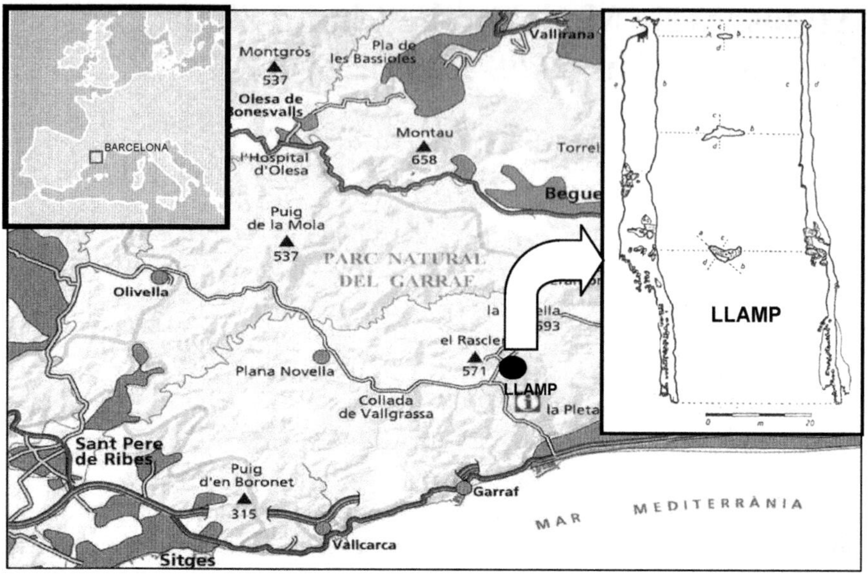


**Fig. 1** Situation of the Llamp shaft

a well shaft with an oval cross section which was formed from the joining together of two spindle well shafts.

## 2 Methodology

In the Llamp shaft, measurements of gas levels were taken periodically from 2002 until July 2009, although the continuous record of these measurements runs from July 2008 until July 2009. The campaign began using a detector of $O_2$, $CO_2$, $SH_2$ and $CH_4$, and after establishing that the significant parameters for variation and continuation were for $O_2$ and $CO_2$, two pairs of $CO_2$–$O_2$ detectors were used (DRÄGER PAC 7000). These were fitted at −5 m and at −50 m. From the hydro-geochemical analysis (in situ) the following have been determined: conductivity, temperature, pH, alkalinity, hardness, nitrites, nitrates and iron.

## 3 Results

The results have demonstrated a difference between the underground atmosphere and the exterior (Table 1). The average concentration of $O_2$ at the depth of 50 m is 18.6%. Consequently the percentage of $CO_2$ rises with the decrease of $O_2$, its

**Table 1** Statistical outline of the atmospheric details of Llamp

| | % $O_2$ (−5 m) | % $O_2$ (−50 m) | % $CO_2$ (−5 m) | % $CO_2$ (−50 m) |
|---|---|---|---|---|
| Sampling period | Jul 08–Jul 09 | Jul 08–Jul 09 | Jul 08–Feb 09 | Jul 08–Feb 09 |
| n (days) | 160 | 142 | 45 | 102 |
| n (30 min) | 7071 | 6329 | 1734 | 4490 |
| Average (%) | 20.0 | 18.6 | 0.7 | 1.1 |
| Percentile 25 (%) | 19.4 | 17.8 | 0.0 | 0.0 |
| Percentile 75 (%) | 20.9 | 19.2 | 1.1 | 1.9 |
| Desv. St. (%) | 1.2 | 1.2 | 0.8 | 1.1 |
| Max (%) | 20.9 | 20.9 | 3.9 | 4.0 |
| Min (%) | 15.7 | 15,3 | 0,0 | 0,0 |

average value at the depth of 50 m being 1.1%. In none of the tests were gases of anthropoid or entirely organic origin such as $SH_2$ and $CH_4$ found. The fluctuations are noticeable, varying with depth; all $CO_2$ concentrations increase in depth (Benavente 2007), as well as seasonal and daily fluctuations.

The relationship between the percentages of $O_2$ and $CO_2$ provides strong indications. The CAI (Cave Air Index) – formula proposed by Halbert (Smith 1999) suggests that: CAI $= \%\ CO_2/(20.9 - \%\ O_2)$. The average CAI obtained in Llamp is 0.8, indicating an organic origin, but a low percentage of organic material in the ground and the absence of other organic gases ($SH_2$ and $CH_4$) were recorded. While it could cautiously be deduced that the production of $CO_2$ could result from processes of calcium precipitation, meteoric drag and organic oxidation, correlation between percentage of $O_2$ and $CO_2$ is $r = -0.89$. The series of $O_2$ are more continual and longer. It has been decided to extend these parameters to be able to present a special and temporal analysis.

In all the series of measurements, a decrease of $O_2$ with depth (Fig. 2) has been detected with a consequential increase in $CO_2$. In this way vertical monitoring has enabled the observation of where exactly the decrease in percentage of oxygen takes place first, confirming that variations always occur primarily in the lower zones and with several hours delay, changes are later detected in the higher parts (Fig. 3). This behavior contradicts the theory of atmospheric sedimentation.

The signal of $O_2$ presents a periodic behaviour with two daily minima and two daily maxima. The two minima are at 4.00 h and 16.00 h. The maxima are at 10.00 h and 22.00 h (Fig. 3). This result is repeated throughout the year. For this reason an atmospheric parameter that shows this same spectrum was looked for and found in the atmospheric pressure, due to the natural process of the atmospheric tides (Berry 1945), which present a daily reading of two maxima and two minima, of the same form as the $O_2$ spectrum, thus establishing an adequate correlation between both parameters. To be able to compare the recordings of $O_2$ with the readings of atmospheric pressure, the monthly and daily breakdown of the pressure reading has to be undertaken.

By using moving averages, it is possible to reduce the time oscillations and the accidental oscillations. This application requires the prior decision of the period

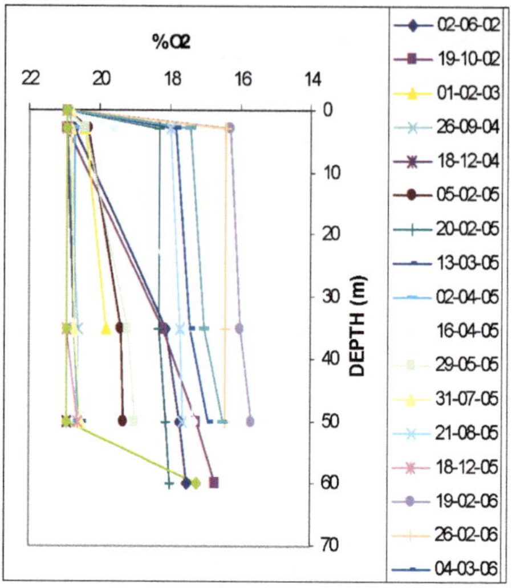


**Fig. 2** $O_2$ decrease in depth

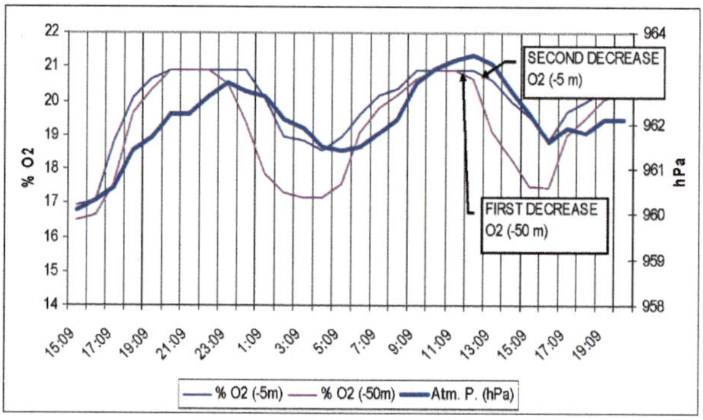


**Fig. 3** It is noted that the decrease in oxygen first occurs in the $-50$ m zone and several hours later the decreases are seen in the $-5$ m zone

in which the behaviour pattern is repeated. In this way, the daily component for atmospheric pressure has been extracted (Fig. 4).

- The moving average is calculated by taking the period of 12-hour grouping that corresponds with the length of the wave of the barometrical tides.
- The daily component is calculated by taking the difference between the average values and the moving average.

The indexes of correlation (r) between the daily component of the atmospheric pressure and the percentage of $O_2$ at $-5$ m and at $-50$ m are superior to 0.8 (during February 2009) in this way validating the lineal dependence between both parameters.

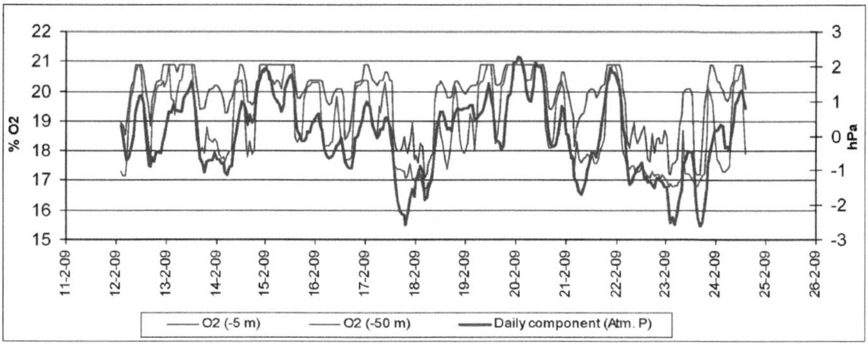


**Fig. 4** Representation of the seasonality at hourly intervals of the atmospheric pressure, the percentage of $O_2$ at $-5$ m and at $-50$ m. The barometric tides are clearly defined by the recording of percentages of $O_2$

In this way, the baro-atmospheric behaviour is explained. In moments of low pressure, the air is charged with $CO_2$, present in the dense fabric of fissures at a pressure higher than the exterior. This $CO_2$ enriched air then migrates to the greater spaces occupying the base zones of the shaft.

The study of the correlograms of $O_2$ at $-50$ m shows the presence of a seasonal timetable in winter and the absence of this in summer (Fig. 5).

There is a strong relationship between $CO_2$ gas, dissolved $CO_2$ $HCO_3^-$, $CO_3^{2-}$ and pH. The vadose zone is where the atmospheric effects are measured and where the correlation between atmosphere and vadose water are established. By using the Henry formula, a relation is shown between $CO_2$ gas and $CO_2$ dissolved in water. At the same time, the amount of $CO_2$ dissolved in the water among other parameters determines its corrosive or depositional potential. To determine the type of behaviour of the water, the saturation index (SI) is used. The values obtained at $-50$ m are outlined as: First quartile $-0.32$ (moderately aggressive), average $-0.03$. The verification of moderately aggressive episodes at $-50$ m is backed up by the morphologies at Llamp. At this depth dissolved formations appear, i.e., wall troughs have developed due to drop-water corrosion – erosion effect on rock surface.

## 4 Synthesis and Interpretation

**Climatic model.** The production of $CO_2$ could result from calcium precipitation, meteoric drag and organic oxidation processes. With a decrease in atmospheric pressure outside, the air in the cavity will escape through all fissures. When a high pressure appears outside, it causes a massive influx of air to balance the system (Fig. 4). These $CO_2$ exhalations are manifested from inside to outside, as verified with the stations located at different depths (Fig. 3).

A barometric model is defined where the atmospheric pressure acts as a motor to the dynamic atmosphere and the temperature conditions the homogenization or

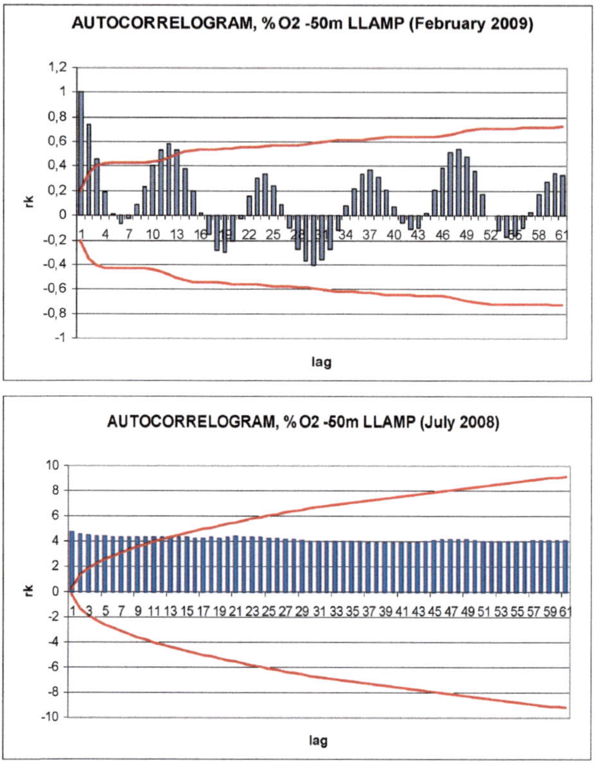


**Fig. 5** The representation of the correlograms of winter and summer verify the different atmospheric behaviour of the base zone at LLamp. In winter the convective movements of the air in the caves homogenizes the internal atmosphere, and in summer the atmosphere is seen to be stratified, exaggerating considerably the differences between the superior and inferior parts

stratification of the cave air. In summer, atmospheric differentiation is more usual. The decreases in pressure (effects of storms and tides) provoke a rapid influx of $CO_2$ charged air, entering first in the deeper parts. Here the air can show up to 3% $CO_2$, a richness due to the fact that the influx of $CO_2$ rich air is faster than the processes of atmospheric homogenization. In winter a homogenization of the atmosphere is produced, due to a thermal convective effect (Fig. 5), ventilation in the cavities is more efficient (Bourges et al. 2001).

## References

Benavente J, Vadillo I, Carrasco F, Liñán C (2007) Distribución vertical del contenido en $CO_2$ en la zona no saturada de un karst mediterráneo (Nerja, España): resultados preliminares. Geogaceta 41 (2007) 23–26.

Berry F A, Bollay E, Beers N (1945) Handbook of Meteorology (New York: McGraw-Hill, 1945). pp. 483–498 and 746–749. Atmospheric tides.

Bourges F, Mangin A, d'Hulst, D (2001) Le gaz carbonique dans la dynamique de l'atmosphère des cavités karstiques: l'exemple de l'Aven d'Orgnac (Ardèche). Earth and Planetary Sciences 333 (2001) 685–692.

Smith GK (1999) Foul air in limestone caves and its effect on cavers. ASF 22nd Biennial Conference proceedings 1999, 48–58.

# Physics of Condensation Corrosion in Caves

F. Gabrovšek, W. Dreybrodt, and M. Perne

**Abstract** Condensation of water from warm, humid air to cold rock walls in caves is regarded to play a significant role in speleogenesis. Water condensing to the cave walls quickly attains equilibrium with the carbon dioxide in the surrounding air, and consequently dissolves limestone or gypsum. Heat released by condensation and heat flux from the air to the cave wall raises the temperature of the cave walls and reduces the condensation rates. For constant air temperature initial condensation rates are high but then drop rapidly by orders of magnitude during the first few days until constant condensation rates are attained, when the heat flux into the rock is fully transmitted to the surface of the karst plateau. When diurnal or seasonal variations of the air temperature are active as is the case close to cave entrances, condensation rates can become quite significant, up to about $10^{-6}$ m/year.

## 1 Introduction

Water vapor from a cave atmosphere condensing to the walls of a cave creates a water film in equilibrium with the partial pressure $pCO_2$ of the cave atmosphere. This solution is therefore aggressive to limestone and the dissolution process based on what has been termed condensation corrosion. Condensation is possible only if the temperature of the cave walls is below the dew point of the air. The heat released by condensation increases the temperature of the cave wall and reduces the condensation rate. In this paper, from a theoretical point of view, the following question is addressed: What are the rates of condensation and what average annual retreat of bedrock follows as a consequence?

W. Dreybrodt
Karst Processes Research Group, University of Bremen, Germany, e-mail: dreybrodt@t-online.de

F. Gabrovšek, W. Dreybrodt, M. Perne
Karst Research Institute ZRC SAZU, Postojna, Slovenia, e-mail: gabrovsek@zrc-sazu.si; perne@zrc-sazu.si

B. Andreo et al. (Eds.), *Advances in Research in Karst Media*
DOI 10.1007/978-3-642-12486-0, © Springer 2010

## 2 Basic Theory and Condensation at Stationary Boundary Conditions

The rate $F\,[\mathrm{g\,m^{-2}\,s^{-1}}]$ of water condensing to a rock surface determines the retreat of rock by dissolution. It is assumed that the rock is covered by a thin film of water, about $10^{-2}$ cm thick. Such a thin film will quickly come to thermal equilibrium with the temperature $T_f$ of the cave wall. It keeps its constant depth as there is flow from the rock surface down to the cave floor. Water condenses to the film, if the partial pressure $P_a$ of vapor in the cave air exceeds the vapor pressure $P_f$ at the temperature $T_f$ of the water film. In the following, it is assumed that the cave air is well mixed, but close to the rock a diffusion boundary of thickness $\varepsilon D\,[\mathrm{m}]$ exists.

For condensation, water molecules are transported through this layer by molecular diffusion. Combining Fick's law with the ideal gas law and the relation $P_a - P_f = \Delta P/\Delta T(T_a - T_f)$, the flux of water condensing to the cave wall is obtained:

$$F_v = \frac{D_m}{\varepsilon_D} \cdot \left(\frac{\Delta P}{\Delta T}\right) \cdot \left(\frac{T_a - T_f}{RT_a}\right) \cdot M = \frac{h_v}{\varepsilon_D}(T_a - T_f)\,, \quad h_v = 1.8 \cdot 10^{-5}\,\mathrm{gm^{-1}\,K^{-1}} \tag{1}$$

$D_m$ is molecular diffusion constant, $R$ is the universal gas constant and $M$ the molar mass of air. The vapor condensing to the water film releases heat of condensation $q_c = 2450\,\mathrm{J/g}$. The flux of heat $F_q$ released by condensation is given by

$$F_q = q_c \cdot F_v = q_c \cdot \frac{D_m}{\varepsilon_D} \cdot \frac{M}{RT_a} \cdot \frac{\Delta P}{\Delta T} \cdot (T_a - T_f) = \frac{4.42 \cdot 10^{-2}}{\varepsilon_D}(T_a - T_f) \quad [\mathrm{Wm^{-2}}] \tag{2}$$

An additional flux $F_c$ from the warm cave air to the water film is given by

$$F_c = \frac{k_a}{\varepsilon T}(T_a - T_f) = \frac{2.6 \cdot 10^{-2}}{\varepsilon T}(T_a - T_f) \quad [\mathrm{Wm^{-2}}] \tag{3}$$

$k_a$ is the thermal conductivity of air ($2.6 \cdot 10^{-2}\,\mathrm{W\,m^{-1}\,K^{-1}}$) and $\varepsilon_T$ is the thickness of the thermal boundary layer, which is related to $\varepsilon_D$ by $\varepsilon_D = \varepsilon_T\,(Sc/Pr)^{1/3}$.

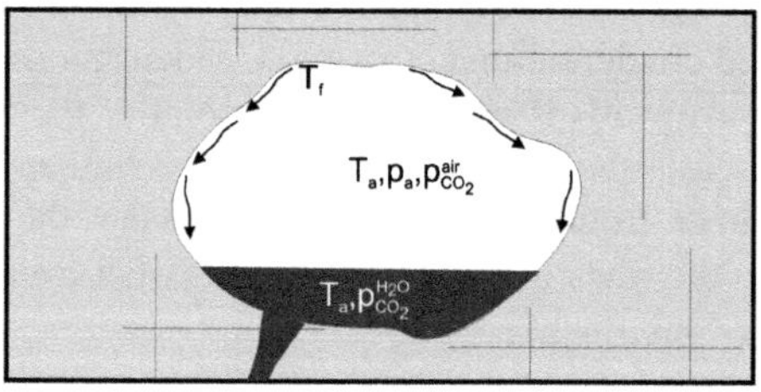


**Fig. 1** Water evaporates from an open surface of elevated temperature $T_a$ to the cave air in thermal equilibrium with the water. Vapor pressure at temperature $T_a$ is given by $P_a$. The partial pressure in the air is in equilibrium with in the water. The vapor condenses at the cave wall with temperature $T_f < T_a$ and flows back as a thin film, designated by *arrows*. These water films become saturated with respect to the mineral composing the rock (limestone or gypsum)

Sc is the Schmidt number for diffusion and $P_r$ the Prandtl number for heat convection (Beek and Muttzall 1975). For air, $S_c = P_r = 1$. Therefore $\varepsilon_D = \varepsilon_T = \varepsilon$.

The total heat flux to the water film is the sum of Eqs. 2 and 3

$$F_{tot} = F_q + F_c = \frac{k}{\varepsilon}(T_a - T_f)\,, \quad k = 0.0702\,\mathrm{W\,m^{-1}\,K^{-1}} \tag{4}$$

This heat flux increases the temperature of the water film, until the heat flux into the rock equals the heat flux from the cave air into the water film.

In a first approach, the problem was reduced to a simple one-dimensional setting, as shown on Fig. 2. An extended large cave is located at depth $Z$ below the surface. Only the vertical heat flux towards the surface is considered. At $z = 0$ the temperature $T_0$ is kept constant for $t > 0$. The cave roof at depth $Z$ experiences a heat flow $F_{tot}$ given by Eq. 4. A solution to this problem is given in Carslaw and Jaeger (1959,

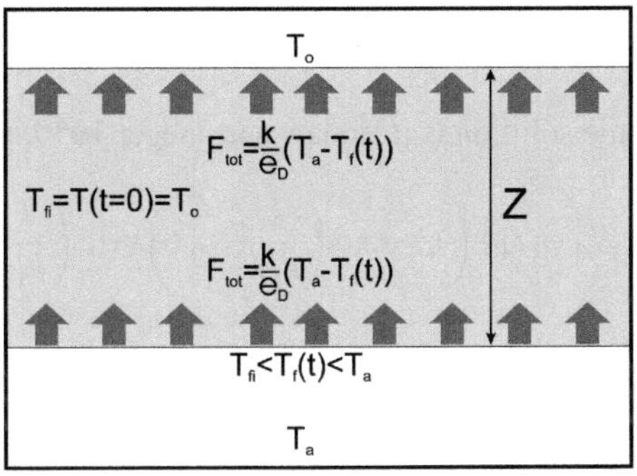


**Fig. 2** A large extended hall at depth $Z$ below ground is filled with air of 100% humidity at constant temperature $T_a$. The temperature of the covering rock is the average annual temperature $T_0$ of the surface. As soon as heating by condensation starts the wall temperature $T_f$ increases. Steady state is reached when the heat flow (*gray arrows*) to the cave ceiling equals that at the surface

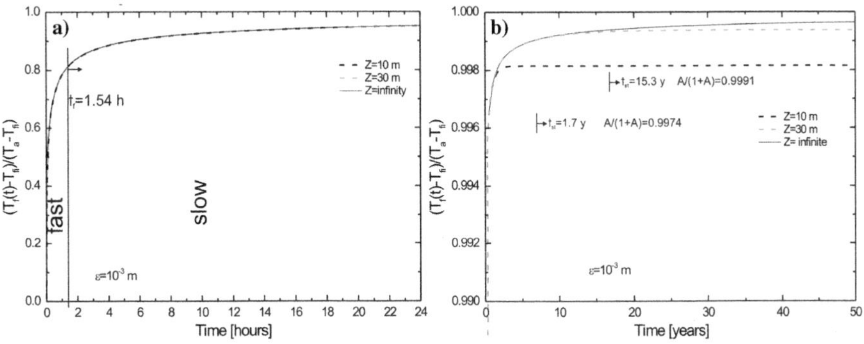


**Fig. 3a,b** Time dependence of the temperature difference $T_a - T_f(t)$ normalized to initial temperature difference $T_a - T_{fi}$. **a** Results for the first 24 hours. **b** Results for long times

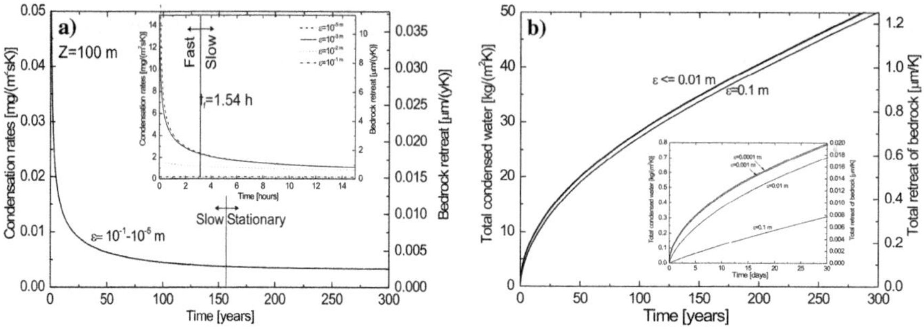


**Fig. 4** **a** Rates of condensation and retreat of bedrock as function of time divided by the initial temperature difference $T_a - T_{fi}$. Values of $\varepsilon$ are written at the *curves*. For large times all curves coincide, i.e the rates become independent of $\varepsilon$. **b** Amount of condensed water per $m^2$ and K of temperature difference and total retreat of bedrock for the cases of Fig. 4a. Note that although the variation of $\varepsilon$ covers two orders of magnitude, the amount of condensed water only weakly depends on $\varepsilon$

p.125). For $Z = \infty$ the solution is (Carslaw and Jaeger 1959, page 70).

$$T_f - T_{fi} = (T_a - T_{fi}) \left(1 - \exp\left(\frac{k^2}{k_r^2 \varepsilon^2} \cdot \kappa t\right) \operatorname{erfc}\left(\frac{k}{k_r \varepsilon} \sqrt{\kappa t}\right)\right), \tag{5}$$

The condensation rate for large times ($t > \tau_f = 9(k_r \varepsilon / k)^2 / \kappa$) is given by:

$$F_V^{slow} = D_m \left(\frac{\Delta P}{\Delta T}\right) \frac{M}{RT_a} \left(\frac{1}{\sqrt{\pi\kappa}} \frac{k_r}{k} \frac{1}{\sqrt{t}}\right) \cdot (T_a - T_{fi}) \quad \left[gm^{-2}s^{-1}\right] \tag{6}$$

The condensation rate continuously drops with the square root of time. However, for a finite $Z$ stationary temperature $T_f^{stat}$ and condensation rate $F_V^{stac}$ are reached after an exponential approach of the wall temperature with the time constant $\tau_S = Z^2/(\beta^2 \kappa)$. The minimum rate of condensation is reached after time $5 \cdot \tau_s$:

$$F_V^{stac} = \frac{D_m}{\varepsilon} \left(\frac{\Delta P}{\Delta T}\right) \frac{M}{RT_a} \cdot \frac{A}{1 + A} (T_a - T_{fi}) = 1.8 \cdot 10^{-5} \frac{T_a - T_{fi}}{\varepsilon + \frac{k}{k_r} Z} \quad \left[gm^{-2}s^{-1}\right] \tag{7}$$

$k_r$ is the thermal conductivity of the rock and $\kappa$ is the thermal diffusivity of the rock. $A = Zk/(\varepsilon k_r)$. The value of $\beta$ is a function of $A$ and is listed by Carslaw and Jaeger (1959). Note that the stationary rates drop approximately with $1/Z$ as can be also seen from Fig. 5d.

Figure 3 shows the exact temperature dependence for limestone with $Z = 10$ m, 30 m, and $\infty$; $\varepsilon = 0.001$ m. Note that the initial rise of temperature is independent of $Z$. Figure 4a shows the rates of condensation for $Z = 100$ m and various values of $\varepsilon$. Figure 4b depicts the amount of condensed water after time $t$, which is obtained by integration of the curves in Fig. 4a.

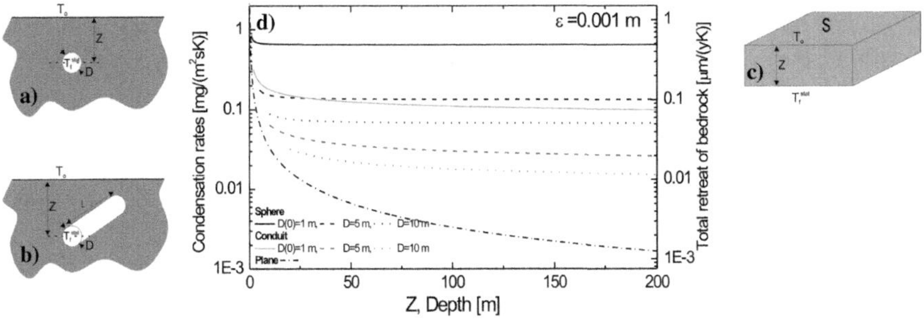


**Fig. 5 a–c** Geometrical configurations for the shape factors in Table 1. **d** Condensation rates and retreat of bedrock in the stationary state in dependence on depth $Z$ for spheres and cylindrical conduits with various diameters. $D(0)$ is the initial diameter. The *blue curve* represents one-dimensional scenario (slab with thickness $Z$) of Fig. 5c. See also Fig. 2

The one-dimensional model is idealistic, therefore other geometries are more suitable, e.g. a spherical room with diameter $D_s$ or a cylindrical conduit of diameter $D_c$ and length $L$, both buried at depth $Z$. These are shown in Fig. 5a–c. For such situations exact analytical results are not available. However, the general behavior is similar to the one-dimensional case, and it is possible to obtain the temperature at stationary state, using the theory of conduction shape factors (Incropera and DeWitt 2002). At stationary state, the wall of the rock is at temperature $T_f^{stat}$ and the surface temperature at $z = 0$ is $T_0$. See Fig. 2. The total amount of heat flowing from the conduit to the surface is then given by

$$Q_r = S \cdot k_r \left(T_f^{stat} - T_0\right) \quad [\mathrm{W}] \tag{8}$$

where $S$ [m] is the shape factor. Some of these are given in Table 1.

The total heat transferred to the wall of the cave is $Q_w = \sigma \cdot (k/\varepsilon)\left(T_a - T_f^{stat}\right)$, where $\sigma$ is the surface area of the cave. Conservation of energy requires $Q_w = Q_r$, which gives:

$$\left(T_a - T_f^{stat}\right) = \frac{1}{1 + \frac{k \cdot \sigma}{k_r \cdot \varepsilon \cdot S}} \left(T_a - T_0\right) \tag{9}$$

From this the rate of condensed water can be calculated by use of Eq. 1.

**Table 1** Shape factors for some typical geometry as shown in Figs. 5a–c

| System | Restriction | Shape factor |
|---|---|---|
| Sphere in a semi infinite medium (Fig. 5a) | $Z > D/2$ | $2\pi D/(1 - D/4Z)$ |
| Cylinder in a semi infinite medium (Fig. 5b) | $L \gg D,\ Z > 3D/2$ | $2\pi L/\ln(4Z/D)$ |
| Slab (Fig. 5c) | | $\sigma/Z$ |

Figure 5d shows these rates per degree K for circular cave rooms and cylindrical conduits, buried at depth $Z$. Note that for $\varepsilon \ll 0.038\sigma/S$ the rates are independent of $\varepsilon$, which is true for all practical applications.

Spheres show the highest condensation rates, almost independent of their depth $Z$ below ground, but dependent on their diameter $D_S$. Dreybrodt et al. (2005) have calculated that a sphere with initial diameter of 0.1 m needs about 530 ky to reach a diameter of 1 m for $T_a - T_{fi} = 1$ K. For $T_a - T_{fi} = 10$ K these times are lower by a factor of ten. For a cylindrical conduit buried at $Z = 100$ m growth from initially 1 m to 10 m takes $2 \cdot 10^8$ years for $T_a - T_{fi} = 1$ K.

## 3 Conclusions and Further Perspectives

The presented scenarios are rather idealistic. However, the basic demand stays the same for all possible cases: the heat of condensation has to be transported away for the condensation to last. Diurnal and yearly temperature variations can enhance condensation rates effectively as the walls cool down during the cold period. Such scenarios were investigated by Dreybrodt et al. (2005). They also envisaged further scenarios with 2D conduits and condensation on stalagmites. They also compared theoretical results to the results of an experimental set-up. Lismonde (2002) presents different scenarios and processes that effectively enhance the condensation. Further work should include field measurements of condensation and evaporation rates, similar to those presented by De Freitas and Schmekal (2003).

## References

Beek WJ, Muttzall KMK (1975) Transport phenomena. Wiley, London

Carslaw HS, Jaeger JC (1959) Conduction of heat in solids. Oxford University Press, Oxford

De Freitas CR, Schmekal AA (2003) Condensation as a microclimate process: Measurement and prediction in the Glowworm Tourist Cave, New Zealand. International Journal of Climatology 23(5):557–575

Dreybrodt W, Gabrovšek F, Perne M (2005) Condensation Corrosion: A Theoretical Approach. Acta Carsologica 34/2:317–347

Incropera FP, DeWitt DP (2002) Fundamentals of heat and mass transfer. J. Wiley, New York

Lismonde B (2002) Climatologie du Monde Souterrain, Aérologie des systèmes karstiques. CDS Isère.

# $\delta^{13}C$ Values from a Stalagmite at the Nerja Cave, South Spain

C. Jiménez de Cisneros and E. Caballero

**Abstract** Carbon isotopes in speleothems may serve as indicators of vegetative change, climatic conditions, and karst processes. Carbon isotope variations in speleothems arise from a variety of causes resulting from changes in both the surface environment (local climate, bioproductivity, vegetation type, etc) and the subsurface environment in which rain water passes through the soil zone and then the epikarst into cave voids. Even within the cave, local environmental controls including evaporation, drip rates and changes in cave air $pCO_2$ controlled by seasonal ventilation can influence the final $\delta^{13}C$ values recorded in the speleothem record; although interpretation of $\delta^{13}C$ records may involve several processes, these potentially contain a useful environmental signal. The $\delta^{13}C$ record of a stalagmite from Nerja Cave was examined and the roles of environmental processes and local cave conditions on speleothem $\delta^{13}C$ values were evaluated.

## 1 Introduction

Climatic conditions, particularly temperature and precipitation, affect the type and density of vegetation above a cave and the vegetative productivity. During periods of greater precipitation, plant cover and biological activity are increased, which raises $pCO_2$ and lowers $\delta^{13}C$ values of soil $CO_2$ (Hou et al. 2003). When conditions are drier, reduced plant cover and biological activity result in lower $pCO_2$ and higher $\delta^{13}C$ values of soil $CO_2$. The carbon isotopic composition of speleothems can be used to infer the isotopic composition of atmospheric $CO_2$ (Baskaran and Krishnamurthy 1993), the type of vegetation growing above the cave (Dorale et al. 1992), and the relative contributions of soils vs. aquifer limestone to the dissolved inorganic

C. Jiménez de Cisneros, E. Caballero
Instituto Andaluz de Ciencias de la Tierra, CSIC, Facultad de Ciencias, Universidad de Granada, Campus Fuentenueva s/n 18002 Granada, Spain, e-mail: concepcion.cisneros@eez.csic.es

B. Andreo et al. (Eds.), *Advances in Research in Karst Media*
DOI 10.1007/978-3-642-12486-0, © Springer 2010

carbon that forms the speleothem (Hendy 1971; Genty et al. 1998). Speleothem $\delta^{13}C$ values reflect the degree of mixing between the bedrock (isotopically heavier) and biogenic (isotopically lighter) $CO_2$.

$\delta^{13}C$ in stalagmites is an additional and important environmental proxy, although $\delta^{13}C$ is more difficult to interpret in climatic terms. Recently, speleothem $\delta^{13}C$ has been linked to kinetic fractionation associated with the degassing of $CO_2$ from the cave dripwater, which in turn varies in response to ambient cave $CO_2$ levels (Mattey et al. 2008).

In this study, the $\delta^{13}C$ record of a stalagmite from Nerja Cave, Southern Spain was examined, and the roles of climatic processes and local cave conditions were evaluated.

## 2 Site Description

The Nerja Cave is situated in the south of the Iberian Peninsula (Málaga, Southern Spain). Geologically, the study regions form part of the Betic Cordillera (Fig. 1). The cave of Nerja is a cavity in the outcrops of Alpujarride marbles that make up Sierra Almijara. This cave is an excellent fossil record of its own history and the paleoclimatic and neo-seismotectonic evolution of the area where the cave is located. Nerja Cave consists of numerous halls with a north-south orientation. The focus of this study is a stalagmite, which was collected in the Montaña Hall.

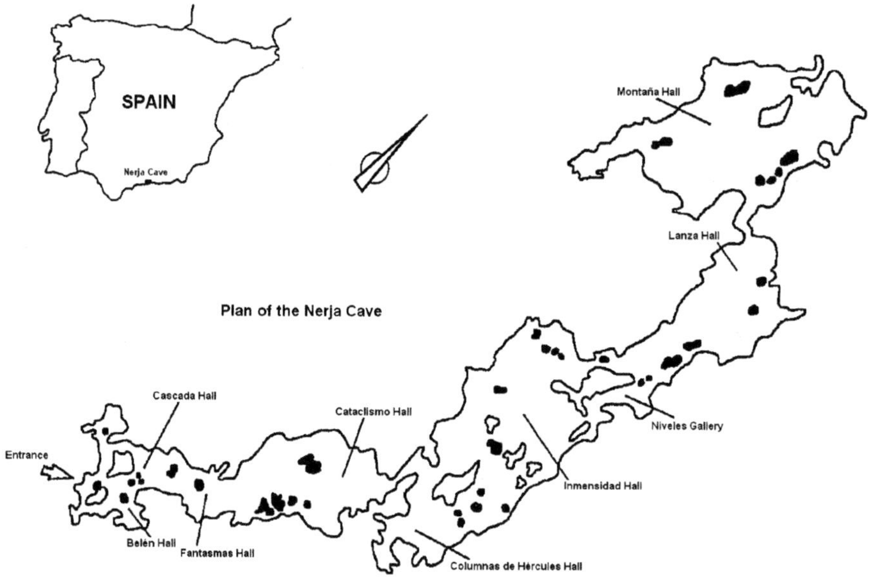


**Fig. 1** Situation of the Nerja Cave in Málaga, southern Spain. Plan of cave showing the location of the halls

## 3 Method

The stalagmite was cut along its growth axis. One longitudinal section was polished and crystallographic observations were performed with a binocular microscope. The sample studied was between 60 and 90 mm long. A regular lamination with alternation of thin layers (<0.2 mm) and thick is present. The stalagmite diameter is variable up to 70 mm. The light and dark laminae visible in hand specimen are related to changes in the fabrics. The stalagmite was divided into two parts, the first clear is composed of beige translucent crystals, in the middle of which lies a thin brown layer which was interpreted as a clayey inclusion, although this does not necessarily represent a long interruption in deposition; in the second part grey-white crystals are observed. Based upon optical and crystallographic features, the crystals display mainly microcrystalline fabrics and subordinately columnar ones (Frisia et al. 2000). The stalagmite consists of aragonite crystals and the mineralogical composition appears uniform throughout, with no observed evidence to indicate recrystallization of aragonite to calcite or the presence of alternating mineral laminae. $^{230}Th/^{234}U$ date indicate that the sampled stalagmite was deposited between 90,000 and 70,000 years ago (Durán 1996).

Stable isotope analyses were taken along the growth axis of stalagmite. The $\delta^{13}C$ values are given in permil relative to PDB and the standard deviations are 0.06‰ based on replicate measurements of an internal standard.

## 4 Results and Discussion

The $\delta^{13}C$ values in the studied stalagmite show large variations. These fluctuations are often explained by changes in vegetation (Dorale et al. 1998), the most obvious being related to photosynthetic pathways. $\delta^{13}C$ measurements range from –0.2‰ to –8.6‰ with an average value of –2.6‰. The complete $\delta^{13}C$ time-series areplotted in Fig. 2. The $\delta^{13}C$ record roughly follows the same trend as the $\delta^{18}O$ record. When the correlation between $\delta^{13}C$ and $\delta^{18}O$ is good, as it is in this case where $r = 0.913$, there is a clear correlation between hydrology and vegetation above the cave.

The stalagmite shows two parts, the first stage featuring relatively high carbon signature (mean value of −1.1‰) and an upper part characterized with relatively decreasing values (mean value –4.8‰). The higher $\delta^{13}C$ values may reflect, at least in part, an enrichment of the heavier $^{13}C$ isotope in the aragonite crystals. Differences in vegetation and local processes also contributed to the higher $\delta^{13}C$ values. Furthermore, some degree of isotopic disequilibrium during aragonite precipitation may have contributed to elevated $\delta^{13}C$ values. The stalagmite was deposited during the time interval 70–90 ky ago, in the ending isotopic stage 5, when the phreatic level of the waters in the cave decreased. This period in the cave is characterized by maximum growth of speleothems.

Paleoenvironmental conditions expressed in the $\delta^{13}C$ time-series record define two intervals: the first stage characterized by decreased vegetation associated with

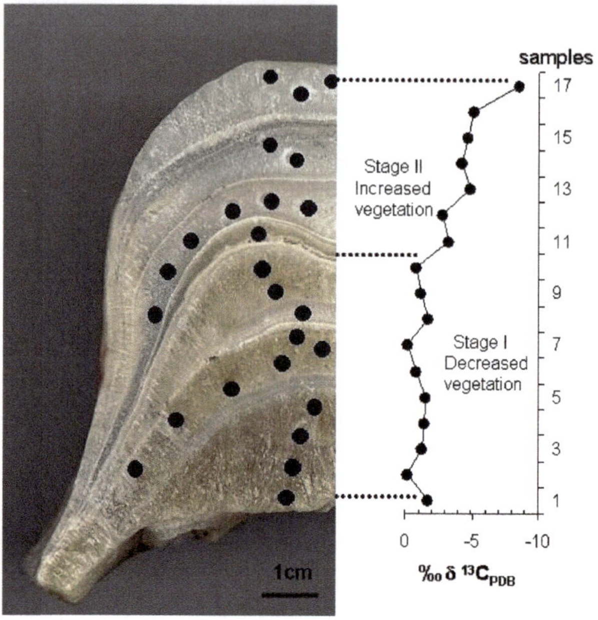


**Fig. 2** Photograph of the cross-section of the stalagmite. Sampling sites for $\delta^{13}C$ analyses are indicated. $\delta^{13}C$ record divided into two stages

a cool-dry conditions during this period, the high values suggest that vegetative productivity declined. These values may also reflect lower drip rates, consistent with decreased precipitation, and degassing in the epikarst. The cold and dry conditions favoured enhanced cave air ventilation, which enhanced degassing and the precipitation of higher speleothem $\delta^{13}C$ values. The second stage with lower $\delta^{13}C$ values suggests more biological productivity associated with warm conditions. The low $\delta^{13}C$ values reflect relatively high drip rates fed by abundant precipitation and a lower degree of mixing between atmospheric and biogenic $CO_2$. Under these conditions, weakened cave air ventilation resulted in reduced degassing and lower speleothem $\delta^{13}C$ values.

## 5 Conclusions

$\delta^{13}C$ values of stalagmite from Nerja Cave provide information about the history of vegetation and climatic conditions. Changes in biological productivity and inorganic processes in response to climatic conditions associated with a mediterranean climate were the origin of the variations in the $\delta^{13}C$ values. During a cool-dry period the vegetative productivity declined which resulted in higher $\delta^{13}C$ values. When the climatic conditions are warm the low $\delta^{13}C$ values reflect relatively high drip rates fed by abundant precipitation and a lower degree of mixing between atmospheric and biogenic $CO_2$.

**Acknowledgements** This research was supported by Project CGL2007-61876/BTE. The authors thank the Foundation of Nerja Cave for access to the sites.

## References

Baskaran M, Krishnamurthy RV (1993) Speleothems as proxy for the carbon isotope composition of atmospheric $CO_2$. Geophys Res Lett 20:2905–2908

Dorale JA, Gonzalez LA, Reagan MK, Pickett DA, Murrell MT, Baker RG (1992) A high-resolution record of Holocene climate change in speleothem calcite from Cold Water Cave, North-east Iowa. Science 258:1626–1630

Durán JJ (1996) Los sistemas kársticos de la provincia de Málaga y su evolución: contribución al conocimiento paleoclimático del Cuaternario en el Mediterráneo occidental. PhD Thesis, University of Madrid

Frisia S, Borsato A, Fairchild I, McDermott F (2000) Calcite fabrics, growth mechanisms and environments of formation in speleothems from the Italian Alps and Southwestern Ireland. Jour Sed Res 70:1183–1196

Genty D, Vokal B, Obelic B, Massault M (1998) Bomb $^{14}C$ time history recorded in two modern stalagmites; importance for soil organic matter dynamics and bomb $^{14}C$ distribution over continents. Earth Planet Sci Lett 160:795–809

Hendy CH (1971) The isotopic geochemistry of speleothems; I, The calculation of the effects of different modes of formation on the isotopic composition of speleothems and their applicability as paleoclimatic indicators. Geochim Cosmochim Acta 35:801–824

Hou J, Tan M, Cheng H, Liu T (2003) Stable isotope records of plant cover change and monsoon variation in the past 2200 years: evidence from laminated stalagmites in Beijing, China. Boreas 32:304–313

Mattey D, Lowry D, Duffet J, Fisher R, Hodge E, Frisia S (2008) A 53 year seasonally resolved oxygen and carbon isotope record from a modern Gibraltar speleothem: reconstructed drip water and relationship to local precipitation. Earth Planet Sci Lett 269:80–95

# Mineral-Forming Processes at Canelobre Cave (Alicante, SE Spain)

J. Cuevas-González, A. Fernández-Cortés, M.C. Muñoz-Cervera, D. Benavente, M.A. García del Cura, J.M. Andreu, and J.C. Cañaveras

**Abstract** Canelobre cave presents a remarkable collection of speleothems, such as stalactites, coralloids, stalagmites, flowstones, draperies, columns, helictites, gours, spars, crusts and flowers. Most of them are calcitic in composition, although sulphate speleothems have been found in some lower chambers in the cave. Gypsum speleothems appear as white uniform crusts or as flowers (antholites). The younger speleothems recognized in the cave corresponds to soda-straw stalactites, helictites and some active stalagmites. Vadose speleothems clearly prevail the phreatic ones, which are represented mainly by calcite spar deposits. Different mechanisms of speleothem formation can be recognized in the cave: dripping or flowing water, seepage, capillarity, etc. The parameters that control the speleothem deposition are the characteristics of karstic waters and the cave microclimate.

## 1 Introduction

Canelobre Cave is situated in the Sierra del Cabeçó d'Or (Alicante, SE Spain). The cave mainly consists of a single huge chamber with a length of about 100 m and a very high roof (approximately 60 m), resembling that of a cathedral. From their

J. Cuevas-González, M.C. Muñoz-Cervera, D. Benavente, J.M. Andreu, J.C. Cañaveras
Dpto. Ciencias de la Tierra y del Medio Ambiente, Univ. Alicante, Campus San Vicente del Raspeig, 03080 Alicante, Spain, e-mail: jaime.cuevas@ua.es

J. Cuevas-González, M.C. Muñoz-Cervera, D. Benavente, M.A. García del Cura, J.C. Cañaveras
Laboratorio Petrología Aplicada, Unidad asociada UA-CSIC, 03080 Alicante, Spain

A. Fernández-Cortés
Dpto. Geología, Museo Nacional de Ciencias Naturales, CSIC, 28006 Madrid, Spain

M.A. García del Cura
Instituto de Geología Económica, CSIC, 28040 Madrid, Spain

B. Andreo et al. (Eds.), *Advances in Research in Karst Media*
DOI 10.1007/978-3-642-12486-0, 

ceiling and walls a spectacular display of speleothems emerges, some of them looking like a 'candelabra', for what is thought that the cave takes its name. Its natural entrance is high (700 m above sea level) on the southwest slopes of the sierra, although visits are made via a man-made tunnel at a lower level. Currently, more than 60,000 people visit this cave each year.

Sierra del Cabeçó d'Or is composed of a 650 m-thick sequence of Jurassic and Lower Cretaceous massive limestones with interbedded marls, sandstones and calcarenites (Andreu 1997). This sequence in the zone behaves as an aquifer, in which the principal recharge comes from direct infiltration of the rainfall on the outcrops of permeable rocks.

## 2 Speleothems and Cave Minerals

Canelobre Cave shows a remarkable collection of speleothems. Most of them are calcitic in composition, although gypsum and celestite-bearing calcite speleothems have been found in some lower chambers (Cañaveras et al. 2005). Traces of fluorite, barite, opal and phosphate minerals have been also detected in different types of speleothems.

### *2.1 Calcite Speleothems*

The younger speleothems recognized in the cave correspond to calcitic soda-straw stalactites, helictites and some active stalagmites. Both younger and ancient carbonate speleothems are low-magnesian calcite (LMC) with an $MgCO_3$ content ranging from 0.5 to 2 mol%.

#### 2.1.1 Speleothems Formed by Dripping Water

Soda straws are incipient, translucent stalactites, with a characteristic interior conduit (Fig. 1a) and occasionally with a banded structure. Growth layers of 0.05–0.5 mm thickness corresponds usually to annual cycles and are defined by inclusions oriented parallel to the long direction of calcite crystals. Different textures have been distinguished in both soda straw and stalactites (Fig. 1a): columnar, fibrous and microcrystalline. Columnar and fibrous textures consist on parallel disposition of euhedral calcite crystals, with uniform extinction and a height-width ratio $< 6{:}1$ (columnar) or $> 6{:}1$ (fibrous). The alignment of impurities or inclusions generally marks lines of growth. Microcrystalline texture shows an alternating clear-dark lamination marked by the size of crystals ($< 10\,\mu m$) and the high proportion of impurities (clays, iron oxides). Macrocrystalline texture is characterized by inclusion-free, big ($> 200\,\mu m$) idiotopic to subidiotopic crystals.

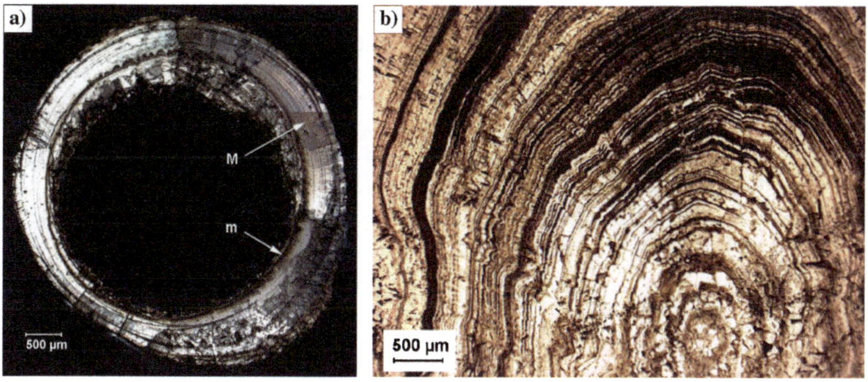


**Fig. 1** **a** Transversal section of a soda-straw. M, macrocrystalline (columnar and fibrous) texture; m, microcrystalline texture. **b** Section of a spherulite composed of an alternation of macrocrystalline and microcrystalline bands (see text). Banding is also outlined by erosive discontinuities and inclusion-rich layers. Bands are commonly 10 µm to 1 mm thick

### 2.1.2 Speleothems Formed by Flowing Water

Flowstones are one of the most common speleothems in the cave. They present a variety of forms, from near horizontal subtypes varieties to cascades and canopies. Crenulations, grading into microgours, are common on flowstone surface. In Canelobre cave this type of speleothem is frequently associated to fallen blocks (ceiling breakdown) that are in the floor of the cavity. Flowstones generate flat surfaces on walls or blocks and are associated with permeable fractures or discontinuities that permit the entrance of infiltration (or dripping) waters inside to the cavity. Flowstones present a very fine lamination, with alternating bands of yellowish, orange, brown and black colour. Calcite crystals form macrocrystalline (> 200 µm) columnar to fibrous aggregates or microcrystalline layers with abundant impurities (clays, iron oxides, terrigenous grains).

Draperies or curtains, a composite flowstone-dripstone speleothems, are common in the inclined part of the ceilings. Also, it is common that flowstones are covered by other types of speleothems, such as coralloids.

### 2.1.3 Speleothems Formed by Seeping Water

Coralloids are also an abundant type of speleothem in Canelobre cave. Coralloid is here used as a generic term that embraces a wide variety of nodular, globular and botryoidal speleothems. Globules or protuberances constitutes the basic element of this speleothem type and can present diverse sizes, morphologies and microstructures. Three main subtypes can be recognized in the cave: dendritic (coral-like) (Fig. 2), crystalline (rugose) and nodular (smooth, cave-popcorn) coralloids. The colour of these speleothems varies from white in globular coatings to orange

**Fig. 2** Coralloid speleothem that shows ramified growth (dendritic or coral-like subtype)

in ramified morphologies. Globular coatings thickness oscillates from some few millimeters to 1–2 cm. On the other, the ramified structures can reach a height of 15–25 cm (Fig. 2). Protuberances or individual globules present a regular concentric structure and vary in width from 2 to 5 mm in globular coatings, from 5 to 10 mm in the botryoidal aggregates and from 5 to 25 mm in ramified subtypes. In cave-popcorn subtypes, a spherulitic to spheoridal crystalline nucleus can be distinguished. This is composed of elongate (sometimes curved) subcrystals (or domains) that grow starting from a center (Fig. 1b). This nucleus attains 5 mm of maximum diameter (perpendicular to growth direction). When the spherulite is composed of curved subindividuals and display asymmetric growth and growth layers of unequal thickness it is termed "spheroidalites" (Self and Hill 2003).

Vermiform helictites growing on cave ceilings, walls or other speleothems are other type of speleothems formed by seeping. Some helictites forms are also observed associated with capillary waters. Their length varies from 1 to 5 cm and usually possesses a tiny central capillary canal.

### 2.1.4 Speleothems Formed by Pool (Thermal) Water

Subaqueous speleothems such as gours, cave rafts and shelfstones are present in the cave, although the main phreatic speleothem in the cave is calcite spar. This deposit consists of rhombohedral or scalenohedral (dogtooth) crystal encustrations covering the walls of big blocks or collapses. Spar crystals are as much as 5 cm long and vary in width from 0.5 to 3 cm. In this cave calcite spar speleothems are usually in higher areas thatnsome vadose forms.

## *2.2 Sulphate Speleothems*

Gypsum speleothems consist in both gypsum flowers (antholites) and 2–4 cm thick crusts covering both limestone bedrock and older carbonate speleothems (coralloids,

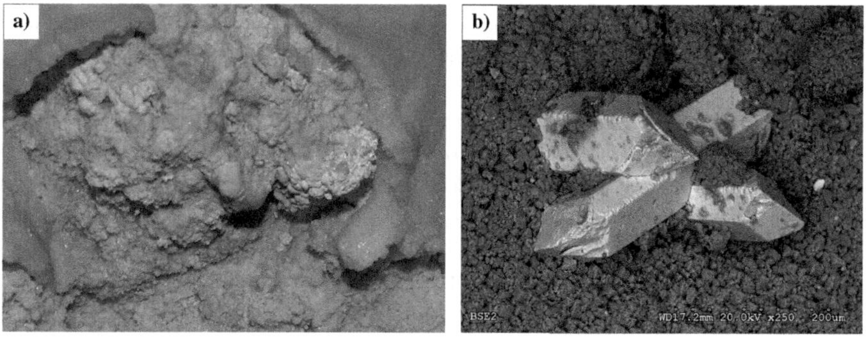


**Fig. 3** **a** Detail of broken gypsum globular crust coating corralloid (calcite) speleothems; **b** prismatic crystal of celestite in association with carbonate speleothems

flowstones) (Fig. 3a). These crusts are constituted of tabular or acicular crystal aggregates. Fibrous gypsum crystals are usually arranged perpendicular to substrate. Occasionally, gypsum crusts show globular or hemispherical mound-like forms with an inner earthy part of calcite, gypsum and celestite crystals (Fig. 3). Some of these morphologies appear broken, probably indicating they flake and collapse from wall as crusts grow thicker and heavier. Celestite is present as tiny ($< 0.5$ mm in length) prismatic colourless crystals in association with calcite coralloids (Fig. 3b) and/or earthy masses below gypsum crusts.

## 3 Discussion

Different mechanisms and environments of mineral formation can be recognized in Canelobre cave. Vadose speleothems clearly prevail over phreatic ones, which are represented mainly by spar deposits and usually appear covering walls of big collapsed blocks. Several cycles of carbonate speleothem formation can be distinguished since gravitational forms (stalactites, stalagmites, curtains, etc.) become overgrown by capillary forms such as knobbly popcorn concretions and phreatic forms are located in upper areas than some vadose forms. Development of different textures of recent speleothems is related to mechanisms of growth and with quantity and composition of karstic waters. This fact is specially highlighted in the case of the speleothems formed from dripping or flowing water. Elongated fabrics (columnar, fibrous) recognized in these speleothems are probably generated from waters in continuous, rather fast, movement. As a consequence, it contributes to numerous nuclei and an almost continuous contribution of reactants to nucleation points. Likewise, soda straws require slow, steady feed water without suspended fines to block the canal.

Subaerial coralloids can grow from thin films of splashing water or seeping where water is controlled by capillary forces within the pore spaces of bedrock. Clear-dark lamination associates with differences in the growth rate, where clear sheets indicate rates of higher growth than dark ones. Each couple of sheets indicates a "cycle" of low-high speed, so that whole aggregate (coralloid) microsequences can be recognized. Capillary forces are also involved in the genesis of other speleothems. A high capillary pressure maintained in their central canals is responsible for the fact that helictites grow without regard to the force of gravity or the direction best suited for evaporation in capillary film environment (Self and Hill 2003). Differences in speleothem subtypes, such as crystalline coralloids (crystallictite subtype) and spherullitic coralloids (cave popcorn subtype), corresponds to variations in the mechanisms of growth. In coralloids, crystallictites grow when growth rate is low meanwhile crystals start splitting and corallites form for a high growth rate. Intermediate forms also occur in cave, such as dendrites consisting of curviedged crystals. Calcite spar deposits, the main phreatic deposits in cave, are associated with degassing of thermal waters (Ford and Williams 2007). Growth of larger spar crystal requires a calcite-saturated solution in a quiet environment during enough time for crystals to grow large.

Flowing and seeping water evaporation appears to be the main mechanism for gypsum crust formation in Canelobre Cave. Microclimate oscillations, specifically short-period (seasonal) variations in humidity, could also be considered as a factor in their genesis (Maltsev 1996). Likewise, globular forms may be related to zones of preferential capillarity flows in the substrate and with the alternation of solution–precipitation processes from seepage water in walls and ceilings (Calaforra 1998). The dissolution of limestone hostrock in which celestite veins have also been recognized in the area (Andreu 1997) is the most probable source of strontium for precipitation of celestite in Canelobre Cave.

Speleothem deposition in Canelobre cave is controlled by the distribution, quantity and chemistry of water percolating through the karstic aquifer and by the cave microclimate. These main factors are therefore controlled by cave geometry, aquifer properties, surface geomorphology, vegetation and external climate.

**Acknowledgements** Financial support has been provided by MCI (CGL2008-05929/BTE). The authors thank Ayto. de Busot and the cave staff for their cooperation.

## References

Andreu JM (1997) Contribución de la sobreexplotación al conocimiento de los acuíferos kársticos de Crevillente, Cid y Cabeçó d'Or (provincia de Alicante) Ph. D. thesis Univ. Alicante, 377 p.

Calaforra JM (1998) Karstología de yesos. Ciencia y Tecnología 3, 384 p.

Cañaveras JC, Andreu JM, Muñoz MC, García del Cura MA (2005) Sulphate speleothems in Canelobre Cave (Alicante, SE Spain). Macla 3:59–60.

Ford D, Williams PW (2007) Karst hydrology and geomorphology. Wiley, 562 p.

Maltsev VA (1996) Sulphate filamentary crystals and their aggregates in caves. Proc Univ Bristol Spel Soc 20(3):171–185.

Self CA, Hill CA (2003) How speleothems grow: An introduction to the ontogeny of cave minerals. Jour Cave and Karst Studies, 65(2):130–151.

# Petrological Study as a Tool to Evaluate the Degradation of Speleothems in Touristic Caves, Castañar de Ibor Cave, Cáceres, Spain

R. Martín-García, A. Martín-Pérez, and A.M. Alonso-Zarza

**Abstract** In Castañar cave the surface of most of the speleothems present dissolution and corrosion features. In touristic caves, this process has usually been related to the acidification of the atmospheric moisture caused by $CO_2$ from the breath of visitors. However, in Castañar cave the process of corrosion has been also observed in rooms that are not visited by tourists. Petrological studies were carried out in the speleothems affected by surface corrosion in Castañar cave. The results indicate that this process occurs not only at present time, but over a period of thousands of years, as evidenced by the presence of corrosion lines inside and on the surface of the speleothems. All this and the fact that Castañar cave recovers very quickly from changes in the environmental parameters, indicates that the dissolution–corrosion process is slow and hence it is not related to the presence of visitors.

## 1 Introduction

The study of degradation processes in caves is a widely studied phenomenom. In touristic caves such studies have been focused on the impact of the visitors and the cave improvements (such as the lighting) over the cave environment and the speleothems. But there are also natural degradation processes, observed in touristic and non-touristic caves that are inherent to the dynamics of the cave systems.

In this contribution a case study is presented from Castañar de Ibor speleothems, in which degradation processes can be observed not only in the outer parts of the speleothems, but at different stages of their formation. In doing so, the processes that cause the alteration that change the surface appearance of the speleothems are envisioned.

R. Martín-García, A. Martín-Pérez, A.M. Alonso-Zarza
Instituto de Geología Económica (CSIC-UCM), Departamento de Petrología y Geoquímica, Facultad de Ciencias Geológicas, 28040, Madrid, Spain, e-mail: rmartingarcia@geo.ucm.es

B. Andreo et al. (Eds.), *Advances in Research in Karst Media* 509
DOI 10.1007/978-3-642-12486-0, © Springer 2010

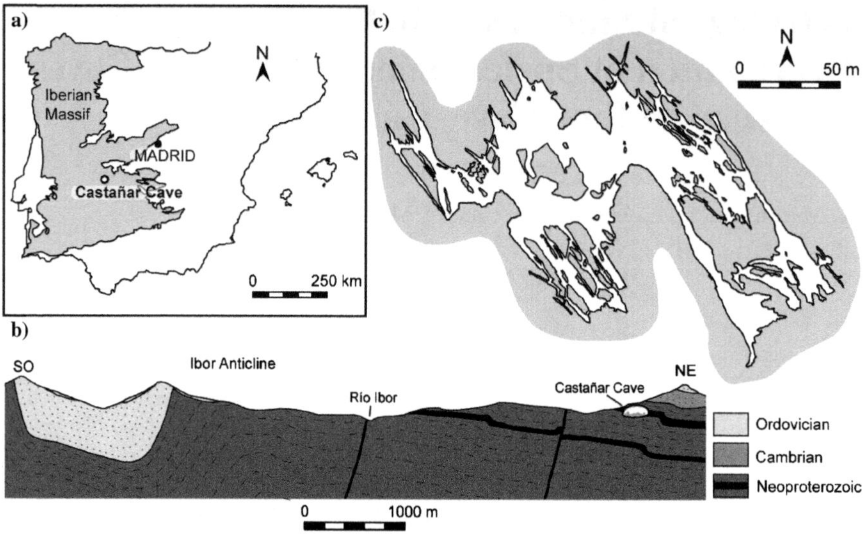


**Fig. 1** **a** Geographical setting of Castañar Cave within the Iberian Peninsula. The *gray zone* corresponds to the Iberian Massif. **b** Map of Castañar Cave. **c** Geological section through the Ibor anticline. The cave is developed in the dolomitic layers of the Neoproterozoic series (*black lines*)

## 2 Geological Setting

Castañar cave is on the SE part or the Iberian Massif (Fig. 1a). The area where it has been developed is mainly siliciclastic, consisting of shales and greywackes with a few interbedded dolomite and magnesite layers (Fig. 1b), all of Neoproterozoic age (Díez-Balda et al. 1990). The formation of the cave is due to the dissolution of the carbonates and the later collapse of the siliciclastic materials.

The development of the cave is horizontal (Fig. 1c) and it has just one entrance that is opened only during the visits, so there is little interchange with the surface atmosphere. The enviromental parameters inside the cave are constant throughout the year: 17°C temperature, and humidity close to 100%. The nature of the host rock determines the composition of the infiltration cave waters, which have a Ca/Mg ratio of 0.5 to 0.1 and are usually saturated with respect to calcite, aragonite and dolomite (Sánchez-Moral et al. 2006). There is another source of water in the condensation of moisture. This moisture is undersaturated in all minerals and it is usually acidic.

The main minerals in Castañar cave are carbonates, mainly calcite and aragonite, but due to the high concentrations of Mg in the cave waters, Mg-carbonates such as dolomite and huntite are also found (Alonso-Zarza and Martín-Pérez 2008).

## 3 Materials and Methods

Conventional optical petrography studies were performed on corroded speleothems. Thin sections were taken from different forms, including stalactites, stalagmites, and draperies. To make the thin sections the samples were immersed in a resin containing Epofer EX 401 and Epofer E 432 in a vacuum system before cutting and polishing. Scanning electron microscopy observations were performed on gold-coated samples using a JEOL 6400 electron microscope working at 20 kV and with a resolution of 35 Å.

## 4 Characterization of the Corrosion Process in Speleothems

The process of surface corrosion on speleothems is a common process in caves. It has been described all over the world in touristic and non-touristic caves (Sarbu and Lascu 1997; Dublyansky and Dublyansky 1998; Tarhule-Lips and Ford 1998; Zupan-Hajna 2002; Auler and Smart 2004; de Freitas and Schmekal 2006; Martín-García et al. 2009).

The corrosion occurs when undersaturated or acidic waters (with high concentrations of $CO_2$ or $H_2S$) contact the surface of the bedrock and speleothems dissolving them. There are different mechanisms of corrosion; the most common is the condensation-corrosion that consists in the presence of aggressive condensed moisture over the rocks.

When a speleothem is corroded, the surface appears as a white matt powdery wrap making the speleothems appear chalky (Fig. 2). If the corrosion layer is thick enough, it usually falls off so the inner fresh part of the speleothems can be seen (Fig. 2a).

When polishing the samples, the thickness of the corrosion layer can be measured. Thinner lines of the same kind appear in the inner zones (Fig. 3a), all of which are concentric and appear between layers of fresh clean crystals (Fig. 3b,c). The thinner lines sometimes are invisible to the naked eye, but are visible under the

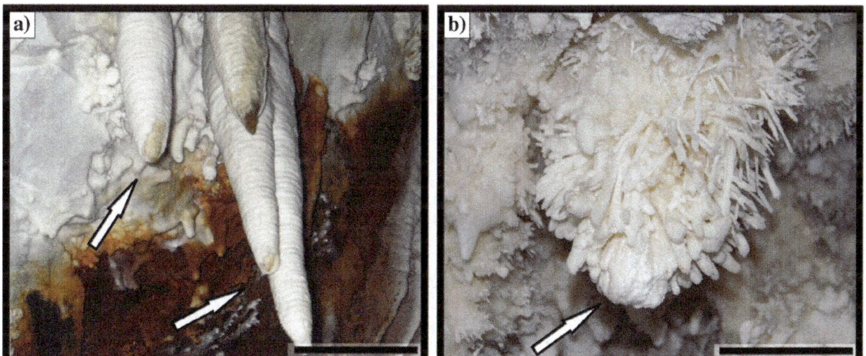


**Fig. 2a,b** Two examples of corroded speleothems in Castañar cave. The overall aspect is chalky and matt. **a** Calcitic stalactites. The tips of these stalactites have lost their chalky coating (*white arrows*). **b** Aragonitic anthodites corroded specially on the tips (*white arrows*)

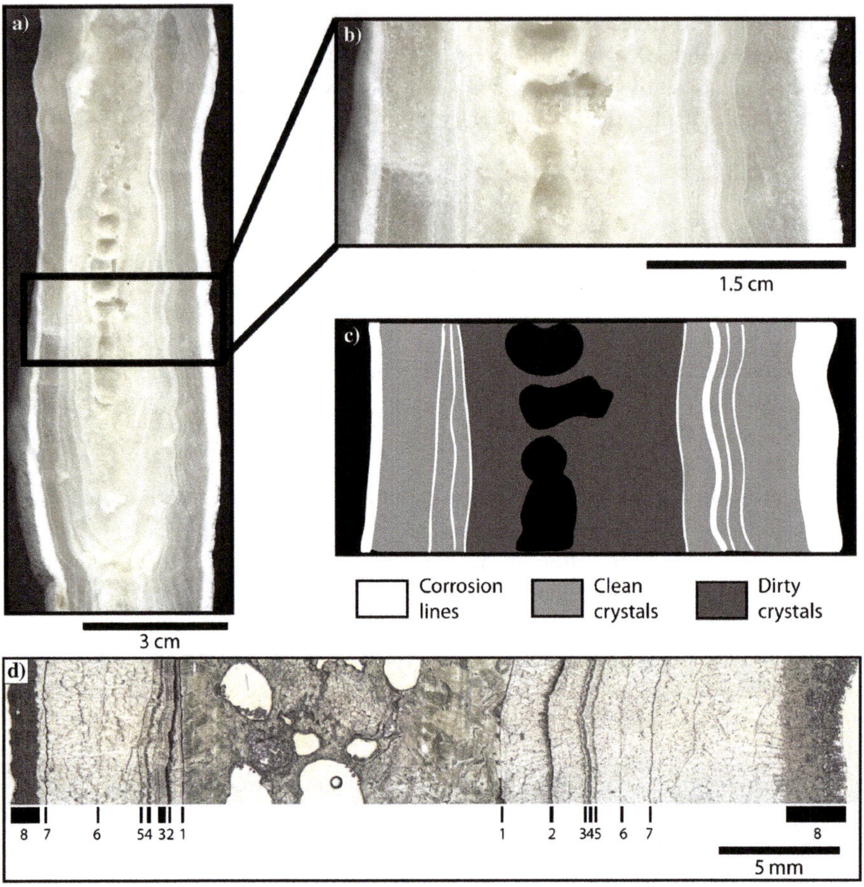


**Fig. 3** **a** Polished sample of a stalactite. **b** Detailed view of a. From the surface to the inner zone it shows: a superficial corrosion layer (*white*), clean crystals (*gray*) and a wide whitish zone that contains concentric *white corrosion lines* of different thickness. The *holes in the middle* are the relics of the central channel that has been partially cemented. **c** Sketch based on b. **d** Microphotograph with plain polarised light. The *white lines* in hand sample here correspond to the *darker ones*. The number of corrosion lines observed under the microscope is larger than the observed in the hand sample. Also, the number of lines at each side of the central channel is the same

microscope (Fig. 3d). At both sides of the central channel the number of lines is the same but the distance between them, that represent growing stages, is different.

Under the microscope, the white layers look like a dark opaque structureless mass (Fig. 4a). This mass is constituted by small crystals formed by the dissolution of the larger ones that form the body of the speleothems (Martín-García el al. 2009) leading to micrite formation (Jones and Kahle 1995). In an initial stage previous stage of corrosion, when the small crystals are still not formed, the surfaces of the crystals appear partially dissolved and etched (Fig. 4b). The contact of this micritic mass with the fresh crystals is usually the crystal boundary, but if the corrosion is

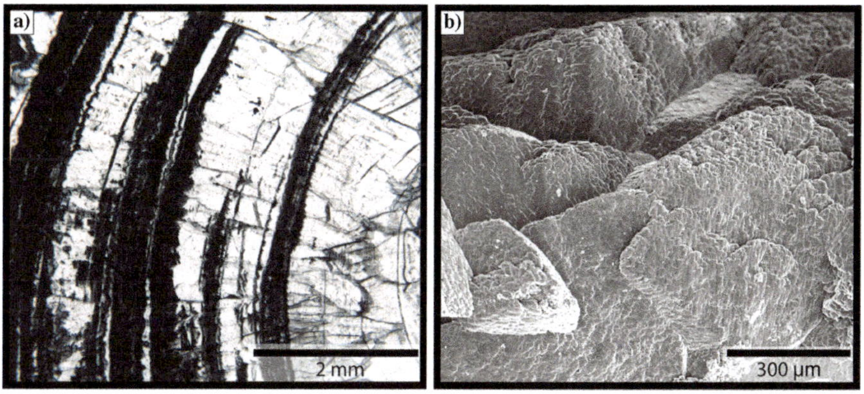


**Fig. 4** **a** Microphotograph showing concentric corrosion lines between the crystal boundaries of the same growth band. Plane polarised light. **b** SEM image of calcite crystals showing etched surfaces

very aggressive the contact does not follow any boundary but they cut across the crystals.

## 5 Discussion

This corrosion process occurs only on the surfaces of the speleothems, thus the inner lines may correspond to ancient surfaces. This may indicate that the process is episodic in the area and that corrosion and growth alternate rather than occurring at the same time within the same speleothem.

The relation between the visits and the chemical degradation of the speleothems is not clear. The acidification of the atmospheric moisture caused by human breath may amount to the same as the natural process, but the inner lines indicate that this process has been occurring for thousand of years, long before the cave was firstly discovered and visited.

The most probable mechanism of corrosion in the case of Castañar cave is the one explained by Palmer (2007) that is applicable to caves where little or no circulation of air and little running water are common features, as in the example of this study. The main idea of this model is that water condenses in the upper parts of the cave and then seeps down through the rock pores or over the rock surfaces dissolving the materials.

## 6 Conclusions

The corrosion of the speleothems is a natural process that occurs in touristic and non-touristic caves causing the loss of brightness and general disfigurement of the formations. It occurs only on the surface of the speleothems and the host rock and it is recorded in the speleothems during their formation providing information about the growing mechanisms.

The contribution of the visits to this degradation process is not clear, but is likely to be pronounced, as evidenced by the existence of old lines preserved.

**Acknowledgements** This research is a contribution to IGCP 513: Global Study of Karst Aquifers and Water Resources from UNESCO and has been supported by: FEOGA-ORIENTACION-FEDER funds and projects CGL-2008-05584-C02-02 from the MCINN and UCM-910404. R. M-G was supported by a JAEPredoc-CSIC grant and A. M-P by an I3P-CSIC grant.

## References

Alonso-Zarza AM, Martín-Pérez A (2008) Dolomite in caves: recent dolomite formation in oxic, non-sulfate environments. Castañar Cave, Spain. Sediment Geol 205:160–164.

Auler AS, Smart PL (2004) Rates of condensation corrosion in speleothems of semi-arid northeastern Brazil. Speleogenesis and Evolution of Karst Aquifers, 2.

de Freitas CR, Schmekal A (2006) Studies of condensation/evaporation processes in the Glowworm Cave, New Zealand. Int J Speleol 35:75–81.

Díez-Balda MA, Vegas R, González-Lodeiro F (1990) Central-Iberian Zone. Autochthonous Sequences: Structure. In: Dallmeyer RD, Martínez-García E (eds) Pre-Mesozoic Geology of Iberia. Springer-Verlag, Berlin.

Dublyansky VN, Dublyansky YV (1998) The problem of condensation in karst studies. J Cave Karst Stud 60:3–17.

Jones B, Kahle CF (1995) Origin of endogenetic micrite in karst terrains: a case study from the Cayman Islands. J Sediment Res A65:283–293.

Martín-García R, Alonso-Zarza AM, Martín-Pérez A (2009) Loss of primary texture and geochemical signatures in speleothems due to diagenesis: Evidences from Castañar Cave, Spain. Sediment Geol 221:141–149.

Palmer AN (2007) Cave Geology. Cave Books, Ohio.

Sánchez-Moral S, Cuezva S, Lario J et al (2006) Hydrochemistry of karstic waters in a low-energy cave (Castañar de Ibor, Spain). In: Durán JJ, Andreo D, Carrasco F (eds) Karst, cambio climático y aguas subterráneas. Hidrogeología y aguas subterráneas. IGME, Madrid.

Sarbu SM, Lascu C (1997) Condensation corrosion in Movile Cave, Romania. J Cave Karst Stud 59:99–102.

Tarhule-Lips RFA, Ford DC (1998) Condensation corrosion in caves of Cayman Brac and Isla de Mona. J Cave Karst Stud 60:84–95.

Zupan-Hajna N (2002) Chemical weathering of limestones and dolomites in a cave environment. In: Gabrovšek F (ed) Evolution of karst: from prekarst to cessation. Zalozba ZRC, Postojna-Ljubljana.

# Features and Origin of Red Clays in Castañar Cave: A Touch of Colour

A. Martín-Pérez, R. Martín-García, A.M. Alonso-Zarza, and M.J. Herrero

**Abstract** In Castañar Cave (Cáceres, Spain), coatings of red clays cover the walls of the chambers, coexisting with diverse speleothems of aragonite, calcite, huntite and dolomite. The mineralogy of the clays is mainly illite, chlorite, kaolinite, smectite, quartz and Fe oxides and hydroxides such as goethite. They can be transported into the cave by infiltration waters or form by in situ alteration of the host rock: layers of dolomite rich in Fe and magnesite interbedded with greywackes and shales. Present-day hydrological conditions in the cave and conditions during the formation of speleothems have determined that the clays have not been transported by any flooding or seepage, but mostly staying in situ, and not included into carbonate crystal forms. Thus, most of the well-preserved speleothems are white and not stained, conforming an interesting chromatic contrast with the red clays that represent an additional attraction in this show cave.

## 1 Introduction

When speaking about clays in caves, there is an image of wet slippery clayey floors. Clays are common deposits in caves, and they give valuable information to different scientist from geomorphologists to archaeologists. In show caves, when clay covers the speleothems it can spoil the aesthetic appearance of speleothems.

In Castañar Cave, clay deposits are not very abundant, but they appear in most parts of the cave. The clays in the cave have an intense brown-red colour although they usually do not stain the speleothems. The varied calcite, aragonite huntite and dolomite speleothems of Castañar Cave are clean, bright and white, and these characteristics are enhanced by the dark colours of the clays.

A. Martín-Pérez, R. Martín-García, A.M. Alonso-Zarza, M.J. Herrero
Instituto de Geología Económica (CSIC-UCM), Departamento de Petrología y Geoquímica, Facultad de Ciencias Geológicas, C/ Jose Antonio Novais, 2, 28040, Madrid, Spain,
e-mail: andreamartin@geo.ucm.es; rmartingarcia@geo.ucm.es; alonsoza@geo.ucm.es; mjherrer@pdi.ucm.es

B. Andreo et al. (Eds.), *Advances in Research in Karst Media* 
DOI 10.1007/978-3-642-12486-0, 

In this study, the features of the red clays and their close relationship with the host rock are described to attempt to determine which processes have been involved in their formation.

## 2 Geological Setting

Castañar Cave is in the town of Castañar de Ibor, in Caceres province, Spain. The study area forms part of the southeastern part of the Iberian Massif, in the Domain of vertical folds (Díez-Balda et al. 1990), the Ibor anticline being the main structure in the area. In the core of this anticline, a mixed carbonate-siliciclastic succession of Neoproterozoic age crops out (Fig. 1a,b). Castañar Cave formed by dissolution and collapse of these materials, following the NW-SE pattern of the main folds and fractures in the area (Alonso-Zarza et al. 2005).

The Neoproterozoics succession that form the host rock consist of shales, greywackes, dolostones and magnesites formed well-defined beds of centimetric to metric thickness (Fig. 1c). In many of the outcrops all these rocks, especially dolostone and magnesites, appear strongly weathered, partially losing their original texture and showing a brown reddish colour and powdery texture. This alteration is also visible inside the cave.

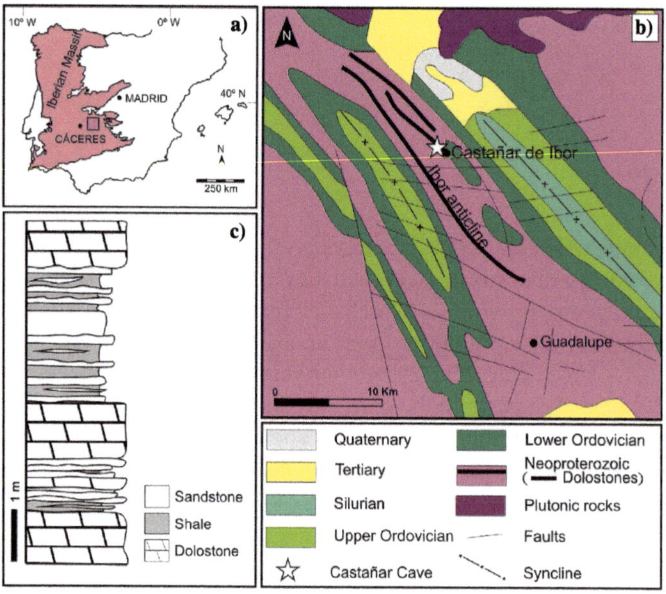


**Fig. 1a–c** Geological setting of Castañar Cave. **a** Location of the studied area in the Iberian Massif. **b** Simplified geological map of the region. **c** Stratigraphic section of the Neoproterozoic rocks near Castañar de Ibor

## 3 Methods

The mineralogical composition of the red clays and the host rock was determined by X-ray diffraction using a Philips PW-1710 XRD system between 2 to 65° $2\theta$. For clay mineralogy, fractions > 20 μm were analyzed using oriented air dried slides that were ethylene glycol solvated and heated at 550°C. Host rock textures were studied under conventional optical petrography in double polished thin sections. Scanning electron microscopy observations were performed on gold-coated samples using a JEOL 6400 electron microscope working at 20 kV and with a resolution of 35 Å.

## 4 Mineralogy and Petrography of the Host Rock

The siliciclastic rocks in the Castañar area are composed of quartz, K-feldspar, plagioclase, clay minerals and, in some cases, authigenic pyrite (Fig. 2). The sandstones are classified as greywackes because they contain up to 40% of clayey matrix (epimatrix and pseudomatrix) mainly consisting of illite, chlorite, smectite and kaolinite. There is also a noticeable amount of dolomite cements replacing part of the matrix.

Carbonate beds (Fig. 3a) are dolostones, magnesites or mixtures of both. In the dolostones Fe-dolomite crystals form non-planar mosaics. Some quartz cements are also present. Magnesite appears as euhedral crystals with shapes varying from prismatic to hexagonal. Minor amounts of siderite have also been found. Dissolution of dolostones and magnesites leads to formation of soft and very porous material (Fig. 3b) mainly composed of goethite (Fig. 3c). Goethite probably formed by oxidation and hydration of Fe present in the carbonates and is seen under SEM as small star-shape aggregates of fibres of about 5 μm in size (Fig. 3d).

## 5 Red Clays in Castañar Cave

Red clays in Castañar Cave appear forming coatings from 2 mm to 3–4 cm thick over the ceiling and walls of most of the chambers. In some parts of the cave, they also cover the breakdowns and the floor. These coatings have formed on the host rock and it is possible to recognise a gradual transition between the texture of the host rock and the clays (Fig. 3e). Only in very few cases the clays have been washed off by dripping water and deposited over speleothems. No lamination or other depositional features that could suggest deposition by water floods have been found. The clays in the coatings are moist and show porous structure under the SEM. They can be rather homogeneous or have a lumpy texture, forming globular aggregates of 1–2 mm diameter and smooth surface. In some cases the clays act as the substrate for the nucleation of delicate aragonite frostwork, providing a beautiful chromatic contrast (Fig. 3f).

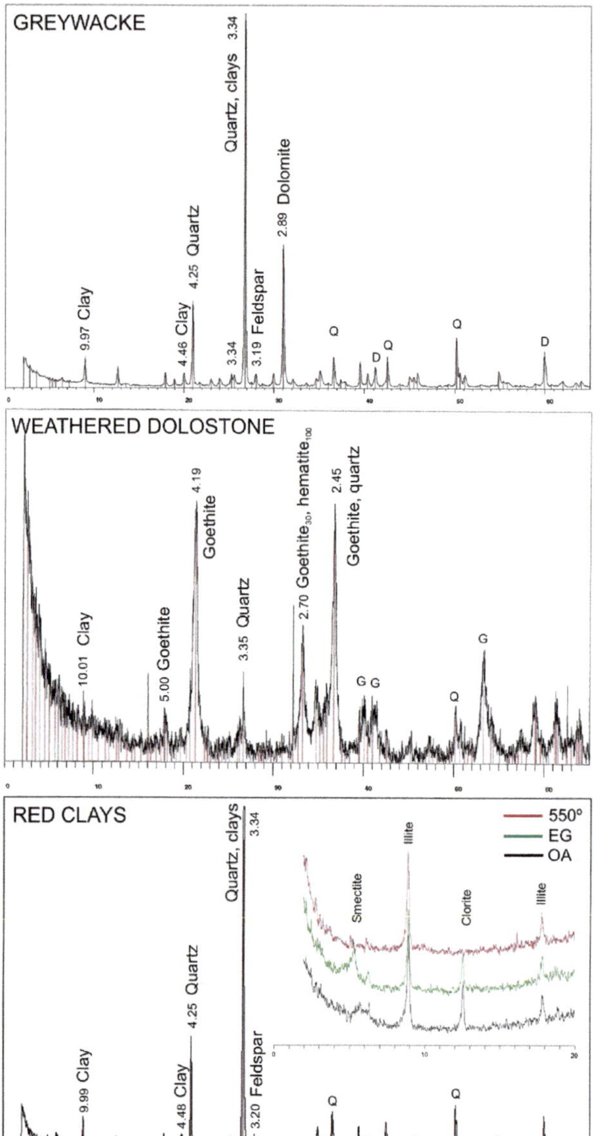


**Fig. 2** XRD patterns showing mineralogy of the greywackes, weathered dolostone and clays

The clays are mainly formed by quartz, clay minerals and goethite having smaller amounts of feldspars, dolomite, magnesite and hematite. Clay minerals are illite, chlorite, kaolinite and smectite. Goethite appears in variable amounts, and it is the main mineral responsible for the red-brown colour of the clays (Fig. 2).

**Fig. 3** **a** Weathered dolostones and greywackes in an outcrop near the cave; **b** The thin bedded host rocks are being altered also inside the cave; **c** Photomicrograph of the euhedral crystals of magnesite largely transformed into goethite. Plane polarized light; **d** SEM image of the fibres of goethite growing over dolomite crystals; **e** Gradual transition between the host rock and the red clays in the cave; **f** Aragonite frostwork nucleating over the red clays

# 6 Discussion and Conclusions

There are two possible origins for the clay deposits in caves (Ford and Williams 2007): allochthonous, as fluvial flows or infiltration from soils overhead, or autochthonous, coming from the weathering of the walls. In Castañar Cave there is no evidence of major streams transporting clastic sediments, and host rock-clay relationships point towards a predominant autochthonous origin. Further evidence of this formation mechanism is the similar mineralogy of the host rock and the clay deposits. Clay minerals, quartz and feldspars can come directly from disgregation of shales and greywackes and from the dissolution of dolomite cements. Clays can also represent weathering remains produced by hydrolysis of silicates (Velde 1992). Iron oxides and hydroxides in caves are generally considered to form by oxidation of iron sulphides, such as pyrite (Onac 2005, Palmer 2007). In Castañar Cave, pyrite is not very common, so it is considered that goethite and hematite formed by oxidation of Fe present in dolomite, magnesite and siderite.

Such in situ transformation of the host rock into red clays is produced by water-rock interaction. In present days, water in the cave comes mainly from the surface by percolation. There are no streams and only small pools. However, climatic and hydrological conditions may have changed during time, as shown by the diversity of morphologies and mineralogies of the speleothems, and the diagenetic processes that affect them (Martín-García et al. 2009).

As can be seen, the varied lithologies of the host rock, and the distribution and composition of cave waters through time are responsible for the remarkable features of the red clays. These special features, such as the colour and distribution, cause the clays to be an attractive element to the tourists that visit Castañar Cave.

**Acknowledgements** This research has been supported by: FEOGA-ORIENTACION-FEDER funds and projects CGL-2008-05584-C02-02 from the MCINN and UCM-910404 and it is a contribution to IGCP 513: Global Study of Karst Aquifers and Water Resources from UNESCO. A. M-P was supported by an I3P-CSIC grant and R. M-G by a JAE Predoc-CSIC grant. Adrijan Košir is thanked for interesting discussions in the field and comments on drafts of this paper. Helpful suggestions from an anonymous reviewer contributed to improve the quality of the manuscript.

## References

Alonso-Zarza AM, Gil Peña I, Martínez Flores E, Muñoz Barco P (2005) La cueva de Castañar. In: Muñoz Barco P and Martínez Flores E (eds) Patrimonio Geológico de Extremadura: geodiversidad y lugares de interés geológico. Junta de Extremadura, Mérida.

Díez-Balda MA, Vegas R, González-Lodeiro F (1990) Central-Iberian Zone. Autochthonous Sequences: Structure. In: Dallmeyer RD and Martínez-García E (eds) Pre-Mesozoic Geology of Iberia. Springer-Verlag, Berlin.

Ford DC, Williams PW (2007) Karst Hydrogeology and Geomorphology. Wiley, Chichester, England.

Martín-García R, Alonso-Zarza AM, Martín-Pérez A (2009) Loss of primary texture and geochemical signatures in speleothems due to diagenesis: Evidences from Castañar Cave, Spain. Sediment Geol 221:141–149.

Onac BP (2005) Minerals. In: Culver DC, White WB (eds) Encyclopedia of Caves. Elsevier Academic Press, London.

Palmer AN (2007) Cave Geology. Cave Books, Ohio.

Velde B (1992) Introduction to clay minerals. Chapman & Hall, London.

# Impact of the Expected Climate Change on the Stability of Underground Cavities in France

P. Gombert, A. Charmoille, D. Christophe, and R. D'hotelans

**Abstract** The different prospective scenarios studied worldwide predict a climate change, with first effects which should be seen before the end of the 21st century. The question is knowing the predictable impact of this climate change on the stability of about a hundred thousand natural and artificial underground cavities which undermine France, notably in karstic regions (limestone, chalk, gypsum). An attempt will be made to describe the different parameters influenced by the climate change and likely to interfere with the stability of underground cavities. The instrumentations set up in situ in artificial cavities as well as the first results obtained and their application to natural cavities, are then presented.

## 1 Introduction

For about one century and a half, human activities have generated large emissions of greenhouse gases in the atmosphere increasing the natural greenhouse effect. In France, the climatic models simulate a change at 2100 in temperatures (+ 3.5°C) and in precipitations (+ 5 to 8%) on a global scale (INSU 2007).

Therefore it seemed necessary to initiate studies in order to anticipate the consequences of this climate change on various themes including the stability of underground cavities. Indeed, France has several hundred of thousands of underground cavities which are sensitive to variations of hydrogeotechnical parameters likely to be affected by the climate change (Al Heib et al. 2008). There are:

- natural cavities which are hollowed out by dissolution in geological formations sensitive to the action of water (limestone, chalk, gypsum, etc.);
- artificial cavities (mines and quarries).

P. Gombert, A. Charmoille, D. Christophe, R. D'hotelans
INERIS (French National Institute for Industrial Environment and Risks), BP 2, 60550 Verneuil-en-Halatte, France, e-mail: philippe.gombert@ineris

B. Andreo et al. (Eds.), *Advances in Research in Karst Media*
DOI 10.1007/978-3-642-12486-0, © Springer 2010

Underground mines and quarries have generally been exploited above the piezometric level or sheltered from groundwater by high flow rate pumping operations. As most of them are today abandoned, they may have been subject to total or partial, temporary or permanent flooding called "rebound". This phenomenon, well studied in the case of artificial cavities, is thus at the origin of irreversible deformations of the rock medium which may cause disorders right up to the surface.

## 2 Expected Impact of Climate Change

In the case of $CO_2$ level doubling, the models predict winter warming by 1 to 2°C in France (MIES 2000). In spring, the warming would exceed 2°C on the southern France and then extend in summer and in autumn to the remainder of the country. These models also predict an increase in winter rainfall and, in the eastern France, in spring rainfall (up to 50%) but a decrease in the south. In summer and autumn, efficient rainfall would decrease on the whole country by up to −20%.

### *2.1 Quantitative Impact on Groundwater*

Some aquifers are particularly sensitive to climate conditions and this will be the case in the future if the climate changes. It is thus possible to distinguish:

- the phreatic aquifers, essentially replenished in winter, such as the limestone of the Paris basin or the alluvium of the great rivers; the increase in winter precipitation should amplify their replenishment and therefore raise their piezometric level; on the other hand, in summer and in autumn, the lack of precipitation will led to lowering of piezometric level;
- the deep aquifers, generally captive, will only be slightly affected in a first phase, because they are subject to pluriannual infiltration and weakly interact with the surface (Albian of Paris basin, Tertiary of Aquitaine basin).

### *2.2 Qualitative Impact on Groundwater*

The increase in rainfall and temperature will probably modify the aggressivity of rain water. Moreover, the pedogenetic activity will also be changed due to temperature and humidity, which will have an impact on the dissolved $CO_2$ flow in percolation water. These elements will perturb calcocarbonic equilibrium which plays a major role in karstification of carbonated rocks. The risks for subterranean cavities relate to an increase in the degradation of walls, and pillars in the case of artificial cavities. The result would be an embrittlement of the surrounding rock and one of the possible responses would be an increase in the frequency of openings at the surface of these cavities.

# 3 Role of Water in the Instability of Artificial Underground Cavities

## 3.1 Laboratory Tests

After their abandonment, artificial underground cavities are subject to collapses which may develop right up to the surface with the formation of a crown hole or a sinkhole. These cavities may be flooded or subject to flooding/unflooding cycles. It has been shown that water absorption by rock causes a reduction of its compressive strength by up to 50% (Fig. 1). Saturation therefore promotes creep and deformation while desaturation induces shrinkage and generally tensile cracks.

In Lorraine iron mines (NW of France), several collapses have occurred during and after rebound. Experimental work shows that the compressive strength of ore is reduced by 50% when its humidity increases from 80 to 100% (GISOS 2005).

Concerning chalk, this very porous rock quickly looses part of its interstitial water in contact with air that alters its mechanical properties. Laboratory tests have shown the occurrence of cracks during saturation/desaturation cycles such as those produced by hygrometric or piezometric variations (Laouafa et al. 2009).

## 3.2 In Situ Measurements

In 2009, the abandoned underground quarries of Saint-Martin-le-Noeud (Oise) and Crouzilles (Indre-et-Loire) were instrumented by INERIS in order to follow the evolution of piezometry and physicochemical parameters (Fig. 2). The first results

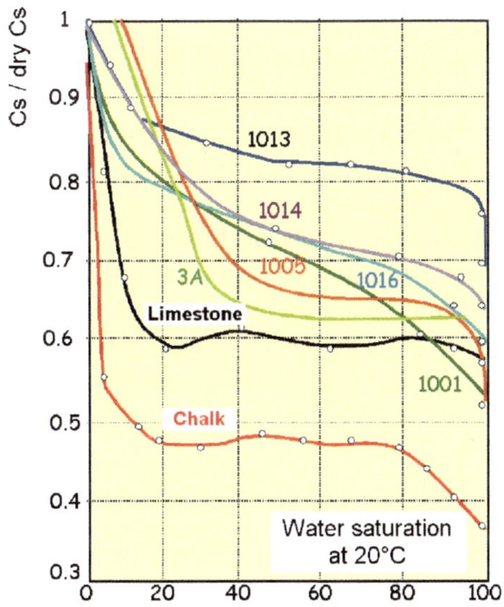


**Fig. 1** Compressive strength of several rocks depending on the saturation level (Watelet 1996) Legend: Cs = compressive strength, 1001 = Quartzite, 1005–1016 = Various sandstone, 3A = Various sandstones, Limestone = coarse limestone from the Paris Basin, Chalk = chalk from the Paris Basin

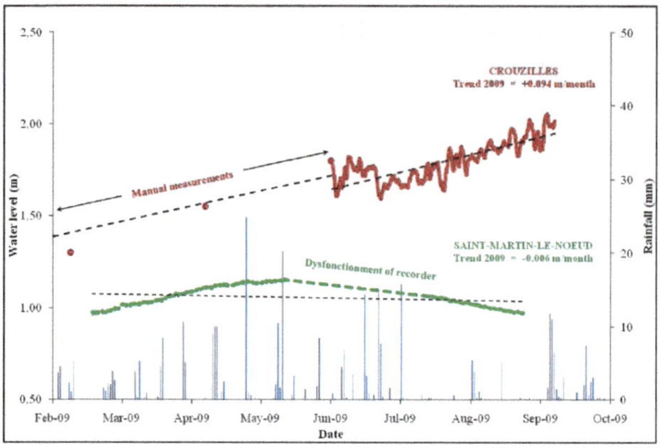


**Fig. 2** Evolution of the flooding water level of the investigated mines

show a stabilization of water level in St-Martin (− 0.07 m/yr) while it increases in Crouzilles (+ 1.13 m/yr). This behaviour expresses a different sensitivity to recharge: in St-Martin, infiltration of effective rainfall is delayed by 4–6 months while in Crouzilles, the response time to rains is three days for piezometry (pressure transfer) and about one week for conductivity (mass transfer). The behaviour of the flooding water at Crouzilles is closer to that of a karstic network.

# 4 Application to the Case of Natural Cavities

## *4.1 Kinetics of Evolution of Karstic Cavities*

The karsts of carbonated formations can be distinguished as those with slow evolution at a human scale and those of evaporitic massifs, the quick evolution of which is close to that of artificial cavities. However, the stability of both of these types of natural cavities will comply with same geomechanical laws as for artificial cavities.

In the carbonated rocks, surface disorders are related to the opening of avens or to paroxysmal evolution of dolines (Salvati and Sasowsky 2002). These generally stable karsts produce isolated disorders with low return frequency except in chalk, but their response to a climate change as rapid as the one announced is unknown.

Karstification of gypsum is due to its quick dissolution in contact with undersaturated waters, hence high risks of occurrence of surface disorders. In France, this is observed in the Tertiary of the Paris basin (Toulemont 1987; Thierry et al. 2008) or the Trias of the southern France (Julian and Nicod 1990).

## *4.2 Response of Karstic Caves to a Climate Change*

**Hydraulic Response.** Amplification of the piezometric fluctuations may be expressed by the collapse of overlying ground because of:

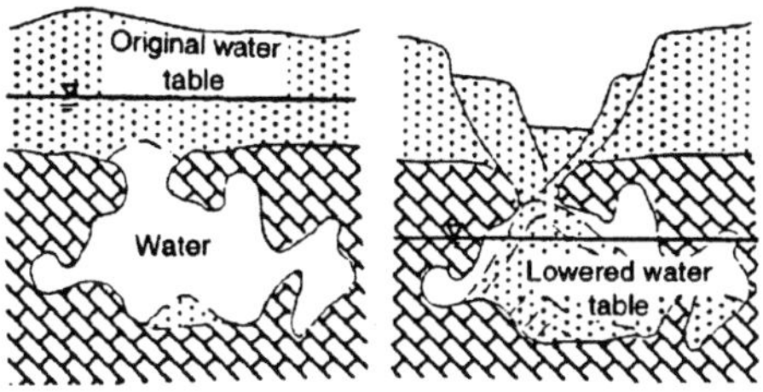


**Fig. 3** Conceptual diagram of a collapse related to a piezometric drop (Ford & Williams, 1989)

- the occurrence of suffosion cavities (Salvati and Sasowsky 2002) following a rise in the piezometric level, notably in the cover of loess and clays of Normandy chalk;
- a piezometric drop as reported by Ford and Williams (1989) and observed in Turkey by Celic and Afsin (1998) in gypsum (Fig. 3).

**Hydromechanical Response.** Rainwater infiltrations may induce karst deformations that may exceed in amplitude all the other deformations and notably earth tides (Lesage 1981). In southern France, Durand (1992) observed a subsidence due to the infiltration of water into the epikarst, followed by a deformation phase related to the piezometric variation. It is difficult to determine the impact of these deformations on the surface disorders, but one may ponder over the cumulative effect in the long run of mechanical fatigue of the rocks. On the other hand, gypsum karsts have not been studied.

**Hydrogeochemical Response.** The increase in rainfall and temperature may modify the karstic dissolution potential of percolation waters. In gypsum, in the Paris region, the active dissolution zones are mainly localized at the transition fringe between normal groundwater and sulphate saturated water in contact with the gypseous levels (Toulemont 1987). Dissolution is stopped as water reaches chemical equilibrium, until the arrival of undersaturated water. Consequently, the higher the renewal rate of groundwater, the less this water will be saturated with regard to gypsum, and the more dissolution will be active; it is therefore possible that the climate change induces new sinkholes in these geological formations with quick kinetics of dissolution.

# 5 Conclusion

Some aquifers will be more affected by the expected climate change: small aquifers, karsts, free and not very deep aquifers. The impact on groundwater, and therefore on the stability of underground cavities, is still poorly known but it should have a dual aspect: 1) quantitatively, groundwater recharge (and level) should increase in winter and strongly decrease in summer and 2) qualitatively, the water may become more aggressive and thereby amplify dissolution of rocks.

In the present state of our knowledge, it is difficult to know how this will be expressed on natural or artificial underground cavities. However, any amplification of the fluctuation of the groundwater level will induce saturation/desaturation cycles which will have an impact on the mechanical properties of surrounding rocks and notably on their single compaction strength.

The evolution of cavities with quick kinetics of dissolution should be more sensitive to climate change and the question is posed of whether this will have an influence at a human scale on the stability of the superficial formations. This is why INERIS has engaged studies on the formation and evolution mechanisms of underground cavities (Thoraval 2005; Laouafa et al. 2009; Al Heib et al. 2008).

## References

Al Heib M, Gombert P, Charmoille A (2008) Impact du changement climatique sur la stabilité des cavités souterraines: Etat des connaissances. Rapport INERIS, DRS-08-95052-11840A, March 2009.

Celik M, Afsin M (1998) The role of hydrogeology in solution-subsidence development and its environmental impacts: a case-study for Sazlika (Nigde, Turkey). Env. Geol., 36:335–342.

Durand V (1992) Structure d'un massif karstique. Relation entre déformations et facteurs hydrométéorologiques. Causse de l'Hortus – Site des sources du Lamalou (Hérault). Doctoral Thesis, USTL.

Ford D, Williams P (1989) Karst geomorphology and hydrology. Unwin Hyman Ed., London.

GISOS (2005) Synthèse des travaux de recherches "après-mines fer". INPL-LAEGO Report, June 2006.

INSU (2007) Les recherches françaises sur le changement climatique. Ministère de l'éducation, de l'enseignement supérieur et de la recherche, MEEDDAT.

Julian M, Nicod J (1990) Catastrophes naturelles et risques afférents aux terrains gypseux (Alpes et Provence). Rev. de Géogr. Alpine, 1990/78-1-3:157–173.

Laouafa F, Degas M, Kazmierczak JB (2009) Influence d'interactions eau-roche sur le comportement à long terme de cavités souterraines dans la craie. INERIS Report, DRS-09-76873-02216A, February 2009.

Lesage P (1981) Développement de méthodes en vue de l'étude des déformations sismotectoniques du sous-sol. Doctoral Thesis, UPMC, Paris.

MIES (2000) Impacts potentiels du changement climatique en France au XXIe siècle. Rapport de la Mission Interministérielle de l'Effet de Serre.

Salvati R, Sasowsky ID (2002) Development of collapse sinkholes in areas of groundwater discharge. J. of Hydrol., 264:1–11.

Thierry P (2008) Geological modelling at turban scale and mapping of ground movement susceptibility from gypsum dissolution: The Paris example (France). Eng. Geol., doi: 10.1016/j.enggeo.2008.12.010:1–11.

Thoraval A (2005) Analyse du comportement hydromécanique des massifs rocheux fracturés à partir d'essais in-situ et de modélisation. INERIS Report, DRS05-66388/RN0I.

Toulemont M (1987) Les risques d'instabilité liés au karst gypseux lutétien de la région parisienne – Prévision en cartographie. Bull. de liaison P et Ch, 150/151, réf. 3192:109–116.

Watelet JM (1996) Méthode d'analyse et diagnostic des conditions de stabilité des carrières souterraines. Engineering Diploma Work. ISAM.SNAM, Paris.

GPSR Compliance

*The European Union's (EU) General Product Safety Regulation (GPSR) is a set of rules that requires consumer products to be safe and our obligations to ensure this.*

*If you have any concerns about our products, you can contact us on ProductSafety@springernature.com*

In case Publisher is established outside the EU, the EU authorized representative is:

Springer Nature Customer Service Center GmbH
Europaplatz 3
69115 Heidelberg, Germany

**Batch number: 10755034**

Printed by Printforce, the Netherlands